Bioprocess Engineering
Basic Concepts
Second Edition

ISBN 0-13-081908-5

90000

9 780130 819086

PRENTICE HALL PTR INTERNATIONAL SERIES IN THE PHYSICAL AND CHEMICAL ENGINEERING SCIENCES

BIOPROCESS ENGINEERING
Basic Concepts
Second Edition

Michael L. Shuler
School of Chemical Engineering
Cornell University

Fikret Kargi
Department of Environmental Engineering
Dokuz Eylūl University
Izmir, Turkey

Pearson Education International

Editorial/production supervision: *Patti Guerrieri*
Acquisitions editor: *Bernard Goodwin*
Marketing manager: *Dan DePasquale*
Manufacturing manager: *Alexis R. Heydt-Long*
Editorial assistant: *Michelle Vincenti*
Cover design director: *Jerry Votta*
Cover designer: *Talar Boorujy*

© 2002 by Prentice Hall PTR
Prentice-Hall, Inc.
Upper Saddle River, NJ 07458

Pearson Education LTD.
Pearson Education Australia PTY, Limited
Pearson Education Singapore, Pte. Ltd.
Pearson Education North Asia Ltd.
Pearson Education Canada, Ltd.
Pearson Educación de Mexico, S.A. de C.V.
Pearson Education—Japan
Pearson Education Malaysia, Pte. Ltd.

For Andy, Kristin, Eric, and Kathy
and for Karen

Contents

11   RECOVERY AND PURIFICATION OF PRODUCTS   329

Part 4   Applications to Nonconventional Biological Systems   385

12   BIOPROCESS CONSIDERATIONS IN USING ANIMAL CELL CULTURES   385

Contents   **xiii**

Preface to the Second Edition

In the decade since the first edition of *Bioprocess Engineering: Basic Concepts*, biotechnology has undergone several revolutions. Currently, the ability to sequence the genome of whole organisms presents opportunities that could be hardly envisioned ten years ago. Many other technological advances have occurred that provide bioprocess engineers with new tools to serve society better. However, the principles of bioprocess engineering stated in the first edition remain sound.

The goals of this revision are threefold. We want to capture for students the excitement created by these advances in biology and biotechnology. We want to inform students about these tools. Most importantly, we want to demonstrate how the principles of bioprocess engineering can be applied in concert with these advances.

This edition contains a new section in the first chapter alerting students to the regulatory issues that constrain bioprocess design and modification. We believe students need to be aware of these industrially critical issues. Part 2, "An Overview of Biological Basics," has been updated throughout and expanded. Greater emphasis is given now to posttranslational processing of proteins, as this is a key issue in choice of bioprocessing strategies to make therapeutic proteins. Basic processes in animal cells are more completely described, since animal cell culture is now an established commercial bioprocess technology. Chapter 5 is made more complete by introduction of a section on noncarbohydrate metabolism. Key concepts in functional genomics have been added to prepare students to understand the impact of these emerging ideas and technologies on bioprocesses.

In Part 3, "Engineering Principles for Bioprocesses," greater attention is given to issues associated with animal cell bioreactors. The discussion of chromatographic processes is expanded. In Part 4, "Applications to Nonconventional Biological Systems," the material has been rearranged and updated and a new chapter added. These changes are evident in the chapters on animal and plant cell culture. Particularly important is the expanded discussion on choice of host-vector systems for production of proteins from recombinant DNA technology. Coverage of two areas of increasing importance to bioprocess engineers, metabolic and protein engineering, has been expanded. A new chapter on biomedical applications illustrates how approaches to bioprocess engineering are relevant to problems typically considered to be biomedical engineering. The chapter on mixed cultures has been extended to cover advanced waste-water treatment processes. An appendix providing descriptive overviews of some traditional bioprocesses is now included.

The suggestions for further reading at the end of each chapter have been updated. We are unable in this book to provide in-depth treatment of many vital topics. These readings give students an easy way to begin to learn more about these topics.

Teaching a subject as broad as bioprocess engineering in the typical one-semester, three-credit class has never been easy. Although some material in the first edition has been removed or condensed, the second edition is longer than the first. For students with no formal background in biology, coverage of all of the material in this book would require a four-credit class. In a three-credit class we suggest that the instructor cover Chapters 1 to 11 (with 7 being optional) and then decide on subsequent chapters based on course goals. A course oriented toward biopharmaceuticals will want to include careful coverage of Chapters 12 and 14 and some coverage of 13 and 15. A course oriented toward utilization of bioresources would emphasize Chapter 16 and the Appendix and selected coverage of topics in Chapters 13 and 14.

Many students now enter a bioprocess engineering course with formal, college-level instruction in biology and biochemistry. For such students Chapters 2, 4, 5, 7, and 8 can be given as reading assignments to refresh their memories and to insure a uniform, minimal level of biological knowledge. Lecture time can be reserved for material in other chapters or for supplementary material. For these five chapters study questions are provided for self-testing. Under these circumstances the instructor should be able to cover the rest of the material in the book.

Once again we have been assisted by comments from many colleagues across the world. These comments have included suggestions for new material to be incorporated and for corrections. While the list is too long to include here, specific contributions deserve special recognition. Mohammad Ataai provided us a summary of IMAC (immobilized metal affinity chromatography) that has been incorporated into the revision. Kelvin Lee provided the paragraph describing 2-D gel electrophoresis. We thank Laura Palomares for an excellent job in updating and revising a first draft of the Appendix on traditional bioprocesses. The first edition of this book was translated into Korean, and Yoon-Mo Koo, Jin Ho Seo, Yong Keun Chang, and Tai Hyun Park provided us an extensive list of corrections, which has been very helpful in preparation of this revision.

We wish to also thank our families for their support during the process of this revision.

Preface to the First Edition

Bioprocess engineering is the application of engineering principles to design, develop, and analyze processes using biocatalysts. These processes may result in the formation of desirable compounds or in the destruction of unwanted or hazardous substances. The tools of the engineer, particularly the chemical engineer, will be essential to the successful exploitation of bioprocesses.

This book's main purpose is to introduce the essential concepts of bioprocessing to traditional chemical engineers. No background in biology is assumed. The material in this book has been used as a basis for a course at Cornell University. Although it was designed primarily for seniors and entering graduate students from chemical engineering, students from agricultural engineering, environmental engineering, food science, soil science, microbiology, and biochemistry have successfully completed this course.

Parts 1 and 2 outline basic biological concepts. These eight chapters are not intended to be a replacement for good courses in microbiology, biochemistry, and genetics. They simply provide sufficient information to make the rest of the book accessible to the reader. A reader who desires a more in-depth understanding of the key biological concepts is referred to the suggested readings at the end of each chapter. Chapters 3 and 6 differ in that they are more detailed and introduce concepts not normally found in the standard biological textbooks.

In Part 3, "Engineering Principles for Bioprocesses," we focus on the generic components of bioprocessing that do not depend on the type of cell used in the process. In

Part 4 we discuss applications to special systems and the particular characteristics of mixed cultures, genetically engineered cells, plant cells, and animal cells. These application chapters reinforce the previous engineering and biological concepts while providing more detailed information about important new biological systems.

This book reflects useful and important suggestions from many people. Students in Chemical Engineering 643 ("Introduction to Bioprocess Engineering") at Cornell University in Fall 1989 provided written critiques on the chapters. This input has been invaluable and is deeply appreciated. The two external reviewers of this text also provided invaluable suggestions. Harvey Blanch, Charles D. Scott, and Octave Levenspiel have provided one of us (FK) with suggested homework problems, which have greatly enhanced the potential usefulness of this book. Finally, the superb technical skills of Bonnie Sisco have been instrumental in converting nearly incomprehensible written scrawls into a readable text.

Finally, both authors wish to thank their families for their patience and support during this process.

Michael L. Shuler
Fikret Kargi

1

What Is a Bioprocess Engineer?

1.1. INTRODUCTORY REMARKS

We can now manipulate life at its most basic level—the genetic. For thousands of years people have practiced genetic engineering at the level of selection and breeding. But now it can be done in a purposeful, predetermined manner with the molecular-level manipulation of DNA. We now have a tool to probe the mysteries of life in a way unimaginable 25 years ago.

With this intellectual revolution emerge new visions and new hopes: new medicines, semisynthetic organs grown in large vats, abundant and nutritious foods, computers based on biological molecules rather than silicon chips, superorganisms to degrade pollutants, and a wide array of consumer products and industrial processes.

These dreams will remain dreams without hard work. Engineers will play an essential role in converting these visions into reality. Biological systems are very complex and beautifully constructed, but they obey the rules of chemistry and physics and they are susceptible to engineering analysis. Living cells are predictable, and the processes to use them can be rationally constructed on commercial scales. Doing this is the job of the bioprocess engineer.

Probably the reason you are reading this book is your desire to participate in this intellectual revolution and to make an important contribution to society. You *can* do it, but it is not easy to combine the skills of the engineer with those of the biologist. Our intent is to help you begin to develop these skills. This book and a one-term course are not enough to make you a complete bioprocess engineer, but we intend to help you form the necessary foundation.

1.2. BIOTECHNOLOGY AND BIOPROCESS ENGINEERING

When new fields emerge from new ideas, old words are usually not adequate to describe these fields. Biotechnology and what constitutes engineering in this field are best described with examples rather than single words or short phrases.

Biotechnology usually implies the use or development of methods of direct genetic manipulation for a socially desirable goal. Such goals might be the production of a particular chemical, but they may also involve the production of better plants or seeds, or gene therapy, or the use of specially designed organisms to degrade wastes. The key element for many workers is the use of sophisticated techniques outside the cell for genetic manipulation. Others interpret biotechnology in a much broader sense and equate it with applied biology; they may include engineering as a subcomponent of biotechnology.

Many words have been used to describe engineers working with biotechnology. *Bioengineering* is a broad title and would include work on medical and agricultural systems; its practitioners include agricultural, electrical, mechanical, industrial, environmental and chemical engineers, and others. *Biological engineering* is similar but emphasizes applications to plants and animals. *Biochemical engineering* has usually meant the extension of chemical engineering principles to systems using a biological catalyst to bring about desired chemical transformations. It is often subdivided into bioreaction engineering and bioseparations. *Biomedical engineering* has been considered to be totally separate from biochemical engineering, although the boundary between the two is increasingly vague, particularly in the areas of cell surface receptors and animal cell culture. Another relevant term is *biomolecular engineering,* which has been defined by the National Institutes of Health as ". . . research at the interface of biology and chemical engineering and is focused at the molecular level."

There is a difference between bioprocess engineering and biochemical engineering. In addition to chemical engineering, *bioprocess engineering* would include the work of mechanical, electrical, and industrial engineers to apply the principles of their disciplines to processes based on using living cells or subcomponents of such cells. The problems of detailed equipment design, sensor development, control algorithms, and manufacturing strategies can utilize principles from these disciplines. Biochemical engineering is more limited in the sense that it draws primarily from chemical engineering principles and broader in the sense that it is not restricted to well-defined artificially constructed processes, but can be applied to natural systems.

We will focus primarily on the application of chemical engineering principles to systems containing biological catalysts, but with an emphasis on those systems making use of biotechnology. The rapidly increasing ability to determine the complete sequence of genes in an organism offers new opportunities for bioprocess engineers in the design and monitoring of bioprocesses. The cell, itself, is now a designable component of the overall process.

1.3. BIOLOGISTS AND ENGINEERS DIFFER IN THEIR APPROACH TO RESEARCH

The fundamental trainings of biologists and engineers are distinctly different. In the development of knowledge in the life sciences, unlike chemistry and physics, mathematical theories and quantitative methods (except statistics) have played a secondary role. Most progress has been due to improvements in experimental tools. Results are qualitative and descriptive models are formulated and tested. Consequently, biologists often have incomplete backgrounds in mathematics but are very strong with respect to laboratory tools and, more importantly, with respect to the interpretation of laboratory data from complex systems.

Engineers usually possess a very good background in the physical and mathematical sciences. Often a theory leads to mathematical formulations, and the validity of the theory is tested by comparing predicted responses to those in experiments. Quantitative models and approaches, even to complex systems, are strengths. Biologists are usually better at the formation of testable hypotheses, experimental design, and data interpretation from complex systems. Engineers are typically unfamiliar with the experimental techniques and strategies used by life scientists.

The skills of the engineer and life scientist are complementary. To convert the promises of molecular biology into new processes to make new products requires the integration of these skills. To function at this level, the engineer needs a solid understanding of biology and its experimental tools. In this book we provide sufficient biological background for you to understand the chapters on applying engineering principles to biosystems. However, if you are serious about becoming a bioprocess engineer, you will need to take further courses in microbiology, biochemistry, and cell biology, as well as more advanced work in biochemical engineering. If you already have these courses, these chapters can be used for review.

1.4. THE STORY OF PENICILLIN: HOW BIOLOGISTS AND ENGINEERS WORK TOGETHER

In September 1928, Alexander Fleming at St. Mary's Hospital in London was trying to isolate the bacterium, *Staphylococcus aureus*, which causes boils. The technique in use was to grow the bacterium on the surface of a nutrient solution. One of the dishes had been contaminated inadvertently with a foreign particle. Normally, such a contaminated plate would be tossed out. However, Fleming noticed that no bacteria grew near the invading substance (see Fig. 1.1).

Fleming's genius was to realize that this observation was meaningful and not a "failed" experiment. Fleming recognized that the cell killing must be due to an antibacterial agent. He recovered the foreign particle and found that it was a common mold of the *Penicillium* genus (later identified as *Penicillium notatum*). Fleming nurtured the mold to grow and, using the crude extraction methods then available, managed to obtain a tiny quantity of secreted material. He then demonstrated that this material had powerful antimicrobial properties and named the product penicillin. Fleming carefully preserved the culture, but the discovery lay essentially dormant for over a decade.

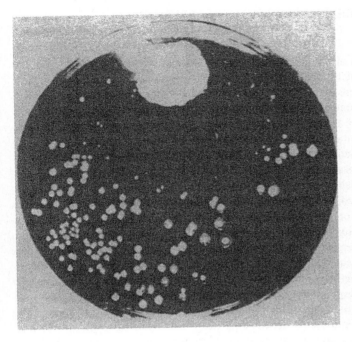

Figure 1.1. Photograph of Alexander Fleming's original plate showing the growth of the mold *Penicillium notatum* and its inhibitory action on bacterial growth. (With permission, from Corbis Corporation.)

World War II provided the impetus to resurrect the discovery. Sulfa drugs have a rather restricted range of activity, and an antibiotic with minimal side effects and broader applicability was desperately needed. Howard Florey and Ernst Chain of Oxford decided to build on Fleming's observations. Norman Heatley played the key role in producing sufficient material for Chain and Florey to test the effectiveness of penicillin. Heatley, trained as a biochemist, performed as a bioprocess engineer. He developed an assay to monitor the amount of penicillin made so as to determine the kinetics of the fermentation, developed a culture technique that could be implemented easily, and devised a novel back-extraction process to recover the very delicate product. After months of immense effort, they produced enough penicillin to treat some laboratory animals.

Eighteen months after starting on the project, they began to treat a London bobby for a blood infection. The penicillin worked wonders initially and brought the patient to the point of recovery. Most unfortunately, the supply of penicillin was exhausted and the man relapsed and died. Nonetheless, Florey and Chain had demonstrated the great potential for penicillin, if it could be made in sufficient amount. To make large amounts of penicillin would require a process, and for such a process development, engineers would be needed, in addition to microbial physiologists and other life scientists.

The war further complicated the situation. Great Britain's industrial facilities were already totally devoted to the war. Florey and his associates approached pharmaceutical firms in the United States to persuade them to develop the capacity to produce penicillin, since the United States was not at war at that time.

Many companies and government laboratories, assisted by many universities, took up the challenge. Particularly prominent were Merck, Pfizer, Squibb, and the USDA Northern Regional Research Laboratory in Peoria, Illinois.

The first efforts with fermentation were modest. A large effort went into attempts to chemically synthesize penicillin. This effort involved hundreds of chemists. Consequently, many companies were at first reluctant to commit to the fermentation process, beyond the pilot-plant stage. It was thought that the pilot-plant fermentation system could produce sufficient penicillin to meet the needs of clinical testing, but large-scale production would soon be done by chemical synthesis. At that time, U.S. companies had achieved a great deal of success with chemical synthesis of other drugs, which gave the companies a great deal of control over the drug's production. The chemical synthesis of penicillin proved to be exceedingly difficult. (It was accomplished in the 1950s, and the synthesis route is still not competitive with fermentation.) However, in 1940 fermentation for the production of a pharmaceutical was an unproved approach, and most companies were betting on chemical synthesis to ultimately dominate.

The early clinical successes were so dramatic that in 1943 the War Production Board appointed A. L. Elder to coordinate the activities of producers to greatly increase the supply of penicillin. The fermentation route was chosen. As Elder recalls, "I was ridiculed by some of my closest scientific friends for allowing myself to become associated with what obviously was to be a flop—namely, the commercial production of penicillin by a fermentation process" (from Elder, 1970). The problems facing the fermentation process were indeed very formidable.

The problem was typical of most new fermentation processes: a valuable product made at very low levels. The low rate of production per unit volume would necessitate very large and inefficient reactors, and the low concentration (titer) made product recovery and purification very difficult. In 1939 the final concentration in a typical penicillin fermentation broth was one part per million (ca. 0.001 g/l); gold is more plentiful in sea water. Furthermore, penicillin is a fragile and unstable product, which places significant constraints on the approaches used for recovery and purification.

Life scientists at the Northern Regional Research Laboratory made many major contributions to the penicillin program. One was the development of a corn steep liquor–lactose based medium. This medium increased productivity about tenfold. A worldwide search by the laboratory for better producer strains of *Penicillium* led to the isolation of a *Penicillium chrysogenum* strain. This strain, isolated from a moldy cantaloupe at a Peoria fruit market, proved superior to hundreds of other isolates tested. Its progeny have been used in almost all commercial penicillin fermentations.

The other hurdle was to decide on a manufacturing process. One method involved the growth of the mold on the surface of moist bran. This bran method was discarded because of difficulties in temperature control, sterilization, and equipment size. The surface method involved growth of the mold on top of a quiescent medium. The surface method used a variety of containers, including milk bottles, and the term "bottle plant" indicated such a manufacturing technique. The surface method gave relatively high yields, but had a long growing cycle and was very labor intensive. The first manufacturing plants were bottle plants because the method worked and could be implemented quickly.

However, it was clear that the surface method would not meet the full need for penicillin. If the goal of the War Production Board was met by bottle plants, it was estimated that the necessary bottles would fill a row stretching from New York City to San Francisco. Engineers generally favored a submerged tank process. The submerged process presented challenges in terms of both mold physiology and in tank design and operation. Large volumes of absolutely clean, oil- and dirt-free sterile air were required. What were then very large agitators were required, and the mechanical seal for the agitator shaft had to be designed to prevent the entry of organisms. Even today, problems of oxygen supply and heat removal are important constraints on antibiotic fermenter design. Contamination by foreign organisms could degrade the product as fast as it was formed, consume nutrients before they were converted to penicillin, or produce toxins.

Figure 1.2(a). Series of large-scale antibiotic fermenters. (With permission, from T.D. Brock, K.M. Brock, and D.M. Ward. *Basic Microbiology with Applications*, 3d ed., Pearson Education, Upper Saddle River, NJ, 1986, p. 507.)

In addition to these challenges in reactor design, there were similar hurdles in product recovery and purification. The very fragile nature of penicillin required the development of special techniques. A combination of pH shifts and rapid liquid-liquid extraction proved useful.

Soon processes using tanks of about 10,000 gal were built. Pfizer completed in less than six months the first plant for commercial production of penicillin by submerged fermentation (Hobby, 1985). The plant had 14 tanks each of 7000-gal capacity. By a combination of good luck and hard work, the United States had the capacity by the end of World War II to produce enough penicillin for almost 100,000 patients per year (see Figs. 1.2 and 1.3).

This accomplishment required a high level of multidisciplinary work. For example, Merck realized that men who understood both engineering and biology were not available. Merck assigned a chemical engineer and microbiologist together to each aspect of the problem. They planned, executed, and analyzed the experimental program jointly, "almost as if they were one man" (see the chapter by Silcox in Elder, 1970).

Progress with penicillin fermentation has continued, as has the need for the interaction of biologists and engineers. From 1939 to now, the yield of penicillin has gone from 0.001 g/l to over 50 g/l of fermentation broth. Progress has involved better understanding of mold physiology, metabolic pathways, penicillin structure, methods of mutation and selection of mold genetics, process control, and reactor design.

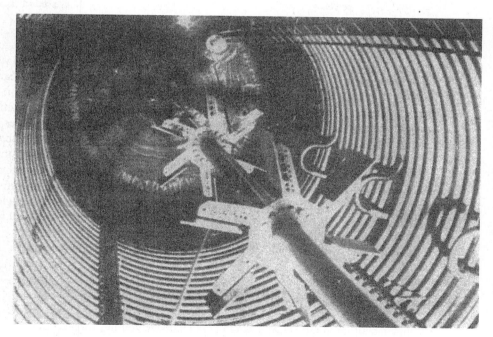

Figure 1.2(b). Inside view of a large antibiotic fermenter. (From *Trends in Biotechnology 3* (6), 1985. Used with permission of Elsevier Science Publishers.)

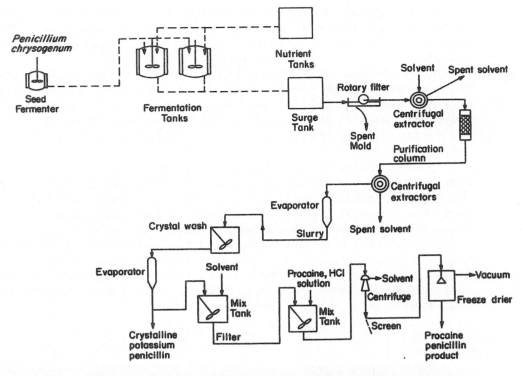

Figure 1.3. Schematic of penicillin production process.

Before the penicillin process, almost no chemical engineers sought specialized training in the life sciences. With the advent of modern antibiotics, the concept of a bioprocess engineer was born. The penicillin process also established a paradigm for bioprocess development and biochemical engineering. This paradigm still guides much of our profession's thinking. The mind set of bioprocess engineers was cast with the penicillin experience. It is for this reason that we have focused on the penicillin story, rather than on an example for production of a protein from a genetically engineered organism. Although many parallels can be made between the penicillin process and our efforts to use recombinant DNA, no similar paradigm has yet emerged from our experience with genetically engineered cells. We must continually reexamine the prejudices the field has inherited from the penicillin experience.

It is you, the student, who will best be able to challenge these prejudices.

1.5. BIOPROCESSES: REGULATORY CONSTRAINTS

To understand the mind set of a bioprocess engineer you must understand the regulatory climate in which many bioprocess engineers work. The U.S. FDA (Food and Drug Administration) and its equivalents in other countries must insure the safety and efficacy of medicines. For bioprocess engineers working in the pharmaceutical or biotechnology industry the primary concern is not reduction of manufacturing cost (although that is still a very desirable goal), but the production of a product of consistently high quality in amounts to satisfy the medical needs of the population.

Consider briefly the process by which a drug obtains FDA approval. A typical drug undergoes 6.5 years of development from the discovery stage through preclinical testing in animals. Human clinical trials are conducted in three phases. Phase 1 clinical trials (about 1 year) are used to test safety; typically 20 to 80 volunteers are used. Phase II clinical trials (about 2 years) use 100 to 300 patients and the emphasis is on efficacy (i.e., does it help the patient) as well as further determining which side effects exist. Compounds that are still promising enter phase III clinical trials (about 3 years) with 1000 to 3000 patients. Since individuals vary in body chemistry, it is important to test the range of responses in terms of both side effects and efficacy by using a representative cross section of the population. Data from clinical trials is presented to the FDA for review (about 18 months). If the clinical trials are well designed and demonstrate statistically significant improvements in health with acceptable side effects, the drug is likely to be approved. Even after this point there is continued monitoring of the drug for adverse effects. The whole drug discovery-through-approval process takes 15 years on the average and costs about $400 million (in 1996). Only one in ten drugs that enter human clinical trials receives approval. Recent FDA reforms have decreased the time to obtain approval for life-saving drugs in treatment of diseases such as cancer and AIDS, but the overall process is still lengthy.

This process greatly affects a bioprocess engineer. FDA approval is for the product *and* the process together. There have been tragic examples where a small process change has allowed a toxic trace compound to form or become incorporated in the final product, resulting in severe side effects, including death. Thus, process changes may require new clinical trials to test the safety of the resulting product. Since clinical trials are very expensive, process improvements are made under a limited set of circumstances. Even during clinical trials it is difficult to make major process changes.

Drugs sold on the market or used in clinical trials must come from facilities that are certified as GMP. GMP stands for *good manufacturing practice*. GMP concerns the actual manufacturing facility design and layout, the equipment and procedures, training of production personnel, control of process inputs (e.g., raw materials and cultures), and handling of product. The plant layout and design must prevent contamination of the product and dictates the flow of material, personnel, and air. Equipment and procedures must be *validated*. Procedures include not only operation of a piece of equipment, but also cleaning and sterilization. Computer software used to monitor and control the process must be validated. Off-line assays done in laboratories must satisfy *good laboratory practices* (GLP). Procedures are documented by SOPs (*standard operating procedures*).

The GMP guidelines stress the need for documented procedures to validate performance. "Process validation is establishing documented evidence which provides a high degree of assurance that a specific process will consistently produce a product meeting its predetermined specifications and quality characteristics" and "There shall be written procedures for production and process-control to assure that products have the identity, strength, quality, and purity they purport or are represented to possess."

The actual process of doing validation is often complex, particularly when a whole facility design is considered. The FDA provides extensive information and guidelines which are updated regularly. If students become involved in biomanufacturing for pharmaceuticals, they will need to consult these sources. However, certain key concepts do not change. These concepts are written documentation, consistency of procedures, consistency of product, and demonstrable measures of product quality, particularly purity and safety. These tasks are demanding and require careful attention to detail. Bioprocess engineers will often find that much of their effort will be to satisfy these regulatory requirements.

The key point is that process changes cannot be made without considering their considerable regulatory impact.

SUGGESTIONS FOR FURTHER READING

A. History of Penicillin

ELDER, A. L., ed., *The History of Penicillin Production*, Chem. Eng. Prog. Symp. Ser. *66* (#100), American Institute of Chemical Engineers, New York, 1970.

HOBBY, G. L., *Penicillin. Meeting the Challenge*, Yale University Press, New Haven, CT, 1985.

MOBERG, C. L., Penicillin's Forgotten Man: Norman Heatley, *Science* 253: 734–735, 1991.

MATELES, R. I., ed., *Penicillin: A Paradigm for Biotechnology*, Candida Corp., Chicago, IL, 1998. (This contains the work by Elder which is no longer in print plus two additional chapters on current pratice.)

SHEEHAN, J. C., *The Enchanted Ring. The Untold Story of Penicillin*, MIT Press, Cambridge, MA, 1982.

B. Regulatory Issues

DURFOR, C. N., AND SCRIBNER, C. L., An FDA perspective of manufacturing changes for products in human use, *Ann. NY Acad. Sci.* 665:356–363, 1992.

NAGLAK, T. J., KEITH, M. G. AND OMSTEAD, D. R., Validation of fermentation processes, *Biopharm* (July/Aug.):28–36, 1994.

REISMAN, H. B., Problems in scale-up of biotechnology production processes, *Crit. Rev. Biotechnol.* *13*:195–253, 1993.

PROBLEMS

1.1. What is GMP and how does it relate to the regulatory process for pharmaceuticals?

1.2. When the FDA approves a process, it requires *validation* of the process. Explain what validation means in the FDA context.

1.3. Why does the FDA approve the process and product together?

The Basics of Biology:
An Engineer's Perspective

—————————— 2 ——————————

An Overview
of Biological Basics

2.1. ARE ALL CELLS THE SAME?

2.1.1. Microbial Diversity

Life is very tenacious and can exist in extreme environments. Living cells can be found almost anywhere that water is in the liquid state. The right temperature, pH, and moisture levels vary from one organism to another.

Some cells can grow at −20°C (in a brine to prevent freezing), while others can grow at 120°C (where water is under high enough pressure to prevent boiling). Cells that grow best at low temperatures (below 20°C) are usually called *psychrophiles*, while those with temperature optima in the range of 20° to 50°C are *mesophiles*. Organisms that grow best at temperatures greater than 50°C are *thermophiles*.

Many organisms have pH optima far from neutrality; some prefer pH values down to 1 or 2, while others may grow well at pH 9. Some organisms can grow at low pH values and high temperatures.

Although most organisms can grow only where water activity is high, others can grow on barely moist solid surfaces or in solutions with high salt concentrations.

Some cells require oxygen for growth and metabolism. Such organisms can be termed *aerobic*. Other organisms are inhibited by the presence of oxygen and grow only *anaerobically*. Some organisms can switch the metabolic pathways to allow them to grow under either circumstance. Such organisms are *facultative*.

Often organisms can grow in environments with almost no obvious source of nutrients. Some *cyanobacteria* (formerly called blue-green algae) can grow in an environment with only a little moisture and a few dissolved minerals. These bacteria are photosynthetic and can convert CO_2 from the atmosphere into the organic compounds necessary for life. They can also convert N_2 into NH_3 for use in making the essential building blocks of life. Such organisms are important in colonizing nutrient-deficient environments.

Organisms from these extreme environments (*extremophiles*) often provide the human race with important tools for processes to make useful chemicals and medicinals. They are also key to the maintenance of natural cycles and can be used in the recovery of metals from low-grade ores or in the desulfurization of coal or other fuels. The fact that organisms can develop the capacity to exist and multiply in almost any environment on earth is extremely useful.

Not only do organisms occupy a wide variety of habitats, but they also come in a wide range of sizes and shapes. Spherical, cylindrical, ellipsoidal, spiral, and pleomorphic cells exist. Special names are used to describe the shape of bacteria. A cell with a spherical or elliptical shape is often called a *coccus* (plural, *cocci*); a cylindrical cell is a rod or *bacillus* (plural, *bacilli*); a spiral-shaped cell is a *spirillum* (plural, *spirilla*). Some cells may change shape in response to changes in their local environment.

Thus, organisms can be found in the most extreme environments and have evolved a wondrous array of shapes, sizes, and metabolic capabilities. This great diversity provides the engineer with an immense variety of potential tools. We have barely begun to learn how to exploit these tools.

2.1.2. Naming Cells

The situation is complicated by the bewildering variety of organisms present. A systematic approach to classifying these organisms is an essential aid to their intelligent use. *Taxonomy* is the development of approaches to organize and summarize our knowledge about the variety of organisms that exist. Although a knowledge of taxonomy may seem remote from the needs of the engineer, it is necessary for efficient communication among engineers and scientists working with living cells. Taxonomy can also play a critical role in patent litigation involving bioprocesses.

While taxonomy is concerned with approaches to classification, *nomenclature* refers to the actual naming of organisms. For microorganisms we use a dual name (binary nomenclature). The names are given in Latin or are Latinized. A genus is a group of related species, while a species includes organisms that are substantially alike. A common gut organism that has been well studied is *Escherichia coli*. *Escherichia* is the genus and *coli* the species. When writing a report or paper, it is common practice to give the full name when the organism is first mentioned, but in subsequent discussion to abbreviate the *genus* to the first letter. In this case we would use *E. coli*. Although organisms that belong to the same species all share the same major characteristics, there are subtle and often technologically important variations within species. An *E. coli* used in one laboratory may differ from that used in another. Thus, various strains and substrains are designated by the addition of letters and numbers. For example, *E. coli* B/r A will differ in growth and physiological properties from *E. coli* K12.

Now that we know how to name organisms, we could consider broader classification up to the level of kingdoms. There is no universal agreement on how to classify microorganisms at this level. Such classification is rather arbitrary and need not concern us. However, we must be aware that there are two primary cell types: *eucaryotic* and *procaryotic*. The primary difference between them is the presence or absence of a membrane around the cell's genetic information.

Procaryotes have a simple structure with a single chromosome. Procaryotic cells have no nuclear membrane and no organelles, such as the mitochondria and endoplasmic reticulum. Eucaryotes have a more complex internal structure, with more than one chromosome (DNA molecule) in the nucleus. Eucaryotic cells have a true nuclear membrane and contain mitochondria, endoplasmic reticulum, golgi apparatus, and a variety of specialized organelles. We will soon describe each of these components (Section 2.1.5). A detailed comparison of procaryotes and eucaryotes is presented in Table 2.1. Structural differences between procaryotes and eucaryotes are discussed later.

Recently, it has become obvious that the situation is even more complicated. Evidence suggests that a common or universal ancestor gave rise to three distinctive branches of life: eucaryotes, eubacteria (or "true" bacteria), and archaebacteria. Table 2.2 summarizes some of the distinctive features of these groups. The ability to sequence the genes of whole organisms will have a great impact on our understanding of how these families evolved and are related.

Viruses cannot be classified under any of these categories, as they are not free-living organisms. Let's consider first some of the characteristics of these rather simple "organisms."

TABLE 2.1   A Comparison of Procaryotes with Eucaryotes

Characteristic	Procaryotes	Eucaryotes
Genome		
No. of DNA molecules	One	More than one
DNA in organelles	No	Yes
DNA observed as chromosomes	No	Yes
Nuclear membrane	No	Yes
Mitotic and meiotic division of the nucleus	No	Yes
Formation of partial diploid	Yes	No
Organelles		
Mitochondria	No	Yes
Endoplasmic reticulum	No	Yes
Golgi apparatus	No	Yes
Photosynthetic apparatus	Chlorosomes	Chloroplasts
Flagella	Single protein, simple structure	Complex structure, with microtubules
Spores	Endospores	Endo- and exospores
Heat resistance	High	Low

With permission, from N. F. Millis in *Comprehensive Biotechnology*, M. Moo-Young, ed., Vol. 1, Elsevier Science, 1985.

TABLE 2.2 Primary Subdivisions of Cellular Organisms That Are Now Recognized

Group	Cell structure	Properties	Constituent groups
Eucaryotes	Eucaryotic	Multicellular; extensive differentiation of cells and tissues Unicellular, coenocytic or mycelial; little or no tissue differentiation	Plants (seed plants, ferns, mosses) Animals (vertebrates, invertebrates) Protists (algae, fungi, protozoa)
Eubacteria	Procaryotic	Cell chemistry similar to eucaryotes	Most bacteria
Archaebacteria	Procaryotic	Distinctive cell chemistry	Methanogens, halophiles, thermo-acidophiles

With permission, from R. Y. Stainer et al., *The Microbial World*, 5th ed., Pearson Education, Upper Saddle River, NJ, 1986.

2.1.3. Viruses

Viruses are very small and are obligate parasites of other cells, such as bacterial, yeast, plant, and animal cells. Viruses cannot capture or store free energy and are not functionally active except when inside their host cells. The sizes of viruses vary from 30 to 200 nanometers (nm). Viruses contain either DNA (DNA viruses) or RNA (RNA viruses) as genetic material. DNA stands for deoxyribonucleic acid, and RNA is ribonucleic acid; we will soon discuss these molecules in more detail. In free-living cells, all genetic information is contained on the DNA, whereas viruses can use either RNA or DNA to encode such information. This nuclear material is covered by a protein coat called a *capsid*. Some viruses have an outer envelope of a lipoprotein and some do not.

Almost all cell types are susceptible to viral infections. Viruses infecting bacteria are called *bacteriophages*. Some bacteriophages have a hexagonal head, tail, and tail fibers. Bacteriophages attach to the cell wall of a host cell with tail fibers, alter the cell wall of the host cell, and inject the viral nuclear material into the host cell. Figure 2.1 describes the attachment of a virus onto a host cell. Bacteriophage nucleic acids reproduce inside the host cells to produce more phages. At a certain stage of viral reproduction, host cells lyse or break apart and phage particles are released, which can infect new host cells. This mode of reproduction of viruses is called the *lytic cycle*. In some cases, phage DNA may be incorporated into the host DNA, and the host may continue to multiply in this state, which is called the *lysogenic cycle*.

Viruses are the cause of many diseases, and antiviral agents are important targets for drug discovery. Additionally, viruses are directly important to bioprocess technology. For example, a phage attack on an *E. coli* fermentation to make a recombinant protein product can be extremely destructive, causing the loss of the whole culture in vessels of many thousands of liters. However, phages can be used as agents to move desired genetic material into *E. coli*. Modified animal viruses can be used as vectors in genetically engineering animal cells to produce proteins from recombinant DNA technology. In some cases a killed virus preparation is used as a vaccine. In other cases genetic engineering allows production of viruslike particles that are empty shells; the shell is the capsid and all nucleic acid is removed. Such particles can be used as vaccines without fear of viral replication, since all of the genetic material has been removed. For gene therapy one approach

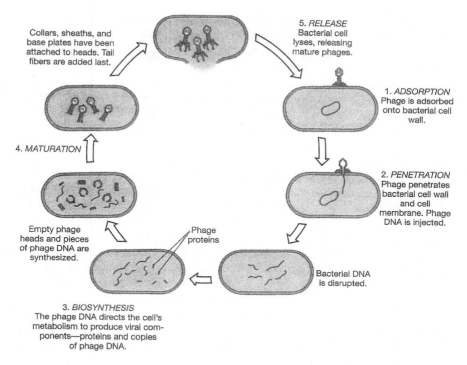

Figure 2.1. Replication of a virulent bacteriophage. A virulent phage undergoes a lytic cycle to produce new phage particles within a bacterial cell. *Cell lysis* releases new phage particles that can infect more bacteria. (With permission, from J. G. Black, *Microbiology: Principles and Applications*, 3d ed., p. 282. This material is used by permission of John Wiley & Sons, Inc.)

is to use a virus where viral genetic material has been replaced with the desired gene to be inserted into the patient. The viral capsid can act as a Trojan Horse to protect the desired gene in a hostile environment and then to deliver it selectively to a particle cell type. Thus, viruses can do great harm but also are important biotechnological tools.

2.1.4. Procaryotes

The sizes of most procaryotes vary from 0.5 to 3 micrometers (μm) in equivalent radius. Different species have different shapes, such as spherical or coccus (e.g., *Staphylococci*), cylindrical or bacillus (*E. coli*), or spiral or spirillum (*Rhodospirillum*). Procaryotic cells grow rapidly, with typical doubling times of one-half hour to several hours. Also, procaryotes can utilize a variety of nutrients as carbon source, including carbohydrates, hydrocarbons, proteins, and CO_2.

2.1.4.1. Eubacteria. The Eubacteria can be divided into several different groups. One distinction is based on the gram stain (developed by Hans Christian Gram in 1884). The staining procedure first requires fixing the cells by heating. The basic dye, crystal violet, is added; all bacteria will stain purple. Next, iodine is added, followed by the addition of ethanol. *Gram-positive* cells remain purple, while *gram-negative* cells become colorless. Finally, counterstaining with safranin leaves gram-positive cells purple, while gram-negative cells are red. This ability to react with the gram stain reveals intrinsic differences in the structure of the cell envelope.

A typical gram-negative cell is *E. coli* (see Fig. 2.2). It has an *outer membrane* supported by a thin peptidoglycan layer. *Peptidoglycan* is a complex polysaccharide with amino acids and forms a structure somewhat analogous to a chain-link fence. A second membrane (the inner or *cytoplasmic membrane*) exists and is separated from the outer membrane by the *periplasmic space*. The cytoplasmic membrane contains about 50% protein, 30% lipids, and 20% carbohydrates. The cell envelope serves to retain important cellular compounds and to preferentially exclude undesirable compounds in the environment. Loss of membrane integrity leads to *cell lysis* (cells breaking open) and cell death. The cell envelope is crucial to the transport of selected material in and out of the cell.

A typical gram-positive cell is *Bacillus subtilis*. Gram-positive cells do not have an outer membrane. Rather they have a very thick, rigid cell wall with multiple layers of peptidoglycan. Gram-positive cells also contain *teichoic acids* covalently bonded to the peptidoglycan. Because gram-positive bacteria have only a cytoplasmic membrane, they are often much better suited to excretion of proteins. Such excretion can be technologically advantageous when the protein is a desired product.

Some bacteria are not gram-positive or gram-negative. For example, the *Mycoplasma* have no cell walls. These bacteria are important not only clinically (e.g., primary atypical pneumonia), but also because they commonly contaminate media used industrially for animal cell culture.

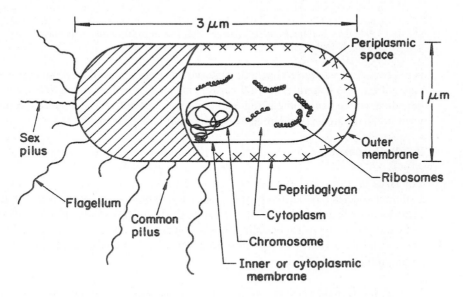

Figure 2.2. Schematic of a typical gram-negative bacterium (*E. coli*). A gram-positive cell would be similar, except that it would have no outer membrane, its peptidoglycan layer would be thicker, and the chemical composition of the cell wall would differ significantly from the outer envelope of the gram-negative cell.

Actinomycetes are bacteria, but, morphologically, actinomycetes resemble molds with their long and highly branched hyphae. However, the lack of a nuclear membrane and the composition of the cell wall require classification as bacteria. Actinomycetes are important sources of antibiotics. Certain Actinomycetes possess amylolytic and cellulolytic enzymes and are effective in enzymatic hydrolysis of starch and cellulose. *Actinomyces, Thermomonospora,* and *Streptomyces* are examples of genera belonging to this group.

Other distinctions within the eubacteria can be made based on cellular nutrition and energy metabolism. One important example is photosynthesis. The cyanobacteria (formerly called blue-green algae) have chlorophyll and fix CO_2 into sugars. Anoxygenic photosynthetic bacteria (the purple and green bacteria) have light-gathering pigments called *bacteriochlorophyll.* Unlike true photosynthesis, the purple and green bacteria do not obtain reducing power from the splitting of water and do not form oxygen.

When stained properly, the area occupied by the procaryotic cell's DNA can be easily seen. Procaryotes may also have other visible structures when viewed under the microscope, such as *ribosomes, storage granules, spores,* and *volutins.* Ribosomes are the site of protein synthesis. A typical bacterial cell contains approximately 10,000 ribosomes per cell, although this number can vary greatly with growth rate. The size of a typical ribosome is 10 to 20 nm and consists of approximately 63% RNA and 37% protein. Storage granules (which are not present in every bacterium) can be used as a source of key metabolites and often contain polysaccharides, lipids, and sulfur granules. The sizes of storage granules vary between 0.5 and 1 μm.

Some bacteria make intracellular spores (often called endospores in bacteria). Bacterial spores are produced as a resistance to adverse conditions such as high temperature, radiation, and toxic chemicals. The usual concentration is 1 spore per cell, with a spore size of about 1 μm. Spores can germinate under favorable growth conditions to yield actively growing bacteria.

Volutin is another granular intracellular structure, made of inorganic polymetaphosphates, that is present in some species. Some photosynthetic bacteria, such as *Rhodospirillum,* have chromatophores that are large inclusion bodies (50 to 100 nm) utilized in photosynthesis for the absorption of light.

Extracellular products can adhere to or become incorporated within the surface of the cell. Certain bacteria have a coating or outside cell wall called *capsule,* which is usually a polysaccharide or sometimes a polypeptide. Extracellular polymers are important to biofilm formation and response to environmental challenges (e.g., viruses). Table 2.3 summarizes the architecture of most bacteria.

2.1.4.2. Archaebacteria. The archaebacteria appear under the microscope to be nearly identical to many of the eubacteria. However, these cells differ greatly at the molecular level. In many ways the archaebacteria are as similar to the eucaryotes as they are to the eubacteria. Some examples of differences between archaebacteria and eubacteria are as follows:

1. Archaebacteria have no peptidoglycan.
2. The nucleotide sequences in the ribosomal RNA are similar within the archaebacteria but distinctly different from eubacteria.
3. The lipid composition of the cytoplasmic membrane is very different for the two groups.

TABLE 2.3 Characteristics of Various Components of Bacteria

Part	Size	Composition and comments
Slime layer		
Microcapsule	5–10 nm	Protein–polysaccharide–lipid complex responsible for the specific antigens of enteric bacteria and other species.
Capsule	0.5–2.0 μm	Mainly polysaccharides (e.g., *Streptococcus*); sometimes polypeptides (e.g., *Bacillus antracis*).
Slime	Indefinite	Mainly polysaccharides (e.g., *Leuconostoc*); sometimes polypeptides (e.g., *Bacillus subtilis*).
Cell wall		
Gram-positive species	10–20 nm	Confers shape and rigidity upon the cell. 20% dry weight of the cell. Consists mainly of macromolecules of a mixed polymer of *N*-acetyl muramic-peptide, teichoic acids, and polysaccharides.
Gram-negative species	10–20 nm	Consists mostly of a protein–polysaccharide–lipid complex with a small amount of the muramic polymer.
Cell membrane	5–10 nm	Semipermeable barrier to nutrients. 5% to 10% dry weight of the cell, consisting of 50% protein, 28% lipid, and 15% to 20% carbohydrate in a double-layered membrane.
Flagellum	10–20 nm by 4–12 μm	Protein of the myosin–keratin–fibringen class, MW of 40,000. Arises from the cell membrane and is responsible for motility.
Pilus (fimbria)	5–10 nm by 0.5–2.0 μm	Rigid protein projections from the cell. Especially long ones are formed by *Escherichia coli*.
Inclusions		
Spore	1.0–1.5 μm by 1.6–2.0 μm	One spore is formed per cell intracellularly. Spores show great resistance to heat, dryness, and antibacterial agents.
Storage granule	0.5–2.0 μm	Glycogenlike, sulfur, or lipid granules may be found in some species.
Chromatophore	50–100 nm	Organelles in photosynthetic species. *Rhodospirillum rubrum* contains about 6000 per cell.
Ribosome	10–30 nm	Organelles for synthesis of protein. About 10,000 ribosomes per cell. They contain 63% RNA and 37% protein.
Volutin	0.5–1.0 μm	Inorganic polymetaphosphates that stain metachromatically.
Nuclear material		Composed of DNA that functions genetically as if the genes were arranged linearly on a single endless chromosome, but that appears by light microscopy as irregular patches with no nuclear membrane or distinguishable chromosomes. Autoradiography confirms the linear arrangement of DNA and suggests a MW of at least 1000×10^6.

With permission, from S. Aiba, A. E. Humphrey, and N. F. Millis, *Biochemical Engineering*, 2d ed., University of Tokyo Press, Tokyo, 1973.

The archaebacteria usually live in extreme environments and possess unusual metabolism. Methanogens, which are methane-producing bacteria, belong to this group, as well as the thermoacidophiles. The thermoacidophiles can grow at high temperatures and low pH values. The halobacteria, which can live only in very strong salt solutions, are members of this group. These organisms are important sources for catalytically active proteins (enzymes) with novel properties.

2.1.5. Eucaryotes

Fungi (yeasts and molds), algae, protozoa, and animal and plant cells constitute the eucaryotes. Eucaryotes are five to ten times larger than procaryotes in diameter (e.g., yeast about 5 μm, animal cells about 10 μm, and plants about 20 μm). Eucaryotes have a true nucleus and a number of cellular organelles inside the cytoplasma. Figure 2.3 is a schematic of two typical eucaryotic cells.

In cell wall and cell membrane structure, eucaryotes are similar to procaryotes. The plasma membrane is made of proteins and phospholipids that form a bilayer structure. Major proteins of the membrane are hydrophobic and are embedded in the phospholipid matrix. One major difference is the presence of sterols in the cytoplasmic membrane of eucaryotes. Sterols strengthen the structure and make the membrane less flexible. The cell wall of eucaryotic cells shows considerable variations. Some eucaryotes have a peptidoglycan layer in their cell wall; some have polysaccharides and cellulose (e.g., algae). The plant cell wall is composed of cellulose fibers embedded in pectin aggregates, which impart strength to the cell wall. Animal cells do not have a cell wall but only a cytoplasmic membrane. For this reason, animal cells are very shear-sensitive and fragile. This factor significantly complicates the design of large-scale bioreactors for animal cells.

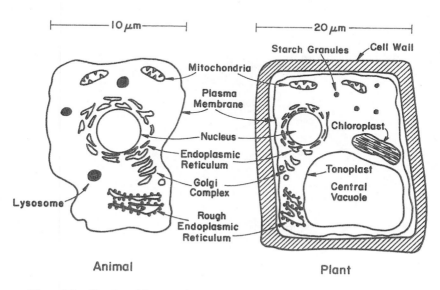

Figure 2.3. Sketches of the two primary types of higher eucaryotic cells. Neither sketch is complete, but each summarizes the principal differences and similarities of such cells.

The *nucleus* of eucaryotic cells contains *chromosomes* as nuclear material (DNA molecules with some closely associated small proteins), surrounded by a membrane. The nuclear membrane consists of a pair of concentric and porous membranes. The nucleolus is an area in the nucleus that stains differently and is the site of ribosome synthesis. However, many chromosomes contain small amounts of RNA and basic proteins called *histones* attached to the DNA. Each chromosome contains a single linear DNA molecule on which the histones are attached.

Cell division (asexual) in eucaryotes involves several major steps, such as DNA synthesis, nuclear division, cell division, and cell separation. Sexual reproduction in eucaryotic cells involves the conjugation of two cells called *gametes* (egg and sperm cells). The single cell formed from the conjugation of gametes is called a *zygote*. The zygote has twice as many chromosomes as does the gamete. Gametes are *haploid* cells, while zygotes are *diploid*. For humans, a haploid cell contains 23 chromosomes, and diploid cells have 46. The cell-division cycle (asexual reproduction) in a eucaryotic cell is depicted in Fig. 2.4.

The cell-division cycle is divided into four phases. The M phase consists of *mitosis* where the nucleous divides, and *cytokinesis* where the cell splits into separate daughter cells. All of the phases between one M phase and the next are known collectively as the *interphase*. The interphase is divided into three phases: G_1, S, and G_2. The cell increases in size during the interphase period. In the S phase the cell replicates its nuclear DNA. There are key checkpoints in the cycle when the cell machinery must commit to entry to the next phase. Checkpoints exist for entry into the S and M phases and exit from M phase. Cells may also be in a G_0 state, which is a resting state where there is no growth.

The *mitochondria* are the powerhouses of a eucaryotic cell, where respiration and oxidative phosphorylation take place. Mitochondria have a nearly cylindrical shape 1 μm in diameter and 2 to 3 μm in length. The typical structure of a mitochondrion is shown in

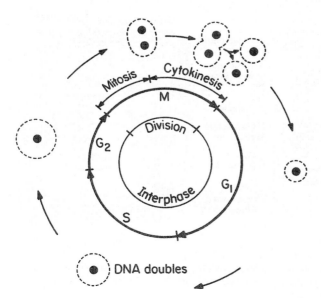

DNA doubles

Figure 2.4. Schematic of cell division cycle in an eucaryote. (See text for details.)

Fig. 2.5. The external membrane is made of a phospholipid bilayer with proteins embedded in the lipid matrix. The mitochondria contain a complex system of inner membranes called *cristae*. A gellike matrix containing large amounts of protein fills the space inside the cristae. Some enzymes of oxidative respiration are bound to the cristae. A mitochondrion has its own DNA and protein-synthesizing machinery and reproduces independently.

The *endoplasmic reticulum* is a complex, convoluted membrane system leading from the cell membrane into the cell. The rough endoplasmic reticulum contains ribosomes on the inner surfaces and is the site of protein synthesis and modifications of protein structure after synthesis. The smooth endoplasmic reticulum is more involved with lipid synthesis.

Lysosomes are very small membrane-bound particles that contain and release digestive enzymes. Lysosomes contribute to the digestion of nutrients and invading substances.

Peroxisomes are similar to lysosomes in their structure, but not in function. Peroxisomes carry out oxidative reactions that produce hydrogen peroxide.

Glyoxysomes are also very small membrane-bound particles that contain the enzymes of the glyoxylate cycle.

Golgi bodies are very small particles composed of membrane aggregates and are responsible for the secretion of certain proteins. Golgi bodies are sites where proteins are modified by the addition of various sugars in a process called *glycosylation*. Such modifications are important to protein function in the body.

Vacuoles are membrane-bound organelles of low density and are responsible for food digestion, osmotic regulation, and waste-product storage. Vacuoles may occupy a large fraction of cell volume (up to 90% in plant cells).

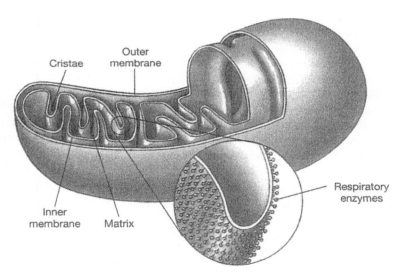

Figure 2.5. Diagram of a mitochondrion. Respiratory enzymes that make ATP are located on the surfaces of the inner membrane and the cristae, which are infoldings of the inner membrane. (With permission, from J. G. Black, *Microbiology Principles and Applications*, 3d ed., 1996, p. 94. This material is used by permission of John Wiley & Sons, Inc.)

Chloroplasts are relatively large, chlorophyll-containing, green organelles that are responsible for photosynthesis in photosynthetic eucaryotes, such as algae and plant cells. Every chloroplast contains an outer membrane and a large number of inner membranes called *thylakoids*. Chlorophyll molecules are associated with thylakoids, which have a regular membrane structure with lipid bilayers. Chloroplasts are autonomous units containing their own DNA and protein-synthesizing machinery.

Certain procaryotic and eucaryotic organisms contain *flagella*—long, filamentous structures that are attached to one end of the cell and are responsible for the motion of the cell. Eucaryotic flagella contain two central fibers surrounded by eighteen peripheral fibers, which exist in doublets. Fibers are in a tube structure called a *microtubule* and are composed of proteins called tubulin. The whole fiber assembly is embedded in an organic matrix and is surrounded by a membrane.

The *cytoskeleton* (in eucaryotic cells) refers to filaments that provide an internal framework to organize the cell's internal activities and control its shape. These filaments are critical in cell movement, transduction of mechanical forces into biological responses, and separation of chromosomes into the two daughter cells during cell division. Three types of fibers are present: *actin filaments, intermediate filaments*, and *microtubules*.

Cilia are flagellalike structures, but are numerous and shorter. Only one group of protozoa, called *ciliates,* contains cilia. *Paramecium* species contain nearly 10^4 cilia per cell. Ciliated organisms move much faster than flagellated ones.

This completes our summary of eucaryotic cell structure. Now let us turn our attention to the microscopic eucaryotes.

Fungi are heterotrophs that are widespread in nature. Fungal cells are larger than bacterial cells, and their typical internal structures, such as nucleus and vacuoles, can be seen easily with a light microscope. Two major groups of fungi are yeasts and molds.

Yeasts are single small cells of 5- to 10-µm size. Yeast cells are usually spherical, cylindrical, or oval. Yeasts can reproduce by asexual or sexual means. Asexual reproduction is by either budding or fission. In budding, a small bud cell forms on the cell, which gradually enlarges and separates from the mother cell. Asexual reproduction by *fission* is similar to that of bacteria. Only a few species of yeast can reproduce by fission. In fission, cells grow to a certain size and divide into two equal cells. Sexual reproduction of yeasts involves the formation of a *zygote* (a diploid cell) from fusion of two haploid cells, each having a single set of chromosomes. The nucleus of the diploid cells divides several times to form *ascospores*. Each ascospore eventually becomes a new haploid cell and may reproduce by budding and fission. The life cycle of a typical yeast cell is presented in Fig. 2.6.

The classification of yeasts is based on reproductive modes (e.g., *budding* or *fission*) and the nutritional requirements of cells. The most widely used yeast, *Saccharomyces cerevisiae*, is used in alcohol formation under anaerobic conditions (e.g., in wine, beer and whiskey making) and also for baker's yeast production under aerobic conditions.

Molds are filamentous fungi and have a mycelial structure. The *mycelium* is a highly branched system of tubes that contains mobil cytoplasm with many nuclei. Long, thin filaments on the mycelium are called *hyphae*. Certain branches of mycelium may grow in the air, and asexual spores called *conidia* are formed on these aerial branches. Conidia are nearly spherical in structure and are often pigmented. Some molds reproduce by sexual means and form sexual spores. These spores provide resistance against heat, freezing,

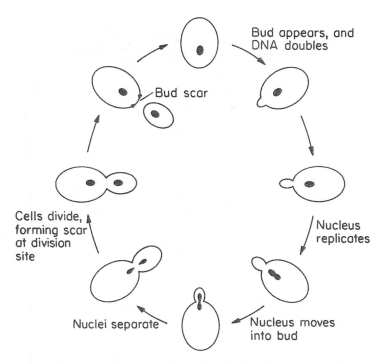

Figure 2.6. Cell-division cycle of a typical yeast, *Saccharomyces cerevisiae*. (With permission, from T. D. Brock, D. W. Smith, and M. T. Madigan, *Biology of Microorganisms*, 4th ed., Pearson Education, Upper Saddle River, NJ, 1984, p. 80.)

drying, and some chemical agents. Both sexual and asexual spores of molds can germinate and form new hyphae. Figure 2.7 describes the structure and asexual reproduction of molds.

Molds usually form long, highly branched cells and easily grow on moist, solid nutrient surfaces. The typical size of a filamentous form of mold is 5 to 20 μm. When grown in submerged culture, molds often form cell aggregates and pellets. The typical size of a mold pellet varies between 50 μm and 1 mm, depending on the type of mold and growth conditions. Pellet formation can cause some nutrient-transfer (mainly oxygen) problems inside the pellet. However, pellet formation reduces broth viscosity, which can improve bulk oxygen transfer.

On the basis of their mode of sexual reproduction, fungi are grouped in four classes.

1. The phycomycetes are algalike fungi; however, they do not possess chlorophyll and cannot photosynthesize. Aquatic and terrestrial molds belong to this category.
2. The ascomycetes form sexual spores called ascospores, which are contained within a sac (a capsule structure). Some molds of the genera *Neurospora* and *Aspergillus* and yeasts belong to this category.
3. The basidiomycetes reproduce by basidiospores, which are extended from the stalks of specialized cells called the basidia. Mushrooms are basidiomycetes.

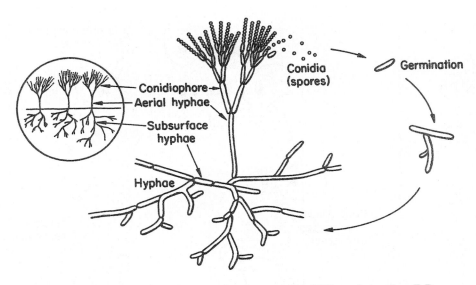

Figure 2.7. Structure and asexual reproduction of molds. (With permission, from T. D. Brock, K. M. Brock, and D. M. Ward, *Basic Microbiology with Applications*, 3d ed., Pearson Education, Upper Saddle River, NJ, 1986, p. 35.)

4. The deuteromycetes (*Fungi imperfecti*) cannot reproduce by sexual means. Only asexually reproducing molds belong to this category. Some pathogenic fungi, such as *Trichophyton*, which causes athlete's foot, belong to the deuteromycetes.

Molds are used for the production of citric acid (*Aspergillus niger*) and many antibiotics, such as penicillin (*Penicillium chrysogenum*). Mold fermentations make up a large fraction of the fermentation industry.

Algae are usually unicellular organisms. However, some plantlike multicellular structures are present in marine waters. All algae are photosynthetic and contain chloroplasts, which normally impart a green color to the organisms. The chloroplasts are the sites of chlorophyll pigments and are responsible for photosynthesis. The size of a typical unicellular alga is 10 to 30 μm. Multicellular algae sometimes form a branched or unbranched filamentous structure. Some algae contain silica or calcium carbonate in their cell wall. Diatoms containing silica in their cell wall are used as filter aids in industry. Some algae, such as *Chlorella, Scenedesmus, Spirullina*, and *Dunaliella*, are used for waste-water treatment with simultaneous single-cell protein production. Certain gelling agents such as agar and alginic acid are obtained from marine algae and seaweeds. Some algae are brown or red because of the presence of other pigments.

Protozoa are unicellular, motile, relatively large (1 mm to 50 mm) eucaryotic cells that lack cell walls. Protozoa usually obtain food by ingesting other small organisms, such as bacteria, or other food particles. Protozoa are usually uninucleate and reproduce by sexual or asexual means. They are classified on the basis of their motion. The *amoebae* move by ameboid motion, whereby the cytoplasm of the cell flows forward to form a pseudopodium (false foot), and the rest of the cell flows toward this lobe. The *flagellates*

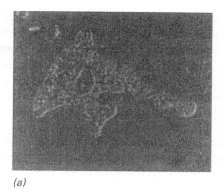

(a)

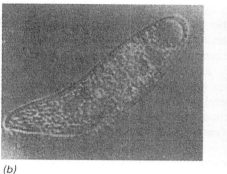

(b)

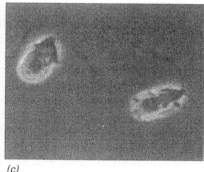

(c)

Figure 2.8. Protozoa. (a) An amoeba, *Amoeba proteus*. Magnification, 125X. (b) A ciliate, *Blepharisma*. Magnification, 120X. (c) A flagellate, *Dunaliella*. Magnification, 1900X. (With permission, from T. D. Brock, K. M. Brock, and D. M. Ward, *Basic Microbiology with Applications*, 3d ed., Pearson Education, Upper Saddle River, NJ, 1986, p. 40.)

move using their flagella. *Trypanosomes* move by flagella and cause a number of diseases in humans. The *ciliates* move by motion of a large number of small appendages on the cell surface called *cilia*. The *sporozoans* are nonmotile and contain members that are human and animal parasites. These protozoa do not engulf food particles, but absorb dissolved food components through their membranes. Protozoa cause some diseases, such as malaria and dysentery. Protozoa may have a beneficial role in removing bacteria from waste water in biological waste-water treatment processes and helping to obtain clean effluents. Microscopic pictures of some protozoa are presented in Fig. 2.8.

2.2. CELL CONSTRUCTION

2.2.1. Introduction

Living cells are composed of high-molecular-weight polymeric compounds such as proteins, nucleic acids, polysaccharides, lipids, and other storage materials (fats, polyhydroxybutyrate, glycogen). These biopolymers constitute the major structural elements of living cells. For example, a typical bacterial cell wall contains polysaccharides, proteins, and lipids; cell cytoplasm contains proteins mostly in the form of enzymes; in eucaryotes, the cell nucleus contains nucleic acids mostly in the form of DNA. In addition to these biopolymers, cells contain other metabolites in the form of inorganic salts (e.g., NH_4^+, PO_4^{3-}, K^+, Ca^{2+}, Na^+, SO_4^{2-}), metabolic intermediates (e.g., pyruvate, acetate), and vitamins. The elemental composition of a typical bacterial cell is 50% carbon, 20% oxygen, 14% nitrogen, 8% hydrogen, 3% phosphorus, and 1% sulfur, with small amounts of K^+, Na^+, Ca^{2+}, Mg^{2+}, Cl^-, and vitamins.

The cellular macromolecules are functional only when in the proper three-dimensional configuration. The interaction among them is very complicated. Each macromolecule is part of an intracellular organelle and functions in its unique microenvironment. Information transfer from one organelle to another (e.g., from nucleus to ribosomes) is mediated by special molecules (e.g., messenger RNA). Most of the enzymes and metabolic intermediates are present in cytoplasm. However, other organelles, such as mitochondria, contain enzymes and other metabolites. A living cell can be visualized as a very complex reactor in which more than 2000 reactions take place. These reactions (metabolic pathways) are interrelated and are controlled in a complicated fashion.

Despite all their complexity, an understanding of biological systems can be simplified by analyzing the system at several different levels: molecular level (molecular biology, biochemistry); cellular level (cell biology, microbiology); population level (microbiology, ecology); and production level (bioprocess engineering). This section is devoted mainly to the structure and function of biological molecules.

2.2.2. Amino Acids and Proteins

Proteins are the most abundant organic molecules in living cells, constituting 40% to 70% of their dry weight. Proteins are polymers built from amino acid monomers. Proteins typically have molecular weights of 6000 to several hundred thousand. The α-amino acids are the building blocks of proteins and contain at least one carboxyl group and one α-amino group, but they differ from each other in the structure of their R groups or side chains.

$$\begin{array}{c} H \\ | \\ H_2N-C-COOH \\ | \\ R \end{array}$$

L-amino acid

Although the sequence of amino acids determines a protein's *primary* structure, the *secondary* and *tertiary* structure are determined by the weak interactions among the various side groups. The ultimate three-dimensional structure is critical to the biological activity of the protein. Two major types of protein conformation are (1) fibrous proteins and (2) globular proteins. Figure 2.9 depicts examples of fibrous and globular proteins. Proteins have diverse biological functions, which can be classified in five major categories:

1. Structural proteins: glycoproteins, collagen, keratin
2. Catalytic proteins: enzymes

α–Helical coil   Supercoiling of α-helical coils to form ropes

Fibrous   Proteins

The tertiary structure
of a single-chain
globular protein

The quaternary structure of a multichain
or oligomeric globular protein

Globular   Proteins

Figure 2.9. Fibrous and globular proteins. (With permission, from A. Lehninger, *Biochemistry*, 2d ed., Worth Publishing, New York, 1975, p. 61.)

3. Transport proteins: hemoglobin, serum albumin
4. Regulatory proteins: hormones (insulin, growth hormone)
5. Protective proteins: antibodies, thrombin

The enzymes represent the largest class of proteins. Over 2000 different kinds of enzymes are known. Enzymes are highly specific in their function and have extraordinary catalytic power. Each enzyme's molecule contains an *active site* to which its specific substrate is bound during catalysis. Some enzymes are regulated and are called *regulatory* enzymes. Most enzymes are globular proteins.

The building blocks of proteins are α-amino acids, and there are 20 common amino acids. Amino acids are named on the basis of the side (R) group attached to the α-carbon. Amino acids are optically active and occur in two isomeric forms.

$$
\begin{array}{cc}
\overset{\displaystyle H}{\underset{\displaystyle R}{H_2N-C-COOH}} & \overset{\displaystyle H}{\underset{\displaystyle R}{HOOC-C-NH_2}} \\[2ex]
\text{L-amino acid} & \text{D-amino acid}
\end{array}
$$

Only L-amino acids are found in proteins. D-amino acids are rare in nature; they are found in the cell walls of some microorganisms and in some antibiotics.

Amino acids have acidic (—COOH) and basic (—NH₂) groups. The acidic group is neutral at low pH (—COOH) and negatively charged at high pH (—COO⁻). At intermediate pH values, an amino acid has positively and negatively charged groups, a dipolar molecule called a *zwitterion*.

$$
\overset{\displaystyle H}{\underset{\displaystyle R}{H_3N^+ - C - COO^-}}
$$

zwitterion

The pH value at which amino acids have no net charge is called the *isoelectric point*, which varies depending on the R group of amino acids. At its isoelectric point, an amino acid does not migrate under the influence of an electric field. Knowledge of the isoelectric point can be used in developing processes for protein purification. A list of 21 amino acids that are commonly found in proteins is given in Table 2.4.

The proteins are amino acid chains. The condensation reaction between two amino acids results in the formation of a *peptide bond*.

$$
\overset{\displaystyle H}{\underset{\displaystyle R_1}{NH_2-C-CO-}} \boxed{OH + H} \overset{\displaystyle H}{\underset{\displaystyle H}{-N}} \overset{}{\underset{\displaystyle R_2}{-C-COOH}}
$$

(2.1)

$$
\rightarrow \ NH_2 - \overset{\displaystyle H}{\underset{\displaystyle R_1}{C}} - \overset{\displaystyle O}{C} - \overset{}{\underset{\displaystyle H}{N}} - \overset{\displaystyle H}{\underset{\displaystyle R_2}{C}} - COOH + H_2O
$$

The peptide bond is planar. Peptides contain two or more amino acids linked by peptide bonds. Polypeptides usually contain fewer than 50 amino acids. Larger amino acid chains are called *proteins*. Many proteins contain organic and/or inorganic components other than amino acids. These components are called *prosthetic groups*, and the proteins containing prosthetic groups are named *conjugated proteins*. Hemoglobin is a conjugated protein and has four heme groups, which are iron-containing organometallic complexes.

TABLE 2.4 Chemical Structure of 21 Amino Acids of the General Structure

$$NH_2-\underset{\underset{R}{|}}{\overset{\overset{COOH}{|}}{C}}-H$$

R Group	Name	Abbreviation	Symbol	Class
—H	Glycine	GLY	G	Aliphatic
—CH_3	Alanine	ALA	A	
—$CH(CH_3)_2$	Valine	VAL	V	
—$CH_2CH(CH_3)_2$	Leucine	LEU	L	
—$CHCH_3CH_2CH_3$	Isoleucine	ILU	I	
—CH_2OH	Serine	SER	S	Hydroxyl or sulfur containing
—$CHOHCH_3$	Threonine	THR	T	
—CH_2SH	Cysteine	CYS	C	
—$(CH_2)_2SCH_3$	Methionine	MET	M	
—CH_2COOH	Aspartic acid	ASP	D	Acids and corresponding amides
—CH_2CONH_2	Asparagine	ASN	N	
—$(CH_2)_2COOH$	Glutamic acid	GLU	E	
—$(CH_2)_2CONH_2$	Glutamine	GLN	Q	
—$(CH_2)_3CH_2NH_2$	Lysine	LYS	K	Basic
—$(CH_2)_3NHCNHNH_2$	Arginine	ARG	R	
(imidazole ring) $-CH_2$	Histidine	HIS	H	
(phenyl ring) $-CH_2$	Phenylalanine	PHE	F	Aromatic
(phenol ring) $-CH_2$—OH	Tyrosine	TYR	Y	
(indole ring) CH_2	Tryptophan	TRP	W	
(pyrrolidine ring) —COOH	Proline	PRO	P	Imino acid
—CH_2—S—S—CH_2—	Cystine	—		Disulfide

The three-dimensional structure of proteins can be described at four different levels.

1. *Primary structure:* The primary structure of a protein is its linear sequence of amino acids. Each protein has not only a definite amino acid composition, but also a unique sequence. The one-dimensional structure of proteins (the amino acid sequence) has a profound effect on the resulting three-dimensional structure and, therefore, on the function of proteins.

2. *Secondary structure:* This is the way the polypeptide chain is extended and is a result of hydrogen bonding between residues not widely separated. Two major types of secondary structure are (a) helixes and (b) sheets. Helical structure can be either α-helical or triple helix. In an α-helical structure, hydrogen bonding can occur between the α-carboxyl group of one residue and the —NH group of its neighbor four units down the chain, as shown in Fig. 2.10. The triple-helix structure present in collagen consists of three α-helixes intertwined in a superhelix. Triple-helix structure is rigid and stretch resistant. The α-helical structure can be easily disturbed, since H bonds are not highly stable. However, the sheet structure (β-pleated sheet) is more stable. The hydrogen bonds between parallel chains stabilize the sheet structure and provide resistance to stretching (Fig. 2.11).

3. *Tertiary structure:* This is a result of interactions between R groups widely separated along the chain. The folding or bending of an amino acid chain induced by interaction between R groups determines the tertiary structure of proteins. R groups may interact by covalent, disulfide, or hydrogen bonds. Hydrophobic and hydrophilic interactions may also be present among R groups. The disulfide bond can cross-link two polypeptide chains (for example, insulin). Disulfide bonds are also critical in proper chain folding, as shown in Fig. 2.12. The tertiary structure of a protein has a profound effect on its function.

4. *Quaternary structure:* Only proteins with more than one polypeptide chain have quaternary structure. Interactions among polypeptide chains determine the quaternary structure (Fig. 2.9). Hemoglobin has four subunits (oligomeric), and interaction among

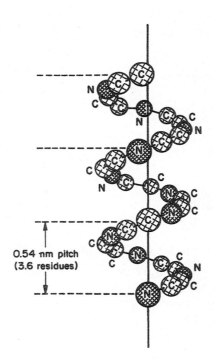

0.54 nm pitch
(3.6 residues)

Figure 2.10. The α-helical structure of fibrous proteins.

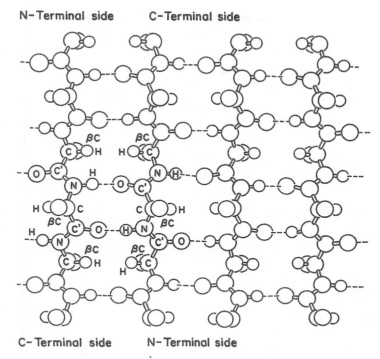

N-Terminal side   **C-Terminal side**

C-Terminal side   **N-Terminal side**

Figure 2.11. Representation of an antiparallel β-pleated sheet. Dashed lines indicate hydrogen bonds between strands.

these subunits results in a quaternary structure. The forces between polypeptide chains can be disulfide bonds or other weak interactions. The subunit structure of enzymes has an important role in the control of their catalytic activity.

Antibodies or *immunoglobulins* are proteins that bind to particular molecules or portions of large molecules with a high degree of specificity. Antibody (Ab) molecules appear in the blood serum and in certain cells of a vertebrate in response to foreign macromolecules. The foreign macromolecule is called the *antigen (Ag)*. The specific antibody molecules can combine with the antigen to form an *antigen–antibody complex*. The complex formation between Ag and Ab is called the *immune response*. In addition to their obvious clinical importance, antibodies are important industrial products for use in diagnostic kits and protein separation schemes. Antibodies may also become a key element in the delivery of some anticancer drugs. Antibodies have emerged as one of the most important products of biotechnology.

Antibody molecules have binding sites that are specific for and complementary to the structural features of the antigen. Antibody molecules usually have two binding sites and can form a three-dimensional lattice of alternating antigen and antibody molecules. This complex precipitates from the serum and is called *precipitin*. Antibodies are highly specific for the foreign proteins that induce their formation.

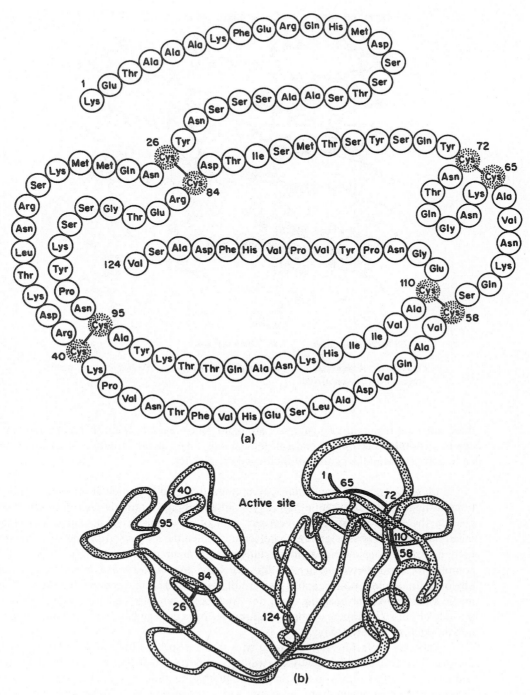

Figure 2.12. Structure of the enzyme ribonuclease. (a) Primary amino acid sequence, showing how sulfur-sulfur bonds between cysteine residues cause folding of the chain. (b) The three-dimensional structure of ribonuclease, showing how the macromolecule folds so that a site of enzymatic activity is formed, the active site. (With permission, from T. D. Brock, K. M. Brock, and D. M. Ward, *Basic Microbiology with Applications*, 3d ed., Pearson Education, Upper Saddle River, NJ, 1986, p. 56.)

The five major classes of immunoglobins in human blood plasma are: IgG, IgA, IgD, IgM, and IgE, of which the IgG globulins are the most abundant and the best understood. Molecular weights of immunoglobulins are about 150 kilodaltons (kD) except for IgM, which has a molecular weight of 900 kD. A *dalton* is a unit of mass equivalent to a hydrogen atom. Immunoglobulins have four polypeptide chains: two heavy (H) chains (about 430 amino acids) and two light (L) chains (about 214 amino acids). These chains are linked together by disulfide bonds into a Y-shaped, flexible structure (Fig. 2.13). The heavy chains contain a covalently bound oligosaccharide component. Each chain has a region of constant amino acid sequence and a variable-sequence region. The Ab molecule has two binding sites for the antigen; the variable portions of the L and H chains

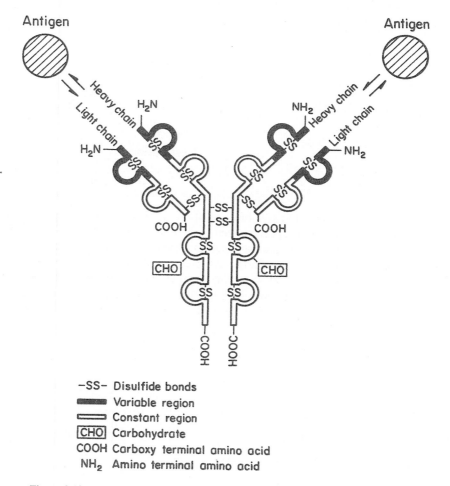

−SS− Disulfide bonds
▬▬ Variable region
▭ Constant region
CHO Carbohydrate
COOH Carboxy terminal amino acid
NH₂ Amino terminal amino acid

Figure 2.13. Structure of immunoglobulin G (IgG). Structure showing disulfide linkages within and between chains and antigen binding site. (With permission, adapted from T. D. Brock, D. W. Smith, and M. T. Madigan, *Biology of Microorganisms*, 4th ed., Pearson Education, Upper Saddle River, NJ, 1984, p. 524.)

contribute to these binding sites. The variable sections have *hypervariable* regions in which the frequency of amino acid replacement is high. The study of how cells develop and produce antibodies is being actively pursued worldwide. Recent developments have also led to insights on how to impart catalytic activities to antibodies. These molecules have been called *abzymes*. Coupling new developments in protein engineering to antibodies promises the development of extremely specific catalytic agents.

2.2.3. Carbohydrates: Mono- and Polysaccharides

Carbohydrates play key roles as structural and storage compounds in cells. They also appear to play critical roles in modulating some aspects of chemical signaling in animals and plants. Carbohydrates are represented by the general formula $(CH_2O)_n$, where $n \geq 3$, and are synthesized through photosynthesis.

The gases CO_2 and H_2O are converted through photosynthesis into sugars in the presence of sunlight and are then polymerized to yield polysaccharides, such as cellulose or starch.

Monosaccharides are the smallest carbohydrates and contain three to nine carbon atoms. Common monosaccharides are presented in Table 2.5. Common monosaccharides are either aldehydes or ketones. For example, glucose is an aldohexose. Glucose may be present in the form of a linear or ring structure. In solution, D-glucose is in the form of a ring (pyranose) structure. The L-form plays a minor role in biological systems.

TABLE 2.5 Common Monosaccharides

I. Aldoses
   a. D-Hexoses

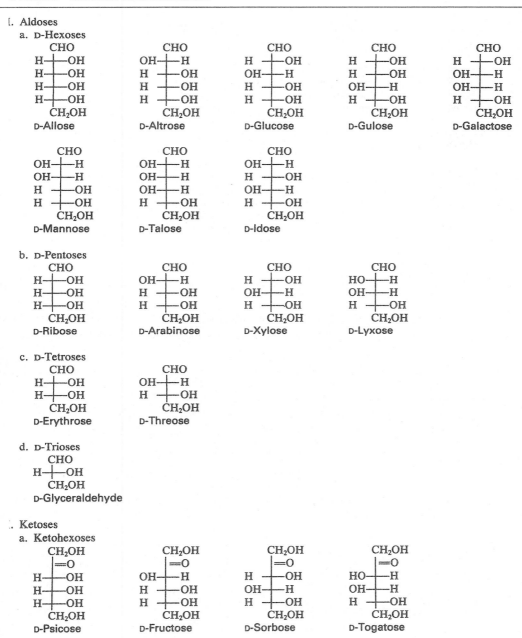

CHO	CHO	CHO	CHO	CHO
H—OH	OH—H	H—OH	H—OH	H—OH
H—OH	H—OH	OH—H	H—OH	OH—H
H—OH	H—OH	H—OH	OH—H	OH—H
H—OH	H—OH	H—OH	H—OH	H—OH
CH$_2$OH	CH$_2$OH	CH$_2$OH	CH$_2$OH	CH$_2$OH
D-Allose	D-Altrose	D-Glucose	D-Gulose	D-Galactose

CHO	CHO	CHO
OH—H	OH—H	OH—H
OH—H	OH—H	H—OH
H—OH	OH—H	OH—H
H—OH	H—OH	H—OH
CH$_2$OH	CH$_2$OH	CH$_2$OH
D-Mannose	D-Talose	D-Idose

   b. D-Pentoses

CHO	CHO	CHO	CHO
H—OH	OH—H	H—OH	HO—H
H—OH	H—OH	OH—H	OH—H
H—OH	H—OH	H—OH	H—OH
CH$_2$OH	CH$_2$OH	CH$_2$OH	CH$_2$OH
D-Ribose	D-Arabinose	D-Xylose	D-Lyxose

   c. D-Tetroses

CHO	CHO
H—OH	OH—H
H—OH	H—OH
CH$_2$OH	CH$_2$OH
D-Erythrose	D-Threose

   d. D-Trioses

CHO
H—OH
CH$_2$OH
D-Glyceraldehyde

2. Ketoses
   a. Ketohexoses

CH$_2$OH	CH$_2$OH	CH$_2$OH	CH$_2$OH
=O	=O	=O	=O
H—OH	OH—H	H—OH	HO—H
H—OH	H—OH	OH—H	OH—H
H—OH	H—OH	H—OH	H—OH
CH$_2$OH	CH$_2$OH	CH$_2$OH	CH$_2$OH
D-Psicose	D-Fructose	D-Sorbose	D-Togatose

TABLE 2.5 (*Continued*)

b. **Ketopentoses**

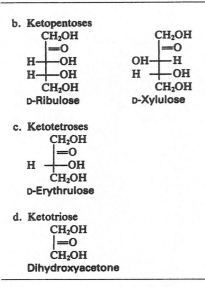

CH_2OH
$|=O$
H——OH
H——OH
CH_2OH
D-Ribulose

CH_2OH
$|=O$
OH——H
H ——OH
CH_2OH
D-Xylulose

c. **Ketotetroses**

CH_2OH
$|=O$
H ——OH
CH_2OH
D-Erythrulose

d. **Ketotriose**

CH_2OH
$|=O$
CH_2OH
Dihydroxyacetone

A particularly important group of monosaccharides are D-ribose and deoxyribose. These are five carbon ring-structured sugar molecules and are essential components of DNA and RNA.

D-Ribose

Deoxyribose

Disaccharides are formed by the condensation of two monosaccharides. For example, maltose is formed by the condensation of two glucose molecules via α-1,4 glycosidic linkage.

α-D-Glucose + α-D-Glucose ⟶ α-Maltose $+ H_2O$

Sucrose is a disaccharide of α-D-glucose and β-D-fructose. Lactose is a disaccharide of β-D-glucose and β-D-galactose.

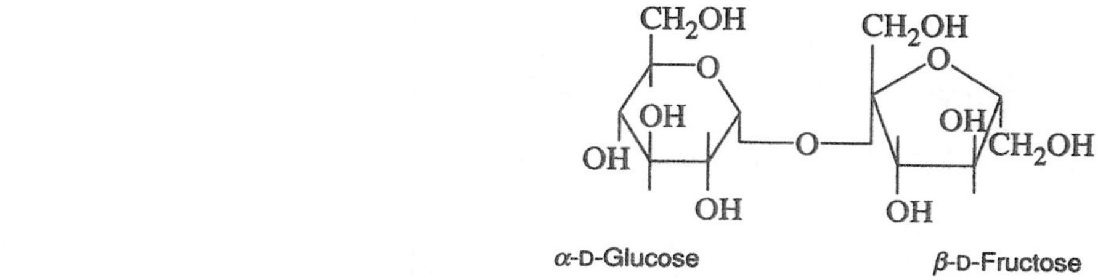

α-D-Glucose   β-D-Fructose

Sucrose

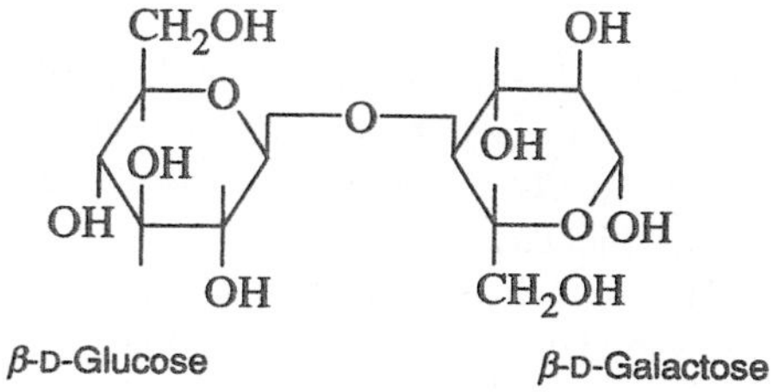

β-D-Glucose   β-D-Galactose

Lactose

Lactose is found in milk and whey, while sucrose is the major sugar in photosynthetic plants. Whey utilization remains an important biotechnological challenge, and sucrose is often a major component in artificial growth media.

Polysaccharides are formed by the condensation of more than two monosaccharides by glycosidic bonds. The polysaccharide processing industry makes extensive use of enzymatic processing and biochemical engineering.

Amylose is a straight chain of glucose molecules linked by α-1,4 glycosidic linkages. The molecular weight (MW) of amylose is between several thousand and one-half million daltons.

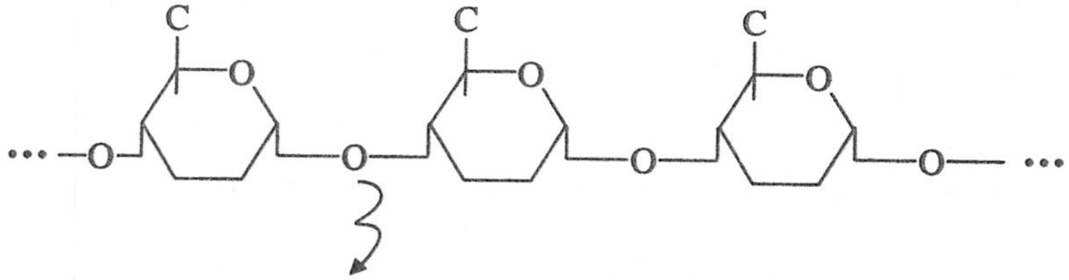

α-1,4-Glycosidic linkages

Amylose is water insoluble and constitutes about 20% of starch.

Amylopectin is a branched chain of D-glucose molecules. Branching occurs between the glycosidic—OH of one chain and the —6 carbon of another glucose, which is called α-1,6 glycosidic linkage.

Sec. 2.2   Cell Construction

Amylopectin molecules are much larger than those of amylose, with a MW of 1 to 2 million daltons. Amylopectin is water soluble. Partial hydrolysis of starch (acidic or enzymatic) yields glucose, maltose, and dextrins, which are branched sections of amylopectin. Dextrins are used as thickeners.

Glycogen is a branched chain of glucose molecules that resembles amylopectin. Glycogen is highly branched and contains about 12 glucose units in straight-chain segments. The MW of a typical glycogen molecule is less than 5×10^6 daltons.

Cellulose is a long, unbranched chain of D-glucose with a MW between 50,000 and 1 million daltons. The linkage between glucose monomers in cellulose is a β-1,4 glycosidic linkage.

The β-1,4 glycosidic bond is resistant to enzymatic hydrolysis. Only a few microorganisms can hydrolyze β-1,4 glycosidic bonds of cellulose. α-1,4 glycosidic bonds in starch or glycogen are relatively easy to break by enzymatic or acid hydrolysis. Efficient cellulose hydrolysis remains one of the most challenging problems in attempts to convert cellulosic wastes into fuels or chemicals.

2.2.4. Lipids, Fats, and Steroids

Lipids are hydrophobic biological compounds that are insoluble in water, but soluble in nonpolar solvents such as benzene, chloroform, and ether. They are usually present in the nonaqueous biological phases, such as plasma membranes. Fats are lipids that can serve as biological fuel-storage molecules. Lipoproteins and lipopolysaccharides are other types of lipids, which appear in the biological membranes of cells. Cells can alter the mix of lipids in their membranes to compensate (at least partially) for changes in temperature or to increase their tolerance to the presence of chemical agents such as ethanol.

The major component in most lipids is *fatty acids*, which are made of a straight chain of hydrocarbon (hydrophobic) groups, with a carboxyl group (hydrophilic) at the end. A typical fatty acid can be represented as

$$CH_3 - (CH_2)_n - COOH$$

An Overview of Biological Basics   Chap. 2

The value of n is typically between 12 and 20. Unsaturated fatty acids contain double —C=C— bonds, such as oleic acid.

$$\text{Oleic acid:} \quad CH_3 - (CH_2)_7 - HC = CH - (CH_2)_7 - COOH$$

A list of common fatty acids is presented in Table 2.6. The hydrocarbon chain of a fatty acid is hydrophobic (water insoluble), but the carboxyl group is hydrophilic (water soluble).

Fats are esters of fatty acids with glycerol. The formation of a fat molecule can be represented by the following reaction:

TABLE 2.6 Examples of Common Fatty Acids

Acid	Structure
Saturated fatty acids	
Acetic acid	CH_3COOH
Propionic acid	CH_3CH_2COOH
Butyric acid	$CH_3(CH_2)_2COOH$
Caproic acid	$CH_3(CH_2)_4COOH$
Decanoic acid	$CH_3(CH_2)_8COOH$
Lauric acid	$CH_3(CH_2)_{10}COOH$
Myristic acid	$CH_3(CH_2)_{12}COOH$
Palmitic acid	$CH_3(CH_2)_{14}COOH$
Stearic acid	$CH_3(CH_2)_{16}COOH$
Arachidic acid	$CH_3(CH_2)_{18}COOH$
Behenic acid	$CH_3(CH_2)_{20}COOH$
Lignoceric acid	$CH_3(CH_2)_{22}COOH$
Monoenoic fatty acids	
Oleic acid	$CH_3(CH_2)_7CH \overset{cis}{=\!=} CH(CH_2)_7COOH$
Dienoic fatty acid	
Linoleic acid	$CH_3(CH_2)_4(CH \overset{cis}{=\!=} CHCH_2)_2(CH_2)_6COOH$
Trienoic fatty acids	
α-Linolenic acid	$CH_3CH_2(CH \overset{cis}{=\!=} CHCH_2)_3(CH_2)_6COOH$
γ-Linolenic acid	$CH_3(CH_2)_4(CH \overset{cis}{=\!=} CHCH_2)_3(CH_2)_3COOH$
Tetraenoic fatty acid	
Arachidonic acid	$CH_3(CH_2)_4(CH \overset{cis}{=\!=} CHCH_2)_4(CH_2)_2COOH$
Unusual fatty acids	
Tariric acid	$CH_3(CH_2)_{10}C \equiv C(CH_2)_4COOH$
Lactobacillic acid	$CH_3(CH_2)_5CH\text{---}CH(CH_2)_9COOH$ (cyclopropane CH$_2$ bridge)
Prostaglandin (PGE$_2$)	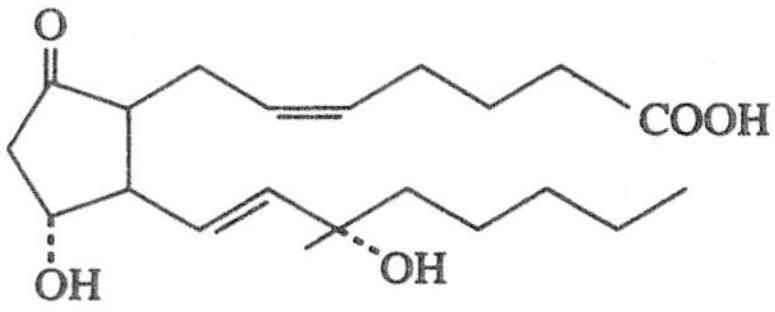

$$
\begin{array}{c}
\text{CH}_2\text{OH} \\[4pt]
| \\[2pt]
\text{CHOH} \\[2pt]
| \\[2pt]
\text{CH}_2\text{OH}
\end{array}
\;+\;
\begin{array}{c}
\quad\quad\overset{\displaystyle O}{\overset{\|}{\text{HO}-\text{C}}}-(\text{CH}_2)n_1-\text{CH}_3 \\[6pt]
\quad\quad\overset{\displaystyle O}{\overset{\|}{\text{HO}-\text{C}}}-(\text{CH}_2)n_2-\text{CH}_3 \\[6pt]
\quad\quad\overset{\displaystyle O}{\overset{\|}{\text{HO}-\text{C}}}-(\text{CH}_2)n_3-\text{CH}_3
\end{array}
\;\longrightarrow\;
\begin{array}{c}
\overset{\displaystyle O}{\overset{\|}{\text{CH}_2\text{O}-\text{C}}}-(\text{CH}_2)n_1-\text{CH}_3 \\[6pt]
\overset{\displaystyle O}{\overset{\|}{\text{CHO}-\text{C}}}-(\text{CH}_2)n_2-\text{CH}_3 \\[6pt]
\overset{\displaystyle O}{\overset{\|}{\text{CH}_2\text{O}-\text{C}}}-(\text{CH}_2)n_3-\text{CH}_3 + 3\text{H}_2\text{O}
\end{array}
$$

glycerol   fatty acids   fat

Phosphoglycerides have similar structures to fats, the only difference being that phosphoric acid replaces a fatty acid and is esterified at one end to glycerol.

Membranes with selective permeability are key to life. Cells must control the entry and exit of molecules. *Phospholipids* are key components, but membranes contain large amounts of proteins. Biological membranes are based on a lipid bilayer. The hydrophobic tails of the phospholipids associate with each other in the core of the membrane. The hydrophilic heads form the outsides of the membrane and associate with the aqueous cytosol or the aqueous extracellular fluid. Some proteins span across the membrane, while others are attached to one of the surfaces. Membranes are dynamic structures, and lipids and proteins can diffuse rapidly. Typical membrane phospholipids include phosphatidylcholine, phosphatidlyserine, phosphatidyl glycerol, and phosphatidyl inositol.

Another class of lipids of increasing technological importance is the polyhydroxy-alkanoates (PHA). In particular, *polyhydroxybutyrate* (PHB) is a good example. It can be used to form a clear, biodegradable polymeric sheet. Polymers with a variety of PHAs are being commercially developed. In some cells PHB is formed as a storage product.

Steroids can also be classified as lipids. Naturally occurring steroids are hormones that are important regulators of animal development and metabolism at very low concentrations (for example, $10^{-8}\,M$). A well-known steroid, cholesterol, is present in membranes of animal tissues. Figure 2.14 depicts the structures of some important steroids. Cortisone is an anti-inflammatory used to treat rheumatoid arthritis and some skin diseases. Derivatives of estrogens and progesterone are used as contraceptives. The commercial production of steroids is very important and depends on microbial conversions. Because of the large number of asymmetric centers, the total synthesis of steroids is difficult. Plants provide a source of abundant lipid precursors for these steroids, but the highly specific hydroxylation of these substrates at positions 11 (and 16) or dehydrogenations at position 1 are necessary to convert the precursors into compounds similar to those made in the adrenal gland. This cannot be done easily with chemical means and is done commercially using microbes that contain enzymes mediating specific hydroxylations or dehydrogenations.

2.2.5. Nucleic Acids, RNA, and DNA

Nucleic acids play the central role in reproduction of living cells. *Deoxyribonucleic acid* (DNA) stores and preserves genetic information. *Ribonucleic acid* (RNA) plays a central role in protein synthesis. Both DNA and RNA are large polymers made of their corresponding nucleotides.

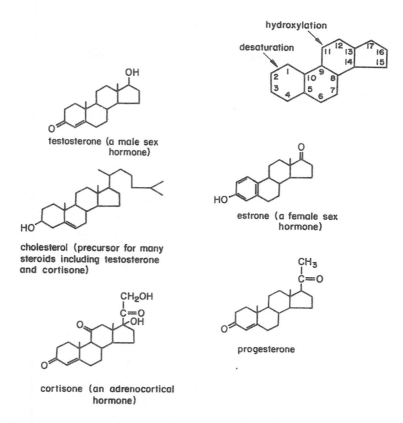

Figure 2.14. Examples of important steroids. The basic numbering of the carbon atoms in these molecules is also shown.

Nucleotides are the building blocks of DNA and RNA and also serve as molecules to store energy and reducing power. The three major components in all nucleotides are phosphoric acid, pentose (ribose or deoxyribose), and a base (purine or pyrimidine). Figure 2.15 depicts the structure of nucleotides and purine-pyrimidine bases. Two major purines present in nucleotides are adenine (A) and guanine (G), and three major pyrimidines are thymine (T), cytosine (C), and uracil (U). Deoxyribonucleic acid (DNA) contains A, T, G, and C, and ribonucleic acid (RNA) contains A, U, G, and C as bases. It is the base sequence in DNA that carries genetic information for protein synthesis. In Chapters 4 and 8 we discuss how this information is expressed and passed on from one generation to another.

In Chapter 5 we will discuss further the role of nucleotides in cellular energetics. The triphosphates of adenosine and to a lesser extent guanosine are the primary energy currency of the cell. The phosphate bonds in *ATP* (*adensosine triphosphate*) and *GTP* (*guanosine triphosphate*) are high-energy bonds. The formation of these phosphate bonds or their hydrolysis is the primary means by which cellular energy is stored or used. For example, the synthesis of a compound that is thermodynamically unfavorable can be

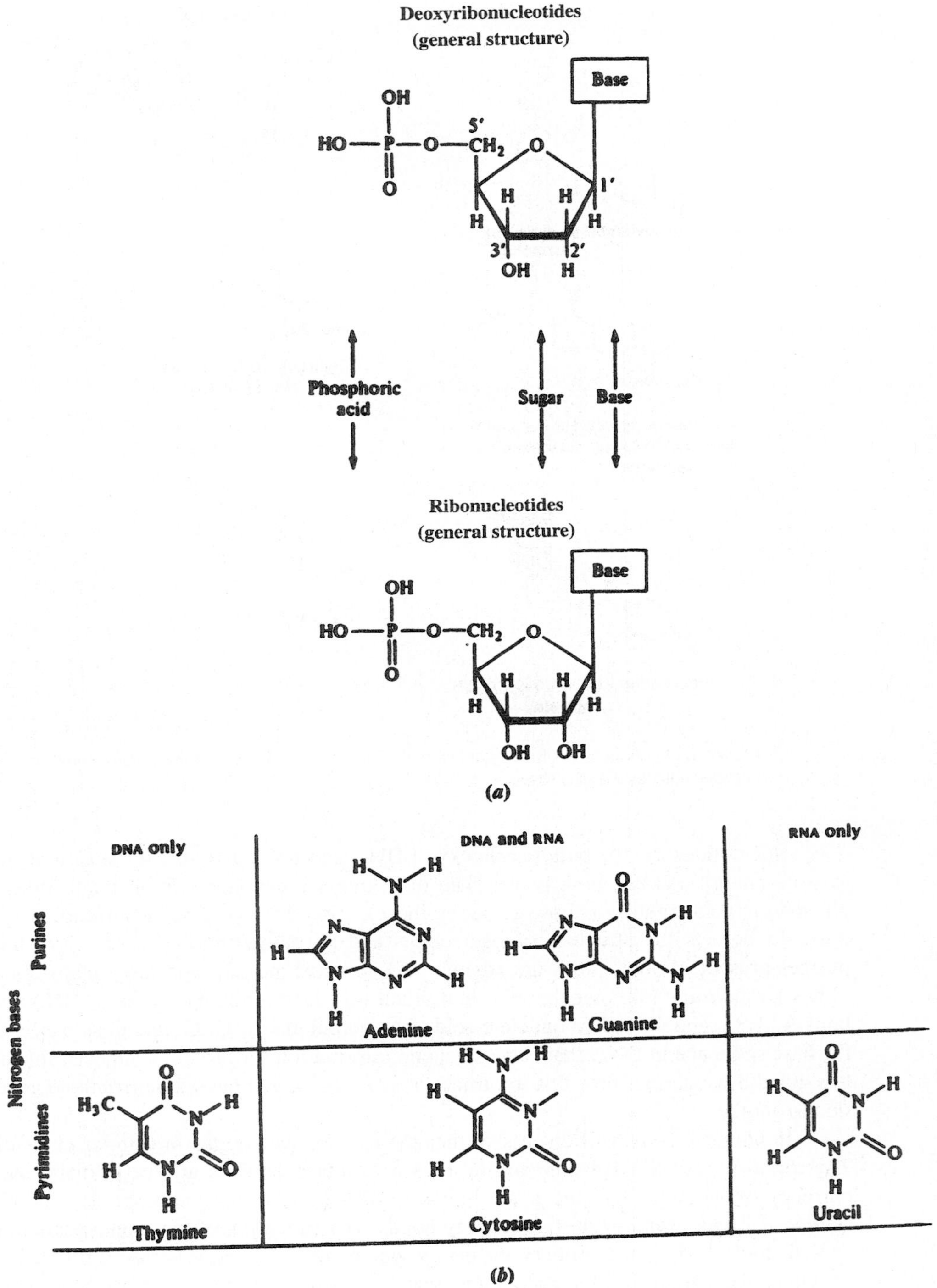

Figure 2.15. (a) General structure of ribonucleotides and deoxyribonucleotides. (b) Five nitrogenous bases found in DNA and RNA. (With permission, from J. E. Bailey and D. F. Ollis, *Biochemical Engineering Fundamentals*, 2d ed., McGraw-Hill Book Co., New York, 1986, p. 43.)

coupled to ATP hydrolysis to ADP (the diphosphate) or AMP (the monophosphate). The coupled reaction can proceed to a much greater extent, since the free-energy change becomes much more negative. In reactions that release energy (for example, oxidation of a sugar), the energy is "captured" and stored by the formation of a phosphate bond in a coupled reaction where ADP is converted into ATP.

In addition to using ATP to store energy, the cell stores and releases hydrogen atoms from biological oxidation-reduction reactions by using nucleotide derivatives. The two most common carriers of reducing power are *nicotinamide adenine dinucleotide* (*NAD*) and *nicotinamide adenine dinucleotide phosphate* (*NADP*).

In addition to this important role in cellular energetics, the nucleotides are important monomers. The polynucleotides (DNA and RNA) are formed by the condensation of nucleotides. The nucleotides are linked together between the 3' and 5' carbons' successive sugar rings by phosphodiester bonds. The structures of DNA and RNA are illustrated in Fig. 2.16.

DNA is a very large threadlike macromolecule (MW, 2×10^9 D in *E. coli*) and has a double-helical three-dimensional structure. The sequence of bases (purines and pyrimidines) in DNA carries genetic information, whereas sugar and phosphate groups perform a structural role. The base sequence of DNA is written in the $5' \rightarrow 3'$ direction, such as pAGCT. The double-helical structure of DNA is depicted in Fig. 2.17. In this structure, two helical polynucleotide chains are coiled around a common axis to form a double-helical DNA, and the chains run in opposite directions, $5' \rightarrow 3'$ and $3' \rightarrow 5'$. The main features of double-helical DNA structure are as follows:

1. The phosphate and deoxyribose units are on the outer surface, but the bases point toward the chain center. The planes of the bases are perpendicular to the helix axis.

2. The diameter of the helix is 2 nm. The helical structure repeats after ten residues on each chain, at an interval of 3.4 nm.

3. The two chains are held together by hydrogen bonding between pairs of bases. *Adenine is always paired with thymine* (two H bonds); *guanine is always paired with cytosine* (three H bonds). *This feature is essential to the genetic role of DNA.*

4. The sequence of bases along a polynucleotide is not restricted in any way, although each strand must be complementary to the other. The precise sequence of bases carries the genetic information.

The large number of H bonds formed between base pairs provides molecular stabilization. Regeneration of DNA from original DNA segments is known as *DNA replication*. When DNA segments are replicated, one strand of the new DNA segment comes directly from the parent DNA, and the other strand is newly synthesized using the parent DNA segment as a template. Therefore, DNA replication is semiconservative, as depicted in Fig. 2.18. The replication of DNA is discussed in more detail in Chapter 4.

Some cells contain circular DNA segments in cytoplasm that are called *plasmids*.† Plasmids are nonchromosomal, autonomous, self-replicating DNA segments. Plasmids are easily moved in and out of the cells and are often used for genetic engineering. Naturally occurring plasmids can encode factors that protect cells from antibiotics or harmful chemicals.

---

†Linear rather than circular plasmids can be found in some yeasts and other organisms.

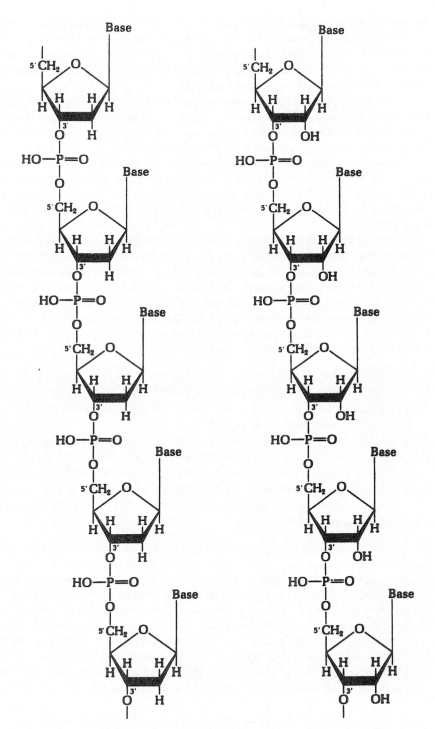

Figure 2.16. Structure of DNA and RNA chains. Phosphodiester bonds are formed between 3' and 5' carbon atoms. (With permission, from A. Lehninger, *Biochemistry*, 2d ed., Worth Publishing, New York, 1975, p. 319.)

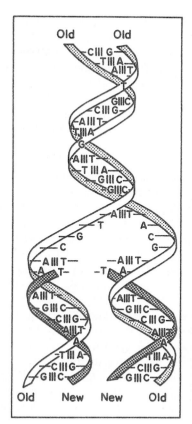

Figure 2.17. Double-helical structure of DNA, showing overall process of replication by complementary base pairing. (With permission, from T. D. Brock, D. W. Smith, and M. T. Madigan, *Biology of Microorganisms*, 4th ed., Pearson Education, Upper Saddle River, NJ, 1984, p. 276.)

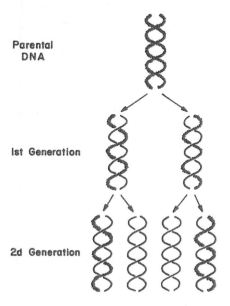

Figure 2.18. Semiconservative replication of DNA.

The major function of DNA is to carry genetic information in its base sequence. The genetic information in DNA is transcribed by RNA molecules and translated in protein synthesis. The templates for RNA synthesis are DNA molecules, and RNA molecules are the templates for protein synthesis. The formation of RNA molecules from DNA is known as DNA *transcription*, and the formation of peptides and proteins from RNA is called *translation*.

Certain RNA molecules function as the genetic information-carrying intermediates in protein synthesis (*messenger*, m-*RNA*), whereas other RNA molecules [*transfer* (t-*RNA*) and *ribosomal* (r-*RNA*)] are part of the machinery of protein synthesis. The ribosomal r-RNA is located in ribosomes which are small particles made of protein and RNA. *Ribosomes* are cytoplasmic organelles (usually attached on the inner surfaces of endoplasmic reticulum in eucaryotes) and are the sites of protein synthesis.

RNA is a long, unbranched macromolecule consisting of nucleotides joined by 3'–5' phosphodiester bonds. An RNA molecule may contain from 70 to several thousand nucleotides. RNA molecules are usually single stranded, except some viral RNA. However, certain RNA molecules contain regions of double-helical structure, like hairpin loops. Figure 2.19 describes the cloverleaf structure of t-RNA (transfer RNA). In double-helical regions of t-RNA, A pairs with U and G pairs with C. The RNA content of cells is usually two to six times higher than the DNA content.

Let us summarize the roles of each class of RNA species:

Messenger RNA (m-RNA) is synthesized on the chromosome and carries genetic information from the chromosome for synthesis of a particular protein to the ribosomes. The m-RNA molecule is a large one with a short half-life.

Transfer RNA (t-RNA) is a relatively small and stable molecule that carries a specific amino acid from the cytoplasm to the site of protein synthesis on ribosomes. t-RNAs contain 70 to 90 nucleotides and have a MW range of 23 to 28 kD. Each one of 20 amino acids has at least one corresponding t-RNA.

Ribosomal RNA (r-RNA) is the major component of ribosomes, constituting nearly 65%. The remainder is various ribosomal proteins. Three distinct types of r-RNAs present in the *E. coli* ribosome are specified as 23S, 16S, and 5S, respectively, on the basis of their sedimentation coefficients (determined in a centrifuge). The symbol S denotes a Svedberg unit. The molecular weights are 35 kD for 5S, 550 kD for 16S, and 1,100 kD for 23S. These three r-RNAs differ in their base sequences and ratios. Eucaryotic cells have larger ribosomes and four different types of r-RNAs: 5S, 7S, 18S, and 28S. Ribosomal RNAs make up a large fraction of total RNA. In *E. coli*, about 85% of the total RNA is r-RNA, while t-RNA is about 12% and m-RNA is 2% to 3%.

2.3. CELL NUTRIENTS

2.3.1. Introduction

A cell's composition differs greatly from its environment. A cell must selectively remove desirable compounds from its extracellular environment and retain other compounds within itself. A semipermeable membrane is the key to this selectivity. Since the cell dif-

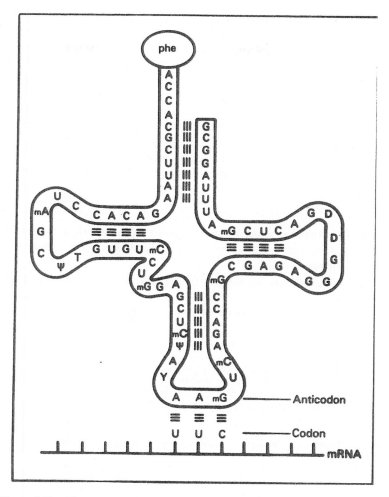

Figure 2.19. The structure of the transfer RNA (tRNA) molecule and the manner in which the anticodon of tRNA associates with the codon on mRNA by complementary base pairing. The amino acid corresponding to this codon (UUC) is phenylalanine which is bound to the opposite end of the tRNA molecule. Many tRNA molecules contain unusual bases, such as methyl cytosine (mC) and pseudouridine (ψ). (With permission, from T. D. Brock, K. M. Brock, and D. M. Ward, *Basic Microbiology with Applications*, 3d ed., Pearson Education, Upper Saddle River, NJ, 1986, p. 138.)

fers so greatly in composition from its environment, it must expend energy to maintain itself away from thermodynamic equilibrium. Thermodynamic equilibrium and death are equivalent for a cell.

All organisms except viruses contain large amounts of water (about 80%). About 50% of dry weight of cells is protein, and the proteins are largely enzymes (proteins that act as catalysts). The nucleic acid content (which contains the genetic code and machinery to make proteins) of cells varies from 10% to 20% of dry weight. However, viruses may contain nucleic acids up to 50% of their dry weight. Typically, the lipid content of most cells varies between 5% to 15% of dry weight. However, some cells accumulate PHB up to 90% of the total mass under certain culture conditions. In general, the intracellular composition of cells varies depending on the type and age of the cells and the composition of the nutrient media. Typical compositions for major groups of organisms are summarized in Table 2.7.

Most of the products formed by organisms are produced as a result of their response to environmental conditions, such as nutrients, growth hormones, and ions. The qualitative and quantitative nutritional requirements of cells need to be determined to optimize growth and product formation. Nutrients required by cells can be classified in two categories:

1. *Macronutrients* are needed in concentrations larger than 10^{-4} *M*. Carbon, nitrogen, oxygen, hydrogen, sulfur, phosphorus, Mg^{2+}, and K^+ are major macronutrients.
2. *Micronutrients* are needed in concentrations of less than 10^{-4} *M*. Trace elements such as Mo^{2+}, Zn^{2+}, Cu^{2+}, Mn^{2+}, Ca^{2+}, Na^+, vitamins, growth hormones, and metabolic precursors are micronutrients.

TABLE 2.7 Chemical Analyses, Dry Weights, and the Populations of Different Microorganisms Obtained in Culture

Organism	Composition (% dry weight)			Typical population in culture (numbers/ml)	Typical dry weight of this culture (g/100 ml)	Comments
	Protein	Nucleic acid	Lipid			
Viruses	50–90	5–50	<1	10^8–10^9	0.0005[a]	Viruses with a lipoprotein sheath may contain 25% lipid.
Bacteria	40–70	13–34	10–15	2×10^8–2×10^{11}	0.02–2.9	PHB content may reach 90%
Filamentous fungi	10–25	1–3	2–7		3–5	Some *Aspergillus* and *Penicillium* sp. contain 50% lipid.
Yeast	40–50	4–10	1–6	1–4×10^8	1–5	Some *Rhodotorula* and *Candida* sp. contain 50% lipid.
Small unicellular algae	10–60 (50)	1–5 (3)	4–80 (10)	4–8×10^7	0.4–0.9	Figure in ( ) is a commonly found value but the composition varies with the growth conditions.

[a]For a virus of 200 nm diameter.

With permission, from S. Aiba, A. E. Humphrey, and N. F. Millis, *Biochemical Engineering*, 2d ed., University of Tokyo Press, Tokyo, 1973.

2.3.2. Macronutrients

Carbon compounds are major sources of cellular carbon and energy. Microorganisms are classified in two categories on the basis of their carbon source: (1) *Heterotrophs* use organic compounds such as carbohydrates, lipids, and hydrocarbons as a carbon and energy source. (2) *Autotrophs* use carbon dioxide as a carbon source. Mixotrophs concomitantly grow under both autotrophic and heterotrophic conditions; however, autotrophic growth is stimulated by certain organic compounds. Facultative autotrophs normally grow under autotrophic conditions; however, they can grow under heterotrophic conditions in the absence of CO_2 and inorganic energy sources. *Chemoautotrophs* utilize CO_2 as a carbon source and obtain energy from the oxidation of inorganic compounds. *Photoautotrophs* use CO_2 as a carbon source and utilize light as an energy source.

The most common carbon sources in industrial fermentations are molasses (sucrose), starch (glucose, dextrin), corn syrup, and waste sulfite liquor (glucose). In laboratory fermentations, glucose, sucrose, and fructose are the most common carbon sources. Methanol, ethanol, and methane also constitute cheap carbon sources for some fermentations. In aerobic fermentations, about 50% of substrate carbon is incorporated into cells and about 50% of it is used as an energy source. In anaerobic fermentations, a large fraction of substrate carbon is converted to products and a smaller fraction is converted to cell mass (less than 30%).

Nitrogen constitutes about 10% to 14% of cell dry weight. The most widely used nitrogen sources are ammonia or the ammonium salts [NH_4Cl, $(NH_4)_2SO_4$, NH_4NO_3], proteins, peptides, and amino acids. Nitrogen is incorporated into cell mass in the form of proteins and nucleic acids. Some organisms such as *Azotobacter sp.* and the cyanobacteria fix nitrogen from the atmosphere to form ammonium. Urea may also be used as a nitrogen source by some organisms. Organic nitrogen sources such as yeast extract and peptone are expensive compared to ammonium salts. Some carbon and nitrogen sources utilized by the fermentation industry are summarized in Table 2.8.

Oxygen is present in all organic cell components and cellular water and constitutes about 20% of the dry weight of cells. Molecular oxygen is required as a terminal electron acceptor in the aerobic metabolism of carbon compounds. Gaseous oxygen is introduced into growth media by sparging air or by surface aeration.

TABLE 2.8 Some Carbon and Nitrogen Sources Utilized by the Fermentation Industry

Carbon sources	Nitrogen sources
Starch waste (maize and potato)	Soya meal
Molasses (cane and beet)	Yeast extract
Whey	Distillers solubles
n-Alkanes	Cottonseed extract
Gas oil	Dried blood
Sulfite waste liquor	Corn steep liquor
Domestic sewage	Fish solubles and meal
Cellulose waste	Groundnut meal
Carbon bean	

With permission, from G. M. Dunn in *Comprehensive Biotechnology*, M. Moo-Young, ed., Vol. 1, Elsevier Science, 1985.

Hydrogen constitutes about 8% of cell dry weight and is derived primarily from carbon compounds, such as carbohydrates. Some bacteria such as methanogens can utilize hydrogen as an energy source.

Phosphorus constitutes about 3% of cell dry weight and is present in nucleic acids and in the cell wall of some gram-positive bacteria such as teichoic acids. Inorganic phosphate salts, such as KH_2PO_4 and K_2HPO_4, are the most common phosphate salts. Glycerophosphates can also be used as organic phosphate sources. Phosphorus is a key element in the regulation of cell metabolism. The phosphate level in the media should be less than 1 mM for the formation of many secondary metabolites such as antibiotics.

Sulfur constitutes nearly 1% of cell dry weight and is present in proteins and some coenzymes. Sulfate salts such as $(NH_4)_2SO_4$ are the most common sulfur source. Sulfur-containing amino acids can also be used as a sulfur source. Certain autotrophs utilize S^{2+} and S^0 as energy sources.

Potassium is a cofactor for some enzymes and is required in carbohydrate metabolism. Cells tend to actively take up K^+ and Mg^{2+} and exclude Na^+ and Ca^{2+}. The most commonly used potassium salts are K_2HPO_4, KH_2PO_4, and K_3PO_4.

Magnesium is a cofactor for some enzymes and is present in cell walls and membranes. Ribosomes specifically require Mg^{2+} ions. Magnesium is usually $MgSO_4 \cdot 7H_2O$ supplied as $MgSO_4 \cdot 7H_2O$ or $MgCl_2$.

Table 2.9 lists the eight major macronutrients and their physiological role.

2.3.3. Micronutrients

Trace elements are essential to microbial nutrition. Lack of essential trace elements increases the lag phase (the time from inoculation to active cell replication in batch culture) and may decrease the specific growth rate and the yield. The three major categories of micronutrients are discussed next.

TABLE 2.9 The Eight Macronutrient Elements and Some Physiological Functions and Growth Requirements

Element	Physiological function	Required concentration (mol l^{-1})
Carbon	Constituent of organic cellular material. Often the energy source.	$>10^{-2}$
Nitrogen	Constituent of proteins, nucleic acids, and coenzymes.	10^{-3}
Hydrogen	Organic cellular material and water.	—
Oxygen	Organic cellular material and water. Required for aerobic respiration.	—
Sulfur	Constituent of proteins and certain coenzymes	10^{-4}
Phosphorus	Constituent of nucleic acids, phospholipids, nucleotides, and certain coenzymes	10^{-4} to 10^{-3}
Potassium	Principal inorganic cation in the cell and cofactor for some enzymes.	10^{-4} to 10^{-3}
Magnesium	Cofactor for many enzymes and chlorophylls (photosynthetic microbes) and present in cell walls and membranes.	10^{-4} to 10^{-3}

With permission, from G. M. Dunn in *Comprehensive Biotechnology,* M. Moo-Young, ed., Vol. I, Elsevier Science, 1985.

1. Most widely needed trace elements are Fe, Zn, and Mn. Iron (Fe) is present in ferredoxin and cytochrome and is an important cofactor. Iron also plays a regulatory role in some fermentation processes (e.g., iron deficiency is required for the excretion of riboflavin by *Ashbya gosypii* and iron concentration regulates penicillin production by *Penicillium chrysogenum).* Zinc (Zn) is a cofactor for some enzymes and also regulates some fermentations such as penicillin fermentation. Manganese (Mn) is also an enzyme cofactor and plays a role in the regulation of secondary metabolism and excretion of primary metabolites.

2. Trace elements needed under specific growth conditions are Cu, Co, Mo, Ca, Na, Cl, Ni, and Se. Copper (Cu) is present in certain respiratory-chain components and enzymes. Copper deficiency stimulates penicillin and citric acid production. Cobalt (Co) is present in corrinoid compounds such as vitamin B_{12}. Propionic bacteria and certain methanogens require cobalt. Molybdenum (Mo) is a cofactor of nitrate reductase and nitrogenase and is required for growth on NO_3 and N_2 as the sole source of nitrogen. Calcium (Ca) is a cofactor for amylases and some proteases and is also present in some bacterial spores and in the cell walls of some cells, such as plant cells.

Sodium (Na) is needed in trace amounts by some bacteria, especially by methanogens for ion balance. Sodium is important in the transport of charged species in eucaryotic cells. Chloride (Cl^-) is needed by some halobacteria and marine microbes, which require Na^+, too. Nickel (Ni) is required by some methanogens as a cofactor and Selenium (Se) is required in formate metabolism of some organisms.

3. Trace elements that are rarely required are B, Al, Si, Cr, V, Sn, Be, F, Ti, Ga, Ge, Br, Zr, W, Li, and I. These elements are required in concentrations of less than 10^{-6} M and are toxic at high concentrations, such as 10^{-4} M.

Some ions such as Mg^{2+}, Fe^{3+}, and PO_4^{3-} may precipitate in nutrient medium and become unavailable to the cells. *Chelating agents* are used to form soluble compounds with the precipitating ions. Chelating agents have certain groups termed *ligands* that bind to metal ions to form soluble complexes. Major ligands are carboxyl (—COOH), amine (—NH$_2$), and mercapto (—SH) groups. Citric acid, EDTA (ethylenediaminetetraacetic acid), polyphosphates, histidine, tyrosine, and cysteine are the most commonly used chelating agents. Na_2 EDTA is the most common chelating agent. EDTA may remove some metal ion components of the cell wall, such as Ca^{2+}, Mg^{2+}, and Zn^{2+} and may cause cell wall disintegration. Citric acid is metabolizable by some bacteria. Chelating agents are included in media in low concentrations (e.g., 1 mM).

Growth factors stimulate the growth and synthesis of some metabolites. Vitamins, hormones, and amino acids are major growth factors. Vitamins usually function as coenzymes. Some commonly required vitamins are thiamine (B_1), riboflavin (B_2), pyridoxine (B_6), biotin, cyanocobalamine (B_{12}), folic acid, lipoic acid, *p*-amino benzoic acid, and vitamin K. Vitamins are required at a concentration range of 10^{-6} M to 10^{-12} M. Depending on the organism, some or all of the amino acids may need to be supplied externally in concentrations from 10^{-6} M to 10^{-13} M. Some fatty acids, such as oleic acid and sterols, are also needed in small quantities by some organisms. Higher forms of life, such as animal and plant cells, require hormones to regulate their metabolism. Insulin is a common hormone for animal cells, and auxin and cytokinins are plant-growth hormones.

2.3.4. Growth Media

Two major types of growth media are defined and complex media. *Defined media* contain specific amounts of pure chemical compounds with known chemical compositions. A medium containing glucose, $(NH_4)_2SO_4$, KH_2PO_4, and $MgCl_2$ is a defined medium. *Complex media* contain natural compounds whose chemical composition is not exactly known. A medium containing yeast extracts, peptone, molasses, or corn steep liquor is a complex medium. A complex medium usually can provide the necessary growth factors, vitamins, hormones, and trace elements, often resulting in higher cell yields, compared to the defined medium. Often, complex media are less expensive than defined media. The primary advantage of defined media is that the results are more reproducible and the operator has better control of the fermentation. Further, recovery and purification of a product is often easier and cheaper in defined media. Table 2.10 summarizes typical defined and complex media.

TABLE 2.10 Compositions of Typical Defined and Complex Media

Constituent	Purpose	Concn (g/liter)
Defined medium		
Group A		
Glucose	C, energy	30
KH_2PO_4	K, P	1.5
$MgSO_4 \cdot 7H_2O$	Mg, S	0.6
$CaCl_2$	Ca	0.05
$Fe_2(SO_4)_3$	Fe	15×10^{-4}
$ZnSO_4 \cdot 7H_2O$	Zn	6×10^{-4}
$CuSO_4 \cdot 5H_2O$	Cu	6×10^{-4}
$MnSO_4 \cdot H_2O$	Mn	6×10^{-4}
Group B		
$(NH_4)_2HPO_4$	N	6
$(NH_4)H_2PO_4$	N	5
Group C		
$C_6H_5Na_3O_7 \cdot 2H_2O$	Chelator	4
Group D		
Na_2HPO_4	Buffer	20
KH_2PO_4	Buffer	10

Complex medium used in a penicillin fermentation	
Glucose or molasses (by continuous feed)	10% of total
Corn steep liquor	1–5% of total
Phenylacetic acid (by continuous feed)	0.5–0.8% of total
Lard oil (or vegetable oil) antifoam by continuous addition	0.5% of total
pH to 6.5 to 7.5 by acid or alkali addition	

2.4. SUMMARY

Microbes can grow over an immense range of conditions: temperatures above boiling and below freezing; high salt concentrations; high pressures (>1000 atm); and at low and high pH values (about 1 to 10). Cells that must use oxygen are known as *aerobic*. Cells that find oxygen toxic are *anaerobic*. Cells that can adapt to growth either with or without oxygen are *facultative*.

The two major groups of cells are *procaryotic* and *eucaryotic*. Eucaryotic cells are more complex. The essential demarcation between procaryotic and eucaryotic is the absence (procaryotes) or presence (eucaryotes) of a membrane around the chromosomal or genetic material.

The procaryotes can be divided into two major groups: the *eubacteria* and *archaebacteria*. The archaebacteria are a group of ancient organisms; subdivisions include methanogens (methane producing), halobacteria (live in high-salt environments), and thermoacidophiles (grow best under conditions of high temperature and high acidity). Most eubacteria can be separated into gram-positive and gram-negative cells. *Gram-positive* cells have an inner membrane and strong cell wall. *Gram-negative* cells have an inner membrane and an outer membrane. The outer membrane is supported by cell wall material but is less rigid than in gram-positive cells. The *cyanobacteria* (blue-green algae) are photosynthetic procaryotes classified as a subdivision of the eubacteria.

The eucaryotes contain both single-celled organisms and multicell systems. The fungi and yeasts, the algae, and the protozoa are all examples of single-celled eucaryotes. Plants and animals are multicellular eucaryotes.

Viruses are replicating particles that are obligate parasites. Some viruses use DNA to store genetic information, while others use RNA. Viruses specific for bacteria are called *bacteriophages* or phages.

All cells contain the macromolecules: *protein*, *RNA*, and *DNA*. Other essential components of these cells are constructed from lipids and carbohydrates. Proteins are polymers of amino acids; typically, 20 different amino acids are used. Each amino acid has a distinctive side group. The sequence of amino acids determines the *primary structure* of the protein. Interactions among the side groups of the amino acids (hydrogen bonding, disulfide bonds, regions of hydrophobicity or hydrophilicity) determine the *secondary* and *tertiary structure* of the molecule. If separate polypeptide chains associate to form the final structure, then we speak of *quaternary structure*. The three-dimensional shape of a protein is critical to its function.

DNA and RNA are polymers of nucleotides. DNA contains the cell's genetic information. RNA is involved in transcribing and translating that information into real proteins. Messenger RNA transcribes the code; transfer RNA is an adapter molecule that transports a specific amino acid to the reaction site for protein synthesis; and ribosomal RNAs are essential components of ribosomes, which are the structures responsible for protein synthesis.

In addition to their role as monomers for DNA and RNA synthesis, nucleotides play important roles in cellular energetics. The *high-energy phosphate bonds* in *ATP* can store energy. The hydrolysis of ATP when coupled to otherwise energetically unfavorable reactions can drive the reaction toward completion. *NAD* and *NAPH* are important carriers of *reducing power*.

Carbohydrates consist of sugars, and the polymerized products of sugars are called *polysaccharides*. Sugars represent convenient molecules for the rapid oxidation and release of energy. The polysaccharides play an important structural role (as in cellulose) or can be used as a cellular reserve of carbon and energy (as in starch).

Lipids and related compounds are critical in the construction of cellular membranes. Some fats also form reserve sources. A number of growth factors or hormones involve lipid materials. *Phospholipids* are the primary components of biological membranes.

The maintenance of cellular integrity requires the selective uptake of nutrients. One class of nutrients is the *macronutrients*, and these are used in large amounts. The *micronutrients* and trace nutrients are used in low concentrations; some of these compounds become toxic if present at too high a level.

In a *defined medium*, all components added to the medium are identifiable chemical species. In a *complex medium*, one or more components are not chemically defined (e.g., yeast extract).

SUGGESTIONS FOR FURTHER READING

ALBERTS, B., D. BRAY, A. JOHNSON, J. LEWIS, M. RAFF, K. ROBERTS, AND P. WALTER, *Essential Cell Biology: An Introduction to the Molecular Biology of the Cell*, Garland Publ., Inc., New York, 1998.

BLACK, J. G., *Microbiology: Principles and Applications*, 3d ed. Prentice Hall, Upper Saddle River, NJ, 1996.

MADIGAN, M.T., J. M. MARTINKO, AND J. PARKER, *Brock Biology of Microorganisms*, 8th ed. Prentice Hall, Upper Saddle River, NJ, 1997.

MORAN, L. A., K. G. SCRIMGEOUR, H. R. HORTON, R. S. OCHS, AND J. D. RAWN, *Biochemistry*, 2d ed. Prentice Hall, Upper Saddle River, NJ, 1994.

PACE, N. R., Microbial Ecology & Diversity, *Am. Soc. Microbiol. News* 65:328–333, 1999.

PROBLEMS

2.1. Briefly compare procaryotes with eucaryotes in terms of internal structure and functions.

2.2. What are the major classes of fungi? Cite the differences among these classes briefly.

2.3. Briefly describe distinct features of actinomycetes and their important products.

2.4. Briefly compare protozoa with algae in terms of their cellular structures and functions.

2.5. What are major sources of carbon, nitrogen, and phosphorus in industrial fermentations?

2.6. Explain the functions of the following trace elements in microbial metabolism: Fe, Zn, Cu, Co, Ni, Mn, vitamins.

2.7. What are chelating agents? Explain their function with an example.

2.8. Cite five major biological functions of proteins.

2.9. Briefly describe the primary, secondary, tertiary, and quaternary structure of proteins. What could happen if you substituted a tyrosine for a cysteine in the active site? What might happen if the substitution occurred elsewhere?

2.10. Contrast DNA and RNA. Cite at least four differences.

2.11. Contrast the advantages and disadvantages of chemically defined and complex media.

2.12. You are asked to develop a medium for production of an antibiotic. The antibiotic is to be made in large amounts (ten 100,000 l fermenters) and is relatively inexpensive. The host cell is a soil isolate of a fungal species, and the nutritional requirements for rapid growth are uncertain. Will you try to develop a defined or complex medium? Why?

2.13. You wish to produce a high-value protein using recombinant DNA technology. Would you try to develop a chemical defined medium or a complex medium? Why?

2.14. Explain what semiconservative replication means.

2.15. Give characteristic dimensions for each of these organisms:

> E. coli
> Yeast (S. cerevisiae)
> Liver cell (hepatocyte)
> Plant cell

2.16. What are the differences in cell envelope structure between gram-negative and gram-positive bacteria? These differences become important if you wish to genetically engineer bacteria to excrete proteins into the extracellular fluid.

2.17. True or False

> **a)** An organism that can grow using oxygen as an electron acceptor and can also grow and metabolize in the absence of oxygen is called facultative.
> **b)** Yeasts are procaryotes.
> **c)** A bacteriophage is a virus that infects bacteria.
> **d)** When you supplement growth medium with amino acids, you should use the D-form.

3

Enzymes

3.1. INTRODUCTION

Enzymes are usually proteins of high molecular weight (15,000 < MW < several million daltons) that act as catalysts. Recently, it has been shown that some RNA molecules are also catalytic, but the vast majority of cellular reactions are mediated by protein catalysts. RNA molecules that have catalytic properties are called *ribozymes*. Enzymes are specific, versatile, and very effective biological catalysts, resulting in much higher reaction rates as compared to chemically catalyzed reactions under ambient conditions. More than 2000 enzymes are known. Enzymes are named by adding the suffix *-ase* to the end of the substrate, such as urease, or the reaction catalyzed, such as alcohol dehydrogenase. Some enzymes have a simple structure, such as a folded polypeptide chain (typical of most hydrolytic enzymes). Many enzymes have more than one subunit. Some protein enzymes require a nonprotein group for their activity. This group is either a cofactor, such as metal ions, Mg, Zn, Mn, Fe, or a coenzyme, such as a complex organic molecule, NAD, FAD, CoA, or some vitamins. An enzyme containing a nonprotein group is called a *holoenzyme*. The protein part of this enzyme is the *apoenzyme* (holoenzyme = apoenzyme + cofactor). Enzymes that occur in several different molecular forms, but catalyze the same reaction, are called *isozymes*. Some enzymes are grouped together to form enzyme complexes. Enzymes are substrate specific and are classified according to the reaction they catalyze. Major classes of enzymes and their functions are listed in Table 3.1.

TABLE 3.1 International Classification of Enzymes: Class Names, Code Numbers, and Types of Reactions Catalyzed

1. Oxidoreductases (oxidation-reduction reactions)	3. Hydrolases (hydrolysis reactions)
	3.1 Esters
1.1 Acting on $\diagdown$CH—OH$\diagup$	3.2 Glycosidic bonds
	3.4 Peptide bonds
	3.5 Other C—N bonds
	3.6 Acid anhydrides
	4. Lyases (addition to double bonds)
1.2 Acting on $\diagdown$C=O$\diagup$	4.1 $\diagdown$C=C$\diagup$
1.3 Acting on $\diagdown$C=CH—$\diagup$	4.2 $\diagdown$C=O$\diagup$
1.4 Acting on $\diagdown$CH—NH$_2$$\diagup$	4.3 $\diagdown$C=N—$\diagup$
1.5 Acting on $\diagdown$CH—NH—$\diagup$	
1.6 Acting on NADH; NADPH	
2. Transferases (transfer of functional groups)	5. Isomerases (isomerization reactions)
2.1 One-carbon groups	5.1 Racemases
2.2 Aldehydic or ketonic groups	6. Ligases (formation of bonds with ATP cleavage)
2.3 Acyl groups	6.1 C—O
2.4 Glycosyl groups	6.2 C—S
2.7 Phosphate groups	6.3 C—N
2.8 S-containing groups	6.4 C—C

With permission, from A. Lehninger, *Biochemistry*, 2d edition, Worth Publishers, New York, 1975.

3.2. HOW ENZYMES WORK

Enzymes lower the activation energy of the reaction catalyzed by binding the substrate and forming an enzyme–substrate complex. Enzymes do not affect the free-energy change or the equilibrium constant. Figure 3.1 illustrates the action of an enzyme from the activation-energy point of view. For example, the activation energy for the decomposition of hydrogen peroxide varies depending on the type of catalysis. The activation energy of the uncatalyzed reaction at 20°C is 18 kilocalories per mole (kcal/mol), whereas the ΔE values for chemically catalyzed (by colloidal platinum) and enzymatically catalyzed (catalase) decomposition are 13 and 7 kcal/mol, respectively. That is, catalase accelerates the rate of reaction by a factor of about 10^8. The reader should note that this large change in rate for a relatively small change in activation energy is due to the exponential dependence of rate on activation energy. In this case, the ratio of the rates is $\exp(-7000/2 \cdot 293) \div \exp(-18,000/2 \cdot 293)$.

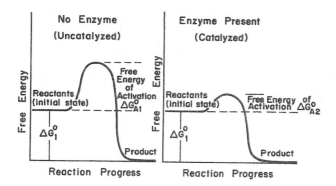

Figure 3.1. Activation energies of enzymatically catalyzed and uncatalyzed reactions. Note that $|\Delta G°_{A2}| < |\Delta G°_{A1}|$.

The molecular aspects of enzyme–substrate interaction are not yet fully understood. This interaction varies from one enzyme–substrate complex to another. Various studies using x-ray and Raman spectroscopy have revealed the presence of the enzyme–substrate (ES) complex. The interaction between the enzyme and its substrate is usually by weak forces. In most cases, van der Waals forces and hydrogen bonding are responsible for the formation of ES complexes. The substrate binds to a specific site on the enzyme known as the *active site*. The substrate is a relatively small molecule and fits into a certain region on the enzyme molecule, which is a much larger molecule. The simplest model describing this interaction is the lock-and-key model, in which the enzyme represents the lock and the substrate represents the key, as described in Fig. 3.2.

In multisubstrate enzyme-catalyzed reactions, enzymes can hold substrates such that reactive regions of substrates are close to each other and to the enzyme's active site, which is known as the *proximity effect*. Also, enzymes may hold the substrates at certain positions and angles to improve the reaction rate, which is known as the *orientation effect*. In some enzymes, the formation of an enzyme–substrate complex causes slight changes in the three-dimensional shape of the enzyme. This induced fit of the substrate to the enzyme molecule may contribute to the catalytic activity of the enzyme, too. The enzymes lysozyme and carboxypeptidase A have been observed to change their three-dimensional structure upon complexing with the substrate. Enzyme catalysis is affected not only by the primary structure of enzymes but also by the secondary, tertiary, and quaternary structures. The properties of the active site of enzymes and the folding characteristics have a profound effect on the catalytic activity of enzymes. Certain enzymes require coenzymes and cofactors for proper functioning. Table 3.2 lists some enzymes and their cofactors and coenzymes.

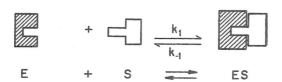

Figure 3.2. Schematic of the lock-and-key model of enzyme catalysis.

TABLE 3.2 Cofactors (Metal Ions) and Coenzymes of Some Enzymes

	Coenzyme	*Entity transferred*
Zn^{2+}	Nicotinamide adenine dinucleotide	Hydrogen atoms (electrons)
Alcohol dehydrogenase	Nicotinamide adenine dinucleotide	Hydrogen atoms (electrons)
Carbonic anhydrase	phosphate	Hydrogen atoms (electrons)
Carboxypeptidase	Flavin mononucleotide	Hydrogen atoms (electrons)
Mg^{2+}	Flavin adenine dinucleotide	Hydrogen atoms (electrons)
Phosphohydrolases	Coenzyme Q	Aldehydes
Phosphotransferases	Thiamin pyrophosphate	Acyl groups
Mn^{2+}	Coenzyme A	Acyl groups
Arginase	Lipoamide	Alkyl groups
Phosphotransferases	Cobamide coenzymes	Carbon dioxide
Fe^{2+} or Fe^{3+}	Biocytin	Amino groups
Cytochromes	Pyridoxal phosphate	Methyl, methylene, formyl,
Peroxidase	Tetrahydrofolate coenzymes	or formimino groups
Catalase		
Ferredoxin		
Cu^{2+} (Cu^+)		
Tyrosinase		
Cytochrome oxidase		
K^+		
Pyruvate kinase (also requires Mg^{2+})		
Na^+		
Plasma membrane ATPase (also requires K^+ and Mg^{2+})		

With permission, from A. Lehninger, *Biochemistry*, 2d ed., Worth Publishers, New York, 1975.

3.3. ENZYME KINETICS

3.3.1. Introduction

A mathematical model of the kinetics of single-substrate-enzyme-catalyzed reactions was first developed by V. C. R. Henri in 1902 and by L. Michaelis and M. L. Menten in 1913. Kinetics of simple enzyme-catalyzed reactions are often referred to as Michaelis–Menten kinetics or *saturation* kinetics. The qualitative features of enzyme kinetics are similar to Langmuir–Hinshelwood kinetics (see Fig. 3.3). These models are based on data from batch reactors with constant liquid volume in which the initial substrate, $[S_0]$, and enzyme, $[E_0]$, concentrations are known. More complicated enzyme-substrate interactions such as multisubstrate–multienzyme reactions can take place in biological systems. An enzyme solution has a fixed number of active sites to which substrates can bind. At high substrate concentrations, all these sites may be occupied by substrates or the enzyme is *saturated*. Saturation kinetics can be obtained from a simple reaction scheme that involves a reversible step for enzyme–substrate complex formation and a dissociation step of the ES complex.

$$E + S \underset{k_{-1}}{\overset{k_1}{\rightleftharpoons}} ES \overset{k_2}{\longrightarrow} E + P \tag{3.1}$$

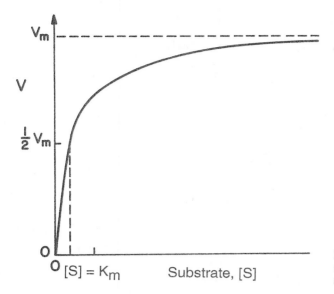

Figure 3.3. Effect of substrate concentration on the rate of an enzyme-catalyzed reaction.

It is assumed that the ES complex is established rather rapidly and the rate of the reverse reaction of the second step is negligible. The assumption of an irreversible second reaction often holds only when product accumulation is negligible at the beginning of the reaction. Two major approaches used in developing a rate expression for the enzyme-catalyzed reactions are (1) rapid-equilibrium approach and (2) quasi-steady-state approach.

3.3.2. Mechanistic Models for Simple Enzyme Kinetics

Both the quasi-steady-state approximation and the assumption of rapid equilibrium share the same few initial steps in deriving a rate expression for the mechanism in eq. 3.1, where the rate of product formation is

$$v = \frac{d[P]}{dt} = k_2[ES] \qquad (3.2)$$

where v is the rate of product formation or substrate consumption in moles/l-s.

The rate constant k_2 is often denoted as k_{cat} in the biological literature. The rate of variation of the ES complex is

$$\frac{d[ES]}{dt} = k_1[E][S] - k_{-1}[ES] - k_2[ES] \qquad (3.3)$$

Since the enzyme is not consumed, the conservation equation on the enzyme yields

$$[E] = [E_0] - [ES] \qquad (3.4)$$

At this point, an assumption is required to achieve an analytical solution.

3.3.2.1. The rapid equilibrium assumption. Henri and Michaelis and Menten used essentially this approach. Assuming a rapid equilibrium between the enzyme and substrate to form an [ES] complex, we can use the equilibrium coefficient to express [ES] in terms of [S].

The equilibrium constant is

$$K'_m = \frac{k_{-1}}{k_1} = \frac{[E][S]}{[ES]} \tag{3.5}$$

Since $[E] = [E_0] - [ES]$ if enzyme is conserved, then

$$[ES] = \frac{[E_0][S]}{(k_{-1}/k_1) + [S]} \tag{3.6}$$

$$[ES] = \frac{[E_0][S]}{K'_m + [S]} \tag{3.7}$$

where $K'_m = k_{-1}/k_1$, which is the dissociation constant of the ES complex. Substituting eq. 3.7 into eq. 3.2 yields

$$v = \frac{d[P]}{dt} = k_2 \frac{[E_0][S]}{K'_m + [S]} = \frac{V_m[S]}{K'_m + [S]} \tag{3.8}$$

where $V_m = k_2[E_0]$.

In this case, the maximum forward velocity of the reaction is V_m. V_m changes if more enzyme is added, but the addition of more substrate has no influence on V_m. K'_m is often called the Michaelis–Menten constant, and the prime reminds us that it was derived by assuming rapid equilibrium in the first step. A low value of K'_m suggests that the enzyme has a high affinity for the substrate. Also, K'_m corresponds to the substrate concentration, giving the half-maximal reaction velocity.

An equation of exactly the same form as eq. 3.8 can be derived with a different, more general assumption applied to the reaction scheme in eq. 3.1.

3.3.2.2. The quasi-steady-state assumption.
In many cases the assumption of rapid equilibrium following mass-action kinetics is not valid, although the enzyme–substrate reaction still shows saturation-type kinetics.

G. E. Briggs and J. B. S. Haldane first proposed using the quasi-steady-state assumption. In most experimental systems a closed system (batch reactor) is used in which the initial substrate concentration greatly exceeds the initial enzyme concentration. They suggest that since $[E_0]$ was small, $d[ES]/dt \approx 0$. (This logic is flawed. Do you see why?) Computer simulations of the actual time course represented by eqs. 3.2, 3.3, and 3.4 have shown that *in a closed system the quasi-steady-state hypothesis holds* after a brief transient *if* $[S_0] \gg [E_0]$ (for example, 100×). Figure 3.4 displays one such time course.

By applying the quasi-steady-state assumption to eq. 3.3, we find

$$[ES] = \frac{k_1[E][S]}{k_{-1} + k_2} \tag{3.9}$$

Substituting the enzyme conservation eq. 3.4 in eq. 3.9 yields

$$[ES] = \frac{k_1([E_0] - [ES])[S]}{k_{-1} + k_2} \tag{3.10}$$

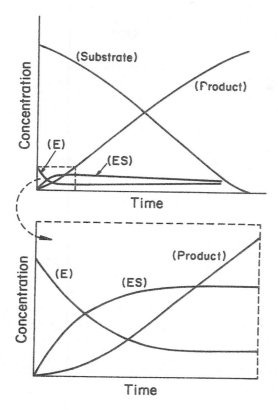

Figure 3.4. Time course of the formation of an enzyme/substrate complex and initiation of the steady state, as derived from computer solutions of data obtained in an actual experiment on a typical enzyme. The portion in the dashed box in the top graph is shown in magnified form on the lower graph. (With permission, adapted from A. Lehninger, *Biochemistry*, 2d ed., Worth Publishers, New York, 1975, p. 191.)

Solving eq. 3.10 for [ES],

$$[ES] = \frac{[E_0][S]}{\dfrac{k_{-1} + k_2}{k_1} + [S]} \tag{3.11}$$

Substituting eq. 3.11 into eq. 3.2 yields

$$v = \frac{d[P]}{dt} = \frac{k_2[E_0][S]}{\dfrac{k_{-1} + k_2}{k_1} + [S]} \tag{3.12a}$$

$$v = \frac{V_m[S]}{K_m + [S]} \tag{3.12b}$$

where K_m is $(k_{-1} + k_2)/k_1$ and V_m is $k_2[E_0]$. Under most circumstances (simple experiments), it is impossible to determine whether K_m or K_m' is more suitable. Since K_m results from the more general derivation, we will use it in the rest of our discussions.

Sec. 3.3   Enzyme Kinetics

3.3.3. Experimentally Determining Rate Parameters for Michaelis–Menten Type Kinetics

The determination of values for K_m and V_m with high precision can be difficult. Typically, experimental data are obtained from *initial-rate experiments*. A batch reactor is charged with a known amount of substrate $[S_0]$ and enzyme $[E_0]$. The product (or substrate concentration) is plotted against time. The initial slope of this curve is estimated (i.e., $v = d[P]/dt|_{t=0} = -d[S]/dt|_{t=0}$). This value of v then depends on the values of $[E_0]$ and $[S_0]$ in the charge to the reactor. Many such experiments can be used to generate many pairs of v and $[S]$ data. These could be plotted as in Fig. 3.3, but the accurate determination of K_m from such a plot is very difficult. Consequently, other methods of analyzing such data have been suggested.

3.3.3.1. Double-reciprocal plot (Lineweaver–Burk plot). Equation 3.12b can be linearized in double-reciprocal form:

$$\frac{1}{v} = \frac{1}{V_m} + \frac{K_m}{V_m}\frac{1}{[S]} \tag{3.13}$$

A plot of $1/v$ versus $1/[S]$ yields a linear line with a slope of K_m/V_m and y-axis intercept of $1/V_m$, as depicted in Fig. 3.5. A double-reciprocal plot gives good estimates on V_m, but not necessarily on K_m. Because the error about the reciprocal of a data point is not symmetric, the reader should be cautious in applying regression analysis (least squares) to such plots. Data points at low substrate concentrations influence the slope and intercept more than those at high substrate concentrations.

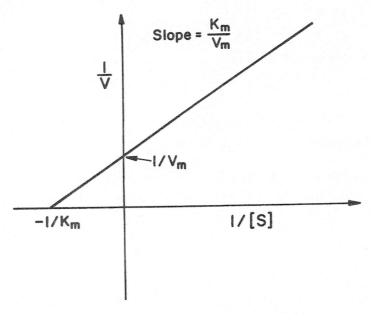

Figure 3.5. Double-reciprocal (Lineweaver–Burk) plot.

3.3.3.2. Eadie–Hofstee plot.

Equation 3.12b can be rearranged as

$$v = V_m - K_m \frac{v}{[S]} \tag{3.14}$$

A plot of v versus $v/[S]$ results in a line of slope $-K_m$ and y-axis intercept of V_m, as depicted in Fig. 3.6. Eadie–Hofstee plots can be subject to large errors since both coordinates contain v, but there is less bias on points at low $[S]$.

3.3.3.3. Hanes–Woolf plot.

Rearrangement of eq. 3.12b yields

$$\frac{[S]}{v} = \frac{K_m}{V_m} + \frac{1}{V_m}[S] \tag{3.15}$$

A plot of $[S]/v$ versus $[S]$ results in a line of slope $1/V_m$ and y-axis intercept of K_m/V_m, as depicted in Fig. 3.7. This plot is used to determine V_m more accurately.

3.3.3.4. Batch kinetics.

The time course of variation of $[S]$ in a batch enzymatic reaction can be determined from

$$v = -\frac{d[S]}{dt} = \frac{V_m[S]}{K_m + [S]} \tag{3.12b}$$

by integration to yield

$$V_m t = [S_0] - [S] + K_m \ln \frac{[S_0]}{[S]} \tag{3.16}$$

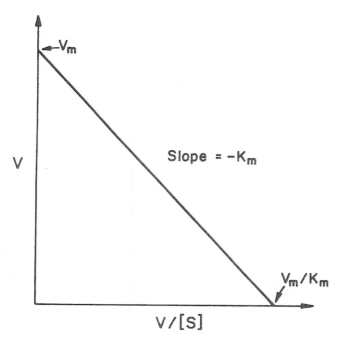

Figure 3.6. Eadie–Hofstee plot.

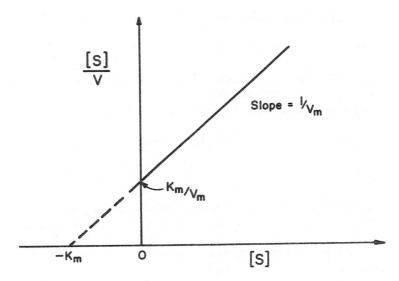

Figure 3.7. Hanes–Woolf plot.

or

$$V_m - \frac{[S_0]-[S]}{t} = \frac{K_m}{t} \ln \frac{[S_0]}{[S]} \tag{3.17}$$

A plot of $1/t \ln[S_0]/[S]$ versus $\{[S_0] - [S]\}/t$ results in a line of slope $-1/K_m$ and intercept of V_m/K_m.

3.3.3.5. Interpretation of K_m and V_m.

While K_m (or K_m') is an intrinsic parameter, V_m is not. K_m is solely a function of rate parameters and is expected to change with temperature or pH. However, V_m is a function of the rate parameter k_2 and the initial enzyme level, $[E_0]$. As $[E_0]$ changes, so does V_m. Of course, k_2 can be readily calculated if $[E_0]$ is known. For highly purified enzyme preparations it may be possible to express $[E_0]$ in terms of mol/l or g/l.

When the enzyme is part of a crude preparation, its concentration is in terms of "units." A "unit" is the amount of enzyme that gives a predetermined amount of catalytic activity under specific conditions. For example, one unit would be formation of one μmol product per minute at a specified pH and temperature with a substrate concentration much greater than the value of K_m. The *specific activity* is the number of units of activity per amount of total protein. For example, a crude cell lysate might have a specific activity of 0.2 units/mg protein which upon purification may increase to 10 units/mg protein. Only enzyme that remains catalytically active will be measured. The enzyme may be *denatured* if it unfolds or has its three-dimensional shape altered by pH extremes or temperature during purification. The denatured enzyme will have no activity.

Example 3.1.

To measure the amount of glucoamylase in a crude enzyme preparation, 1 ml of the crude enzyme preparation containing 8 mg protein is added to 9 ml of a 4.44% starch solution. One unit of activity of glucoamylase is defined as the amount of enzyme which produces 1 μmol of glucose per min in a 4% solution of Lintner starch at pH 4.5 and at 60°C. Initial rate experiments show that the reaction produces 0.6 μmol of glucose/ml-min. What is the specific activity of the crude enzyme preparation?

Solution   The total amount of glucose made is 10 ml × 0.6 μmol glucose/ml-min or 6 μmol glucose per min. The specific activity is then:

$$\text{specific activity} = \frac{6 \text{ units}}{1 \text{ ml protein solution} \cdot 8 \text{ mg/ml}}$$
$$= 6 \text{ units/8 mg protein}$$
$$= 0.75 \text{ units/mg protein}$$

V_m must have units such as μmol product/ml-min. Since $V_m = k_2 E_0$, the dimensions of k_2 must reflect the definition of units in E_0. In the above example we had a concentration of enzyme of 8 mg protein/10 ml solution · 0.75 units/mg protein or 0.6 units/ml. If, for example, $V_m = 1$ μmol/ml-min, then $k_2 = 1$ μmol/ml-min ÷ 0.6 units/ml or $k_2 = 1.67$ μmol/unit-min.

3.3.4. Models for More Complex Enzyme Kinetics

3.3.4.1. Allosteric enzymes.   Some enzymes have more than one substrate binding site. The binding of one substrate to the enzyme facilitates binding of other substrate molecules. This behavior is known as *allostery* or *cooperative binding,* and regulatory enzymes show this behavior. The rate expression in this case is

$$v = -\frac{d[S]}{dt} = \frac{V_m[S]^n}{K_m'' + [S]^n} \tag{3.18}$$

where n = cooperativity coefficient and $n > 1$ indicates positive cooperativity. Figure 3.8 compares Michaelis–Menten kinetics with allosteric enzyme kinetics, indicating a sigmoidal shape of v –[S] plot for allosteric enzymes.

The cooperativity coefficient can be determined by rearranging eq. 3.18 as

$$\ln \frac{v}{V_m - v} = n \ln[S] - \ln K_m'' \tag{3.19}$$

and by plotting $\ln v/(V_m - v)$ versus $\ln[S]$ as depicted in Fig. 3.9.

3.3.4.2. Inhibited enzyme kinetics.   Certain compounds may bind to enzymes and reduce their activity. These compounds are known to be enzyme inhibitors. Enzyme inhibitions may be irreversible or reversible. Irreversible inhibitors such as heavy metals (lead, cadium, mercury, and others) form a stable complex with enzyme and reduce enzyme activity. Such enzyme inhibition may be reversed only by using chelating agents such as EDTA (ethylenediaminetetraacetic acid) and citrate. Reversible inhibitors may dissociate more easily from the enzyme after binding. The three major classes of re-

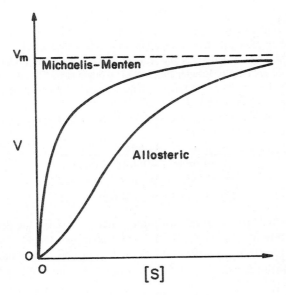

Figure 3.8. Comparison of Michaelis–Menten and allosteric enzyme kinetics.

versible enzyme inhibitions are competitive, noncompetitive, and uncompetitive inhibitions. The substrate may act as an inhibitor in some cases.

Competitive inhibitors are usually substrate analogs and compete with substrate for the active site of the enzyme. The competitive enzyme inhibition scheme can be described as

$$E + S \underset{k_{-1}}{\overset{k_1}{\rightleftharpoons}} ES \overset{k_2}{\longrightarrow} E + P$$

$$+$$

$$I$$

$$\Big\Updownarrow K_1 \qquad\qquad (3.20)$$

$$EI$$

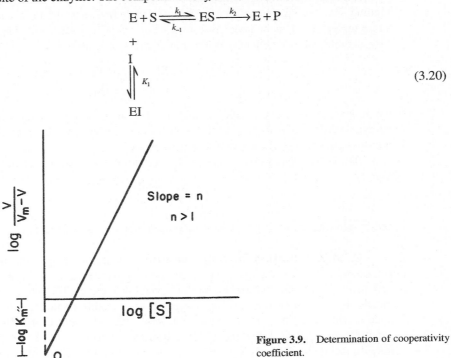

Figure 3.9. Determination of cooperativity coefficient.

Assuming rapid equilibrium and with the definition of

$$K'_m = \frac{[E][S]}{[ES]}, \qquad K_I = \frac{[E][I]}{[EI]}$$

$$[E_0] = [E] + [ES] + [EI] \quad \text{and} \quad v = k_2[ES] \tag{3.21}$$

we can develop the following equation for the rate of enzymatic conversion:

$$v = \frac{V_m[S]}{K'_m \left[1 + \dfrac{[I]}{K_I}\right] + [S]} \tag{3.22}$$

or

$$v = \frac{V_m[S]}{K'_{m,\,app} + [S]} \tag{3.23}$$

where $K'_{m,app} = K'_m \left(1 + \dfrac{[I]}{K_I}\right)$.

The net effect of competitive inhibition is an increased value of $K'_{m,\,app}$ and, therefore, reduced reaction rate. Competitive inhibition can be overcome by high concentrations of substrate. Figure 3.10 describes competitive enzyme inhibition in the form of a double-reciprocal plot.

Noncompetitive inhibitors are not substrate analogs. Inhibitors bind on sites other than the active site and reduce enzyme affinity to the substrate. Noncompetitive enzyme inhibition can be described as follows:

$$\begin{array}{ccc}
E + S & \underset{K'_m}{\rightleftharpoons} ES \xrightarrow{k_2} E + P \\
+ & & + \\
I & & I \\
K_I \big\updownarrow & & \big\updownarrow \\
EI + S & \underset{K'_m}{\rightleftharpoons} ESI
\end{array} \tag{3.24}$$

With the definition of

$$K'_m = \frac{[E][S]}{[ES]} = \frac{[EI][S]}{[ESI]}, \qquad K_I = \frac{[E][I]}{[EI]} = \frac{[ES][I]}{[ESI]}$$

$$[E_0] = [E] + [ES] + [EI] + [ESI] \quad \text{and} \quad v = k_2[ES] \tag{3.25}$$

we can develop the following rate equation:

$$v = \frac{V_m}{\left(1 + \dfrac{[I]}{K_I}\right)\left(1 + \dfrac{K'_m}{[S]}\right)} \tag{3.26}$$

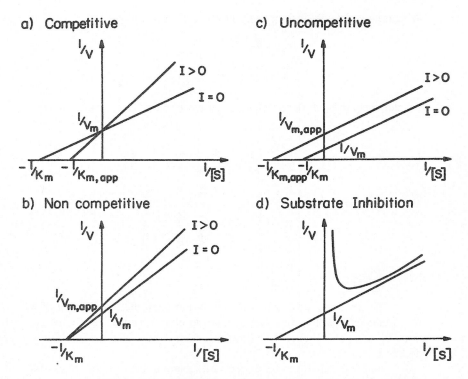

Figure 3.10. Different forms of inhibited enzyme kinetics.

or

$$v = \frac{V_{m,\text{app}}}{\left(1 + \dfrac{K'_m}{[S]}\right)} \tag{3.27}$$

where $V_{m,\text{app}} = \dfrac{V_m}{\left(1 + \dfrac{[I]}{K_1}\right)}$

The net effect of noncompetitive inhibition is a reduction in V_m. High substrate concentrations would not overcome noncompetitive inhibition. Other reagents need to be added to block binding of the inhibitor to the enzyme. In some forms of noncompetitive inhibition V_m is reduced and K'_m is increased. This occurs if the complex ESI can form product.

Uncompetitive inhibitors bind to the ES complex only and have no affinity for the enzyme itself. The scheme for uncompetitive inhibition is

$$E + S \xrightleftharpoons{K'_m} ES \xrightarrow{k_2} E + P$$

$$+$$

$$I$$

$$\Big\updownarrow K_1 \qquad \qquad \text{(3.28)}$$

$$ESI$$

With the definition of

$$K'_m = \frac{[E][S]}{[ES]}, \quad K_1 = \frac{[ES][I]}{[ESI]} \qquad \text{(3.29)}$$

$$[E_0] = [E] + [ES] + [ESI] \quad \text{and} \quad v = k_2[ES]$$

we can develop the following equation for the rate of reaction:

$$v = \frac{\dfrac{V_m}{\left(1 + \dfrac{[I]}{K_1}\right)}[S]}{\dfrac{K'_m}{\left(1 + \dfrac{[I]}{K_1}\right)} + [S]} \qquad \text{(3.30)}$$

or

$$v = \frac{V_{m,\text{app}}[S]}{K'_{m,\text{app}} + [S]} \qquad \text{(3.31)}$$

The net effect of uncompetitive inhibition is a reduction in both V_m and K'_m values. Reduction in V_m has a more pronounced effect than the reduction in K'_m, and the net result is a reduction in reaction rate. Uncompetitive inhibition is described in Fig. 3.10 in the form of a double-reciprocal plot.

High substrate concentrations may cause inhibition in some enzymatic reactions, known as *substrate inhibition*. Substrate inhibition is graphically described in Fig. 3.11.

The reaction scheme for uncompetitive substrate inhibition is

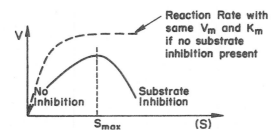

Figure 3.11. Comparison of substrate-inhibited and uninhibited enzymatic reactions.

$$E + S \underset{K'_m}{\overset{}{\rightleftharpoons}} ES \xrightarrow{k_2} E + P$$

$$+$$

$$S$$

$$\Big\Updownarrow K_{S_1}$$ (3.32)

$$ES_2$$

With the definitions of

$$K_{S_1} = \frac{[S][ES]}{[ES_2]}, \quad K'_m = \frac{[S][E]}{[ES]}$$ (3.33)

the assumption of rapid equilibrium yields

$$v = \frac{V_m[S]}{K'_m + [S] + \dfrac{[S]^2}{K_{S_1}}}$$ (3.34)

A double-reciprocal plot describing substrate inhibition is given in Fig. 3.10.

At low substrate concentrations, $[S]^2/K_{S_1} \ll 1$, and inhibition effect is not observed. The rate is

$$v = \frac{V_m}{\left[1 + \dfrac{K'_m}{[S]}\right]}$$ (3.35)

or

$$\frac{1}{v} = \frac{1}{V_m} + \frac{K'_m}{V_m}\frac{1}{[S]}$$ (3.36)

A plot of $1/v$ versus $1/[S]$ results in a line of slope K'_m/V_m and intercept of $1/V_m$.

At high substrate concentrations, $K'_m/[S] \ll 1$, and inhibition is dominant. The rate in this case is

$$v = \frac{V_m}{\left(1 + \dfrac{[S]}{K_{S_1}}\right)}$$ (3.37)

or

$$\frac{1}{v} = \frac{1}{V_m} + \frac{[S]}{K_{S_1}V_m}$$ (3.38)

A plot of $1/v$ versus $[S]$ results in a line of slope $1/K_{s_1} \cdot V_m$ and intercept of $1/V_m$.

The substrate concentration resulting in the maximum reaction rate can be determined by setting $dv/d[S] = 0$. The $[S]_{max}$ is given by

$$[S]_{max} = \sqrt{K_m' K_{S_1}}$$ (3.39)

Example 3.2

The following data have been obtained for two different initial enzyme concentrations for an enzyme-catalyzed reaction.

$v([E_o] = 0.015$ g/l) (g/l-min)	[S] (g/l)	$v([E_o] = 0.00875$ g/l) (g/l-min)
1.14	20.0	0.67
0.87	10.0	0.51
0.70	6.7	0.41
0.59	5.0	0.34
0.50	4.0	0.29
0.44	3.3	
0.39	2.9	
0.35	2.5	

a. Find K_m.
b. Find V_m for $[E_0] = 0.015$ g/l.
c. Find V_m for $[E_0] = 0.00875$ g/l.
d. Find k_2.

Solution A Hanes–Woolf plot (Fig. 3.12) can be used to determine V_m and K_m.

$$\frac{[S]}{v} = \frac{K_m}{V_m} + \frac{1}{V_m}[S]$$

[S]/v (E_o = 0.015) (min)	[S]/ v (E_o = 0.00875) (min)	[S] (g/l)
17.5	30	20.0
11.5	20	10.0
9.6	16	6.7
8.5	15	5.0
8.0	14	4.0
7.6		3.3
7.3		2.9
7.1		2.5

From a plot of $[S]/v$ versus $[S]$ for $E_0 = 0.015$ g/l, the slope is found to be 0.6 min/g/l and $V_m = 1/0.6 = 1.7$ g/l min. The y-axis intercept is $K_m/V_m = 5.5$ min and $K_m = 9.2$ g $[S]/l$.

Also, $V_m = k_2E_0$ and $k_2 = 1.7/0.015 = 110$ g/g enzyme-min. The Hanes–Woolf plot for $E_0 = 0.00875$ g/l gives a slope of 1.0 min/g/l and $V_m = 1.0$ g/l-min; $k_2 = V_m/E_0 = 1.0/0.00875 = 114$ g/g enzyme-min.

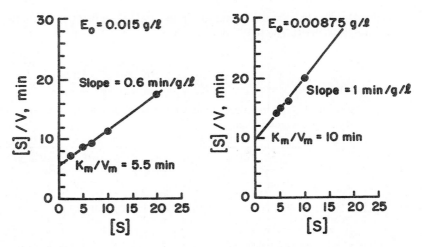

Figure 3.12. Hanes–Woolf plots for $E_0 = 0.015$ g/l and $E_0 = 0.00875$ g/l (Example 3.1).

Example 3.3

The hydrolysis of urea by urease is an only partially understood reaction and shows inhibition. Data for the hydrolysis of the reaction are given next.

Substrate concentration:	0.2 M		0.02 M	
	$1/v$	I	$1/v$	I
	0.22	0	0.68	0
	0.33	0.0012	1.02	0.0012
	0.51	0.0027	1.50	0.0022
	0.76	0.0044	1.83	0.0032
	0.88	0.0061	2.04	0.0037
	1.10	0.0080	2.72	0.0044
	1.15	0.0093	3.46	0.0059

where v = moles/l-min and I is inhibitor molar concentration.

    a. Determine the Michaelis–Menten constant (K'_m) for this reaction.

    b. What type of inhibition reaction is this? Substantiate the answer.

    c. Based on the answer to part b, what is the value of K_i?

Solution A double-reciprocal plot of $1/v$ versus $1/[S]$ for inhibitor concentrations I = 0, 0.0012, 0.0044, and 0.006 indicates that the inhibition is noncompetitive (Fig. 3.13). From the x-axis intercept of the plot, $-1/K'_m = -13$ and $K'_m \approx 7.77 \times 10^{-2}$ M. For [S] = 0.2 M and I = 0 from the intercept of $1/V$ versus 1/S, $1/v_m = 0.2$ and $V_m \cong 5$ moles/l-min. For I = 0.0012 M and [S] = 0.2 M, $v = 3$ moles/l-min. Substituting these values in

$$v = \frac{V_m}{\left(1 + \frac{[I]}{K_I}\right)\left(1 + \frac{K'_m}{[S]}\right)}$$

gives $K_I = 6 \times 10^{-3}$ M.

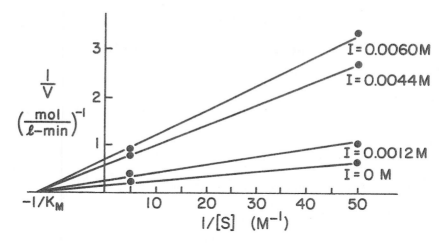

Figure 3.13. Double-reciprocal plot for different inhibitor concentrations (Example 3.2).

3.3.5. Effects of pH and Temperature

3.3.5.1. pH effects. Certain enzymes have ionic groups on their active sites, and these ionic groups must be in a suitable form (acid or base) to function. Variations in the pH of the medium result in changes in the ionic form of the active site and changes in the activity of the enzyme and hence the reaction rate. Changes in pH may also alter the three-dimensional shape of the enzyme. For these reasons, enzymes are only active over a certain pH range. The pH of the medium may affect the maximum reaction rate, K_m, and the stability of the enzyme. In some cases, the substrate may contain ionic groups, and the pH of the medium affects the affinity of the substrate to the enzyme.

The following scheme may be used to describe pH dependence of the enzymatic reaction rate for ionizing enzymes.

$$
\begin{array}{c}
\mathrm{E^- + H^+} \\
K_2 \Big\Updownarrow \\
\mathrm{EH + S} \underset{}{\overset{K'_m}{\rightleftharpoons}} \mathrm{EHS} \xrightarrow{k_2} \mathrm{EH + P} \\
+ \\
\mathrm{H^+} \\
K_1 \Big\Updownarrow \\
\mathrm{EH_2^+}
\end{array}
\tag{3.40}
$$

With the definition of

$$K'_m = \frac{[EH][S]}{[EHS]}$$

$$K_1 = \frac{[EH][H^+]}{[EH_2^+]} \tag{3.41}$$

$$K_2 = \frac{[E^-][H^+]}{[EH]}$$

$$[E_0] = [E^-] + [EH] + [EH_2^+] + [EHS], \quad v = k_2[EHS]$$

We can derive the following rate expression:

$$v = \frac{V_m[S]}{K'_m\left[1 + \frac{K_2}{[H^+]} + \frac{[H^+]}{K_1}\right] + [S]} \tag{3.42}$$

or

$$v = \frac{V_m[S]}{K'_{m,\text{app}} + [S]} \tag{3.43}$$

where $K'_{m,\text{app}} = K'_m\left[1 + \frac{K_2}{[H^+]} + \frac{[H^+]}{K_1}\right]$

As a result of this behavior, the pH optimum of the enzyme is between pK_1 and pK_2.

For the case of ionizing substrate, the following scheme and rate expression can be developed:

$$\begin{array}{c} SH^+ + E \underset{}{\overset{K'_m}{\rightleftharpoons}} ESH^+ \xrightarrow{\ k_2\ } E + HP^+ \\ \updownarrow K_1 \\ S + H^+ \end{array} \tag{3.44}$$

$$v = \frac{V_m[S]}{K'_m\left(1 + \frac{K_1}{[H^+]}\right) + [S]} \tag{3.45}$$

Theoretical prediction of the pH optimum of enzymes requires a knowledge of the active site characteristics of enzymes, which are very difficult to obtain. The pH optimum for an enzyme is usually determined experimentally. Figure 3.14 depicts variation of enzymatic activity with pH for two different enzymes.

3.3.5.2. Temperature effects. The rate of enzyme-catalyzed reactions increases with temperature up to a certain limit. Above a certain temperature, enzyme activ-

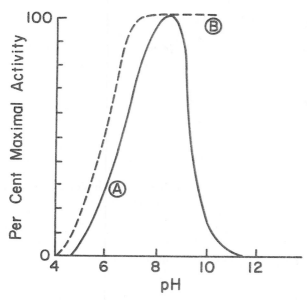

Figure 3.14. The pH-activity profiles of two enzymes. (A) approximate activity for trypsin; (B) approximate activity for cholinesterase.

ity decreases with temperature because of enzyme denaturation. Figure 3.15 depicts the variation of reaction rate with temperature and the presence of an optimal temperature. The ascending part of Fig. 3.15 is known as *temperature activation*. The rate varies according to the Arrhenius equation in this region.

$$v = k_2 [E] \tag{3.46a}$$

$$k_2 = Ae^{-E_a/RT} \tag{3.46b}$$

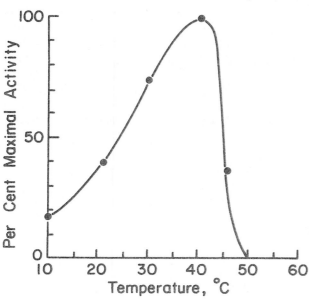

Figure 3.15. Effect of temperature on the activity of an enzyme. Here we have assumed a value of $E_a = 11$ kcal/g-mol and $E_d = 70$ kcal/g-mol. The descending portion of the curve is due to thermal denaturation and is calculated assuming a 10-min exposure to the temperature. Note that the nature of the plot will depend on the length of time the reaction mixture is exposed to the test temperature.

where E_a is the activation energy (kcal/mol) and [E] is the active enzyme concentration. A plot of ln υ versus $1/T$ results in a line of slope $-E_a/R$.

The descending part of Fig. 3.15 is known as *temperature inactivation* or *thermal denaturation*. The kinetics of thermal denaturation can be expressed as

$$-\frac{d[E]}{dt} = k_d[E] \tag{3.47}$$

or

$$[E] = [E_0]e^{-k_dt} \tag{3.48}$$

where $[E_0]$ is the initial enzyme concentration and k_d is the denaturation constant. k_d also varies with temperature according to the Arrhenius equation.

$$k_d = A_de^{-E_a/RT} \tag{3.49}$$

where E_d is the deactivation energy (kcal/mol). Consequently,

$$v = Ae^{-E_a/RT}E_0e^{-k_dt} \tag{3.50}$$

The activation energies of enzyme-catalyzed reactions are within the 4 to 20 kcal/g mol range (mostly about 11 kcal/g mol). Deactivation energies E_d vary between 40 and 130 kcal/g mol (mostly about 70 kcal/g mol). That is, enzyme denaturation by temperature is much faster than enzyme activation. A rise in temperature from 30° to 40°C results in a 1.8-fold increase in enzyme activity, but a 41-fold increase in enzyme denaturation. Variations in temperature may affect both V_m and K_m values of enzymes.

3.3.6. Insoluble Substrates

Enzymes are often used to attack large, insoluble substrates such as wood chips (in biopulping for paper manufacture) or cellulosic residues from agriculture (e.g., cornstalks). In these cases access to the reaction site on these biopolymers by enzymes is often limited by enzyme diffusion. The number of potential reactive sites exceeds the number of enzyme molecules. This situation is opposite that of the typical situation with soluble substrates, where access to the enzyme's active site limits reaction. If we consider initial reaction rates and if the reaction is first order with respect to the concentration of enzyme bound to substrate (i.e., [ES]), then we can derive a rate expression:

$$v = \frac{V_{max,S}[E]}{K_{eq} + [E]} \tag{3.51a}$$

where

$$V_{max,S} = k_2[S_0] \tag{3.51b}$$

and

$$K'_{eq} = k_{des}/k_{ads} \tag{3.51c}$$

The previous equation assumes slow binding of enzyme (i.e., $[E] \approx [E_0]$), S_0 is the number of substrate bonds available initially for breakage, and k_{des} and k_{ads} refer to rates of enzyme desorption and adsorption onto the insoluble matrix, respectively.

3.4 IMMOBILIZED ENZYME SYSTEMS

The restriction of enzyme mobility in a fixed space is known as *enzyme immobilization*. Immobilization of enzymes provides important advantages, such as enzyme reutilization and elimination of enzyme recovery and purification processes, and may provide a better environment for enzyme activity. Since enzymes are expensive, catalyst reuse is critical for many processes. Since some of the intracellular enzymes are membrane bound, immobilized enzymes provide a model system to mimic and understand the action of some membrane-bound intracellular enzymes. Product purity is usually improved, and effluent handling problems are minimized by immobilization.

3.4.1. Methods of Immobilization

Major methods of immobilization are summarized in Fig. 3.16. The two major categories are entrapment and surface immobilization.

3.4.1.1. Entrapment.

Entrapment is the physical enclosure of enzymes in a small space. Matrix entrapment and membrane entrapment, including microencapsulation, are the two major methods of entrapment.

Matrices used for enzyme immobilization are usually polymeric materials such as Ca-alginate, agar, κ-carrageenin, polyacrylamide, and collagen. However, some solid matrices such as activated carbon, porous ceramic, and diatomaceous earth can also be used for this purpose. The matrix can be a particle, a membrane, or a fiber. When immobilizing

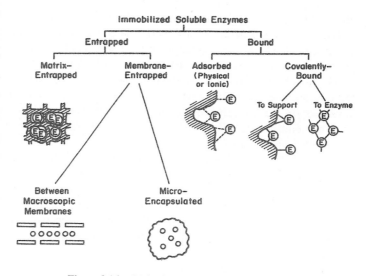

Figure 3.16. Major immobilization methods.

in a polymer matrix, enzyme solution is mixed with polymer solution before polymerization takes place. Polymerized gel-containing enzyme is either extruded or a template is used to shape the particles from a liquid polymer-enzyme mixture. Entrapment and surface attachment may be used in combination in some cases.

Membrane entrapment of enzymes is possible; for example, hollow fiber units have been used to entrap an enzyme solution between thin, semipermeable membranes. Membranes of nylon, cellulose, polysulfone, and polyacrylate are commonly used. Configurations, other than hollow fibers, are possible, but in all cases a semipermeable membrane is used to retain high-molecular-weight compounds (enzyme), while allowing small-molecular-weight compounds (substrate or products) access to the enzyme.

A special form of membrane entrapment is *microencapsulation*. In this technique, microscopic hollow spheres are formed. The spheres contain the enzyme solution, while the sphere is enclosed within a porous membrane. The membrane can be polymeric or an enriched interfacial phase formed around a microdrop.

Despite the aforementioned advantages, enzyme entrapment may have its inherent problems, such as enzyme leakage into solution, significant diffusional limitations, reduced enzyme activity and stability, and lack of control of microenvironmental conditions. Enzyme leakage can be overcome by reducing the MW cutoff of membranes or the pore size of solid matrices. Diffusion limitations can be eliminated by reducing the particle size of matrices and/or capsules. Reduced enzyme activity and stability are due to unfavorable microenvironmental conditions, which are difficult to control. However, by using different matrices and chemical ingredients, by changing processing conditions, and by reducing particle or capsule size, more favorable microenvironmental conditions can be obtained. Diffusion barrier is usually less significant in microcapsules as compared to gel beads.

3.4.1.2. Surface immobilization.
The two major types of immobilization of enzymes on the surfaces of support materials are adsorption and covalent binding.

Adsorption is the attachment of enzymes on the surfaces of support particles by weak physical forces, such as van der Waals or dispersion forces. The active site of the adsorbed enzyme is usually unaffected, and nearly full activity is retained upon adsorption. However, desorption of enzymes is a common problem, especially in the presence of strong hydrodynamic forces, since binding forces are weak. Adsorption of enzymes may be stabilized by cross-linking with glutaraldehyde. Glutaraldehyde treatment can denature some proteins. Support materials used for enzyme adsorption can be inorganic materials, such as alumina, silica, porous glass, ceramics, diatomaceous earth, clay, and bentonite, or organic materials, such as cellulose (CMC, DEAE-cellulose), starch, activated carbon, and ion-exchange resins, such as Amberlite, Sephadex, and Dowex. The surfaces of the support materials may need to be pretreated (chemically or physically) for effective immobilization.

Covalent binding is the retention of enzymes on support surfaces by covalent bond formation. Enzyme molecules bind to support material via certain functional groups, such as amino, carboxyl, hydroxyl, and sulfhydryl groups. These functional groups must not be in the active site. One common trick is to block the active site by flooding the enzyme solution with a competitive inhibitor prior to covalent binding. Functional groups on support material are usually activated by using chemical reagents, such as cyanogen bromide, carbodiimide, and glutaraldehyde. Support materials with various functional groups and the chemical reagents used for the covalent binding of proteins are listed in Table 3.3.

TABLE 3.3 Methods of Covalent Binding of Enzymes to Supports

Supports with —OH

(a) Using cyanogen bromide

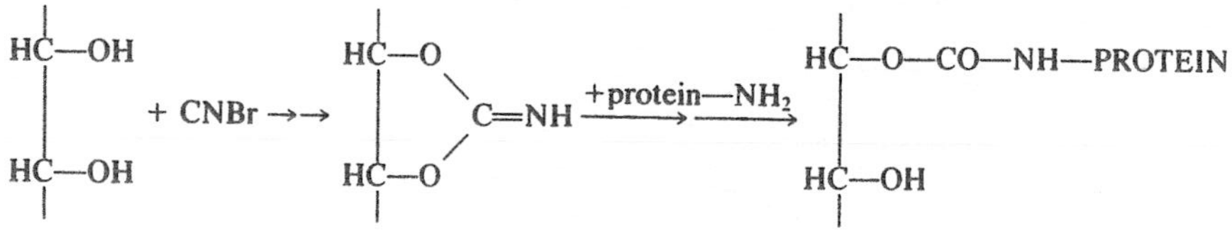

$$\text{HC—OH} \atop \text{HC—OH} \quad + \text{ CNBr} \longrightarrow \quad {\text{HC—O} \atop \text{HC—O}}\!\!\diagdown\!\! \text{C=NH} \xrightarrow[\text{+ protein—NH}_2]{} \quad {\text{HC—O—CO—NH—PROTEIN} \atop \text{HC—OH}}$$

(b) Using S-triazine derivatives

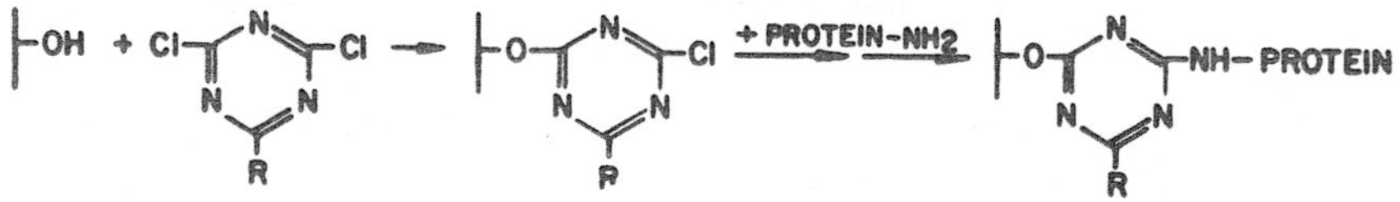

$$\left|\text{OH} + \text{Cl}\!\!-\!\!{\overset{N}{\underset{N}{\diagup\!\!\diagdown}}}\!\!-\!\!{\overset{Cl}{\underset{R}{}}} \longrightarrow \left|\text{O}\!\!-\!\!{\overset{N}{\underset{N}{\diagup\!\!\diagdown}}}\!\!-\!\!{\overset{Cl}{\underset{R}{}}} \xrightarrow{\text{+ PROTEIN—NH}_2} \left|\text{O}\!\!-\!\!{\overset{N}{\underset{N}{\diagup\!\!\diagdown}}}\!\!-\!\!{\overset{NH—PROTEIN}{\underset{R}{}}}\right.$$

Supports with —NH$_2$

(a) By diazotization

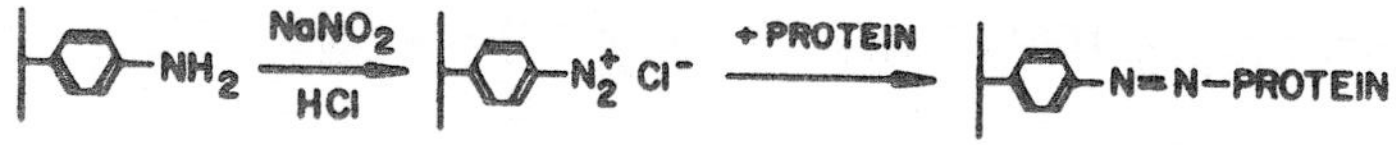

$$\left|\!\!-\!\!\bigcirc\!\!-\!\!\text{NH}_2 \xrightarrow[\text{HCl}]{\text{NaNO}_2} \left|\!\!-\!\!\bigcirc\!\!-\!\!\text{N}_2^+ \text{Cl}^- \xrightarrow{\text{+ PROTEIN}} \left|\!\!-\!\!\bigcirc\!\!-\!\!\text{N}\!\!=\!\!\text{N—PROTEIN}\right.$$

(b) Using glutaraldehyde

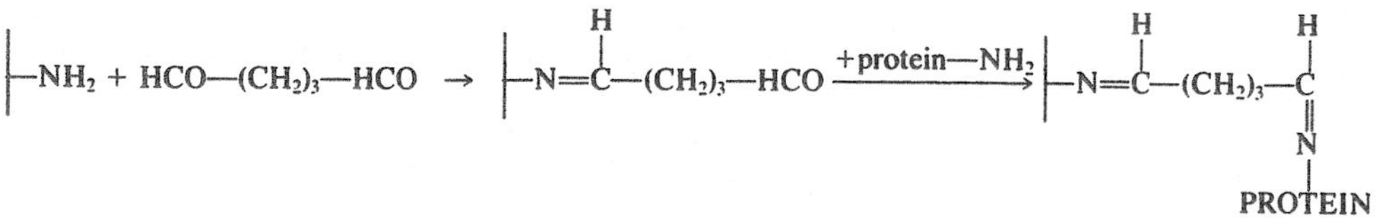

$$\left|\!\!-\!\!\text{NH}_2 + \text{HCO—(CH}_2)_3\text{—HCO} \longrightarrow \left|\!\!-\!\!{\overset{H}{\underset{}{\text{N}\!\!=\!\!\text{C}}}}\text{—(CH}_2)_3\text{—HCO} \xrightarrow{\text{+ protein—NH}_2} \left|\!\!-\!\!{\overset{H}{\underset{}{\text{N}\!\!=\!\!\text{C}}}}\text{—(CH}_2)_3\text{—}{\overset{H}{\underset{N}{\text{C}\!\!=\!\!}}}{\atop\text{PROTEIN}}\right.$$

(continued)

TABLE 3.3 Methods of Covalent Binding of Enzymes to Supports (Continued)

Supports with —COOH

(a) Via azide derivative

1) $\text{—O—CH}_2\text{—COOH} \xrightarrow[\text{H}^+]{\text{CH}_3\text{OH}} \text{—O—CH}_2\text{—COOCH}_3 \xrightarrow{\text{H}_2\text{NNH}_2} \text{—O—CH}_2\text{—CO—NH—NH}_2$

2) $\text{—O—CH}_2\text{—CO—NH—NH}_2 \xrightarrow[\text{HCl}]{\text{NaNO}_2} \text{—O—CH}_2\text{—CON}_3 \xrightarrow{\text{+protein—NH}_2} \text{—O—CH}_2\text{—CO—NH—PROTEIN}$

(b) Using a carbodiimide

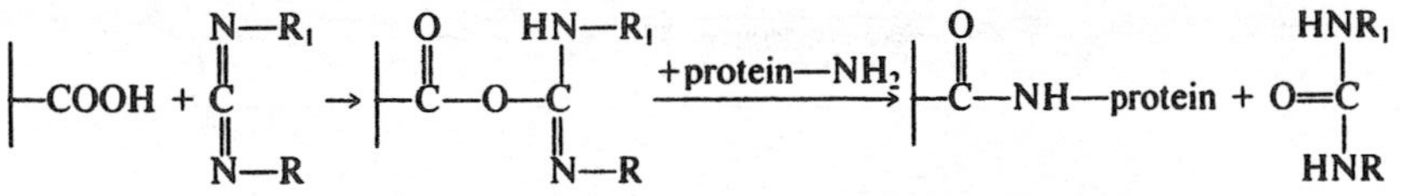

$$\text{COOH} + \begin{array}{c} \text{N—R}_1 \\ \| \\ \text{C} \\ \| \\ \text{N—R} \end{array} \longrightarrow \begin{array}{c} \text{HN—R}_1 \\ \text{O} \quad | \\ \| \quad | \\ \text{—C—O—C} \\ \| \\ \text{N—R} \end{array} \xrightarrow{\text{+protein—NH}_2} \begin{array}{c} \text{O} \\ \| \\ \text{—C—NH—protein} \end{array} + \begin{array}{c} \text{HNR}_1 \\ \text{O=C} \\ \text{HNR} \end{array}$$

Supports containing anhydrides

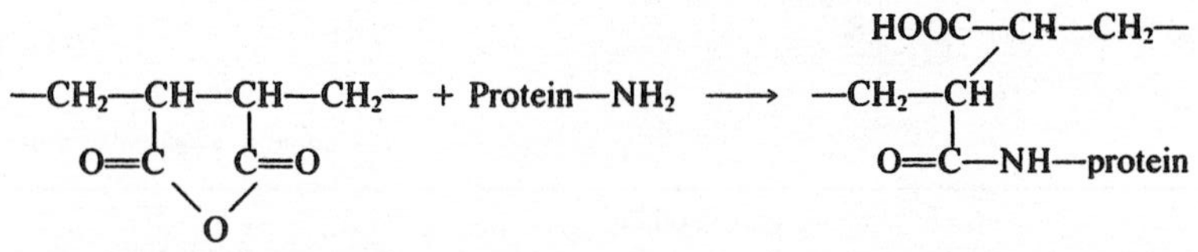

$$\begin{array}{c} \text{—CH}_2\text{—CH—CH—CH}_2\text{—} \\ \quad\quad | \quad\quad | \\ \quad\quad \text{O=C} \quad \text{C=O} \\ \quad\quad\quad \backslash \quad / \\ \quad\quad\quad\quad \text{O} \end{array} + \text{Protein—NH}_2 \longrightarrow \begin{array}{c} \text{HOOC—CH—CH}_2\text{—} \\ \quad\quad\quad | \\ \text{—CH}_2\text{—CH} \\ \quad\quad\quad | \\ \quad\quad \text{O=C—NH—protein} \end{array}$$

With permission, from D. I. C. Wang et al., *Fermentation and Enzyme Technology*, John Wiley & Sons, New York, 1979.

Binding groups on the protein molecule are usually side groups (R) or the amino or carboxyl groups of the polypeptide chain.

The cross-linking of enzyme molecules with each other using agents such as glutaraldehyde, *bis*-diazobenzidine, and 2,2-disulfonic acid is another method of enzyme immobilization. Cross-linking can be achieved in several different ways: enzymes can be cross-linked with glutaraldehyde to form an insoluble aggregate, adsorbed enzymes may be cross-linked, or cross-linking may take place following the impregnation of porous support material with enzyme solution. Cross-linking may cause significant changes in the active site of enzymes, and also severe diffusion limitations may result.

The most suitable support material and immobilization method vary depending on the enzyme and particular application. Two major criteria used in the selection of support material are (1) the binding capacity of the support material, which is a function of charge density, functional groups, porosity, and hydrophobicity of the support surface, and (2) stability and retention of enzymatic activity, which is a function of functional groups on support material and microenvironmental conditions. If immobilization causes some conformational changes on the enzyme, or if reactive groups on the active site of the enzyme are involved in binding, a loss in enzyme activity can take place upon immobilization. Usually, immobilization results in a loss in enzyme activity and stability. However, in some cases, immobilization may cause an increase in enzyme activity and stability due to more favorable microenvironmental conditions. Because enzymes often have more than one functional site that can bind the surface, an immobilized enzyme preparation may be very heterogeneous. Even when binding does not alter enzyme structure, some enzyme can be bound with the active site oriented away from the substrate solution and toward the support surface, decreasing the access of the substrate to the enzyme. Retention of activity varies with the method used. Table 3.4 summarizes the retention of activity of aminoacylase immobilized by different methods.

TABLE 3.4 Effect of Immobilization Methods on the Retention of Enzymatic Activity of Aminoacylase

Support	Method	Observed activity (units)	Enzyme activity immobilized (%)
Polyacrylamide	Entrapment	526	52.6
Nylon	Encapsulation	360	36.0
DEAE-cellulose	Ionic binding	668	55.2
DEAE-Sephadex A-5O	Ionic binding	680	56.2
CM-Sephadex C-5O	Ionic binding	0	0
Iodoacetyl cellulose	Covalent binding	472	39.0
CNBr-activated Sephadex	Covalent binding	12	1.0
AE-cellulose	Cross-linked with glutaraldehyde	8	0.6

With permission, from D. I. C. Wang et al., *Fermentation and Enzyme Technology*, John Wiley & Sons, New York, 1979.

3.4.2. Diffusional Limitations in Immobilized Enzyme Systems

Diffusional resistances may be observed at different levels in immobilized enzymes. These resistances vary depending on the nature of the support material (porous, nonporous), hydrodynamical conditions surrounding the support material, and distribution of the enzyme inside or on the surface of the support material. Whether diffusion resistance has a significant effect on the rate of enzymatic reaction rate depends on the relative rate of the reaction rate and diffusion rate, which is characterized by the Damköhler number (Da).

$$Da = \frac{\text{maximum rate of reaction}}{\text{maximum rate of diffusion}} = \frac{V_{m'}}{k_L [S_b]} \qquad (3.52)$$

where $[S_b]$ is substrate concentration in bulk liquid (g/cm^3) and k_L is the mass-transfer coefficient (cm/s).

The rate of enzymatic conversion may be limited by diffusion of the substrate or reaction, depending on the value of the Damköhler number. If Da >> 1, the diffusion rate is limiting. For Da << 1, the reaction rate is limiting, and for Da ≈ 1, the diffusion and reaction resistances are comparable. Diffusion and enzymatic reactions may be simultaneous, with enzymes entrapped in a solid matrix, or may be two consecutive phenomena for adsorbed enzymes.

3.4.2.1. Diffusion effects in surface-bound enzymes on nonporous support materials.
Assume a situation where enzymes are bound and evenly distributed on the surface of a nonporous support material, all enzyme molecules are equally active, and substrate diffuses through a thin liquid film surrounding the support surface to reach the reactive surfaces, as depicted in Fig. 3.17. Assume further that the process of immobilization has not altered the protein structure, and the intrinsic kinetic parameters (V_m, K_m) are unaltered.

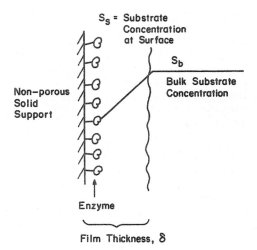

Figure 3.17. Substrate concentration profile in a liquid film around adsorbed enzymes.

At steady state, the reaction rate is equal to the mass-transfer rate:

$$J_s = k_L \left([S_b] - [S_s] \right) = \frac{V_m' [S_s]}{K_m + [S_s]} \qquad (3.53)$$

where V_m' is the maximum reaction rate per unit of external surface area and k_L is the liquid mass-transfer coefficient. This equation is quadratic in $[S_s]$, the substrate concentration at the surface. It can be solved analytically, but the solution is cumbersome. Furthermore, the value of $[S_s]$ is not amenable to direct experimental observation.

Equation 3.53 can be solved graphically as depicted in Fig. 3.18. Such a plot also makes it easy to visualize the effects of parameter changes such as stirring rate, changes in bulk substrate concentration, or enzyme loading.

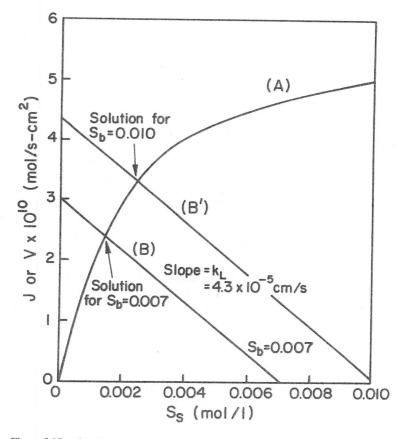

Figure 3.18. Graphical solution for amount of reaction per unit surface area for enzyme immobilized on a nonporous catalyst. Curve A results from a knowledge of the intrinsic solution-based kinetic parameters and the surface loading of enzyme (right side of eq. 3.53). Line B is the mass transfer equation (left side of eq. 3.53). The intersection of the two lines is the reaction rate, υ, that can be sustained in the system. The responses for two different bulk substrate concentrations are shown.

When the system is strongly mass-transfer limited, $[S_s] \approx 0$, since the reaction is rapid compared to mass transfer, and

$$v \approx k_L[S_b], \quad \text{(for Da} \gg 1) \tag{3.54}$$

and the system behaves as pseudo first order.

When the system is reaction limited (Da $\ll$ 1), the reaction rate is often expressed as

$$v = \frac{V'_m [S_b]}{K_{m,\text{app}} + [S_b]} \tag{3.55a}$$

where, with appropriate assumptions,

$$K_{m,\text{app}} = K_m \left\{ 1 + \frac{V'_m}{k_L([S_b] + K_m)} \right\} \tag{3.55b}$$

Under these circumstances, the apparent Michaelis–Menten "constant" is a function of stirring speed. Usually, $K_{m,\text{app}}$ is estimated experimentally as the value of $[S_b]$, giving one-half of the maximal reaction rate.

Example 3.4

Consider a system where a flat sheet of polymer coated with enzyme is placed in a stirred beaker. The intrinsic maximum reaction rate (V_m) of the enzyme is 6×10^{-6} mol/s-mg enzyme. The amount of enzyme bound to the surface has been determined to be 1×10^{-4} mg enzyme/cm^2 of support. In solution, the value of K_m has been determined to be 2×10^{-3} mol/l. The mass-transfer coefficient can be estimated from standard correlations for stirred vessels. We assume in this case a very poorly mixed system where $k_L = 4.3 \times 10^{-5}$ cm/s. What is the reaction rate when (a) the bulk concentration of the substrate is 7×10^{-3} mol/l? (b) $S_b = 1 \times 10^{-2}$ mol/l?

Solution The solution is given in Fig. 3.18. The key is to note that the mass-transfer rate equals the reaction rate at steady state, and as a consequence the right side of eq. 3.53 must equal the left side. In case (a), this occurs at a substrate surface concentration of about 0.0015 mol/l with a reaction rate of 2.3×10^{-10} mol/s-cm^2. By increasing the bulk substrate concentration to 0.01 mol/l, the value of $[S_s]$ increases to 0.0024 mol/l with a reaction rate about 3.3×10^{-10} mol/s-cm^2.

3.4.2.2. Diffusion effects in enzymes immobilized in a porous matrix.

When enzymes are immobilized on internal pore surfaces of a porous matrix, substrate diffuses through the tortuous pathway among pores and reacts with enzyme immobilized on pore surfaces. Diffusion and reaction are simultaneous in this case, as depicted in Fig. 3.19.

Assume that enzyme is uniformly distributed in a spherical support particle; the reaction kinetics are expressed by Michaelis–Menten kinetics, and there is no partitioning of the substrate between the exterior and interior of the support. Then we write the following equation, stating that diffusion rate is equal to reaction rate at steady state:

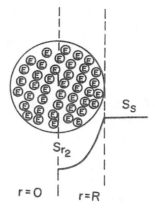

Figure 3.19. Substrate concentration profile in a porous support particle containing immobilized enzymes. Here it is assumed that no external substrate limitation exists so that the bulk and surface concentrations are the same.

$$D_e \left(\frac{d^2[S]}{dr^2} + \frac{2}{r} \frac{d[S]}{dr} \right) = \frac{V_m''[S]}{K_m + [S]} \tag{3.56}$$

with boundary conditions $[S] = [S_s]$ at $r = R$ and $d[S]/dr = 0$ at $r = 0$, where V_m'' is the maximum reaction rate per unit volume of support, and D_e is the effective diffusivity of substrate within the porous matrix.

Equation 3.56 can be written in dimensionless form by defining the following dimensionless variables:

$$\bar{S} = \frac{[S]}{[S_s]}, \qquad \bar{r} = \frac{r}{R}, \qquad \beta = \frac{K_m}{[S_s]}$$

$$\frac{d^2\bar{S}}{d\bar{r}^2} + \frac{2}{\bar{r}} \frac{d\bar{S}}{d\bar{r}} = \frac{R^2 V_m''}{S_s D_e} \left(\frac{\bar{S}}{\bar{S} + \beta} \right) \tag{3.57a}$$

or

$$\frac{d^2\bar{S}}{d\bar{r}^2} + \frac{2}{\bar{r}} \frac{d\bar{S}}{d\bar{r}} = \phi^2 \frac{\bar{S}}{1 + \bar{S}/\beta} \tag{3.57b}$$

where

$$\phi = R \sqrt{\frac{V_m''/K_m}{D_e}} = \text{Thiele modulus} \tag{3.57c}$$

With boundary conditions of $\bar{S} = 1$ at $\bar{r} = 1$ and $d\bar{S}/d\bar{r} = 0$ at $\bar{r} = 0$, eq. 3.57 can be numerically solved to determine the substrate profile inside the matrix. The rate of substrate consumption is equal to the rate of substrate transfer through the external surface of the support particle at steady state into the sphere.

$$r_s = N_s = 4\pi R^2 D_e \left. \frac{d[S]}{dr} \right|_{r=R} \tag{3.58}$$

Under diffusion limitations, the rate per unit volume is usually expressed in terms of the effectiveness factor as follows:

$$r_s = \eta \frac{V''_m \left[S_s \right]}{K_m + \left[S_s \right]} \tag{3.59}$$

The *effectiveness factor* is defined as the ratio of the reaction rate with diffusion limitation (or diffusion rate) to the reaction rate with no diffusion limitation. The value of the effectiveness factor is a measure of the extent of diffusion limitation. For $\eta < 1$, the conversion is diffusion limited, whereas for $\eta \approx 1$ values, conversion is limited by the reaction rate and diffusion limitations are negligible. The factor is a function of ϕ and β as depicted in Figure 3.20.

For a zero-order reaction rate ($\beta \to 0$), $\eta \approx 1$ for a large range of Thiele modulus values such as $1 < \phi < 100$. For a first-order reaction rate ($\beta \to \infty$), $\eta = (\phi, \beta)$ and η is approximated to the following equation for high values of ϕ.

$$\eta = \frac{3}{\phi} \left[\frac{1}{\tanh \phi} - \frac{1}{\phi} \right] \tag{3.60}$$

When internal diffusion limits the enzymatic reaction rate, the rate-constant $V_{m,\text{app}}$ and $K_{m,\text{app}}$ values are not true intrinsic rate constants, but apparent values. To obtain true intrinsic rate constants in immobilized enzymes, diffusion resistances should be eliminated by using small particle sizes, a high degree of turbulence around the particles, and high substrate concentrations.

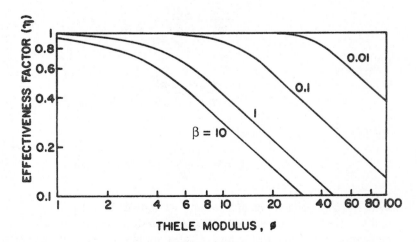

Figure 3.20. Theoretical relationship between the effectiveness factor η and first-order Thiele modulus, ϕ, for a spherical porous immobilized particle for various values of β, where β is the dimensionless Michaelis constant. (With permission, from D. I. C. Wang et al., *Fermentation and Enzyme Technology*, John Wiley & Sons, Inc., New York, 1979, p. 329.)

Enzymes   Chap. 3

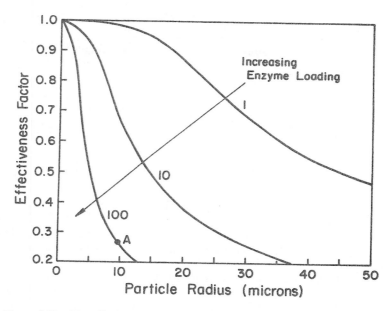

Figure 3.21. The effectiveness factor decreases with increases in enzyme loading or with increases in particle diameter. Point A represents the value of the effectiveness factor for a particle radius of 10 μm with an enzyme loading of 100 mg/cm³, an enzyme activity of 100 μmol/min per mg enzyme, a substrate diffusivity of 5×10^{-6} cm²/s, and a bulk substrate concentration tenfold higher than K_m.

When designing immobilized enzyme systems using a particular support, the main variables are V_m and R, since the substrate concentration, K_m, and D_e are fixed. The particle size (R) should be as small as possible within the constraints of particle integrity, resistance to compression, and the nature of the particle recovery systems. The maximum reaction rate is determined by enzyme activity and concentration in the support. High enzyme content will result in high enzyme activity per unit of reaction volume but low effectiveness factor. On the other hand, low enzyme content will result in lower enzyme activity per unit volume but a high effectiveness factor. For maximum conversion rates, particle size should be small ($D_p \leq 10$ μm) and enzyme loading should be optimized. As depicted in the example in Fig. 3.21, $D_p \leq 10$ μm and enzyme loadings of less than 10 mg/cm³ are required for high values of the effectiveness factor ($\eta \geq 0.8$).

Example 3.5

D. Thornton and co-workers studied the hydrolysis of sucrose at pH = 4.5 and 25°C using crude invertase obtained from baker's yeast in free and immobilized form. The following initial velocity data were obtained with 408 units of crude enzyme (1 unit = quantity of enzyme hydrolyzing 1 μmol of sucrose/min when incubated with 0.29 M sucrose in a buffer at pH 4.5 and 25°C).

V_0 (mmol hydrolyzed/l-min)		
Free enzyme	Immobilized enzyme	S_0 (mol/l)
0.083	0.056	0.010
0.143	0.098	0.020
0.188	0.127	0.030
0.222	0.149	0.040
0.250	0.168	0.050
0.330	0.227	0.100
0.408	0.290	0.290

     a. Determine the K_m and V_m for this reaction using both free and immobilized enzyme.

     b. Do the data indicate any diffusion limitations in the immobilized enzyme preparation?

Solution   From a double-reciprocal plot of $1/v$ versus $1/S$ for free enzyme (Fig. 3.22), $-1/K_m = -20$ and $K_m = 0.05$ M. $1/V_m = 2$ and $V_m = 0.5$ mmol/l min. From a double-reciprocal plot of $1/v$ versus $1/S$ for the immobilized enzyme, $-1/K_m = -20$ and $K_m \doteq 0.05$ M. $1/V_m = 3$ and $V_m = 0.33$ mmol/l-min. Since the K_m values for free and immobilized enzymes are the same, there is no diffusion limitation.

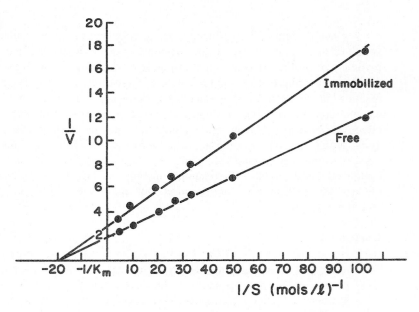

Figure 3.22.   Double-reciprocal plots for free and immobilized enzymes (Example 3.4).

3.4.3. Electrostatic and Steric Effects in Immobilized Enzyme Systems

When enzymes are immobilized in a charged matrix as a result of a change in the microenvironment of the enzyme, the apparent bulk pH optimum of the immobilized enzyme will shift from that of soluble enzyme. The charged matrix will repel or attract substrates, product, cofactors, and H^+ depending on the type and quantity of surface charge. For an enzyme immobilized onto a charged support, the shift in the pH-activity profile is given by

$$\Delta pH = pH_i - pH_e = 0.43 \frac{zF\psi}{RT} \qquad (3.61)$$

where pH_i and pH_e are internal and external pH values, respectively; z is the charge (valence) on the substrate; F is the Faraday constant (96,500 coulomb/eq. g); ψ is the electrostatic potential; and R is the gas constant. Expressions similar to eq. 3.61 apply to other nonreactive charged medium components. The intrinsic activity of the enzyme is altered by the local changes in pH and ionic constituents. Further alterations in the apparent kinetics are due to the repulsion or attraction of substrates or inhibitors.

The activity of an enzyme toward a high-molecular-weight substrate is usually reduced upon immobilization to a much greater extent than for a low-molecular-weight substrate. This is mainly because of steric hindrance by the support. Certain substrates, such as starch, have molecular weights comparable to those of enzymes and may therefore not be able to penetrate to the active sites of immobilized enzymes.

Immobilization also affects the thermal stability of enzymes. Thermal stability often increases upon immobilization due to the presence of thermal diffusion barriers and the constraints on protein unfolding. However, decreases in thermal stability have been noted in a few cases. The pH stability of enzymes usually increases upon immobilization, too.

3.5. LARGE-SCALE PRODUCTION OF ENZYMES

Among various enzymes produced at large scale are proteases (subtilisin, rennet), hydrolases (pectinase, lipase, lactase), isomerases (glucose isomerase), and oxidases (glucose oxidase). These enzymes are produced using overproducing strains of certain organisms. Separation and purification of an enzyme from an organism require disruption of cells, removal of cell debris and nucleic acids, precipitation of proteins, ultrafiltration of the desired enzyme, chromatographic separations (optional), crystallization, and drying. The process scheme varies depending on whether the enzyme is intracellular or extracellular. In some cases, it may be more advantageous to use inactive (dead or resting) cells with the desired enzyme activity in immobilized form. This approach eliminates costly enzyme separation and purification steps and is therefore economically more feasible. Details of protein separations are covered in Chapter 11.

The first step in the large-scale production of enzymes is to cultivate the organisms producing the desired enzyme. Enzyme production can be regulated and fermentation conditions can be optimized for overproduction of the enzyme. Proteases are produced by using overproducing strains of *Bacillus*, *Aspergillus*, *Rhizopus*, and *Mucor*; pectinases are produced by

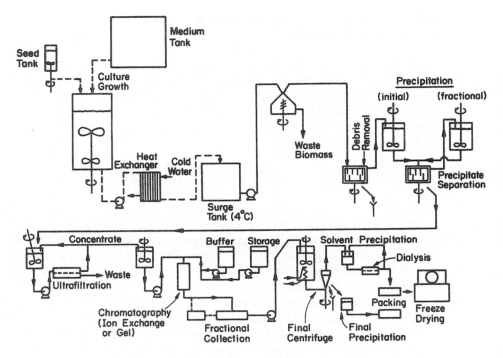

Figure 3.23. A flowsheet for the production of an extracellular enzyme.

Aspergillus niger; lactases are produced by yeast and *Aspergillus;* lipases are produced by certain strains of yeasts and fungi; glucose isomerase is produced by *Flavobacterium arborescens* or *Bacillus coagulans.* After the cultivation step, cells are separated from the media usually by filtration or sometimes by centrifugation. Depending on the intracellular or extracellular nature of the enzyme, either the cells or the fermentation broth is further processed to separate and purify the enzyme. The recovery of intracellular enzymes is more complicated and involves the disruption of cells and removal of cell debris and nucleic acids. Figure 3.23 depicts a schematic of an enzyme plant producing intracellular enzymes.

In some cases, enzyme may be both intracellular and extracellular, which requires processing of both broth and cells. Intracellular enzymes may be released by increasing the permeability of cell membrane. Certain salts such as $CaCl_2$ and other chemicals such as dimethylsulfoxide (DMSO) and pH shift may be used for this purpose. If enzyme release is not complete, then cell disruption may be essential.

The processes used to produce these industrial enzymes have much in common with our later discussions on processes to make proteins from recombinant DNA.

3.6. MEDICAL AND INDUSTRIAL UTILIZATION OF ENZYMES

Enzymes have been significant industrial products for more than a hundred years. However, the range of potential application is increasing rapidly. With the advent of recombinant DNA technology it has become possible to make formerly rare enzymes in large quantities and, hence, reduce cost. Also, in pharmaceutical manufacture the desire to make chirally pure compounds is leading to new opportunities. Chirality is important in a product; in a racemic mixture one enantiomer is often therapeutically useful while the other may cause side effects and add no therapeutic value. The ability of enzymes to recognize chiral isomers and react with only one of them can be a key component in pharmaceutical synthesis. Processes that depend on a mixture of chemical and enzymatic synthesis are being developed for a new generation of pharmaceuticals.

Technological advances have facilitated the use of enzymes over an increasingly broad range of process conditions. Enzymes from organisms that grow in unusual environments (e.g., deep ocean, salt lakes, and hot springs) are increasingly available for study and potential use. New enzymes and better control of reaction conditions allow the use of enzymes in the presence of high concentrations of organics, in high-salt aqueous environments, or at extreme temperatures, pH, or pressures. As we couple new insights into the relationship of enzyme structure to biological function with recombinant DNA technology, we are able to produce enzymes that are human designed or manipulated (see Section 14.9 on protein engineering). We no longer need to depend solely on natural sources for enzymes.

While there are many reasons to be optimistic about increasing use of enzymes, the number of enzymes made at high volume for industrial purposes evolves more slowly. In 1996 the U.S. sales of industrial enzymes were $372 million, and sales are projected to grow to $686 million by 2006. The products made in enzyme processes are worth billions of dollars. Table 3.5 provides a breakdown of projected enzyme sales by industrial sector. Table 3.6 lists some industrially important enzymes.

Proteases hydrolyze proteins into smaller peptide units and constitute a large and industrially important group of enzymes. Proteases constitute about 60% of the total enzyme market. Industrial proteases are obtained from bacteria (*Bacillus*), molds (*Aspergillus, Rhizopus,* and *Mucor*), animal pancreas, and plants. Most of the industrial proteases are endoproteases. Proteases are used in food processing, such as cheese making (rennet), baking, meat tenderization (papain, trypsin), and brewing (trypsin, pepsin); in detergents for the hydrolysis of protein stains (subtilisin Carlsberg); and in tanning and the medical treatment of wounds.

TABLE 3.5. Industrial Enzyme Market*

Application	1996 Sales (U.S. $ in millions)	2006 Projected Sales (U.S. $ in millions)
Food	170	214
Detergent	160	414
Textiles	27	32
Leather	11	13
Paper & Pulp	1	5
Other	3	8
TOTAL	372	686

*Data from C. Wrotnowski, *Genetic & Engineering News,* pp. 14 and 30, Feb. 1, 1997.

TABLE 3.6 Some Industrially Important Enzymes

Name	Example of Source	Application
Amylase	*Bacillus subtilis, Aspergillus niger*	Starch hydrolysis, glucose production
Glucoamylase	*A. niger, Rhizopus niveus, Endomycopsis*	Saccharification of starch, glucose production
Trypsin	Animal pancreas	Meat tenderizer, beer haze removal
Papain	Papaya	Digestive aid, meat tenderizer, medical applications
Pepsin	Animal stomach	Digestive aid, meat tenderizer
Rennet	Calf stomach/recombinant *E. coli*	Cheese manufacturing
Glucose isomerase	*Flavobacterium arborescens, Bacillus coagulans, Lactobacillus brevis*	Isomerization of glucose to fructose
Penicillinase	*B. subtilis*	Degradation of penicillin
Glucose oxidase	*A. niger*	Glucose → gluconic acid, dried-egg manufacture
Lignases	Fungal	Biopulping of wood for paper manufacture
Lipases	*Rhizopus,* pancreas	Hydrolysis of lipids, flavoring and digestive aid
Invertase	*S. cerevisiae*	Hydrolysis of sucrose for further fermentation
Pectinase	*A. oryzae, A. niger, A. flavus*	Clarification of fruit juices, hydrolysis of pectin
Cellulase	*Trichoderma viride*	Cellulose hydrolysis

Pectinases are produced mainly by *A. niger*. The major components in pectinases are pectin esterase, polygalacturonase, and polymethylgalacturonatelyase. Pectinases are used in fruit juice processing and wine making to increase juice yield, reduce viscosity, and clear the juice.

Lipases hydrolyze lipids into fatty acids and glycerol and are produced from animal pancreas, some molds, and yeasts. Lipases may be used to hydrolyze oils for soap manufacture and to hydrolyze the lipid-fat compounds present in waste-water streams. Interesterification of oils and fats may be catalyzed by lipases. Lipases may also be used in the cheese and butter industry to impart flavor as a result of the hydrolysis of fats. Lipase-containing detergents are an important application of lipases.

Amylases are used for the hydrolysis of starch and are produced by many different organisms, including *A. niger* and *B. subtilis*. Three major types of amylases are α-amylase, β-amylase, and glucoamylase. α-amylase breaks α-1,4 glycosidic bonds randomly on the amylose chain and solubilizes amylose. For this reason, α-amylase is known as the starch-liquefying enzyme. β-amylase hydrolyzes α-1,4 glycosidic bonds on the nonreducing ends of amylose and produces maltose residues. β-amylase is known as a saccharifying enzyme. α-1,6 glycosidic linkages in the amylopectin fraction of starch are hydrolyzed by glucoamylase, which is also known as a saccharifying enzyme. In the United States on the average, nearly 1.3×10^9 lb/yr of glucose is produced by the enzymatic hydrolysis of starch. The enzyme pullulanase also hydrolyzes α-1,6 glycosidic linkages in starch selectively.

Cellulases are used in the hydrolysis of cellulose and are produced by some *Trichoderma* species, such as *Trichoderma viride* or *T. reesei*; and by some molds, such as *Aspergillus niger* and *Thermomonospora;* and by some *Clostridium* species. Cellulase is an enzyme complex and its formation is induced by cellulose. *Trichoderma* cellulase hydrolyzes crystalline cellulose, but *Aspergillus* cellulase does not. Cellulose is first hydrolyzed to cellobiose by cellulase, and cellobiose is further hydrolyzed to glucose by β-glucosidase. Both of these enzymes are inhibited by their end products, cellobiose and glucose. Cellulases are used in cereal processing, alcohol fermentation from biomass, brewing, and waste treatment.

Hemicellulases hydrolyze hemicellulose to five-carbon sugar units and are produced by some molds, such as white rot fungi and *A. niger*. Hemicellulases are used in combination with other enzymes in baking doughs, brewing mashes, alcohol fermentation from biomass, and waste treatment.

Lactases are used to hydrolyze lactose in whey to glucose and galactose and are produced by yeast and some *Aspergillus* species. Lactases are used in the fermentation of cheese whey to ethanol.

Other microbial β-1,4 glucanases produced by *Bacillus amyloliquefaciens*, *A. niger*, and *Penicillium emersonii* are used in brewing mashes containing barley or malt. These enzymes improve wort filtration and extract yield.

Penicillin acylase is used by the antibiotic industry to convert penicillin G to 6-aminopenicillanic acid (6-APA), which is a precursor for semisynthetic penicillin derivatives.

Among other important industrial applications of enzymes are the conversion of fumarate to L-aspartate by aspartase. In industry, this conversion is realized in a packed column of immobilized dead *E. coli* cells with active aspartase enzyme. Fumarate solution is passed through the column, and aspartate is obtained in the effluent stream. Aspartate is further coupled with L-phenylalanine to produce aspartame, which is a low-calorie sweetener known as "Nutrasweet®."

The conversion of glucose to fructose by immobilized glucose isomerase is an important industrial process. Fructose is nearly 1.7 times sweeter than glucose and is used as a sweetener in soft drinks. Glucose isomerase is an intracellular enzyme and is produced by different organisms, such as *Flavobacterium arborescens*, *Bacillus licheniformis*, and some *Streptomyces* and *Arthrobacter* species. Immobilized inactive whole cells with glucose isomerase activity are used in a packed column for fructose formation from glucose. Cobalt (Co^{2+}) and magnesium (Mg^{2+}) ions (4×10^{-4} M) enhance enzyme activity. Different immobilization methods are used by different companies. One uses flocculated whole cells of *F. arborescens* treated with glutaraldehyde in the form of dry spherical particles. Entrapment of whole cells in gelatin treated with glutaraldehyde, the use of glutaraldehyde-treated lysed cells in the form of dry particles, and immobilization of the enzyme on inorganic support particles such as silica and alumina are methods used by other companies.

DL-Acylamino acids are converted to a mixture of L- and D-amino acids by immobilized aminoacylase. L-Amino acids are separated from D-acylaminoacid, which is recycled back to the column. L-Amino acids have important applications in food technology and medicine.

Enzymes are commonly used in medicine for diagnosis, therapy, and treatment purposes. Trypsin can be used as an antiinflammatory agent; lysozyme, which hydrolyzes the cell wall of gram-positive bacteria, is used as an antibacterial agent; streptokinase is used as an antiinflammatory agent; urokinase is used in dissolving and preventing blood clots. Asparaginase, which catalyzes the conversion of L-asparagine to L-aspartate, is used as an anticancer agent. Cancer cells require L-asparagine and are inhibited by asparaginase. Asparaginase is produced by *E. coli*. Glucose oxidase catalyzes the oxidation of glucose to gluconic acid and hydrogen peroxide, which can easily be detected. Glucose oxidase is used for the determination of glucose levels in blood and urine. Penicillinases hydrolyze penicillin and are used to treat allergic reactions against penicillin. Tissue plasminogen activator (TPA) and streptokinase are used in the dissolution of blood clots (particularly following a heart attack or stroke).

The development of biosensors using enzymes as integral components is proceeding rapidly. Two examples of immobilized enzyme electrodes are those used in the determination of glucose and urea by using glucose oxidase and urease immobilized on the electrode membrane, respectively. Scarce enzymes (e.g., tissue plasminogen activator) are finding increasing uses, as the techniques of genetic engineering now make it possible to produce usable quantities of such enzymes.

The preceding list of enzymes and uses is not exhaustive, but merely illustrative.

3.7. SUMMARY

Enzymes are protein, glycoprotein, or RNA molecules that catalyze biologically important reactions. Enzymes are very effective, specific, and versatile biocatalysts. Enzymes bind substrate molecules and reduce the activation energy of the reaction catalyzed, resulting in significant increases in reaction rate. Some protein enzymes require a nonprotein group for their activity as a cofactor.

Simple single-enzyme-catalyzed reaction kinetics can be described by Michaelis-Menten kinetics, which has a hyperbolic form in terms of substrate concentration. The activity of some enzymes can be altered by inhibitory compounds, which bind the enzyme molecule and reduce its activity. Enzyme inhibition may be competitive, noncompetitive, and uncompetitive. High substrate and product concentrations may be inhibitory, too.

Enzymes require optimal conditions (pH, temperature, ionic strength) for their maximum activity. Enzymes with an ionizing group on their active site show a distinct optimal pH that corresponds to the natural active form of the enzyme. The activation energy of enzyme-catalyzed reactions is within 4 to 20 kcal/g mol. Above the optimal temperature, enzymes lose their activity, and the inactivation energy is on the order of 40 to 130 kcal/g mol.

Enzymes can be used in suspension or in immobilized form. Enzymes can be immobilized by entrapment in a porous matrix, by encapsulation in a semipermeable membrane capsule or between membranes, such as in a hollow-fiber unit, or by adsorption onto a solid support surface. Enzyme immobilization provides enzyme reutilization, eliminates costly enzyme recovery and purification processes, and may result in increased activity by providing a more suitable microenvironment for the enzyme. Enzyme immobilization may result in diffusion limitations within the matrix. Immobilization may also cause enzyme instability, loss of activity, and a shift in optimal conditions (pH, ionic strength). To obtain maximum reaction rates, the particle size of the support material and enzyme loading need to be optimized, and a support material with the correct surface characteristics must be selected.

Enzymes are widely used in industry and have significant medical applications. Among the most widely used enzymes are proteases (papain, trypsin, subtilisin); amylases (starch hydrolysis); rennet (cheese manufacturing); glucose isomerase (glucose-to-fructose conversion); glucose oxidase (glucose-to-gluconic acid conversion); lipases (lipid hydrolysis), and pectinases (pectin hydrolysis). Enzyme production and utilization are a multibillion-dollar business with a great potential for expansion.

SUGGESTIONS FOR FURTHER READING

ADAMS, M. W. W., AND R. M. KELLY, Enzymes from Microorganisms in Extreme Environments, *Chemical & Engineering News* (Dec. 18), 32–42 (1995).

BAILEY, J .E., AND D. F. OLLIS, *Biochemical Engineering Fundamentals*, 2d ed. McGraw-Hill Book Co., New York, 1986.

BLANCH, H. W., AND D. S. CLARK, *Biochemical Engineering*, Marcel Dekker, Inc., New York, 1996.

KATCHALSKI-KATZIR, E., Immobilized Enzymes—Learning from Past Successes and Failures, *Trends in Biotechnology 11*: 471–478, 1993.

MORAN, L. A., K. G. SCRIMGEOUR, H. R. HORTON, R. S. OCHS, AND J. D. RAWN, *Biochemistry*, Prentice-Hall, Inc., Upper Saddle River, NJ, 1994.

SELLEK, G. A., AND J. B. CHAUDHURI, Biocatalysis in Organic Media Using Enzymes from Extremophiles, *Enzyme Microbial Technol. 25*: 471–482 (1999).

STINSON, S. C., Counting on Chiral Drugs, *Chemical & Engineering News* (Sept. 21), 83–104, 1998.

PROBLEMS

3.1. Consider the following reaction sequence:

$$S + E \underset{k_2}{\overset{k_1}{\rightleftharpoons}} (ES)_1 \underset{k_4}{\overset{k_3}{\rightleftharpoons}} (ES)_2 \xrightarrow{k_5} P + E$$

Develop a suitable rate expression for production formation $[v = k_5(ES)_2]$ by using (a) the equilibrium approach, and (b) the quasi-steady-state approach.

3.2. Consider the reversible product-formation reaction in an enzyme-catalyzed bioreaction:

$$E + S \underset{k_{-1}}{\overset{k_1}{\rightleftharpoons}} (ES) \underset{k_{-2}}{\overset{k_2}{\rightleftharpoons}} E + P$$

Develop a rate expression for product-formation using the quasi-steady-state approximation and show that

$$v = \frac{d[P]}{dt} = \frac{(v_s/K_m)[S] - (v_p/K_p)[P]}{1 + \dfrac{[S]}{K_m} + \dfrac{[P]}{K_p}}$$

where $K_m = \dfrac{k_{-1} + k_2}{k_1}$ and $K_p = \dfrac{k_{-1} + k_2}{k_{-2}}$ and $V_s = k_2[E_0]$, $V_p = k_{-1}[E_0]$.

3.3. The enzyme, fumarase, has the following kinetic constants:

$$S + E \underset{k_{-1}}{\overset{k_1}{\rightleftharpoons}} ES \xrightarrow{k_2} P + E$$

where $k_1 = 10^9 \, M^{-1} \, s^{-1}$

$\qquad k_{-1} = 4.4 \times 10^4 \, s^{-1}$

$\qquad k_2 = 10^3 \, s^{-1}$

a. What is the value of the Michaelis constant for this enzyme?

b. At an enzyme concentration of $10^{-6} \, M$, what will be the initial rate of product formation at a substrate concentration of $10^{-3} \, M$?

[Courtesy of D. J. Kirwan from "Collected Coursework Problems in Biochemical Engineering" compiled by H. W. Blanch for 1977 Am. Soc. Eng. Educ. Summer School.]

3.4. The hydration of CO_2 is catalyzed by carbonic anhydrase as follows:

$$H_2O + CO_2 \overset{E}{\rightleftharpoons} HCO_3^- + H^+$$

The following data were obtained for the forward and reverse reaction rates at pH 7.1 and an enzyme concentration of $2.8 \times 10^{-9} \, M$.

Hydration		Dehydration	
$1/v, M^{-1}$ $(s \times 10^{-3})$	$[CO_2]$ $(M \times 10^3)$	$1/v, M^{-1}$ $(s \times 10^{-3})$	$[HCO_3^-]$ $(M \times 10^3)$
36	1.25	95	2
20	2.5	45	5
12	5	29	10
6	20	25	15

v is the *initial* reaction rate at the given substrate concentration. Calculate the forward and reverse catalytic and Michaelis constants.

[Courtesy of D. J. Kirwan from "Collected Coursework Problems in Biochemical Engineering" compiled by H. W. Blanch for 1977 Am. Soc. Eng. Educ. Summer School.]

3.5. An inhibitor (I) is added to the enzymatic reaction at a level of 1.0 g/l. The following data were obtained for $K_m = 9.2$ g S/l.

v	S
0.909	20
0.658	10
0.493	6.67
0.40	5
0.333	4
0.289	3.33
0.227	2.5

a. Is the inhibitor competitive or noncompetitive?

b. Find K_1.

3.6. During a test of kinetics of an enzyme-catalyzed reaction, the following data were recorded:

E_0 (g/l)	T (°C)	I (mmol/ml)	S (mmol/ml)	V (mmol/ml-min)
1.6	30	0	0.1	2.63
1.6	30	0	0.033	1.92
1.6	30	0	0.02	1.47
1.6	30	0	0.01	0.96
1.6	30	0	0.005	0.56
1.6	49.6	0	0.1	5.13
1.6	49.6	0	0.033	3.70
1.6	49.6	0	0.01	1.89
1.6	49.6	0	0.0067	1.43
1.6	49.6	0	0.005	1.11
0.92	30	0	0.1	1.64
0.92	30	0	0.02	0.90
0.92	30	0	0.01	0.58
0.92	30	0.6	0.1	1.33
0.92	30	0.6	0.033	0.80
0.92	30	0.6	0.02	0.57

a. Determine the Michaelis–Menten constant for the reaction with no inhibitor present at 30°C and at 49.6°C.

b. Determine the maximum velocity of the uninhibited reaction at 30°C and an enzyme concentration of 1.6 g/l.

c. Determine the K_1 for the inhibitor at 30°C and decide what type of inhibitor is being used.

3.7. An enzyme ATPase has a molecular weight of 5×10^4 daltons, a K_M value of 10^{-4} M, and a k_2 value of $k_2 = 10^4$ molecules ATP/min molecule enzyme at 37°C. The reaction catalyzed is the following:

$$\text{ATP} \xrightarrow{\text{ATPase}} \text{ADP} + P_i$$

which can also be represented as

$$E + S \underset{k_2}{\overset{k_1}{\rightleftharpoons}} ES \xrightarrow{k_2} E + P$$

where S is ATP. The enzyme at this temperature is unstable. The enzyme inactivation kinetics are first order:

$$E = E_0 e^{-k_d t}$$

where E_0 is the initial enzyme concentration and $k_d = 0.1$ min^{-1}. In an experiment with a partially pure enzyme preparation, 10 µg of total crude protein (containing enzyme) is added to a 1 ml reaction mixture containing 0.02 M ATP and incubated at 37°C. After 12 hours the reaction ends (i.e., $t \to \infty$) and the inorganic phosphate (P_i) concentration is found to be 0.002 M, which was initially zero. What fraction of the crude protein preparation was the enzyme? *Hint*: Since $[S] \gg K_m$, the reaction rate can be represented by

$$\frac{d(P)}{dt} = k_2[E]$$

3.8. Assume that for an enzyme immobilized on the surface of a nonporous support material the external mass transfer resistance for substrate is not negligible as compared to the reaction rate. The enzyme is subject to substrate inhibition (eq. 3.34).

    a. Are multiple states possible? Why or why not?

    b. Could the effectiveness factor be greater than one?

3.9. The following data were obtained for an enzyme-catalyzed reaction. Determine V_{max} and K_m by inspection. Plot the data using the Eadie–Hofstee method and determine these constants graphically. Explain the discrepancy in your two determinations. The initial rate data for the enzyme-catalyzed reaction are as follows:

[S] mol/l	v µmol/min
5.0×10^{-4}	125
2.0×10^{-4}	125
6.0×10^{-5}	121
4.0×10^{-5}	111
3.0×10^{-5}	96.5
2.0×10^{-5}	62.5
1.6×10^{-5}	42.7
1.0×10^{-5}	13.9
8.0×10^{-6}	7.50

Do these data fit into Michaelis–Menten kinetics? If not, what kind of rate expression would you suggest? Use graphical methods.

3.10. a. H. H. Weetall and N. B. Havewala report the following data for the production of dextrose from corn starch using both soluble and immobilized (azo-glass beads) glucoamylase in a fully agitated CSTR system.

      1. Soluble data: $T = 60°C$, $[S_0] = 168$ mg starch/ml, $[E_0] = 11{,}600$ units, volume = 1000 ml.

      2. Immobilized data: $T = 60°C$, $[S_0] = 336$ mg starch/ml, $[E_0] = 46{,}400$ units initially, immobilized, volume = 1000 ml.

Time (min)	Product concentration (mg dextrose/ml) Soluble	Immobilized
0	12.0	18.4
15	40.0	135
30	76.5	200
45	94.3	236
60	120.0	260
75	135.5	258
90	151.2	262
105	150.4	266
120	155.7	278
135	160.1	300
150	164.9	310
165	170.0	306
225	—	316
415	—	320

Determine the maximum reaction velocity, V_m (mg/ml-min · unit of enzyme) and the saturation constant, K_M (mg/ml).

b. The same authors studied the effect of temperature on the maximum rate of the hydrolysis of corn starch by glucoamylase. The results are tabulated next. Determine the activation energy (ΔE cal/g mole) for the soluble and immobilized enzyme reaction.

T, °C	V_{max} (m mol/min 10^6)	
	Soluble	Azo-immobilized
25	0.62	0.80
35	1.42	1.40
45	3.60	3.00
55	8.0	6.2
65	16.0	11.0

c. Using these results, determine if immobilized enzyme is diffusion limited.

[Courtesy of A. E. Humphrey from "Collected Coursework Problems in Biochemical Engineering" compiled by H. W. Blanch for 1977 Am. Soc. Eng. Educ. Summer School.]

3.11. Michaelis–Menten kinetics are used to describe intracellular reactions. Yet $[E_0] \approx [S_0]$. In in vitro batch reactors, the quasi-steady-state hypothesis does not hold for $[E_0] \approx [S_0]$. The rapid equilibrium assumption also will not hold. Explain why Michaelis–Menten kinetics and the quasi-steady-state approximation are still reasonable descriptions of intracellular enzyme reactions.

3.12. You are working for company A and you join a research group working on immobilized enzymes. Harry, the head of the lab, claims that immobilization improves the stability of the enzyme. His proof is that the enzyme has a half-life of 10 days in free solution, but under identical conditions of temperature, pH, and medium composition, the measured half-life of a packed column is 30 days. The enzyme is immobilized in a porous sphere 5 mm in diameter. Is Harry's reasoning right? Do you agree with him? Why or why not?

3.13. The following data were obtained from enzymatic oxidation of phenol by phenol oxidase at different phenol concentrations.

S (mg/l)	10	20	30	50	60	80	90	110	130	140	150
v (mg/l-h)	5	7.5	10	12.5	13.7	15	15	12.5	9.5	7.5	5.7

a. What type of inhibition is this ?
b. Determine the constants V_m, K_m, and K_{si}.
c. Determine the oxidation rate at $[S] = 70$ mg/l.

3.14. Uric acid is degraded by uricase enzyme immobilized in porous Ca-alginate beads. Experiments conducted with different bead sizes result in the following rate data:

Bead Diameter, Dp (cm)	0.1	0.2	0.3	0.4	0.5	0.6	0.7	0.8
Rate, v (mg/l.h)	200	198	180	140	100	70	50	30

a. Determine the effectiveness factor for particle sizes Dp = 0.5 cm and Dp = 0.7 cm.
b. The following data were obtained for Dp = 0.5 cm at different bulk uric acid concentrations. Assuming negligible liquid film resistance, calculate V_m and K_s for the enzyme. Assume no substrate or product inhibition.

S_0 (mg UA/l)	10	25	50	100	200	250
v (mg UA/l.h)	10	20	30	40	45	46

3.15. The enzyme, urease, is immobilized in Ca-alginate beads 2 mm in diameter. When the urea concentration in the bulk liquid is 0.5 mM the rate of urea hydrolysis is $v = 10$ mmoles-l-h. Diffusivity of urea in Ca-alginate beads is $D_e = 1.5 \times 10^{-5}$ cm^2/sec, and the Michaelis constant for the enzyme is $K_m' = 0.2$ mM. By neglecting the liquid film resistance on the beads (i.e., $[S_o] = [S_s]$) determine the following:

    a. Maximum rate of hydrolysis V_m, Thiele modulus (ϕ), and effectiveness factor (η).

    b. What would be the V_m, ϕ, and η values for a particle size of Dp $= 4$ mm?

    Hint: Assume $\eta = 3/\phi$ for large values of ϕ ($\phi > 2$).

3.16. Decarboxylation of glyoxalate (S) by mitochondria is inhibited by malonate (I). Using the following data obtained in batch experiments, determine the following:

Glyox,S (mM)	Rate of CO_2 evolution, v (mmoles/l-h)		
	I = 0	I = 1.26 mM	I = 1.95 mM
0.25	1.02	0.73	0.56
0.33	1.39	0.87	0.75
0.40	1.67	1.09	0.85
0.50	1.89	1.30	1.00
0.60	2.08	1.41	1.28
0.75	2.44	1.82	1.39
1.00	2.50	2.17	1.82

    a. What type of inhibition is this?

    b. Determine the constants V_m, K_m', and K_I.

3.17. Urea dissolved in aqueous solution is degraded to ammonia and CO_2 by the enzyme urease immobilized on surfaces of nonporous polymeric beads. Conversion rate is controlled by transfer of urea to the surface of the beads through liquid film, and the conversion takes place on the surfaces of the beads. The following parameters are given for the system.

    $k_L = 0.2$ cm/s; $K_m = 200$ mg/l

    $V_m' = 0.1$ mg urea/cm^2 support surface -s.

    $S_b = 1000$ mg urea/l

    a. Determine the surface concentration of urea.

    b. Determine the rate of urea degradation under mass transfer controlled conditions.

3.18. Two enzymes are both immobilized on the same flat, nonporous surface. For enzyme A the substrate is S_1. For enzyme B the substrate is S_2. The product of the first reaction is S_2. That is:

$$S_1 \xrightarrow[E_A]{} S_2 \xrightarrow[E_B]{} P$$

    a. Figure 3.P1 depicts the rate of the first reaction on the surface as a function of local concentrations of S_1. If the bulk concentration of S_1 is 100 mg/l and the mass transfer coefficient is 4×10^{-5} cm/s, what is the rate of consumption of S_1 for a 1 cm^2 surface? What is the surface concentration of S_1?

    b. The rate of the second reaction is:

$$-d[S_2]/dt = d[P]/dt = \frac{V_m'' S_{2\,surface}}{K_m + S_{2\,surface}}$$

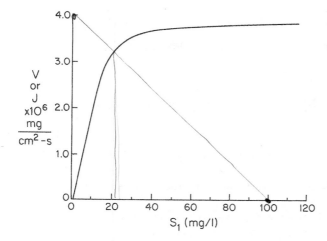

Figure 3P1. Reaction rate data for problem 3.18. Reaction rate dependence on substrate one for reaction catalyzed by E_A.

where $K_m = 5$ mg/l (or 5×10^{-3} mg/cm^3) and $V_m'' = 4.0 \times 10^{-6}$ mg/cm$^2 \cdot$s. The bulk concentration of S_2 [S_{2bulk}] is maintained as 5 mg/l and the mass transfer coefficient is the same for S_1 and S_2. Calculate [$S_{2surface}$] and the rate of formation of P (assuming all stoichiometric coefficients are one).

3.19. Consider the case of two enzymes immobilized on the same nonporous, planar surface. S is a substrate used by both enzymes in the following reactions:

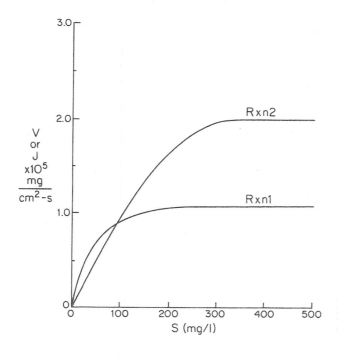

Figure 3P2. Reaction rate data for problem 3.19. The reaction rate data for two different enzymes (E_1 and E_2) are shown.

Problems

103

$$S + E_1 \longrightarrow E_1 + P_1 \tag{1}$$

and

$$S + E_2 \longrightarrow E_2 + P_2 \tag{2}$$

The final product P_3 is formed by the spontaneous reaction of P_1 and P_2:

$$P_1 + P_2 \longrightarrow P_3 \tag{3}$$

Reactions 1 and 2 occur only at the surface and reaction 3 is a homogeneous reaction occurring throughout the bulk liquid phase.

Figure 3.P2 gives the predicted reaction-rate dependence of reaction 1 (bottom curve) alone and reaction 2 (top curve) alone based on the measured amount of each enzyme immobilized and assuming the intrinsic reaction kinetics are not altered in the process of immobilization.

a. If $k_L = 6 \times 10^{-5}$ cm/s and the bulk concentration of substrate is 500 mg/l, what is the *total* rate of substrate disappearance?

b. What is the overall effectiveness factor under the conditions of part a?

c. What will be the ratio of P_2 to P_1 under the conditions of part a?

d. If you want to produce equimolar amounts of P_1 and P_2 and if $k_L = 6 \times 10^{-5}$ cm/s, what value of bulk substrate concentration must you pick?

4

How Cells Work

4.1. INTRODUCTION

So far you have learned something about how cells are constructed and how enzymes function. But a cell is much more than a bag filled with lipids, amino acids, sugars, enzymes, and nucleic acids. The cell must control how these components are made and interact with each other. In this chapter we will explore some examples of metabolic regulation. It is this ability to coordinate a wide variety of chemical reactions that makes a cell a cell.

The key to metabolic regulation is the flow and control of information. The advent of computers has led many to speak of the "Information Age." There are many analogies between human society's need to use and exchange information and a cell's need to use and exchange information between subcellular components. Human society depends primarily on electronic signals for information storage, processing, and transmission; cells use chemical signals for the same purposes. Molecular biology is primarily the study of information flow and control.

4.2. THE CENTRAL DOGMA

Fortunately, almost all living systems have the same core approach to the storage, expression, and utilization of information. Information is stored in the DNA molecule and, like information on a computer tape or disc, can be replicated. It can also be played back or transcribed to produce a message. The message must translate into some action, such as a series of calculations with the ultimate display of the results, for the message to be useful.

Cells operate with an analogous system. Figure 4.1 displays the *central dogma* of biology. Information is stored on the DNA molecule. That information can be *replicated* directly to form a second identical molecule. Further segments of information on the molecule can be *transcribed* to yield RNAs. Using a variety of RNAs, this information is *translated* into proteins. The proteins then perform a structural or enzymatic role, mediating almost all the metabolic functions in the cell. The information content of the DNA molecule is static; changes occur slowly through infrequent mutations or rearrangements. Which species of RNA that are present and in what amount varies with time and with changes in culture conditions. Likewise, the proteins that are present will change with time but on a different time scale than for RNA species. Some of the proteins produced in the cell bind to DNA and regulate the transcriptional process to form RNAs.

The important feature of the central dogma is its universality from the simplest to most complex organisms. One important, although relatively minor, deviation is that some RNA tumor viruses (*retroviruses*) contain an enzyme called *reverse transcriptase*. (The virus that causes AIDS, the human immunodeficiency virus or HIV, is a retrovirus, and

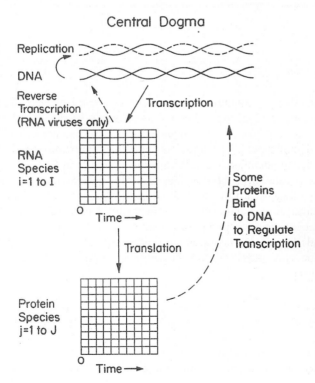

Figure 4.1. The primary tenet of molecular biology is the central dogma, which applies to all organisms. DNA serves as the template for its own replication, as well as transcription to RNA. The information transcribed into the RNA can then be translated into proteins using an RNA template. Note that information on the DNA molecule is relatively time independent, while the information which exists in the form of RNA and protein molecules depends on the history and environment of the cell and is time dependent. Some of the proteins interact with DNA to control which genes are transcribed.

one approach to treatment is to selectively inhibit reverse transcrip'
virus encodes information on a RNA molecule. In the host cell, vir
produces a viral DNA molecule using the viral RNA as a template. Such .
cally important and the enzyme, reverse transcriptase, is an important tool in genec
neering. Nonetheless, the process depicted in Fig. 4.1 is essentially applicable to any cell
of commercial importance.

For information storage and exchange to take place, there must be a language. We
can conceive of all life as using a four-letter alphabet made up of the nucleotides discussed in Chapter 2 (that is, A, T, G, and C in DNA). All words are three letters long; such
words are called *codons*. With four letters and only three-letter words, we have a maximum of 64 words. These words, when expressed, represent a particular amino acid or
"stop" protein synthesis.

When these words are put into a sequence, they can make a "sentence" (i.e., a *gene*),
which when properly transcribed and translated is a protein. Other combinations of words
regulate when the gene is expressed. Carrying the analogy to the extreme, we may look at
the complete set of information in an organism's DNA (i.e., the *genome*) as a book. (For
the human genome it would be more than 1000 books the size of this one.)

This simple language of 64 words is all that is necessary to summarize your total
physical makeup at birth and all your natural capabilities. It is essentially universal, the
same for *E. coli* and *Homo sapiens* (humans). This universality has helped us to make
great strides in understanding life and is a practical tool in genetic engineering and commercial biotechnology.

Each of the steps in information storage and transfer (Fig. 4.1) requires a macromolecular template. Let us examine how these templates are made and how this genetic-level
language is preserved and expressed.

4.3. DNA REPLICATION: PRESERVING AND PROPAGATING THE CELLULAR MESSAGE

The double-helix structure of DNA discussed in Chapter 2 is extremely well suited to its
role of preserving genetic information. Information resides simply in the linear arrangement of the four nucleotide letters (A, T, G, and C). *Because G can hydrogen-bind only to
C and A only to T, the strands must be complementary* if an undistorted double helix is to
result. Replication is *semiconservative* (see Fig. 2.18); each daughter chromosome contains one parental strand and one newly generated strand.

To illustrate the replication process (see Figs. 4.2 and 4.3), let us briefly consider
DNA replication in *E. coli*. The enzyme responsible for covalently linking the monomers
is DNA polymerase. *Escherichia coli* has three DNA polymerases (named Pol I, Pol II,
and Pol III). A DNA polymerase is an enzyme that will link deoxynucleotides together to
form a DNA polymer. Pol III enzymatically mediates the addition of nucleotides to an
RNA primer. Pol I can hydrolyze an RNA primer and duplicates single-stranded regions
of DNA; it is also active in the repair of DNA molecules. The exact role of Pol II is still
unclear.

In addition to the enzyme, the enzymatic reaction requires activated monomer and
the template. The activated monomers are the nucleoside triphosphates. The formation of

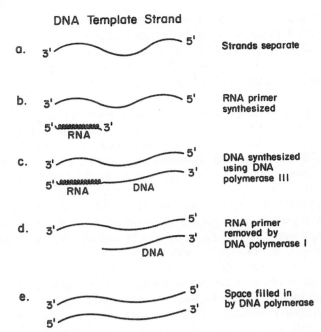

DNA Template Strand

a. 3'——————5' Strands separate

b. 3'——————5' RNA primer
   5'~~~~~~3' synthesized
      RNA

c. 3'——————5' DNA synthesized
   5'~~~~~~————3' using DNA
      RNA   DNA   polymerase III

d. 3'——————5' RNA primer
   ——————3' removed by
        DNA   DNA polymerase I

e. 3'——————5' Space filled in
   5'——————3' by DNA polymerase

Figure 4.2. Initiation of DNA synthesis requires the formation of an RNA primer.

the 5'–3' phosphodiester bond to link a nucleotide with the growing DNA molecule results in the release of a pyrophosphate, which provides the energy for such a biosynthetic reaction. The resulting nucleoside monophosphates are the constituent monomers of the DNA molecule.

Replication of the chromosome normally begins at a predetermined site, the *origin of replication*, which in *E. coli* is attached to the plasma membrane at the start of replication. Initiator proteins bind to DNA at the origin of replication, break hydrogen bonds in the local region of the origin, and force the two DNA strands apart. When DNA replication begins, the two strands separate to form a Y-shaped structure called a *replication fork*. Movement of the fork must be facilitated by the energy-dependent action of *DNA gyrase* and *unwinding enzymes*. In *E. coli* the chromosome is circular. In *E. coli* (but not all organisms) the synthesis of DNA is *bidirectional*. Two forks start at the origin and move in opposite directions until they meet again, approximately 180° from the origin.

To initiate DNA synthesis, an *RNA primer* is required; RNA polymerase requires no primer to initiate the chain-building process, while DNA polymerase does. (We can speculate on why this is so. In DNA replication, it is critical that no mistakes be made in the addition of each nucleotide. The DNA polymerase, Pol III, can *proofread*, in part due to the enzyme's 3'-to-5' exonuclease activity, which can remove mismatches by moving backward. On the other hand, a mistake in RNA synthesis is not nearly so critical, so RNA polymerase lacks this proofreading capacity.) Once a short stretch of RNA complementary to one of the DNA strands is made, DNA synthesis begins with Pol III. Next, the RNA portion is degraded by Pol I, and DNA is synthesized in its place. This process is summarized in Fig. 4.2.

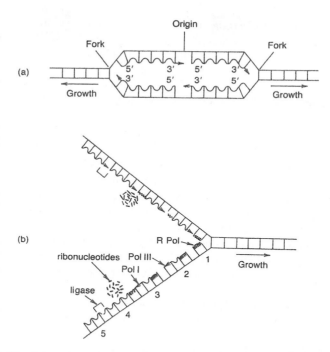

Figure 4.3. Schematic representation of the steps of replication of the bacterial chromosome. Part (a) represents a portion of a replicating bacterial chromosome at a stage shortly after replication has begun at the origin. The newly polymerized strands of DNA (wavy lines) are synthesized in the 5′-to-3′ direction (indicated by the arrows), using the preexisting DNA strands (solid lines) as a template. The process creates two replication forks, which travel in opposite directions until they meet on the opposite side of the circular chromosome, completing the replication process. Part (b) represents a more detailed view of one of the replicating forks and shows the process by which short lengths of DNA are synthesized and eventually joined to produce a continuous new strand of DNA. For purposes of illustration, four short segments of nucleic acid are illustrated at various stages. In (1), primer RNA (thickened area) is being synthesized by an RNA polymerase (R Pol). Then, successively, in (2) DNA is being polymerized to it by DNA polymerase III (Pol III); in (3) a preceding primer RNA is being hydrolyzed, while DNA is being polymerized in its place by the exonuclease and polymerase activities of DNA polymerase I (Pol I); finally, the completed short segment of DNA (4) is joined to the continuous strand (5) by the action of DNA ligase. (With permission, from R. Y. Stanier and others, *The Microbial World*, 5th ed., Pearson Education, Upper Saddle River, NJ, 1986, p. 133.)

DNA polymerase works only in the 5′-to-3′ direction, which means that the next nucleotide is always added to the exposed 3′-OH group of the chain. Thus, one strand (the *leading strand*) can be formed continuously if it is synthesized in the same direction as the replication fork is moving. The other strand (the *lagging strand*) must be synthesized discontinuously. Short pieces of DNA attached to RNA are formed on the lagging strand. These fragments are called *Okazaki fragments*. The whole process is summarized in Fig. 4.3. The enzyme, *DNA ligase*, which joins the two short pieces of DNA on the continuous strand, will be very important in our discussions of genetic engineering.

This brief discussion summarizes the essentials of how one DNA molecule is made from another and thus preserves and propagates the genetic information in the original molecule. Now we turn our attention to how this genetic information can be transferred.

4.4. TRANSCRIPTION: SENDING THE MESSAGE

The primary products of transcription are the three major types of RNA we introduced in Chapter 2: m-RNA, t-RNA, and r-RNA. Their rates of synthesis determine the cell's capacity to make proteins. RNA synthesis from DNA is mediated by the enzyme, *RNA polymerase*. To be functional, RNA polymerase must have two major subcomponents: the *core* enzyme and the *sigma factor*. The core enzyme contains the catalytic site, while the sigma factor is a protein essential to locating the appropriate beginning for the message. The core enzyme plus the sigma factor constitutes the *holoenzyme*.

The student may wonder which of the two strands of DNA is actually transcribed. It turns out that either strand can be read. RNA polymerase always reads in the 3'- to 5'-direction, so the direction of reading will be opposite on each strand. On one part of the chromosome, one strand of DNA may serve as the template or *sense strand*, and on another portion of the chromosome, the other strand may serve as a template.

The processes of initiation, elongation, and termination are summarized in Fig. 4.4. The sigma factor is involved only in initiation. The sigma factor recognizes a specific sequence of nucleotides on a DNA strand. This sequence is the *promoter region*. Promoters can vary somewhat, and this alters the affinity of the sigma factor (and consequently the holoenzyme) for a particular promoter. A *strong promoter* is one with a high affinity for the holoenzyme. The rate of formation of transcripts is determined primarily by the frequency of initiation of transcription, which is directly related to promoter strength. This will be important in our discussions of genetic engineering. Cells usually have one dominant sigma factor that is required to recognize the vast majority of promoters in the cell. However, other sigma factors can play important roles under different growth conditions (particularly stress) and are used to initiate transcription from promoters that encode proteins important to the cell for coping with unusual growth conditions or stress.

After the initiation site is recognized, elongation of the transcript begins. As soon as elongation is established, the sigma factor is released so it can be reused. The synthesis of the growing RNA molecule is energy requiring, so activated triphosphate monomers of the ribonucleotides are required.

The transcript is made until the RNA polymerase encounters a stop signal, or *transcription terminator*. At this point, the RNA polymerase disassociates from the DNA template and the RNA transcript is released. In some cases an additional protein, the *rho* protein, is required for termination. Terminators can be strong or weak. If a weak terminator is coupled with a strong promoter, some of the RNA polymerase will *read through* the terminator, creating an artificially long transcript and possibly disrupting subsequent control regions on that DNA strand. We must consider terminator regions and their strength when constructing recombinant DNA systems.

The transcripts that are formed may be roughly lumped as either stable or unstable RNA species. The stable RNA species are r-RNA and t-RNA. Messenger-RNA is highly unstable (about a 1-min half-life for a typical *E. coli* m-RNA, although m-RNA may be

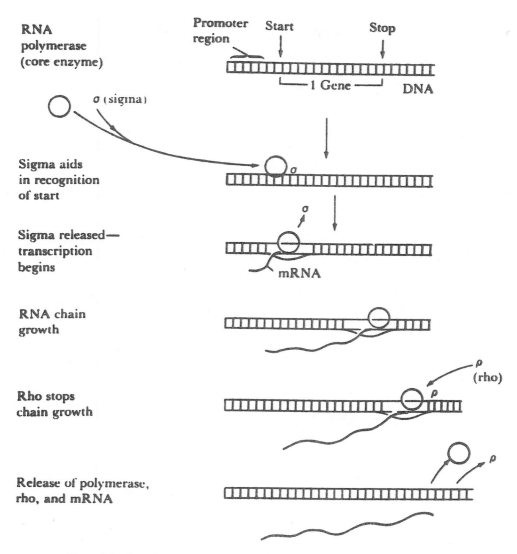

Figure 4.4. Steps in messenger RNA synthesis. The start and stop sites are specific nucleotide sequences on the DNA. RNA polymerase moves down the DNA chain, causing temporary opening of the double helix and transcription of one of the DNA strands. Rho binds to the termination site and stops chain growth; termination can also occur at some sites without rho. (With permission, from T. D. Brock, D. W. Smith, and M. T. Madigan, *Biology of Microorganisms*, 4th ed., Pearson Education, Upper Saddle River, NJ, 1984, p. 285.)

considerably more stable in cells from higher plants and animals). The student should consider for a moment why *m*-RNA is relatively unstable. The answer should become apparent as we discuss translation and regulation.

Although the general features of transcription are universal, there are some significant differences in transcription between procaryotic and eucaryotic cells. One example is

that in procaryotes related proteins are often encoded in a row without interspacing terminators. Thus, transcription from a single promoter may result in a *polygenic* message.
Polygenic indicates many genes; *each single gene encodes a separate protein*. Thus, the
regulation of transcription from a single promoter can provide efficient regulation of functionally related proteins; such a strategy is particularly important for relatively small and
simple cells. On the other hand, eucaryotic cells do not produce polygenic messages.

In procaryotic cells, there is no physical separation of the chromosome from the cytoplasma and ribosomes. Often an *m*-RNA will bind to a ribosome and begin translation
immediately, even while part of it is still being transcribed! However, in eucaryotes,
where the nuclear membrane separates chromosomes and ribosomes, the *m*-RNA is often
subject to processing before translation (see Fig. 4.5). The DNA can encode for a transcript with an intervening sequence (called an *intron*) in the middle of the transcript. This
intron is then cut out of the transcript at two specific sites. The ends of the remaining fragments are joined by a process called *m-RNA splicing*. The spliced message can then be
translated into an actual protein. The part of the transcript forming the intron is degraded
and the monomers recycled. When *m*-RNA is recovered from the cytosol it will be in the
mature form, while *m*-RNA within the nucleus has introns. Many eucaryotic genes contain "nonsense DNA," which encodes for the intronic part of the transcript. The word
"nonsense" denotes that that particular sequence of DNA does not encode for amino

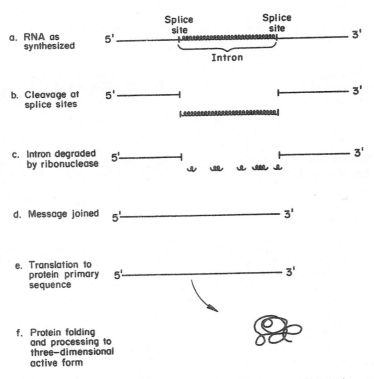

Figure 4.5. In eucaryotes, RNA splicing is important. The presence of introns is a complication in cloning genes from eucaryotes to procaryotes.

acids. The role of introns is not well understood, but they likely play a role in either evolution or cellular regulation or both. The presence of introns complicates the transfer of eucaryotic genes to protein production systems in procaryotes such as *E. coli.*

Two other *m*-RNA processing steps occur in eucaryotic cells that do not occur in procaryotes. One is *RNA capping,* in which the 5′ end is modified by the addition of a guanine nucleotide with a methyl group attached. The other is *polyadenylation,* in which a string of adenine nucleotides are added to the 3′ end. This tail of adenines is often several hundred nucleotides long. These two modifications are thought to increase *m*-RNA stability and facilitate transport across the nuclear membrane.

Once we understand transcription, we can tackle translation. The translation of information on *m*-RNA into proteins occupies a very large fraction of the cells' resources. Like a large automobile plant, the generation of blueprints and the construction of the manufacturing machinery is worthless until product is made.

4.5. TRANSLATION: MESSAGE TO PRODUCT

4.5.1. Genetic Code: Universal Message

The blueprint for any living cell is the genetic code. The code is made up of three-letter words (*codons*) with an alphabet of four letters. Sixty-four words are possible, but many of these words are redundant. Although such a "language" may appear to be ridiculously simple, it is sufficient to serve as a complete blueprint for the "construction" of the reader.

The dictionary for this language is given in Table 4.1, and an illustration of the relationship of nucleotides in the chromosome and *m*-RNA to the final protein product is given in Fig. 4.6. The code is *degenerate* in that more than one codon can specify a particular amino acid (for example, UCU, UCC, UCA, and UCG all specify serine). Three codings, UAA, UAG, and UGA, are called *nonsense codons* in that they do not encode normally for amino acids. These codons act as stop points in translation and are encoded at the end of each gene.

The genetic code is essentially universal, although some exceptions exist (particularly in the mitochondria and for inclusion of rare amino acids). This essential universality greatly facilitates genetic engineering. The language used to make a human protein is understood in *E. coli* and yeast, and these simple cells will faithfully produce the same amino acid sequence as a human cell.

Knowing the genetic language, we may ask more about the mechanism by which proteins are actually made.

4.5.2. Translation: How the Machinery Works

The process of translation consists of three primary steps: *initiation*, *elongation*, and *termination.*

For initiation, *m*-RNA must bind to the ribosome. All protein synthesis begins with a AUG codon (or GUG) on the *m*-RNA. This AUG encodes for a modified methionine, *N*-formylmethionine. In the middle of a protein, AUG encodes for methionine, so the question is how the cell knows that a particular AUG is an initiation codon for

TABLE 4.1 The Genetic Code: Correspondence between Codons and Amino Acids

First base	Second bases							
	U		C		A		G	
U	UUU	phe[a]	UCU	ser	UAU	tyr	UGU	cys
	UUC	phe	UCC	ser	UAC	tyr	UGC	cys
	UUA	leu	UCA	ser	UAA	(none)[b]	UGA	(none)[b]
	UUG	leu	UCG	ser	UAG	(none)[b]	UGG	try
C	CUU	leu	CCU	pro	CAU	his	CGU	arg
	CUC	leu	CCC	pro	CAC	his	CGC	arg
	CUA	leu	CCA	pro	CAA	glu-N	CGA	arg
	CUG	leu	CCG	pro	CAG	glu-N	CGG	arg
A	AUU	ileu	ACU	thr	AAU	asp-N	AGU	ser
	AUC	ileu	ACC	thr	AAC	asp-N	AGC	ser
	AUA	ileu	ACA	thr	AAA	lys	AGA	arg
	AUG	met	ACG	thr	AAG	lys	AGG	arg
G	GUU	val	GCU	ala	GAU	asp	GGU	gly
	GUC	val	GCC	ala	GAC	asp	GGC	gly
	GUA	val	GCA	ala	GAA	glu	GGA	gly
	GUG	val	GCG	ala	GAG	glu	GGG	gly

[a]Amino acids are abbreviated as the first three letters in each case, except for glutamine (glu-N), asparagine (asp-N), and isoleucine (ileu).
[b]The codons UAA, UAG, and UGA are nonsense codons; UAA and UAG are called the ochre codon and the amber codon, respectively.

With permission, from R. Y. Stanier and others, *The Microbial World*, 5th ed., Pearson Education, Upper Saddle River, NJ, 1986, p. 139.

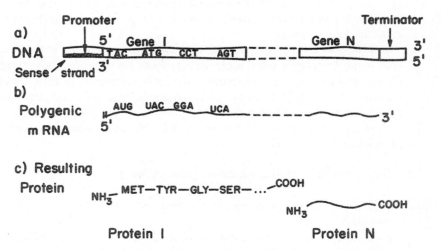

Figure 4.6. Overview of the transfer of information from codons on the DNA template to proteins. In procaryotes, messages are often polygenic, whereas in eucaryotes, polygenic messages are not made.

N-formylmethionine. The answer lies about ten nucleotides upstream of the AUG, where the ribosome binding site (Shine-Delgarno box) is located. Ribosome binding sites can vary in strength and are an important consideration in genetic engineering. The initiation of polymerization in procaryotes requires an *initiation complex* composed of a 30s ribosomal unit with an *N*-formylmethionine bound to its initiation region, a 50s ribosomal unit, three proteins called initiation factors (IF1, IF2, and IF3), and the phosphate bond energy from GTP.

The elongation of the amino acid chain uses *t*-RNAs as decoders. One end of the *t*-RNA contains the *anticodon*, which is complementary to the codon on the *m*-RNA. The other end of the *t*-RNA binds a specific amino acid. The *t*-RNA is called *charged* when it is carrying an amino acid. The binding of an amino acid to the *t*-RNA molecule requires the energy from two phosphate bonds and enzymes known as *aminoacyl-t-RNA synthetases*. Figure 2.19 depicts a *t*-RNA molecule.

The actual formation of the peptide bond between the two amino acids occurs on adjacent sites on the ribosome: the P or *peptidyl* site and the A or *aminoacyl* site (see Fig. 4.7). The growing protein occupies the P site, while the next amino acid to be added occupies the A site. As the peptide bond is formed, the *t*-RNA associated with the P site is released, and a rachet mechanism moves the *m*-RNA down one codon so as to cause the *t*-RNA that was in the A site to be in the P site. Then a charged *t*-RNA with the correct anticodon can be recognized and inserted into the A site. The whole process is then repeated. The cell expends a total of four phosphate bonds to add one amino acid to each growing polypeptide (two to charge the *t*-RNA and two in the process of elongation), and this accounts for most of the cellular energy expenditure in bacteria.

When a nonsense or stop codon is reached, the protein is released from the ribosome with the aid of a protein *release factor* (RF). The 70S ribosome then dissociates into 30S and 50S subunits. An *m*-RNA typically is being read by many (for example, 10 to 20) ribosomes at once; as soon as one ribosome has moved sufficiently far along the message that the ribosome binding site is not physically blocked, another ribosome can bind and initiate synthesis of a new polypeptide chain.

4.5.3. Posttranslational Processing: Making the Product Useful

Often the polypeptide formed from the ribosome must undergo further processing before it can become truly useful. First, the newly formed chain must fold into the proper structure; in some cases, several different chains must associate to form a particular enzyme or structural protein. Additionally, *chaperones* are an important class of proteins that assist in the proper folding of peptides. There are distinct pathways to assist in folding polypeptides. The level of chaperones in a cell increases in response to environmental stresses such as high temperature. Misfolded proteins are subject to degradation if they remain soluble. Often misfolded proteins aggregate and form insoluble particles (i.e., inclusion bodies). High levels of expression of foreign proteins through recombinant DNA technology in *E. coli* often overwhelm the processing machinery, resulting in inclusion bodies. The formation of proteins in inclusion bodies greatly complicates any bioprocess, since in vitro methods to unfold and refold the protein product must be employed. Even when a

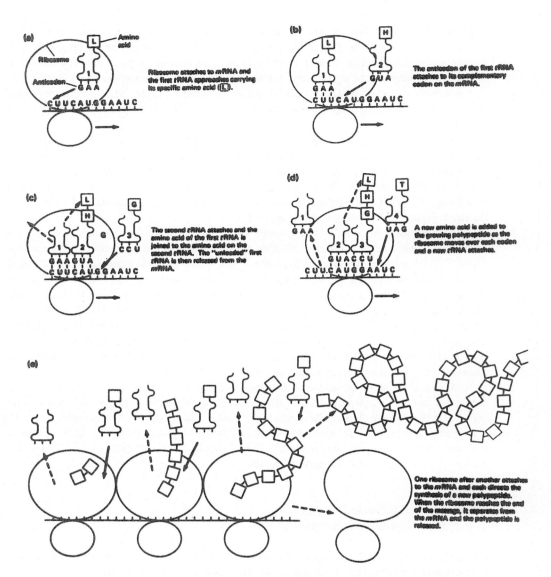

Figure 4.7. Translation of genetic information from a nucleotide sequence to an amino acid sequence. (With permission, from M. W. Jensen and D. N. Wright, *Introduction to Medical Microbiology*, Pearson Education, Upper Saddle River, NJ, 1985, p. 66.)

cell properly folds a protein, additional cellular processing steps must occur to make a useful product.

Many proteins are secreted through a membrane. In many cases, the translocation of the protein across the membrane is done *cotranslationally* (during translation), while in some cases *posttranslation* movement across the membrane occurs. When proteins move across a membrane, they have a *signal sequence* (about 20 to 25 amino acids). This signal sequence is clipped off during secretion. Such proteins exist in a pre-form and mature form. The pre-form is what is made from the *m*-RNA, but the actual active form is the mature form. The pre-form is the signal sequence plus the mature form.

In procaryotes secretion of proteins occurs through the cytoplasmic membrane. In *E. coli* and most gram-negative bacteria the outer membrane blocks release of the secreted protein into the extracellular compartment. In gram-positive cells secreted proteins readily pass the cell wall into the extracellular compartment. Whether a protein product is retained in a cell or released has a major impact on bioprocess design.

In eucaryotic cells proteins are released by two pathways. Both involve *exocytosis,* where *transport vesicles* fuse with the plasma membrane and release their contents. Transport vesicles mediate the transport of proteins and other chemicals from the endoplasmic reticulum (ER) to the Golgi apparatus and from the Golgi apparatus to other membrane-enclosed compartments. Such vesicles bud from a membrane and enclose an aqueous solution with specific proteins, lipids, or other compounds. In the secretory pathway vesicles, carrying proteins bud from the ER, enter the *cis* face of the Golgi apparatus, exit the Golgi *trans* face, and then fuse with the plasma membrane. Only proteins with a signal sequence are processed in the ER to enter the secretory pathway.

Two pathways exist. One is the *constitutive exocytosis pathway,* which operates at all times and delivers lipids and proteins to the plasma membrane. The second is the *regulated exocytosis pathway,* which typically is in specialized secretory cells. These cells secrete proteins or other chemicals only in response to specific chemical signals.

Other modifications to proteins can take place, particularly in higher eucaryotic cells. These modifications involve the addition of nonamino acid components (for example, sugars and lipids) and *phosphorylation. Glycosylation* refers to the addition of sugars. These modifications can be quite complex and are important considerations in the choice of host organisms for the production of proteins. A bioprocess engineer must be aware that many proteins are subject to extensive processing after the initial polypeptide chain is made.

A particularly important aspect of posttranslational processing is *N-linked glycosylation.* The glycosylation pattern can serve to target the protein to a particular compartment or to control its degradation and removal from the organism. For therapeutic proteins injected into the human body these issues are critical ones. A protein product may be ineffective if the N-linked glycosylation pattern is not humanlike, as the protein may not reach the target tissue or may be cleared (i.e., removed) from the body before it exerts the desired action. Further, undesirable immunogenic responses can occur if a protein has a nonhumanlike pattern. Thus, the glycoform of a protein product is a key issue in bioprocesses to make therapeutic proteins (see Chapter 14).

The process of N-linked glycosylation occurs *only* in eucaryotic cells and involves both the ER and Golgi. Thus, the use of procaryotic cells, such as *E. coli,* to serve as hosts for expression of human therapeutic proteins is limited to those proteins where N-linked glycosylation is not present or unimportant. However, not all eucaryotic cells produce proteins with humanlike, N-linked glycosylation. For example, yeasts, lower fungi, and insect cells often produce partially processed products. Even mammalian cells (including human cells) will show altered patterns of glycosylation when cultured in bioreactors, and these patterns can shift upon scale-up in bioreactor size.

The process of N-linked glycosylation is depicted in Fig. 4.8. The pattern shown is "typical," and many variants are possible. The natural proteins in the human body usually

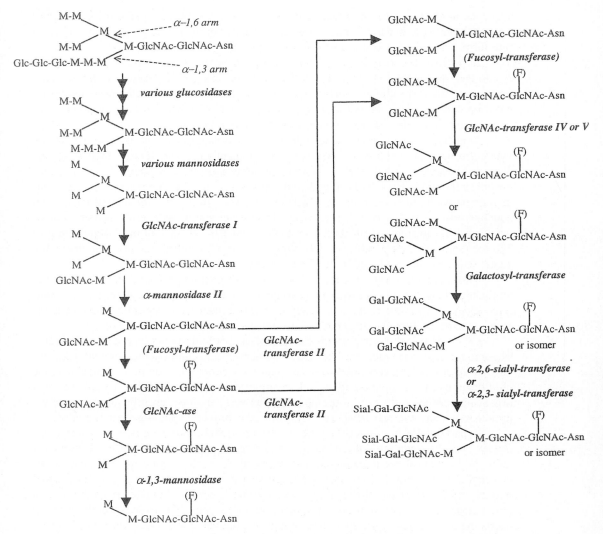

Figure 4.8. Example of a N-linked glycosylation pathway (Glc = glucose, M = mannose, GlcNAc = N-acetylglucosamine, F = fucose, Gal = galactose, Sial = sialic acid). The oligosaccharide side-chain is bound to an asparagine (Asn) of the protein. The upper arm represents the α-1,6 arm and the lower one the α-1,3 arm. The parentheses refer to an optional fucosylation. The GlcNAc-ase step is important in insect cells, but not mammalian cells. (Courtesy of C. Joosten.)

display a range of glycoforms; a single form is not observed. A simple sequence of three amino acids, of which asparagine must be one, is required for attachment of N-linked sugars and amino sugars. The sequence at the attachment site is Asn-Xaa-Ser/Thr, where Xaa is any amino acid and the third amino acid in the sequence must be serine or threonine. The process of N-linked glycosylation begins in the ER, where a preformed branched oligosaccharide (the *dolichol pyrophosphate-oligosaccharide*) with 14 sugars is transferred to the amino group of asparagine.

The 14-sugar residue is first "trimmed" by a set of specific glycosidases. In yeast, oligosaccharide processing often stops in the ER, leading to *simple glycoforms* (or high mannose or oligomannose forms). The initial trimming takes place in the ER, followed by transfer to the Golgi apparatus where final trimming occurs, followed by addition of various sugars or aminosugars. These units are added through the action of various glycosyltransferases using nucleotide-sugar cosubstrates as sugar donors. In insect cells high levels of N-acetylglucosaminidase activity results typically in dead-end structures with a mannose cap. *Complex glycoforms* have sugar residues (N-acetyl glucosamine, galactose, and/or sialic acid) added to all branches of the oligosaccharide structure. *Hybrid glycoforms* have at least one branch modified with one of these sugar residues and one or more with mannose as the terminal residue.

4.6. METABOLIC REGULATION

Metabolic regulation is the heart of any living cell. Regulation takes place principally at the genetic level and at the cellular level (principally, control of enzyme activity and through cell surface receptors). Let us first consider genetic-level changes, as these fit most closely with our discussion of transcription and translation.

4.6.1. Genetic-level Control: Which Proteins Are Synthesized?

As the reader will recall, the formation of a protein requires transcription of a gene. Transcriptional control of protein synthesis is the most common control strategy used in bacteria. Control of protein synthesis in eucaryotes can be more complex, but the same basic concepts hold. In the simplest terms, the cell senses that it has too much or too little of a particular protein and responds by increasing or decreasing the rate of transcription of that gene. One form of regulation is *feedback repression*; in this case, the end product of enzymatic activity accumulates and blocks transcription. Another form of regulation is *induction*; a metabolite (often a substrate for a pathway) accumulates and acts as an *inducer* of transcription. These concepts are summarized in Figs. 4.9 and 4.10. In both cases a repressor protein is required. The repressor can bind to the *operator* region and hinder RNA polymerase binding. For repression, a corepressor (typically the end product of the pathway) is required, and the repressor can block transcription only when bound to the corepressor. For induction, the inducer (typically a substrate for a reaction) will combine with the repressor, and the complex is inactive as a repressor.

Note in Figs. 4.9 and 4.10 that several genes are under the control of a single promoter. A set of contiguous genes, encoding proteins with related functions, under the control of a single promoter–operator is called an *operon*. The *operon* concept is central to understanding microbial regulation.

Control can be even more complex than indicated in Figs. 4.9 and 4.10. The lactose (or lac) operon controls the synthesis of three proteins involved in lactose utilization as a carbon and energy source in *E. coli*. These genes are *lac z*, *lac y*, and *lac a*. Lac z encodes

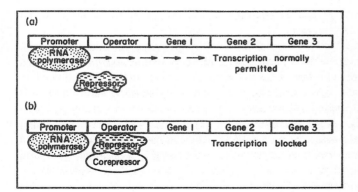

Figure 4.9. Process of enzyme repression. (a) Transcription of the operon occurs because the repressor is unable to bind to the operator. (b) After a corepressor (small molecule) binds to the repressor, the repressor now binds to the operator and blocks transcription. *m*-RNA and the proteins it codes for are not made. (With permission, from T. D. Brock, K. M. Brock, and D. M. Ward, *Basic Microbiology with Applications*, 3d ed., Pearson Education, Upper Saddle River, NJ, 1986, p. 143.)

β-galactosidase (or lactase), which cleaves lactose to glucose and galactose. The lac y protein is a *permease*, which acts to increase the rate of uptake of lactose into the cell. Lactose is modified in the cell to allolactose, which acts as the inducer. The conversion of lactose to allolactose is through a secondary activity of the enzyme β-galactosidase. Repression of transcription in uninduced cells is incomplete, and a low level (*basal level*) of proteins from the operon is made. Allolactose acts as indicated in Fig. 4.10, but induction by allolactose is not both necessary and sufficient for maximum transcription. Further regulation is exerted through *catabolite repression* (also called the *glucose effect*).

When *E. coli* senses the presence of a carbon–energy source preferred to lactose, it will not use the lactose until the preferred substrate (e.g., glucose) is fully consumed. This control mechanism is exercised through a protein called *CAP* (cyclic-AMP-activating protein). Cyclic AMP (cAMP) levels increase as the amount of energy available to the cell decreases. Thus, if glucose or a preferred substrate is depleted, the level of cAMP will increase. Under these conditions, cAMP will readily bind to CAP to form a complex that binds near the lac promoter. This complex greatly enhances RNA polymerase binding to the lac promoter. *Enhancer* regions exist in both procaryotes and eucaryotes.

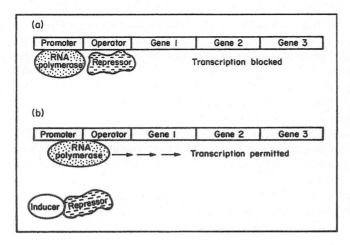

Figure 4.10. Process of enzyme induction. (a) A repressor protein binds to the operator region and blocks the action of RNA polymerase. (b) Inducer molecule binds to the repressor and inactivates it. Transcription by RNA polymerase occurs and an *m*-RNA for that operon is formed. (With permission, from T. D. Brock, K. M. Brock, and D. M. Ward, *Basic Microbiology with Applications*, 3d ed., Pearson Education, Upper Saddle River, NJ, 1986, p. 142.) In the lac operon, gene 1 is lac z, gene 2 is lac y, and gene 3 is lac a. The repressor is made on a separate gene called lac i.

The reader should now see how the cellular control strategy emerges. If the cell has an energetically favorable carbon–energy source available, it will not expend significant energy to create a pathway for utilization of a less favorable carbon–energy source. If, however, energy levels are low, then it seeks an alternative carbon–energy source. If and only if lactose is present will it activate the pathway necessary to utilize it.

Catabolite repression is a global response that affects more than lactose utilization. Furthermore, even for lactose, the glucose effect can work at levels other than genetic. The presence of glucose inhibits the uptake of lactose, even when an active uptake system exists. This is called *inducer exclusion*.

The role of global regulatory systems is still emerging. One concept is that of a *regulon*. Many noncontiguous gene products under the control of separate promoters can be coordinately expressed in a regulon. The best studied regulon is the *heat shock regulon*. The cell has a specific response to a sudden increase in temperature,[†] which results in the elevated synthesis of specific proteins. Evidence now exists that this regulon works by employing the induction of an alternative sigma factor, which leads to high levels of transcription from promoters that do not readily recognize the normal *E. coli* sigma factor. Examples of other regulons involve nitrogen and phosphate starvation, as well as a switch from aerobic to anaerobic conditions.

Although many genes are regulated, others are not. Unregulated genes are termed *constitutive*, which means that their gene products are made at a relatively constant rate irrespective of changes in growth conditions. Constitutive gene products are those that a cell expects to utilize under almost any condition; the enzymes involved in glycolysis are an example.

Example 4.1.

Diauxic growth is a term to describe the sequential use of two different carbon–energy sources. Industrially, diauxic growth is observed when fermenting a mixture of sugars, such as might result from the hydrolysis of biomass. The classic example of diauxic growth is growth of *E. coli* on a glucose–lactose mixture. Observations on this system led to formulation of the operon hypothesis and the basis for a Nobel prize (for J. Monod and F. Jacob). Consider the plot in Fig. 4.11, where the utilization of glucose and lactose and the growth of a culture are depicted for a batch culture (batch reactor). As we will discuss in more detail in Chapter 6, the amount of biomass in a culture, X, accumulates exponentially. Note that at 2 h after inoculation, cells are growing rapidly, glucose is being consumed, and lactose is not being utilized. At 7 h, cell mass accumulation is zero. All the glucose has been consumed. At 10 h the culture is growing and lactose is being consumed, but the rate of growth (cell mass accumulation) is less than at 2 h.

Explain what is happening with intracellular control to account for the observations at 2, 7, and 10 h. What rate of β-galactosidase formation would you expect to find in the culture at these times in comparison to the basal rate (which is < 1% of the maximum rate)?

Solution

At 2 h the lac operon is fully induced, since lactose converted to allolactose combines with the *lac i* repressor protein, inactivating it. With the repressor protein deactivated, RNA polymerase is free to bind to the promoter, but does so inefficiently. Glucose

---

[†]Or other stresses that result in abnormal protein formation or membrane disruption.

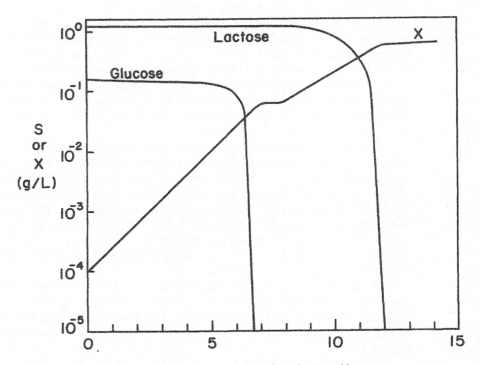

Figure 4.11. Diauxic growth curve for *E. coli* on glucose and lactose.

levels are still high, which results in higher levels of ATP and low levels of AMP and cAMP. Consequently, little cAMP–CAP complex is formed, and the interaction of the lac promoter with RNA polymerase is weak in the absence of cAMP–CAP. The rate of β-galactosidase formation would be slightly increased from the basal level—perhaps 5% of the maximal rate.

At 7 h the glucose has been fully consumed. The cell cannot generate energy, and the level of ATP decreases and cAMP increases. The cAMP–CAP complex level is high, which increases the efficiency of binding RNA polymerase to the lac promoter. This increased binding leads to increased transcription and translation. The rate of β-galactosidase formation is maximal and much higher than the basal rate or the 2 h rate. However, the cells have not yet accumulated sufficient intracellular concentrations of β-galactosidase and lac permease to allow efficient use of lactose and rapid growth.

At 10 h the intracellular content of proteins made from the lac operon is sufficiently high to allow maximal growth on lactose. However, this growth rate on lactose is slower than on glucose, since lactose utilization generates energy less efficiently. Consequently, the cAMP level remains higher than when the cell was growing on glucose. The level of production of β-galactosidase is thus higher than the basal or 2 h level.

Irrespective of whether an enzyme is made from a regulated or constitutive gene, its activity in the cell is regulated. Let us now consider control at the enzymatic level.

4.6.2. Metabolic Pathway Control

In Chapter 3, we learned how enzyme activity could be modulated by inhibitors or activators. Here we discuss how the activities of a group of enzymes (a pathway) can be controlled. The cell will attempt to make the most efficient use of its resources; the fermentation specialist tries to disrupt the cell's control strategy so as to cause the cell to overproduce the product of commercial interest. An understanding of how cells control their pathways is vital to the development of many bioprocesses.

First, consider the very simple case of a linear pathway making a product, P_1. Most often the first reaction in the pathway is inhibited by accumulation of the product (*feedback inhibition* or *end-product inhibition*). The enzyme for the entry of substrate into the pathway would be an allosteric enzyme (as described in Chapter 3), where the binding of the end product in a secondary site distorts the enzyme so as to render the primary active site ineffective. Thus, if the cell has a sufficient supply of P_1 (perhaps through an addition to the growth medium), it will deactivate the pathway so that the substrates normally used to make P_1 can be utilized elsewhere.

This simple concept can be extended to more complicated pathways with many branch points (see Fig. 4.12). Assuming that P_1 and P_2 are both essential metabolites, the cell may use one of several strategies to ensure adequate levels of P_1 and P_2 with efficient utilization of substrates.

One strategy is the use of isofunctional enzymes (*isozymes*). Two separate enzymes are made to carry out the same conversion, while each is sensitive to inhibition by a different end product. Thus, if P_1 is added in excess in the growth medium, it inhibits one of the

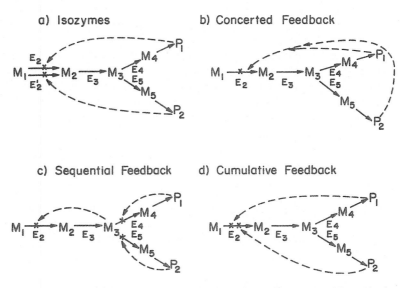

Figure 4.12. Examples of feedback control of branched pathways. P_1 and P_2 are the desired end products. $M_1, M_2, \ldots, M_j$ are intermediates, and E_j is the enzyme involved in converting metabolite M_{j-1} to M_j. Possible paths of inhibition are shown by dashed lines.

isozymes (E_2), while the other enzyme (E_2') is fully active. Sufficient activity remains through isozyme E_2' to ensure adequate synthesis of P_2.

An alternative approach is *concerted feedback inhibition.* Here a single enzyme with two allosteric binding sites (for P_1 and P_2) controls entry into the pathway. A high level of either P_1 or P_2 is not sufficient by itself to inhibit enzyme E_2, while a high level of both P_1 and P_2 will result in full inhibition.

A third possibility is *sequential feedback inhibition,* by which an intermediate at the branch point can accumulate and act as the inhibitor of metabolic flux into the pathway. High levels of P_1 and P_2 inhibit enzymes E_4 and E_5, respectively. If either E_4 or E_5 is blocked, M_3 will accumulate, but not as rapidly as when both E_4 or E_5 are blocked. Thus, intermediate flux levels are allowed if either P_1 or P_2 is high, but the pathway is inactivated if both P_1 and P_2 are high.

Other effects are possible in more complex pathways. A single allosteric enzyme may have effector sites for several end products of a pathway; each effector causes only partial inhibition. Full inhibition is a cumulative effect, and such control is called *cumulative feedback inhibition* or *cooperative feedback inhibition.* In other cases, effectors from related pathways may also act as activators. Typically, this situation occurs when the product of one pathway was the substrate for another pathway. An example of control of a complex pathway (for aspartate) is described in Section 4.9.

The reader should pause to consider the differences between feedback inhibition and repression. Inhibition occurs at the enzyme level and is rapid; repression occurs at the genetic level and is slower and more difficult to reverse. In bacteria where growth rates are high, unwanted enzymes are diluted out by growth. Would such a strategy work for higher cells in differentiated structures? Clearly not, since growth rates would be nearly zero. In higher cells (animals and plants) the control of enzyme levels is done primarily through the control of protein degradation, rather than at the level of synthesis. Most of our discussion has centered on procaryotes; the extension of these concepts to higher organisms must be done carefully.

Another caution is that the control strategy that one organism adopts for a particular pathway may differ greatly from that adopted by even a closely related organism with an identical pathway. Even if an industrial organism is closely related to a well-studied organism, it is prudent to check whether the same regulatory strategy has been adopted by both organisms. Knowing the cellular regulatory strategy facilitates choosing optimal fermenter operating strategy, as well as guiding strain improvement programs.

We have touched on some aspects of cellular metabolic regulation. A related form of regulation that we are just now beginning to appreciate has to do with the cell surface.

4.7. HOW THE CELL SENSES ITS EXTRACELLULAR ENVIRONMENT

4.7.1. Mechanisms to Transport Small Molecules across Cellular Membranes

A cell must take nutrients from its extracellular environment if it is to grow or retain metabolic activity. As we discussed in Example 4.1, which nutrients enter the cell and at what rate can be important in regulating metabolic activity.

Molecules enter the cell through either energy-independent or energy-dependent mechanisms. The two primary examples of energy-independent uptake are *passive diffusion* and *facilitated diffusion*. Energy-dependent uptake mechanisms include *active transport* and *group translocation*.

In passive diffusion, molecules move down a concentration gradient (from high to low concentration) that is thermodynamically favorable. Consequently,

$$J_A = K_p(C_{AE} - C_{AI}) \tag{4.1}$$

where J_A is the flux of species A across the membrane (mol/cm^2 – s), K_p is the permeability (cm/s), C_{AE} is the extracellular concentration of species A (mol/cm^3), and C_{AI} is the intracellular concentration. The cytoplasmic membrane consists of a lipid core with perhaps very small pores. For charged or large molecules, the value K_p is very low and the flow of material across the membrane is negligible. The cellular uptake of water and oxygen appears to be due to passive diffusion. Furthermore, lipids or other highly hydrophobic compounds have relatively high diffusivities (10^{-8} cm^2/s) in cellular membranes, and passive diffusion can be a mechanism of quantitative importance in their transport.

With facilitated transport, a carrier molecule (protein) can combine specifically and reversibly with the molecule of interest. The carrier protein is considered embedded in the membrane. By mechanisms that are not yet understood, the carrier protein, after binding the target molecule, undergoes conformational changes, which result in release of the molecule on the intracellular side of the membrane. The carrier can bind to the target molecule on the intracellular side of the membrane, resulting in the efflux or exit of the molecule from the cell. Thus, the net flux of a molecule depends on its concentration gradient.

The carrier protein effectively increases the solubility of the target molecule in the membrane. Because the binding of the molecule to the carrier protein is saturable (just as the active sites in the enzyme solution can be saturated), the flux rate of the target molecule into the cell depends on concentration differently than indicated in eq. 4.1. A simple equation to represent uptake by facilitated transport is

$$J_A = J_{A\ MAX} \left[\frac{C_{AE}}{K_{MT} + C_{AE}} - \frac{C_{AI}}{K_{MT} + C_{AI}} \right] \tag{4.2}$$

where K_{MT} is related to the binding affinity of the substrate (mol/cm^3) and $J_{A\ MAX}$ is the maximum flux rate of A (mol/cm^2–s). When $C_{AE} > C_{AI}$, the net flux will be into the cell. If $C_{AI} > C_{AE}$, there will be a net efflux of A from the cell. The transport is down a concentration gradient and is thermodynamically favorable. Facilitated transport of sugars and other low-molecular-weight organic compounds is common in eucaryotic cells, but infrequent in procaryotes. However, the uptake of glycerol in enteric bacteria (such as *E. coli*) is a good example of facilitated transport.

Active transport is similar to facilitated transport in that proteins embedded in the cellular membrane are necessary components. The primary difference is that active transport occurs *against a concentration gradient*. The intracellular concentration of a molecule may be a hundredfold or more greater than the extracellular concentration. The movement of a molecule up a concentration gradient is thermodynamically unfavorable and will not occur spontaneously; energy must be supplied. In active transport, several energy sources are possible: (1) the electrostatic or pH gradients of the proton-motive force, and (2) secondary gradients (for example, of Na^+ or other ions) derived from the proton-motive force by other active transport systems and by the hydrolysis of ATP.

The *proton-motive force* results from the extrusion of hydrogen as protons. The respiratory system of cells (see Section 5.4) is configured to ensure the formation of such gradients. Hydrogen atoms, removed from hydrogen carriers (most commonly NADH) on the inside of the membrane, are carried to the outside of the membrane, while the electrons removed from these hydrogen atoms return to the cytoplasmic side of the membrane. These electrons are passed to a final electron acceptor, such as O_2. When O_2 is reduced, it combines with H^+ from the cytoplasm, causing the net formation of OH^- on the inside. Because the flow of H^+ and OH^- across the cellular membrane by passive diffusion is negligible, the concentration of chemical species cannot equilibrate. This process generates a pH gradient and an electrical potential across the cell. The inside of the cell is alkaline compared to the extracellular compartment. The cytoplasmic side of the membrane is electrically negative, and the outside is electrically positive. The proton-motive force is essential to the transport of many species across the membrane, and any defect in the cellular membrane that allows free movement of H^+ and OH^- across the cell boundary can collapse the proton-motive force and lead to cell death.

Some molecules are actively transported into the cell without coupling to the ion gradients generated by the proton-motive force. By a mechanism that is not fully understood, the hydrolysis of ATP to release phosphate bond energy is utilized directly in transport (e.g., the transport of maltose in *E. coli*).

For these mechanisms of active transport, irrespective of energy source, we can write an equation analogous to Michaelis–Menten kinetics to describe uptake:

$$J_A = J_{A\ MAX} \frac{C_{AE}}{K_{MT} + C_{AE}} \tag{4.3}$$

The use of eq. 4.3 is meaningful only when the cell is in an energy-sufficient state.

Another energy-dependent approach to the uptake of nutrients is group translocation. The key factor here is the chemical modification of the substrate during the process of transport. The best-studied system of this type is the *phosphotransferase system*. This system is important in the uptake of many sugars in bacteria. The biological system itself is complex, consisting of four separate phosphate-carrying proteins. The source of energy is phosphoenolpyruvate (PEP).

Effectively, the process can be represented by:

$$\text{sugar}_{(\text{extracellular})} + \text{PEP}_{(\text{intracellular})} \rightarrow \text{sugar-P}_{(\text{intracellular})} + \text{pyruvate}_{(\text{intracellular})} \tag{4.4}$$

By converting the sugar to the phosphorylated form, the sugar is trapped inside the cell. The asymmetric nature of the cellular membrane and this process make the process essentially irreversible. Because the phosphorylation of sugars is a key step in their metabolism, nutrient uptake of these compounds by group translocation is energetically preferable to active transport. In active transport, energy would be expended to move the unmodified substrate into the cell, and then further energy would be expended to phosphorylate it.

Certainly, the control of nutrient uptake is a critical cellular interface with its extracellular environment. In some cases, however, cells can sense their external environment without the direct uptake of nutrients.

4.7.2. Role of Cell Receptors in Metabolism and Cellular Differentiation

Almost all cells have receptors on their surfaces. These receptors can bind a chemical in the extracellular space. Such receptors are important in providing a cell with information about its environment. Receptors are particularly important in animals in facilitating cell-to-cell communication. Animal cell surface receptors are important in transducing signals for growth or cellular differentiation. These receptors are also prime targets for the development of therapeutic drugs. Many viruses mimic certain chemicals (e.g., a growth factor) and use cell surface receptors as a means to entering a cell.

Simpler examples exist with bacteria. Some motile bacteria have been observed to move up concentration gradients for nutrients or down gradients of toxic compounds. This response is called *chemotaxis*. Some microbes also respond to gradients in oxygen (*aerotaxis*) or light (*phototaxis*). Such tactic phenomena are only partially understood. However, the mechanism involves receptors binding to specific compounds, and this binding reaction results in changes in the direction of movement of the flagella. Motile cells move in a random-walk fashion; the binding of an attractant extends the length of time the cell moves on a "run" toward the attractant. Similarly, repellents decrease the length of runs up the concentration gradient. Chemotaxis is described in Fig. 4.13.

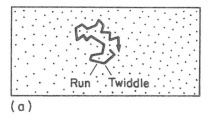

(a)

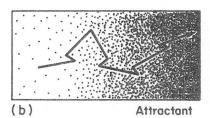

(b)　　　　　　　　Attractant

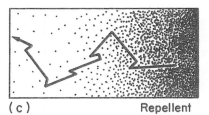

(c)　　　　　　　　Repellent

Figure 4.13. Diagrammatic representation of *Escherichia coli* movement, as analyzed with the tracking microscope. These drawings are two-dimensional projections of the three-dimensional movement. (a) Random movement of a cell in a uniform chemical field. Each run is followed by a twiddle, and the twiddles occur fairly frequently. (b) Directed movement toward a chemical attractant. The runs still go off in random directions, but when the run is up the chemical gradient, the twiddles occur less frequently. The net result is movement toward the chemical. (c) Directed movement away from a chemical repellent. (With permission, adapted from T. D. Brock, D. W. Smith, and M. T. Madigan, *Biology of Microorganisms*, 4th ed., Pearson Education, Upper Saddle River, NJ, 1984, p. 43.)

Microbial communities can be highly structured (e.g., biofilms), and cell-to-cell communication is important in the physical structure of the biofilm. Cell-to-cell communication is also important in microbial phenomena such as bioluminescence, exoenzyme synthesis, and virulence factor production. Basically, these phenomena depend on local cell concentration. How do bacteria count? They produce a chemical known as *quorum sensing molecule*, whose accumulation is related to cell concentration. When the quorum sensing molecule reaches a critical concentration, it activates a response in all of the cells present. A typical quorum sensing molecule is an acylated homoserine lactone. The mechanism of quorum sensing depends on an intracellular receptor protein, while chemotaxis depends on surface receptor proteins.

With higher cells, the timing of events in cellular differentiation and development is associated with surface receptors. With higher organisms, these receptors are highly evolved. Some receptors respond to steroids (*steroid hormone receptors*). Steroids do not act by themselves in cells, but rather the hormone–receptor complex interacts with specific gene loci to activate the transcription of a target gene.

A host of other animal receptors respond to a variety of small proteins that act as *hormones* or *growth factors*. These growth factors are normally required for the cell to initiate DNA synthesis and replication. Such factors are a critical component in the large-scale use of animal tissue cultures. Other cell surface receptors are important in the attachment of cells to surfaces. Cell adhesion can lead to changes in cell morphology, which are often critical to animal cell growth and normal physiological function. The exact mechanism by which receptors work is only now starting to emerge. One possibility for growth factors that stimulate cell division is that binding of the growth factor to the receptor causes an alteration in the structure of the receptor. This altered structure possesses catalytic activity (e.g., tyrosine kinase activity), which begins a cascade of reactions leading to cellular division. Surface receptors are continuously internalized, complexes degraded, and receptors recycled to supplement newly formed receptors. Thus, the ability of cells to respond to changes in environmental signals is continuously renewed. Such receptors will be important in our later discussions on animal cell culture.

4.8. SUMMARY

In this chapter you have learned some of the elementary concepts of how cells control their composition in response to an ever-changing environment. The essence of an organism resides in the chromosome as a linear sequence of nucleotides that form a language (*genetic code*) to describe the production of cellular components. The cell controls the storage and transmission of such information, using macromolecular templates. DNA is responsible for its own *replication* and is also a template for *transcription* of information into RNA species that serve both as machinery and template to *translate* genetic information into proteins. Proteins often must undergo posttranslational processing to perform their intended functions.

The cell controls both the amount and activity of proteins it produces. Many proteins are made on *regulated genes* (e.g., *repressible* or *inducible*), although other genes are *constitutive*. With regulated genes, small effector molecules alter the binding of regulatory proteins to specific sequences of nucleotides in the *operator* or *promoter* regions. Such regulatory proteins can block transcription or in other cases enhance it. A group of contiguous genes under the control of a single promoter–operator is called an *operon*. More global control through *regulons* is also evident. Some gene products are not regulated, and their synthesis is constitutive.

Once a protein is formed, its activity may be continuously modulated through *feedback inhibition*. A number of alternative strategies are employed by the cell to control the flux of material through a pathway. Another form of regulation occurs through the interaction of extracellular compounds with cell surface protein receptors.

4.9. APPENDIX: EXAMPLES OF REGULATION OF COMPLEX PATHWAYS[†]

In *E. coli* the conversion of aspartic acid to aspartyl phosphate is mediated by three isofunctional enzymes, of which two (designated as **a** and **c** in Fig. 4A.1) also mediate the conversion of aspartic acid semialdehyde to homoserine. Enzyme **a**, possessing both these functions, is feedback inhibited, and its synthesis is repressed by threonine. Enzyme **c**, which similarly possesses both functions, is inhibited, and repressed by lysine. The third, aspartokinase (enzyme **b**), is not subject to end-product inhibition, but its synthesis is repressed by methionine (Table 4A.1).

The enzymes of the L-lysine branch (**m–q**) and the L-methionine branch (**r–v**) catalyze reactions leading in each case to a single end product and are subject to specific repression by that end product (L-lysine and L-methionine, respectively).

The third branch of the aspartate pathway is subject to much more complex regulation, for two reasons. First, L-threonine, formed through this branch, is both a component of proteins and an intermediate in the synthesis of another amino acid, L-isoleucine. Second, four of the five enzymes (**e–h**) that catalyze L-isoleucine synthesis from L-threonine, also catalyze analogous steps in the completely separate biosynthetic pathway by which L-valine is synthesized from pyruvic acid. The intermediate of this latter pathway, α-ketoisovaleric acid, is also a precursor of the amino acid L-leucine. These interrelationships are shown in Fig. 4A.1(b).

L-isoleucine is an end-product inhibitor of the enzyme **d**, catalyzing the first step in its synthesis from L-threonine; this enzyme has no other biosynthetic role. L-valine is an end-product inhibitor of an enzyme (**e**) that has a dual metabolic role, since it catalyzes steps in both isoleucine and valine biosynthesis. In certain strains of *E. coli*, this enzyme is extremely sensitive to valine inhibition, with the result that exogenous valine prevents growth, an effect that can be reversed by the simultaneous provision of exogenous isoleucine. The L-leucine branch of the valine pathway is regulated by L-leucine, which is an end-product inhibitor of the first enzyme, **i**, specific to this branch. These interrelationships are shown in Fig. 4A.1(b).

As shown in Table 4A.2, many of the enzymes that catalyze steps in the synthesis of L-isoleucine, L-valine, and L-leucine are subject to repression only by a mixture of the three end products, a phenomenon known as *multivalent repression*. However, the five enzymes specific to L-leucine synthesis are specifically repressed by this amino acid alone.

[†]With permission, from R. Y. Stanier and others, *The Microbial World*, 5th ed., Pearson Education, Upper Saddle River, NJ, 1986.

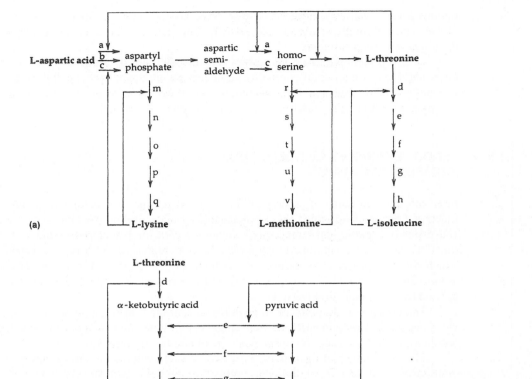

Figure 4A.1 A simplified diagram of the aspartate pathway in *E. coli*. Each solid arrow designates a reaction catalyzed by one enzyme. The biosynthetic products of the pathway (in boldface) are all allosteric inhibitors of one or more reactions. Careful study of this diagram reveals that with a single exception (the inhibition exerted by valine) the inhibition imposed by one amino acid does not cause starvation for a different amino acid. Part (a) shows the regulatory interrelationships of the L-lysine, L-methionine, and L-isoleucine branches of the pathway. Part (b) shows the regulatory interrelationships of the L-isoleucine, L-valine, and L-leucine branches

TABLE 4A.1 Control of the First Step of the Aspartate Pathway, Mediated by Three Different Aspartokinases, in the Bacterium *Escherichia coli*

Enzyme	Corepressor	Allosteric inhibitor
Aspartokinase I	Threonine and isoleucine	Threonine
Aspartokinase II	Methionine	No allosteric control
Aspartokinase III	Lysine	Lysine

TABLE 4A.2 Repressive Control of the Enzymes of the Isoleucine—Valine—Leucine Pathway (See Fig. 4A.1)

Enzymes	Corepressor
d, e, f, g, h	Isoleucine + valine + leucine
i, j, jj, k, l	Leucine

SUGGESTIONS FOR FURTHER READING

ALBERTS, B., AND OTHERS, *Essential Cell Biology,* Garland Publishing, Inc., New York, 1998. (This book is an up-to-date, highly readable, relatively short text. The same authors have written a more detailed text, *Molecular Biology of the Cell.*)

BLACK, J. G., *Microbiology: Principles and Applications* 3d ed., Prentice Hall, Upper Saddle River, NJ, 1996.

COOK, P. R., The Organization of Replication and Transcription, *Science* 284:1790–1795, 1999.

ELGARD, L., M. MOLINARI, AND A. HELENIUS, Setting the Standards: Quality Control in the Secretory Pathway, *Science* 286:1882–1888, 1999. (Also two related articles on quality control in post-translational processing and translation follow.)

KELLY, M. T., AND T. R. HOOVER, Bacterial Enhancers Function at a Distance, *ASM News* 65:484–489, 1999.

KOLTER, R., AND R. LOSICK, One for All and All for One, *Science* 280:226–227, 1998. (This article is an overview related to a more detailed article on quorum sensing in biofilms; see DAVIES, D. G., AND OTHERS, *Science* 280:295–298, 1998.)

MADIGAN, M. T., J. M. MARTINKO, AND J. PARKER, *Brock Biology of Microorganisms,* 8th ed., Prentice Hall, Upper Saddle River, NJ, 1997.

MORAN, L. A., K. G. SCRIMGEOUR, H. R. HORTON, R. S. OCHS, AND J. D. ROWN, *Biochemistry,* 2d ed., Prentice Hall, Upper Saddle River, NJ, 1994.

STANIER, R. Y., AND OTHERS, *The Microbial World*, 5th ed., Prentice Hall, Englewood Cliffs, NJ, 1986.

VON HIPPEL, P. H., An Integrated Model of the Transcription Complex in Elongation, Termination and Editing, *Science* 281:660–665, 1998.

PROBLEMS

4.1. Consider the aspartic acid pathway shown in Fig. 4A.1. Assume you have been asked to develop a high-lysine-producing mutant. What strategy would you pursue? (That is, which steps would you modify by removing feedback inhibition, and what changes in medium composition would you make over a simple mineral salts–glucose base medium?)

4.2. Why is *m*-RNA so unstable in most bacteria (half-life of about 1 min)? In many higher organisms, *m*-RNA half-lives are much longer (> 1 h). Why?

4.3. What would be the consequence of one base deletion at the beginning of the message for a protein?

4.4. How many ribosomes are actively synthesizing proteins at any instant in an *E. coli* cell growing with a 45-min doubling time? The birth size of *E. coli* is 1-μm diameter and 2-μm length. The water content is 75%. About 60% of the dry material is protein, and the rate of amino acid addition per ribosome is 20 amino acids per second. The average molecular weight of free amino acids in *E. coli* is 126.

4.5. Describe simple experiments to determine if the uptake of a nutrient is by passive diffusion, facilitated diffusion, active transport, or group translocation.

4.6. For the *m*-RNA nucleotide code below: (a) Deduce the corresponding sequence of amino acids. (b) What is the corresponding nucleotide sequence on the chromosome? This sequence codes for a part of insulin.

CCG UAU CGA CUU GUA ACA ACG CGC

4.7. Consider the pathway in Fig. 4A.1 for production of lysine, methionine, isoleucine, and theronine. You need to produce lysine. Describe a strategy for making large amounts of lysine. Your strategy can consist of adding various amino acids to the medium and choosing the mutant cells altered in regulation. Say you can identify up to two points of mutation (e.g., removal of feedback inhibition).

4.8. Suggest an experiment to determine if the uptake of a compound is by either facilitated or active transport.

4.9. What is catabolite repression, and how does it affect the level of protein expression from the lac operon?

4.10. Explain the difference between feedback inhibition and feedback repression.

4.11. You are asked by your boss to produce a human protein in *E. coli*. Because you have learned some of the differences in the way that procaryotes and eucaryotes make proteins, you worry about at least two factors that could complicate production of an authentic protein for human use.

  a) What complication might you worry about if the human DNA encoding the protein were placed directly in *E. coli*?

  b) Assume that the correct primary sequence of amino acids has been produced. What post-translational steps do you worry about and why?

4.12. Consider the process of N-linked glycosylation.

  a) What organelles are required?

  b) What is the residual sugar on a glycoprotein that has simple glycosylation?

  c) If glycosylation is complete, what will be the final sugar on the glycoform?

  d) Why may N-linked glycosylation be important?

5

Major Metabolic Pathways

5.1. INTRODUCTION

A major challenge in bioprocess development is to select an organism that can efficiently make a given product. Before about 1980 only naturally occurring organisms were available. With the advent of genetic engineering, it is possible to remove and add genes to an organism to alter its metabolic functions in a predetermined manner (*metabolic engineering*). In any case, the bioprocess developer must understand the metabolic capabilities of natural organisms either to use them directly or to know how to metabolically engineer them to make a desired, perhaps novel, product. Consequently, we turn our focus towards learning about some essential metabolic pathways.

Differences in microbial metabolism can be attributed partly to genetic differences and/or to differences in their responses to changes in their environment. Even the same species may produce different products when grown under different nutritional and environmental conditions. The control of metabolic pathways by nutritional and environmental regulation has become an important consideration in bioprocess engineering. For example, *Saccharomyces cerevisiae* (baker's yeast) produces ethanol when grown under anaerobic conditions. However, the major product is yeast cells (baker's yeast) when growth conditions are aerobic. Moreover, even under aerobic conditions, at high glucose concentrations some ethanol formation is observed, which indicates metabolic regulation not only by oxygen, but also by glucose. This effect is known as the *Crabtree effect*.

133

Ethanol formation during baker's yeast fermentation may be reduced or eliminated by culture with intermittent addition of glucose or by using carbon sources other than sucrose or glucose that support less rapid growth.

The major metabolic pathways and products of various microorganisms will be briefly covered in this chapter. Metabolic pathways are subgrouped as aerobic and anaerobic metabolism.

There are two key concepts in our discussion. *Catabolism* is the intracellular process of degrading a compound into smaller and simpler products (e.g., glucose to CO_2 and H_2O). Catabolism produces energy for the cell. *Anabolism* is involved in the synthesis of more complex compounds (e.g., glucose to glycogen) and requires energy.

5.2. BIOENERGETICS

Living cells require energy for biosynthesis, transport of nutrients, motility, and maintenance. This energy is obtained from the catabolism of carbon compounds, mainly carbohydrates. Carbohydrates are synthesized from CO_2 and H_2O in the presence of light by photosynthesis. The sun is the ultimate energy source for the life processes on earth.[†]

Metabolic reactions are fairly complicated and vary from one organism to another. However, these reactions can be classified in three major categories. A schematic diagram of these reactions is presented in Fig. 5.1. The major categories are (I) degradation of nutrients, (II) biosynthesis of small molecules (amino acids, nucleotides), and (III) biosynthesis of large molecules. These reactions take place in the cell simultaneously. As a result of metabolic reactions, end products are formed and released from the cells. These end products (organic acids, amino acids, antibiotics) are often valuable products for human and animal consumption.

Energy in biological systems is primarily stored and transferred via adenosine triphosphate (ATP), which contains high-energy phosphate bonds.

The active form of ATP is complexed with Mg^{2+}. The standard free-energy charge for the hydrolysis of ATP is 7.3 kcal/mol. The actual free-energy release in the cell may be substantially higher because the concentration of ATP is often much greater than that for ADP.

$$ATP + H_2O \rightleftharpoons ADP + P_i, \quad \Delta G° = -7.3 \text{ kcal/mol} \tag{5.1}$$

Biological energy is stored in ATP by reversing this reaction to form ATP from ADP and P_i. Similarly, ADP dissociates to release energy.

$$ADP + H_2O \rightleftharpoons AMP + P_i, \quad \Delta G° = -7.3 \text{ kcal/mol} \tag{5.2}$$

Analog compounds of ATP, such as guanosine triphosphate (GTP), uridine triphosphate (UTP), and cytidine triphosphate (CTP), also store and transfer high-energy phosphate bonds, but not to the extent of ATP. High-energy phosphate compounds produced during metabolism, such as phosphoenol pyruvate and 1,3-diphosphoglycerate, transfer

---

[†]The only exception is near some thermal vents at the bottom of the ocean, where nonphotosynthetic ecosystems exist independently of sunlight.

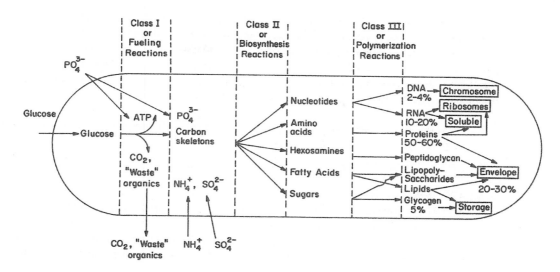

Figure 5.1. Schematic diagram of reactions in a bacterial cell.

their ~P group into ATP. Energy stored in ATP is later transferrded to lower-energy phosphate compounds such as glucose-6-phosphate and glycerol-3-phosphate, as depicted in Fig. 5.2.

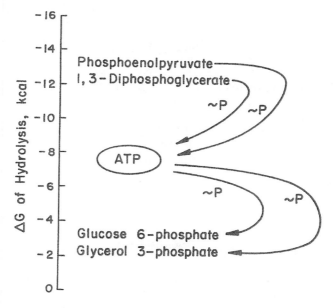

Figure 5.2 Transfer of biological energy from high-energy to low-energy compounds via ATP. Phosphoenolpyruvate and 1,3-diphosphoglycerate are high-energy phosphate compounds and act as phosphate donors. Low-energy compounds such as glucose 6-phosphate and glycerol 3-phosphate are phosphate acceptors.

Hydrogen atoms released in biological oxidation–reduction reactions are carried by nucleotide derivatives, especially by nicotinamide adenine dinucleotide (NAD^+) and nicotinamide adenine dinucleotide phosphate ($NADP^+$) (see Fig. 5.3). The oxidation–reduction reaction described is readily reversible. NADH can donate electrons to certain compounds and accept from others, depending on the oxidation–reduction potential of the compounds. NADH has two major functions in biological systems:

1. *Reducing power:* NADH and NADPH supply hydrogen in biosynthetic reactions, such as CO_2 fixation by autotrophic organisms.

$$CO_2 + 4\,H \longrightarrow CH_2O + H_2O \qquad (5.3)$$

2. *ATP formation in respiratory metabolism:* The electrons (or H atoms) carried by NADH are transferred to oxygen via a series of intermediate compounds (respiratory chain). The energy released from this electron transport results in the information of up to three ATP molecules.[†] ATP can be formed from the reducing power in NADH in the absence of oxygen if an alternative electron acceptor is available (e.g., NO_3^-).

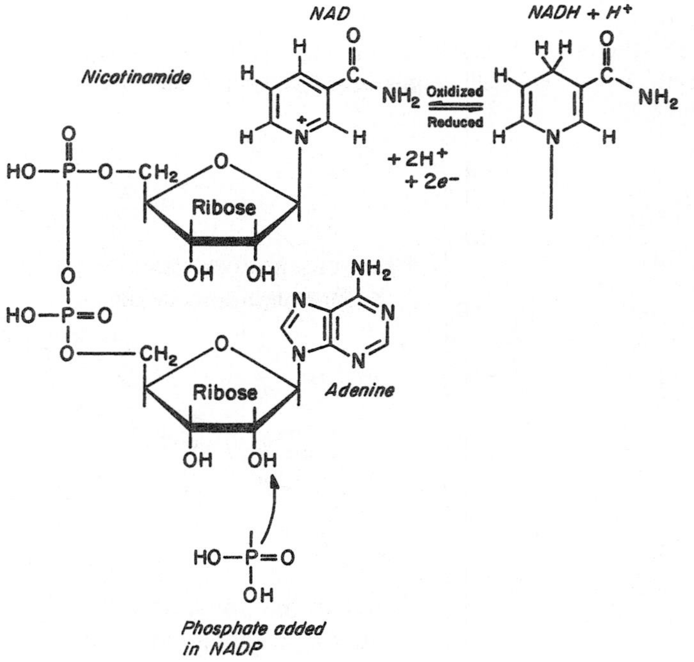

Figure 5.3. Structure of the oxidation–reduction coenzyme nicotinamide adenine dinucleotide (NAD). In NADP, a phosphate group is present, as indicated. Both NAD and NADP undergo oxidation–reduction as shown. (With permission, from T. D. Brock, D. W. Smith, and M. T. Madigan, *Biology of Microorganisms*, 4th ed., Pearson Education, Upper Saddle River, NJ, 1984, p. 104.)

[†]Some research suggests that the theoretical limit may be 2.5 ATP for each NADH, rather than 3.

5.3. GLUCOSE METABOLISM: GLYCOLYSIS AND THE TCA CYCLE

Glucose is a major carbon and energy source for many organisms. Several different metabolic pathways are used by different organisms for the catabolism of glucose. The catabolism of glucose by *glycolysis*, or the *Embden–Meyerhof–Parnas* (EMP) pathway, is the primary pathway in many organisms. Other pathways, such as the *hexose monophosphate* (HMP) and *Entner–Doudoroff* (ED) pathways, will be covered later.

Aerobic catabolism of organic compounds such as glucose may be considered in three different phases:

1. EMP pathway for fermentation of glucose to pyruvate.
2. *Krebs*, *tricarboxylic acid (TCA)*, or *citric acid cycle* for conversion of pyruvate to CO_2 and NADH.
3. Respiratory or *electron transport chain* for formation of ATP by transferring electrons from NADH to an electron acceptor.

The final phase, respiration, changes reducing power into a biologically useful energy form (ATP). *Respiration* may be aerobic or anaerobic, depending on the final electron acceptor. If oxygen is used as final electron acceptor, the respiration is called *aerobic respiration*. When other electron acceptors, such as NO_3^-, SO_4^{3-}, Fe^{3+}, Cu^{2+}, and S^0, are used, the respiration is termed *anaerobic respiration*.

Glycolysis or the EMP pathway results in the breakdown of glucose to two pyruvate molecules. The enzymatic reaction sequence involved in glycolysis is presented in Fig. 5.4. The first step in glycolysis is phosphorylation of glucose to glucose-6-phosphate (G-6P) by hexokinase. Phosphorylated glucose can be kept inside the cell. Glucose-6-phosphate is converted to fructose-6-phosphate (F-6P) by phosphoglucose isomerase, which is converted to fructose 1,6-diphosphate by phosphofructokinase. The first and the third reactions are the only two ATP-consuming reactions in glycolysis. They are irreversible. The breakdown of fructose-1,6-diphosphate into dihydroxyacetone phosphate (DHAP) and glyceraldehyde-3-phosphate (GA-3P) by aldolase is one of the key steps in glycolysis (e.g., C_6 to 2 C_3). DHAP and GA-3P are in equilibrium. As GA-3P is utilized in glycolysis, DHAP is continuously converted to GA-3P. Glyceraldehyde-3-phosphate is first oxidized with the addition of inorganic phosphate (Pi) to 1,3-diphosphoglycerate (1,3-dP-GA) by glyceraldehyde-3-phosphate dehydrogenase. 1,3-dP-GA releases one phosphate group to form ATP from ADP and is converted to 3-phosphoglycerate (3P-GA) by 3-phosphoglycerate kinase. 3P-GA is further converted to 2-phosphoglycerate (2P-GA) by phosphoglyceromutase. Dehydration of 2P-GA to phosphenol pyruvate (PEP) by enolase is the next step. PEP is further dephosphorylated to pyruvate (Pyr) by pyruvate kinase, with the formation of an ATP. Reactions after DHAP and GA-3P formation repeat twice during glycolysis.

The end-product pyruvate is a key metabolite in metabolism. Under anaerobic conditions, pyruvate may be converted to lactic acid, ethanol, or other products, such as acetone, butanol, and acetic acid. Anaerobic conversion of glucose to the aforementioned compounds used to be known as *fermentation*. However, that term today covers a whole

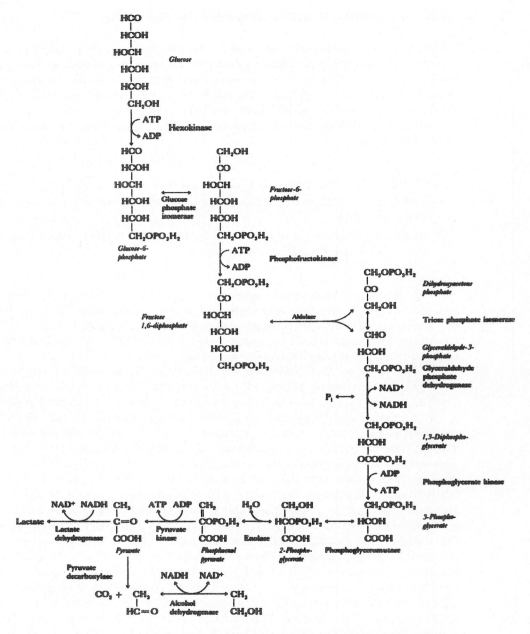

Figure 5.4. Reactions involved in the breakdown of glucose by glycolysis. (With permission, from T. D. Brock, D. W. Smith, and M. T. Madigan, *Biology of Microorganisms*, 4th ed., Pearson Education, Upper Saddle River, NJ, 1984, p. 788.)

range of enzymatic and microbial conversions. Under aerobic conditions, pyruvate is converted to CO_2 and NADH through the TCA cycle.

The overall reaction in glycolysis is

$$glucose + 2\ ADP + 2\ NAD^+ + 2\ Pi \longrightarrow 2\ pyruvate + 2\ ATP + 2(NADH + H^+) \qquad (5.4)$$

The net ATP yield in glycolysis is 2 mol ATP/glucose under anaerobic conditions.

Pyruvate produced in the EMP pathway transfers its reducing power to NAD^+ via the Krebs cycle. Glycolysis takes place in cytoplasm, whereas the site for the Krebs cycle is the matrix of mitochondria in eucaryotes. In procaryotes, these reactions are associated with membrane-bound enzymes. Entry into the Krebs cycle is provided by the acylation of coenyzme-A by pyruvate.

$$pyruvate + NAD^+ + CoA\text{-}SH \xrightarrow{\underset{\text{dehydrogenase}}{\text{pyruvate}}} acetyl\ CoA\ +\ CO_2 + NADH + H^+ \qquad (5.5)$$

Acetyl CoA is transferred through mitochondrial membrane at the expense of the conversion of the two NADHs produced in glycolysis to 2 FADH. Acetyl CoA is a key intermediate in the metabolism of amino acids and fatty acids.

The reactions involved in the TCA cycle are presented in Fig. 5.5. Condensation of acetyl CoA with oxaloacetic acid results in citric acid, which is further converted to isocitric acid and then to α-ketoglutaric acid (α-KGA) with a release of CO_2. α-KGA is decarboxylated and oxidized to succinic acid (SA), which is further oxidized to fumaric acid (FA). Hydration of fumaric acid to malic acid (MA) and oxidation of malic acid to oxaloacetic acid (OAA) are the two last steps of the TCA cycle. For each pyruvate molecule entering the cycle, three CO_2, four $NADH + H^+$, and one $FADH_2$ are produced. The succinate and α-ketoglutarate produced during the TCA cycle are used as precursors for the synthesis of certain amino acids. The reducing power ($NADH + H^+$ and $FADH_2$) produced is used either for biosynthetic pathways or for ATP generation through the electron transport chain.

The overall reaction of the TCA cycle is

$$\begin{aligned} acetyl\text{-}CoA + 3\ NAD^+ + FAD + GDP + Pi + 2H_2O \longrightarrow & \qquad (5.6)\\ CoA + 3(NADH + H^+) + FADH_2 + GTP + 2\ CO_2 \end{aligned}$$

Note from Fig. 5.5 that GTP can be converted easily into ATP; some descriptions of the TCA cycle show directly the conversion of ADP plus phosphorus into ATP, as succinyl CoA is converted to succinate.

The major roles of the TCA cycle are (1) to provide electrons (NADH) for the electron transport chain and biosynthesis, (2) to supply C skeletons for amino acid synthesis, and (3) to generate energy.

Many of the intermediates in the TCA cycle are used extensively in biosynthesis. Removal of these intermediates "short-circuits" the TCA cycle. To maintain a functioning TCA cycle, the cell can fix CO_2 (heterotrophic CO_2 fixation). In some microbes, phosphoenolpyruvate (PEP) can be combined with CO_2 to yield oxalacetate. Three enzymes that can catalyze such a conversion have been found (PEP carboxylase, PEP carboxykinase, and PEP carboxytransphosphorylase). Pyruvate can be combined with CO_2 to yield

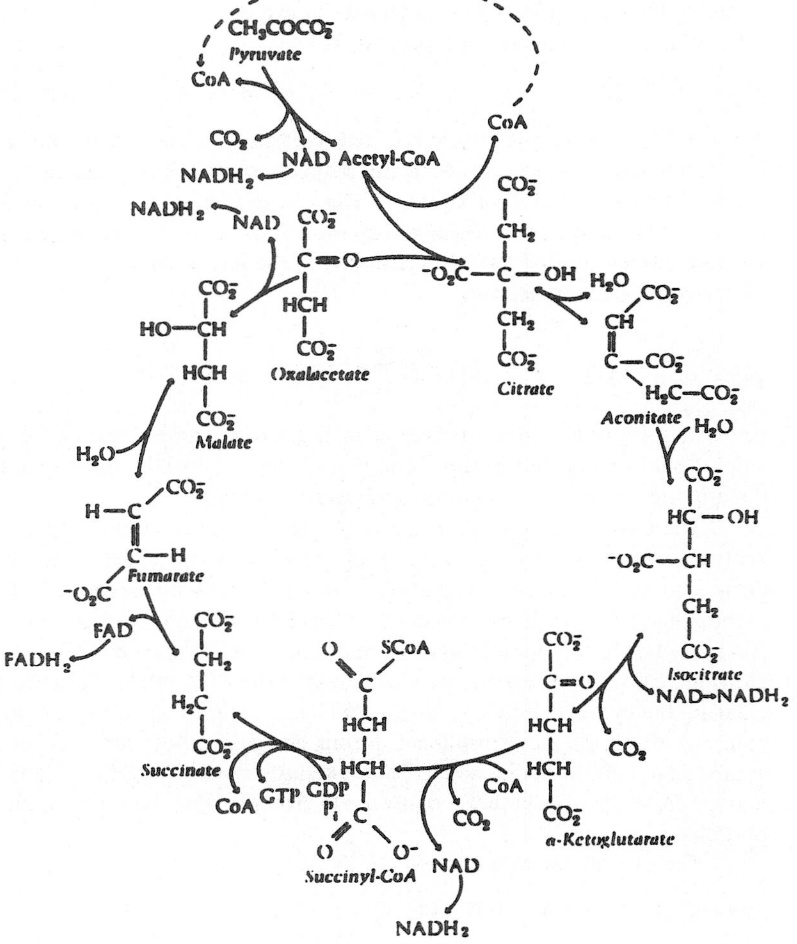

Overall reaction: Pyruvate + 4NAD + FAD → $3CO_2$ + $4NADH_2$ + $FADH_2$

GDP + phosphate → GTP

GTP + ADP → GDP + ATP

Oxidative phosphorylation: $4NADH_2$ ≡ 12ATP; $FADH_2$ ≡ 2ATP — 15ATP

Figure 5.5. Krebs or TCA (tricarboxylic acid) cycle with energy yields based on eucaryotic cells. (With permission, from T. D. Brock, D. W. Smith, and M. T. Madigan, *Biology of Microorganisms*, 4th ed., Pearson Education, Upper Saddle River, NJ, 1984, p. 789.)

oxalacetate, with the expenditure of one ATP, using the enzyme pyruvate carboxylase. Malic enzyme promotes the reversible formation of malate from pyruvate and CO_2, using the reducing power of one molecule of NADPH + H^+.

Heterotrophic CO_2 fixation can be an important factor in culturing microbes. When a culture is initiated at low density (i.e., few cells per unit volume) with little accumulation of intracellular CO_2, or when a gas sparge rate into a fermentation tank is high, then growth may be limited by the rate of CO_2 fixation to maintain the TCA cycle.

5.4. RESPIRATION

The respiration reaction sequence is also known as the electron transport chain. The process of forming ATP from the electron transport chain is known as *oxidative phosphorylation*. Electrons carried by $NADH + H^+$ and $FADH_2$ are transferred to oxygen via a series of electron carriers, and ATPs are formed. Three ATPs are formed from each $NADH + H^+$ and two ATPs for each $FADH_2$ in eucaryotes. The details of the respiratory (cytochrome) chain are depicted in Fig. 5.6. The major role of the electron transport chain is to regenerate NADs for glycolysis and ATPs for biosynthesis. The term P/O ratio is used to indicate the number of phosphate bonds made $(ADP + Pi \rightarrow ATP)$ for each oxygen atom used as an electron acceptor (e.g., $\frac{1}{2}O_2 + NADH + H^+ \rightarrow H_2O + NAD^+$).

Formation of $NADH + H^+$, $FADH_2$, and ATP at different stages of the aerobic catabolism of glucose are summarized in Table 5.1. The overall reaction (assuming 3 ATP/NADH) of aerobic glucose catabolism in eucaryotes[†]

$$\text{glucose} + 36\,Pi + 36\,ADP + 6\,O_2 \longrightarrow 6\,CO_2 + 6\,H_2O + 36\,ATP \qquad (5.7)$$

The energy deposited in 36 mol of ATP is 263 kcal/mol glucose. The free-energy change in the direct oxidation of glucose is 686 kcal/mol glucose. Therefore, the energy efficiency of glycolysis is 38% under standard conditions. With the correction for nonstandard conditions, this efficiency is estimated to be greater than 60%, which is significantly higher than the efficiency of man-made machines. The remaining energy stored in glucose is dissipated as heat. However, in procaryotes the conversion of the reducing power to ATP is less efficient. The number of ATP generated from $NADH + H^+$ is usually ≤ 2, and only one ATP may be generated from $FADH_2$. Thus, in procaryotes a single glucose molecule will yield less than 24 ATP, and the P/O ratio is generally between 1 and 2.

TABLE 5.1 Summary of NADH, $FADH_2$, and ATP Formation during Aerobic Catabolism of Glucose (Based on Consumption of 1 Mole of Glucose)

	NADH	$FADH_2$	ATP	Total ATP[a]
Glycolysis	2	–	2	6[b]
Oxidative decarboxylation of pyruvate	2	–	–	6
TCA cycle	6	2	2	24
Total	10	2	4	36 mol ATP

[a]Assumes a P/O ratio of 3 (i.e., 3 ATP phosphates made for each proton transported by the electron transport chain) for NADH and a P/O ratio of 2 for each $FADH_2$.

[b]The yield is 6 ATP rather than 8 because 2 NADH are converted to 2 $FADH_2$ for the transfer of acetyl CoA into mitochondria.

[†]The yield is about 30 ATP if we assume 2.5 ATP/NADH.

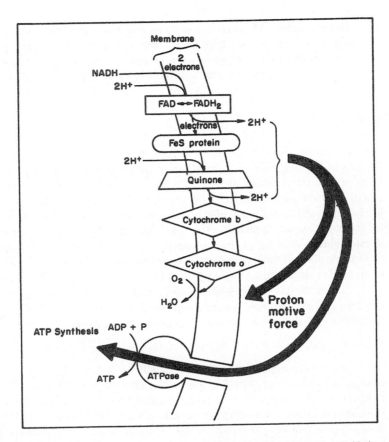

Figure 5.6. Electron transport and electron transport phosphorylation. *Top:* Oxidation of NADH and the flow of electrons through the electron transport system, leading to the transfer of protons (H^+) from the inside to the outside of the membrane. The tendency of protons to return to the inside is called *proton-motive force*. *Bottom:* ATP synthesis occurs as protons reenter the cell. An ATPase enzyme uses the proton-motive force for the synthesis of ATP. The proton-motive force is discussed in Section 4.7. (With permission, from T. D. Brock, K. M. Brock, and D. M. Ward, *Basic Microbiology with Applications*, 3d ed., Pearson Education, Upper Saddle River, NJ, 1986, p. 71.)

5.5. CONTROL SITES IN AEROBIC GLUCOSE METABOLISM

Several enzymes involved in glycolysis are regulated by feedback inhibition. The major control site in glycolysis is the phosphorylation of fructose-6-phosphate by phosphofructokinase.

$$\text{fructose-6-phosphate} + \text{ATP} \longrightarrow \text{fructose-1,6-diphosphate} + \text{ADP} \qquad (5.8)$$

The enzyme phosphofructokinase is an allosteric enzyme activated by ADP and Pi, but inactivated by ATP.

$$\text{phosphofructokinase (active)} \underset{\text{ADP}}{\overset{\text{ATP}}{\rightleftharpoons}} \text{phosphofructokinase (inactive)} \qquad (5.9)$$

At high ATP/ADP ratios, this enzyme is inactivated, resulting in a reduced rate of glycolysis and reduced ATP synthesis.

The concentration of dissolved oxygen or oxygen partial pressure has a regulatory effect on rate of glycolysis, known as the *Pasteur effect*. The rate of glycolysis under anaerobic conditions is higher than that under aerobic conditions. In the presence of oxygen, ATP yield is high, since the TCA cycle and electron transport chain are operating. As a result of high levels of ATP, ADP and Pi become limiting, and phosphofructokinase becomes inhibited. Also, some enzymes of glycolysis with —SH groups are inhibited by high levels of oxygen. A high NADH/NAD+ ratio also reduces the rate of glycolysis.

Certain enzymes of the Krebs cycle are also regulated by feedback inhibition. Pyruvate dehydrogenase is inhibited by ATP, NADH, and acetyl CoA and activated by ADP, AMP, and NAD^+. Similarly, citrate synthase is inactivated by ATP and activated by ADP and AMP; succinyl CoA synthetase is inhibited by NAD^+. In general, high ATP/ADP and NADH/NAD$^+$ ratios reduce the processing rate of the TCA cycle.

Several steps in the electron transport chain are inhibited by cyanide, azide, carbon monoxide, and certain antibiotics, such as amytal. Such inhibition is important due to its potential to alter cellular metabolism.

5.6. METABOLISM OF NITROGENOUS COMPOUNDS

Most of the organic nitrogen compounds have an oxidation level between carbohydrates and lipids. Consequently, nitrogenous compounds can be used as nitrogen, carbon, and energy source. Proteins are hydrolyzed to peptides and further to amino acids by proteases. Amino acids are first converted to organic acids by deamination (removal of amino group). Deamination reaction may be oxidative, reductive, or dehydrative, depending on the enzyme systems involved. A typical oxidative deamination reaction can be represented as follows:

$$\underset{\underset{NH_2}{|}}{R-CH-COOH} + H_2O + NAD^+ \longrightarrow \overset{\overset{O}{\|}}{R-C-COOH} + NH_3 + NADH + H^+ \quad (5.10)$$

Ammonia released from deamination is utilized in protein and nucleic acid synthesis as a nitrogen source, and organic acids can be further oxidized for energy production (ATP).

Transamination is another mechanism for conversion of amino acids to organic acids and other amino acids. The amino group is exchanged for the keto group of α-keto acid. A typical transamination reaction is

$$\text{glutamic acid + oxaloacetic acid} \longrightarrow \alpha\text{-keto glutaric acid + aspartic acid} \qquad (5.11)$$

Nucleic acids can also be utilized by many organisms as carbon, nitrogen, and energy source. The first step in nucleic acid utilization is enzymatic hydrolysis by specific nucleases hydrolyzing RNA and DNA. Nucleases with different specificities hydrolyze different bonds in nucleic acid structure, producing ribose/deoxyribose, phosphoric acid, and purines/pyrimidines. Sugar molecules are metabolized by glycolysis and the TCA cycle, producing CO_2 and H_2O under aerobic conditions. Phosphoric acids are used in ATP, phospholipid, and nucleic acid synthesis.

Purines/pyrimidines are degraded into urea and acetic acid and then to ammonia and CO_2. For example, the hydrolysis of adenine and uracil can be represented as follows:

$$\text{adenine} \longrightarrow CO_2 + NH_3 + \text{acetic acid} + \text{urea} \longrightarrow 5\,NH_3 + 5\,CO_2 \quad (5.12)$$

$$\text{uracil} \longrightarrow \text{alanine} + NH_3 + CO_2 \longrightarrow 2\,NH_3 + 4\,CO_2 \quad (5.13)$$

5.7. NITROGEN FIXATION

Certain microrganisms fix atmospheric nitrogen to form ammonia under reductive or microaerophilic conditions. Organisms capable of fixing nitrogen under aerobic conditions include *Azotobacter, Azotomonas, Azotococcus,* and *Biejerinckia*. Nitrogen fixation is catalyzed by the enzyme "nitrogenase," which is inhibited by oxygen. Typically these aerobic organisms sequester nitrogenase in compartments that are protected from oxygen.

$$N_2 + 6\,H^+ + 6\,e \longrightarrow 2\,NH_3 \quad (5.14)$$

Azotobacter species present in soil provide ammonium for plants by fixing atmospheric nitrogen, and some form associations with plant roots. Some facultative anaerobes such as *Bacillus, Klebsiella, Rhodopseudomonas,* and *Rhodospirillum* fix nitrogen under strict anaerobic conditions as well as strict anaerobes such as *Clostridia* can also fix nitrogen under anaerobic conditions. Certain cyanobacteria, such as *Anabaena sp.,* fix nitrogen under aerobic conditions. The lichens are *associations* of cyanobacteria and fungi. *Cyanobacteria* provide nitrogen to fungi by fixing atmospheric nitrogen. *Rhizobium* species are heterotrophic organisms growing in the roots of leguminous plants. *Rhizobium* fix atmospheric nitrogen under low oxygen pressure and provide ammonium to plants. *Rhizobium* and *Azospirillum* are widely used for agricultural purposes and are bioprocess products.

5.8. METABOLISM OF HYDROCARBONS

The metabolism of aliphatic hydrocarbons is important in some bioprocesses and often critical in applications such as bioremediation. Such metabolism requires oxygen, and only few organisms (e.g., *Pseudomonas, Mycobacteria,* certain yeasts and molds) can metabolize hydrocarbons. The low solubility of hydrocarbons in water is a barrier to rapid metabolism.

The first step in metabolism of aliphatic hydrocarbons is oxygenation by oxygenases. Hydrocarbon molecules are converted to an alcohol by incorporation of oxygen into the end of the carbon skeleton. The alcohol molecule is further oxidized to aldehyde and then to an organic acid which is further converted to acetyl-CoA, which is metabolized by the TCA cycle.

Oxidation of aromatic hydrocarbons takes place by the action of oxygenases and proceeds much slower than those of aliphatic hydrocarbons. Cathecol is the key intermediate in this oxidation sequence and can be further broken down ultimately to acetyl-CoA or TCA cycle intermediates. Aerobic metabolism of benzene is depicted below:

$$
\text{benzene} \longrightarrow \text{cathecol} \longrightarrow \textit{cis-cis-}\text{muconate} \longrightarrow \quad \beta\text{-keto adipate} \qquad (5.15)
$$
$$
\downarrow
$$
$$
\text{acetyl-CoA} + \text{succinate}
$$

Anaerobic metabolism of hydrocarbons is more difficult. Only a few organisms can metabolize hydrocarbons under anaerobic conditions. They cleave the C–C bonds and saturate with hydrogen to yield methane.

5.9. OVERVIEW OF BIOSYNTHESIS

The TCA cycle and glycolysis are critical catabolic pathways and also provide important precursors for the biosynthesis of amino acids, nucleic acids, lipids, and polysaccharides. Although many additional pathways exist, we will describe just two more, one in the context of biosynthesis and the other under anaerobic metabolism.

The first is the *pentose–phosphate pathway* or *hexose–monophosphate pathway* (HMP) (see Fig. 5.7). Although this pathway produces significant reducing power, which could be used, in principle, to supply energy to the cell, its primary role is to provide carbon skeletons for biosynthetic reactions and the reducing power necessary to support anabolism. Normally, NADPH is used in biosynthesis, whereas NADH is used in energy production. This pathway provides an array of small organic compounds with three, four, five, and seven carbon atoms. These compounds are particularly important for the synthesis of ribose, purines, coenzymes, and the aromatic amino acids. The glyceraldehyde-3-phosphate formed can be oxidized to yield energy through conversion to pyruvate and further oxidation of pyruvate in the TCA cycle.

A vital component of biosynthesis, which consumes a large amount of cellular building blocks, is the production of amino acids. Many amino acids are also important commercial products, and the alteration of pathways to induce overproduction (see Chapter 4) is critical to commercial success. The 20 amino acids can be grouped into various families. Figure 5.8 summarizes these families and the compounds from which they are derived. The amino acid, histidine, is not included in Fig. 5.8. Its biosynthesis is fairly complicated and cannot be easily grouped with the others. However, ribose-5-phosphate from HMP is a key precursor in its synthesis.

In addition to the synthesis of amino acids and nucleic acids, the cell must be able to synthesize lipids and polysaccharides. The key precursor is acetyl-CoA (see Fig. 5.5 for the TCA cycle). Fatty acid synthesis consists of the stepwise buildup of acetyl-CoA. Also, CO_2 is an essential component in fatty acid biosynthesis. Acetyl-CoA and CO_2 produce malomyl-CoA, which is a three-carbon-containing intermediate in fatty acid synthesis. This requirement for CO_2 can lengthen the start-up phase (or lag phase; see Chapter 6) for com-

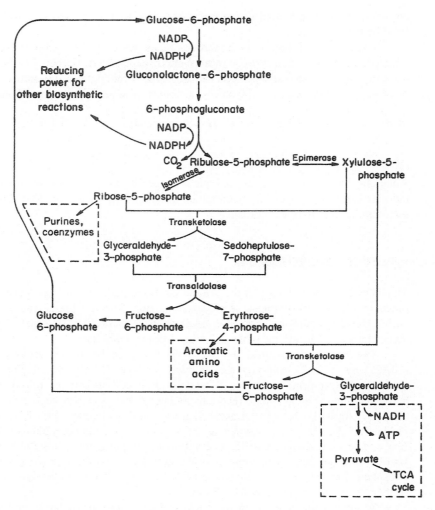

Figure 5.7. Pentose-phosphate pathway (hexose-monophosphate) with relations to other cellular processes. Compounds and reactions in the dashed boxes are not part of the pathway but rather represent connections to other metabolic activities. (With permission, from T. D. Brock, D. W. Smith, and M. T. Madigan, *Biology of Microorganisms*, 4th ed., Pearson Education, Upper Saddle River, NJ, 1984, p. 792.)

mercial fermentations if the system is not operated carefully. The requirement for CO_2 can be eliminated if the medium is formulated to supply key lipids, such as oleic acid.

The synthesis of most of the polysaccharides from glucose or other hexoses is readily accomplished in most organisms. However, if the carbon energy source has less than six carbons, special reactions need to be used. Essentially, the EMP pathway needs to be operated in reverse to produce glucose. The production of glucose is called *gluconeogenesis*. Since several of the key steps in the EMP pathway are irreversible, the cell must cir-

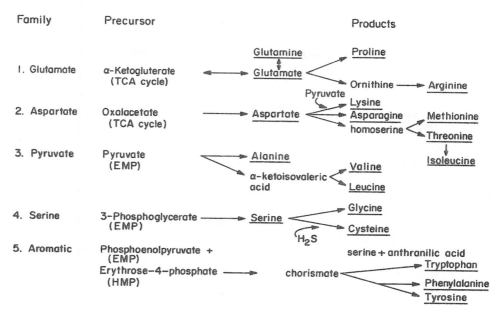

Figure 5.8. Summary of the amino acid families and their synthesis from intermediates in the EMP, TCA, and HMP pathways. The amino acids are underlined.

cumvent these irreversible reactions with energy-consuming reactions. Since pyruvate can be synthesized from a wide variety of pathways, it is the starting point. However, in glycolysis the final step to convert phosphoenolpyruvate (PEP) into pyruvate is irreversible. In gluconeogenesis, PEP is produced from pyruvate from

$$\text{pyruvate} + CO_2 + ATP + H_2O \longrightarrow \text{oxalacetate} + ADP + Pi + 2H^+ \qquad (5.16)$$

and

$$\text{oxalacetate} + ATP \longrightarrow PEP + ADP + CO_2 \qquad (5.17)$$

or a net reaction of

$$\text{pyruvate} + 2\,ATP + H_2O \longrightarrow PEP + 2\,ADP + 2\,H^+ \qquad (5.18)$$

The reactions in glycolysis are reversible (under appropriate conditions) up to the formation of fructose-1,6-diphosphate. To complete glucogenesis, two enzymes (fructose-1, 6-diphosphatase and glucose-6-phosphatase) not in the EMP pathway are required. Thus, an organism with these two enzymes and the ability to complete reaction 5.19 should be able to grow a wide variety of nonhexose carbon–energy sources.

So far we have concentrated on aerobic metabolism. Many of the reactions we have described would be operable under anaerobic conditions. The primary feature of anaerobic metabolism is energy production in the absence of oxygen and in most cases the absence of other external electron acceptors. The cell must also balance its generation and consumption of reducing power. In the next section we show how the pathways we have discussed can be adapted to the constraints of anaerobic metabolism.

5.10. OVERVIEW OF ANAEROBIC METABOLISM

The production of energy in the absence of oxygen can be accomplished by *anaerobic respiration* (see also Section 5.2). The same pathways as employed in aerobic metabolism can be used; the primary difference is the use of an alternative electron acceptor. One excellent example is the use of nitrate NO_3^-, which can act as an electron acceptor. Its product, nitrous oxide (N_2O^-), is also an acceptor leading to the formation of dinitrogen (N_2). This process, *denitrification*, is an important process environmentally. Many advanced biological waste-treatment systems are operated to promote denitrification.

Many organisms grow without using the electron transport chain. The generation of energy without the electron transport chain is called *fermentation*. This definition is the exact and original meaning of the term fermentation, although currently it is often used in a broader context. Since no electron transport is used, the organic substrate must undergo a balanced series of oxidative and reductive reactions. This constraint requires that the rates of conversion of NAD^+ and $NADP^+$ to NADH and NADPH must equal the rates of conversion of NADH and NADPH to NAD^+ and $NADP^+$. For example, with the EMP pathway the 2 mol of NAD reduced in this pathway in the production of pyruvate are reoxidized by oxidation of pyruvate to other products. Two prime examples are lactic acid and ethanol production (see Fig. 5.9). Both lactic acid and ethanol are important

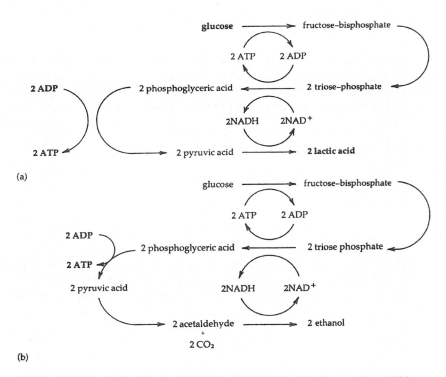

Figure 5.9. Comparison between (a) lactic acid and (b) alcoholic fermentations. (With permission, from R. Y. Stanier and others, *The Microbial World*, 5th ed., Pearson Education, Upper Saddle River, NJ, 1986, p. 94.)

commercial products from bioprocesses. Other partially oxidized by-products from fermentation are or have been commercially important [acetone–butanol fermentation, propionic acid, acetic acid (for vinegar), 2,3-butanediol, isopropanol, and glycerol].

Figure 5.10 summarizes common routes to some of these fermentation end products. Pyruvate is a key metabolite in these pathways. In most cases, pyruvate is formed through glycolysis. However, alternative pathways to form pyruvate exist. The most common of these is the *Entner–Doudoroff pathway* (see Fig. 5.11). This pathway is important in the fermentation of glucose by the bacterium *Zymomonas*. The use of *Zymomonas* to convert glucose into ethanol is of potential commercial interest, because the use of the Entner–Doudoroff pathway produces only 1 mol of ATP per mole of glucose. This low energy yield forces more glucose into ethanol and less into cell mass than for yeast, which uses glycolysis to produce pyruvate, which yields 2 mol ATP per mole of glucose.

No one organism makes all the products indicated in Fig. 5.10. Different organisms will contain different combinations of pathways. Thus, it is important to screen a wide va-

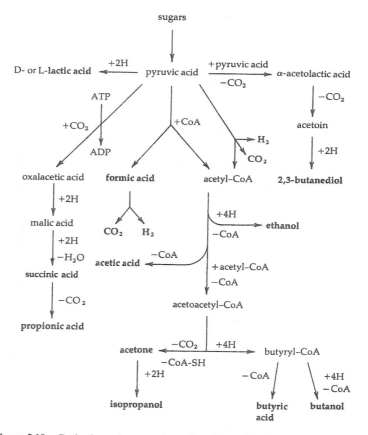

Figure 5.10. Derivations of some major end products of the bacterial fermentations of sugars from pyruvic acid. The end products are shown in boldface type. (With permission, from R. Y. Stanier and others, *The Microbial World*, 5th ed., Pearson Education, Upper Saddle River, NJ, 1986, p. 95.)

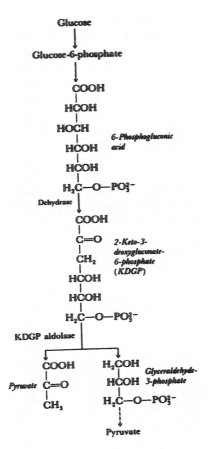

Figure 5.11. Entner–Doudoroff pathway. (With permission, from T. D. Brock. D. W. Smith, and M. T. Madigan, *Biology of Microorganisms*, 4th ed., Pearson Education, Upper Saddle River, NJ, 1984, p. 791.)

riety of organisms to select one that will maximize the yield of a desired product, while minimizing the formation of other by-products.

5.11. OVERVIEW OF AUTOTROPHIC METABOLISM

So far we have been concerned primarily with *heterotrophic growth* (e.g., organic molecules serve as carbon–energy sources). However, *autotrophs* obtain nearly all their carbon from CO_2. Most autotrophs (either photoautotrophs or chemoautotrophs) fix or capture CO_2 by a reaction catalyzed by the enzyme ribulose bisphosphate carboxylase, which converts ribulose-1,5-diphosphate plus CO_2 and H_2O into two molecules of glyceric acid-3-phosphate. This is the key step in the *Calvin cycle* (or Calvin–Benson cycle). This cycle is summarized in Fig. 5.12 and provides the building blocks for autotrophic growth.

Energy for autotrophic growth can be supplied by light (photoautotroph) or chemicals (chemoautotroph). Here we consider the special case of photoautotrophic growth.

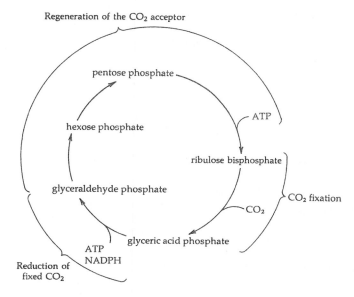

Figure 5.12. Schematic representation of the Calvin–Benson cycle, illustrating its three phases: CO_2 fixation, reduction of fixed CO_2, and regeneration of the CO_2 acceptor. (With permission, from R. Y. Stanier and others, *The Microbial World*, 5th ed., Pearson Education, Upper Saddle River, NJ, 1986, p. 96.)

Photosynthesis takes place in two phases. The overall reaction is

$$6\,CO_2 + 6\,H_2O \xrightarrow{\text{light}} C_6H_{12}O_6 + 6\,O_2 \tag{5.19}$$

The first phase of photosynthesis is known as the *light phase*. Light energy is captured and converted into biochemical energy in the form of ATP and reducing agents, such as NADPH. In this process, hydrogen atoms are removed from water molecules and are used to reduce NADP, leaving behind molecular oxygen. Simultaneously, ADP is phosphorylated to ATP. The light-phase reaction of photosynthesis is

$$H_2O + NADP^+ + P_i + ADP \xrightarrow{\text{light}} \text{oxygen} + NADPH + H^+ + ATP \tag{5.20}$$

In the second phase, the energy-rich products of the first phase, NADPH and ATP, are used as the sources of energy to reduce the CO_2 to yield glucose (see Fig 5.12). Simultaneously, NADPH is reoxidized to $NADP^+$, and the ATP is converted into ADP and phosphate. This *dark phase* is described by the following reaction:

$$CO_2 + NADPH + H^+ + ATP \longrightarrow \tfrac{1}{6}\text{glucose} + NADP^+ + ADP + P_i \tag{5.21}$$

Both procaryotic and eucaryotic cells can fix CO_2 by photosynthesis. In procaryotes (e.g., cyanobacteria), photosynthesis takes place in stacked membranes, whereas in eucaryotes an organelle called the *chloroplast* conducts photosynthesis. Both systems contain chlorophyll to absorb light. Light absorption by chlorophyll molecules results in an electronic excitation. The excited chlorophyll molecule returns to the normal state by emitting light quanta in a process known as *fluorescence*. The excited chlorophyll donates an electron to a sequence of enzymes, and ATP is produced as the electrons travel through the chain. This ATP generation process is called *photophosphorylation*. Electron carriers in this process are ferrodoxin and several cytochromes.

Sec. 5.11   Overview of Autotrophic Metabolism

The light phase of photosynthesis consists of two photosystems. Photosystem I (PS I) can be excited by light of wavelength shorter than 700 nm and generates NADPH. Photosystem II (PS II) requires light of wavelength shorter than 680 nm and splits H_2O into $\frac{1}{2} O_2 + 2 H^+$. ATPs are formed as electrons flow from PS II to PS I.

5.12. SUMMARY

Cellular metabolism is concerned with two primary functions: *catabolism* and *anabolism*. Catabolism involves the degradation of a substrate to more highly oxidized end products for the purpose of generating energy and reducing power. Anabolism is the biosynthesis of more complex compounds from simpler compounds, usually with the consumption of energy and reducing power. The key compound to store and release energy is adenosine triphosphate or *ATP*. Reducing power is stored by nicotinamide adenine dinucleotide (*NADH*) or nicotinamide adenine dinucleotide phosphate (*NADPH*).

Three of the most important pathways in the cell are (1) the *Embden–Meyerhof–Parnas* (EMP) pathway, or *glycolysis*, which converts glucose into pyruvate; (2) the *tricarboxylic acid cycle*, which can oxidize pyruvate through acetyl-CoA into CO_2 and H_2O; and (3) the *pentose–phosphate* or *hexose–monophosphate* (HMP) pathway, which converts glucose-6-phosphate into a variety of carbon skeletons (C_3, C_4, C_5, C_6, and C_7), with glyceraldehyde-3-phosphate as the end product. Although all three pathways can have catabolic and anabolic roles, the EMP pathway and TCA cycle are the primary means for energy generation, and HMP plays a key role in supplying carbon skeletons and reducing power for direct use in biosynthesis. In this chapter we have briefly considered the relationship of these pathways to amino acid, fatty acid, and polysaccharide biosynthesis. The conversion of pyruvate to glucose, necessary for polysaccharide biosynthesis when the carbon source does not have six carbons, is called *glucogenesis*.

Reducing power can be used to generate ATP through the *electron transport chain*. If oxygen is the final electron acceptor for this reducing power, the process is called *aerobic respiration*. If another electron acceptor is used in conjunction with the electron transport chain, then the process is called *anaerobic respiration*. Cells that obtain energy without using the electron transport chain use *fermentation*. Substrate-level phosphorylation supplies ATP. The end products of fermentative metabolism (e.g., ethanol, acetone–butanol, and lactic acid) are important commercially and are formed in response to the cell's need to balance consumption and the production of reducing power.

Autotrophic organisms use CO_2 as their carbon source and rely on the *Calvin* (or *Calvin–Benson*) cycle to incorporate (or fix) carbon from CO_2 into cellular material. Energy is obtained either through light (*photoautotroph*) or oxidation of inorganic chemicals (*chemoautotroph*). Figure 5.13 summarizes the major metabolic pathways and their interrelationship.

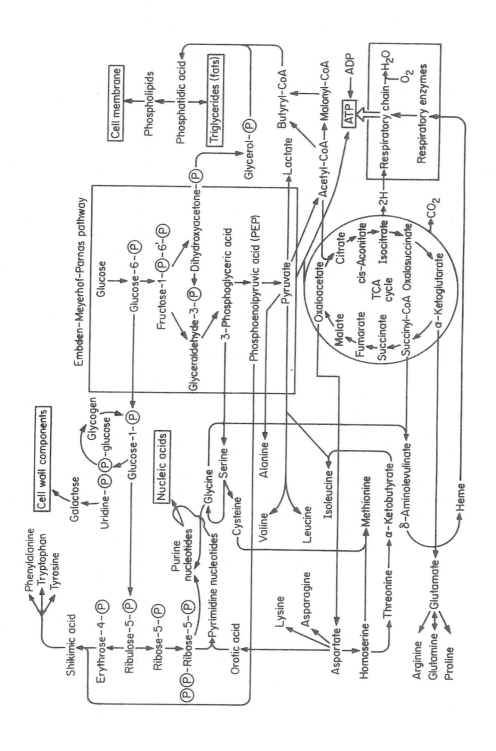

Figure 5.13. Interrelationship of major metabolic pathways in *E. coli*. (Adapted from J. D. Watson, *Molecular Biology of the Gene*, 2d ed., W. A. Benjamin, Inc., New York, 1970.)

SUGGESTIONS FOR FURTHER READING

A. General information on metabolic pathways

ALBERTS, B., D. BRAY, A. JOHNSON, J. LEWIS, M. RAFF, K. ROBERTS, AND P. WALTER, *Essential Cell Biology: An Introduction to the Molecular Biology of the Cell*, Garland Publ., Inc., New York, 1998.

BLACK, J. G., *Microbiology: Principles and Applications,* 3d ed., Prentice Hall, Upper Saddle River, NJ, 1996.

MADIGAN, M. T., J. M. MARTINKO, AND J. PARKER, *Brock Biology of Microorganisms,* 8th ed., Prentice Hall, Upper Saddle River, NJ, 1997.

MORAN, L. A., K. G. SCRIMGEOUR, H. R. HORTON, R. S. OCHS, AND J. D. RAWN, *Biochemistry,* 2d ed., Prentice Hall, Upper Saddle River, NJ, 1994.

B. Specific information on interaction of metabolism and product formation

CRUEGER, W., AND A. CRUEGER, *Biotechnology. A Textbook of Industrial Microbiology* (T. D. Brock, ed., English edition), 2d ed., Sinauer Associates, Inc., Sunderland, MA, 1990.

NEWAY, J. O., *Fermentation Process Development of Industrial Organisms*, Marcel-Dekker, Inc., New York, 1989.

PROBLEMS

5.1. Cite the ATP-consuming and ATP-generating steps in glycolysis.

5.2. Briefly specify major functions of the TCA cycle.

5.3. What are the major control sites in glycolysis?

5.4. What is the Pasteur effect? Explain in terms of regulation of metabolic flow into a pathway.

5.5. How is glucose synthesized from pyruvate?

5.6. Explain the major functions of the dark and light reaction phases in photosynthesis.

5.7. What are the major differences in photosynthesis between microbes and plants?

5.8. What is transamination? Provide an example.

5.9. Briefly explain the Crabtree Effect.

5.10. What are the major steps in aerobic metabolism of hydrocarbons? What are the end products?

5.11. What is nitrogen fixation? Compare the aerobic and anaerobic nitrogen fixation mechanisms.

6

How Cells Grow

6.1. INTRODUCTION

For microbes, growth is their most essential response to their physiochemical environment. Growth is a result of both replication and change in cell size. Microorganisms can grow under a variety of physical, chemical, and nutritional conditions. In a suitable nutrient medium, organisms extract nutrients from the medium and convert them into biological compounds. Part of these nutrients are used for energy production and part are used for biosynthesis and product formation. As a result of nutrient utilization, microbial mass increases with time and can be described simply by

$$\text{substrates } + \text{ cells} \longrightarrow \text{extracellular products } + \text{ more cells}$$
$$\Sigma S \ + \ X \longrightarrow \ \Sigma P + \ nX \tag{6.1}$$

Microbial growth is a good example of an autocatalytic reaction. The rate of growth is directly related to cell concentration, and cellular reproduction is the normal outcome of this reaction.

The rate of microbial growth is characterized by the net *specific growth rate*, defined as

$$\mu_{\text{net}} \equiv \frac{1}{X}\frac{dX}{dt} \tag{6.2a}$$

155

$$\mu_{net} = \mu_g - k_d \tag{6.2b}$$

where X is cell mass concentration (g/l), t is time (h), and μ_{net} is net specific growth rate (h^{-1}). The net specific growth is the difference between a gross specific growth rate, μ_g (h^{-1}), and the rate of loss of cell mass due to cell death or endogenous metabolism, k_d (h^{-1}).

Microbial growth can also be described in terms of cell number concentration, N, as well as X. In that case

$$\mu_R \equiv \frac{1}{N}\frac{dN}{dt} \tag{6.3}$$

where μ_R is the net specific replication rate (h^{-1}). If we ignore cell death, k_d, then we use the symbol μ'_R; and in cases where cell death is unimportant, μ_R will equal μ'_R.

In this chapter we will discuss how the specific growth rate changes with its environment. First, we will consider growth in batch culture, where growth conditions are constantly changing.

6.2. BATCH GROWTH

Batch growth refers to culturing cells in a vessel with an initial charge of medium that is not altered by further nutrient addition or removal. This form of cultivation is simple and widely used both in the laboratory and industrially.

6.2.1. Quantifying Cell Concentration

The quantification of cell concentration in a culture medium is essential for the determination of the kinetics and stoichiometry of microbial growth. The methods used in the quantification of cell concentration can be classified in two categories: direct and indirect. In many cases, the direct methods are not feasible due to the presence of suspended solids or other interfering compounds in the medium. Either cell number or cell mass can be quantified depending on the type of information needed and the properties of the system. Cell mass concentration is often preferred to the measurement of cell number density when only one is measured, but the combination of the two measurements is often desirable.

6.2.1.1. Determining cell number density. A Petroff–Hausser slide or a *hemocytometer* is often used for direct cell counting. In this method, a calibrated grid is placed over the culture chamber, and the number of cells per grid square is counted using a microscope. To be statistically reliable, at least 20 grid squares must be counted and averaged. The culture medium should be clear and free of particles that could hide cells or be confused with cells. Stains can be used to distinguish between dead and live cells. This method is suitable for nonaggregated cultures. It is difficult to count molds under the microscope because of their mycelial nature.

Plates containing appropriate growth medium gelled with agar (Petri dishes) are used for counting viable cells. (The word *viable* used in this context means capable of reproduction.) Culture samples are diluted and spread on the agar surface and the plates are incubated. Colonies are counted on the agar surface following the incubation period. The results are expressed in terms of colony-forming units (CFU). If cells form aggregates, then a single colony may not be formed from a single cell. This method (*plate counts*) is more suitable for bacteria and yeasts and much less suitable for molds. A large number of colonies must be counted to yield a statistically reliable number. Growth media have to be selected carefully, since some media support growth better than others. The *viable count* may vary, depending on the composition of the growth medium. From a single cell, it may require 25 generations to form an easily observable colony. Unless the correct medium and culture conditions are chosen, some cells that are metabolically active may not form colonies.

In an alternative method, an agar–gel medium is placed in a small ring mounted on a microscope slide, and cells are spread on this miniature culture dish. After an incubation period of a few doubling times, the slide is examined with a microscope to count cells. This method has many of the same limitations as plate counts, but it is more rapid, and cells capable of only limited reproduction will be counted.

Another method is based on the relatively high electrical resistance of cells (Fig. 6.1). Commercial *particle counters* employ two electrodes and an electrolyte solution. One electrode is placed in a tube containing an orifice. A vacuum is applied to the inner tube, which causes an electrolyte solution containing the cells to be sucked through the orifice. An electrical potential is applied across the electrodes. As cells pass through the orifice, the electrical resistance increases and causes pulses in electrical voltage. The number of pulses is a measure of the number of particles; particle concentration is known, since the counter is activated for a predetermined sample volume. The height of the pulse is a measure of cell size. Probes with various orifice sizes are used for different cell sizes. This method is suitable for discrete cells in a particulate-free medium and cannot be used for mycelial organisms.

The number of particles in solution can be determined from the measurement of scattered light intensity with the aid of a phototube (nephelometry). Light passes through

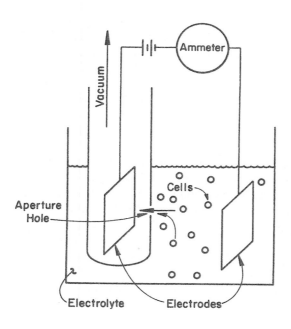

Figure 6.1. Diagram of a particle counter using the electrical resistance method for measuring cell number and cell size distribution. The ratio of volumes of a nonconducting particle to the orifice volume (altered by changing orifice diameter) determines the size of the voltage pulse. (Adapted with permission, from D. I. C. Wang and others, *Fermentation and Enzyme Technology*, John Wiley & Sons, New York, 1979, p. 64.)

the culture sample, and a phototube measures the light scattered by cells in the sample. The intensity of the scattered light is proportional to cell concentration. This method gives best results for dilute cell and particle suspensions.

6.2.1.2. Determining cell mass concentration.   *Direct methods.* Determination of cellular *dry weight* is the most commonly used direct method for determining cell mass concentration and is applicable only for cells grown in solids-free medium. If noncellular solids, such as molasses solids, cellulose, or corn steep liquor, are present, the dry weight measurement will be inaccurate. Typically, samples of culture broth are centrifuged or filtered and washed with a buffer solution or water. The washed wet cell mass is then dried at 80°C for 24 hours; then dry cell weight is measured.

Packed cell volume is used to rapidly but roughly estimate the cell concentration in a fermentation broth (e.g., industrial antibiotic fermentations). Fermentation broth is centrifuged in a tapered graduated tube under standard conditions (rpm and time), and the volume of cells is measured.

Another rapid method is based on the absorption of light by suspended cells in sample culture media. The intensity of the transmitted light is measured using a spectrometer. Turbidity or *optical density* measurement of the culture medium provides a fast, inexpensive, and simple method of estimating cell density in the absence of other solids or light-absorbing compounds. The extent of light transmission in a sample chamber is a function of cell density and the thickness of the chamber. Light transmission is modulated by both absorption and scattering. Pigmented cells give different results than unpigmented ones. Background absorption by components in the medium must be considered, particularly if absorbing dissolved species are taken into cells. The medium should be essentially particle free. Proper procedure entails using a wavelength that minimizes absorption by medium components (600- to 700-nm wavelengths are often used), "blanking" against medium, and the use of a calibration curve. The calibration curve relates optical density (OD) to dry-weight measurements. Such calibration curves can become nonlinear at high OD values (> 0.3) and depend to some extent on the physiological state of the cells.

Indirect methods. In many fermentation processes, such as mold fermentations, direct methods cannot be used. In such cases indirect methods are used, which are based mainly on the measurement of substrate consumption and/or product formation during the course of growth.

Intracellular components of cells such as RNA, DNA, and protein can be measured as indirect measures of cell growth. During a batch growth cycle, the concentrations of these intracellular components change with time. Figure 6.2 depicts the variation of certain intracellular components with time during a batch growth cycle. Concentration of RNA (RNA/cell weight) varies significantly during a batch growth cycle; however, DNA and protein concentrations remain fairly constant. Therefore, in a complex medium, DNA concentration can be used as a measure of microbial growth. Cellular protein measurements can be achieved using different methods. Total amino acids, Biuret, Lowry (folin reagent), and Kjeldahl nitrogen measurements can be used for this purpose. Total amino acids and the Lowry method are the most reliable. Recently, protein determination kits from several vendors have been developed for simple and rapid protein measurements.

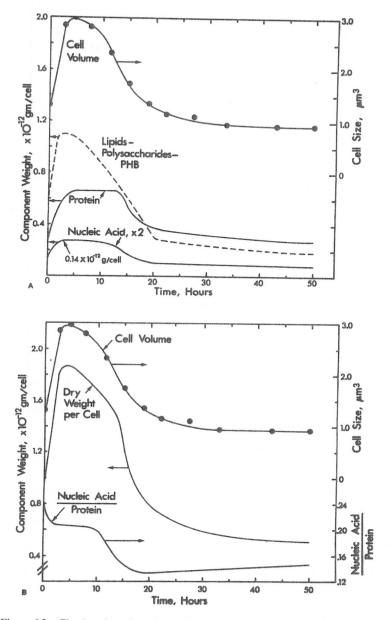

Figure 6.2. The time-dependent changes in cell composition and cell size for *Azotobacter vinelandii* in batch culture are shown. (With permission, from M. L. Shuler and H. M. Tsuchiya, "Cell Size as an Indicator of Changes in Intracellular Composition of *Azotobacter vinelandii*," *Can. J. Microbiol. 31*:927, 1975 and National Research Council of Canada, Ottawa.)

However, many media contain proteins as substrates, which limits the usefulness of this approach.

The intracellular ATP concentration (mg ATP/mg cells) is approximately constant for a given organism. Thus, the ATP concentration in a fermentation broth can be used as a measure of biomass concentration. The method is based on luciferase activity, which catalyzes oxidation of luciferin at the expense of oxygen and ATP with the emission of light.

$$\text{luciferin} + O_2 + ATP \xrightarrow{\text{luciferase}} \text{light} \qquad (6.4)$$

When oxygen and luciferin are in excess, total light emission is proportional to total ATP present in the sample. Photometers can be used to detect emitted light. Small concentrations of biomass can be measured by this method, since very low concentrations of ATP (10^{-12} g ATP/l) can be measured by photometers or scintillation counters. The ATP content of a typical bacterial cell is 1 mg ATP/g dry-weight cell, approximately.

Sometimes, nutrients used for cellular mass production can be measured to follow microbial growth. Nutrients used for product formation are not suitable for this purpose. Nitrate, phosphate, or sulfate measurements can be used. The utilization of a carbon source or oxygen uptake rate can be measured to monitor cellular growth when cell mass is the major product.

The products of cell metabolism can be used to monitor and quantify cellular growth. Certain products produced under anaerobic conditions, such as ethanol and lactic acid, can be related nearly stoichiometrically to microbial growth. Products must be either growth associated (ethanol) or mixed growth associated (lactic acid) to be correlated with microbial growth. For aerobic fermentations, CO_2 is a common product and can be related to microbial growth. In some cases, changes in the pH or acid–base addition to control pH can be used to monitor nutrient uptake and microbial growth. For example, the utilization of ammonium results in the release of hydrogen ions (H^+) and therefore a drop in pH. The amount of base added to neutralize the H^+ released is proportional to ammonium uptake and growth. Similarly, when nitrate is used as the nitrogen source, hydrogen ions are removed from the medium, resulting in an increase in pH. In this case, the amount of acid added is proportional to nitrate uptake and therefore to microbial growth.

In some fermentation processes, as a result of mycelial growth or extracellular polysaccharide formation, the viscosity of the fermentation broth increases during the course of fermentation. If the substrate is a biodegradable polymer, such as starch or cellulose, then the viscosity of the broth decreases with time as biohydrolysis continues. Changes in the viscosity of the fermentation broth can be correlated with the extent of microbial growth. Although polymeric broths are usually non-Newtonian, the apparent viscosity measured at a fixed rate can be used to estimate cell or product concentration.

6.2.2. Growth Patterns and Kinetics in Batch Culture

When a liquid nutrient medium is inoculated with a seed culture, the organisms selectively take up dissolved nutrients from the medium and convert them into biomass. A typical batch growth curve includes the following phases: (1) lag phase, (2) logarithmic or exponential growth phase, (3) deceleration phase, (4) stationary phase, and (5) death phase. Figure 6.3 describes a batch growth cycle.

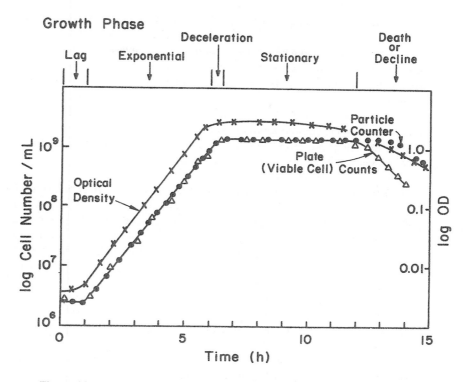

Figure 6.3. Typical growth curve for a bacterial population. Note that the phase of growth (shown here for cell number) depends on the parameter used to monitor growth.

The *lag phase* occurs immediately after inoculation and is a period of adaptation of cells to a new environment. Microorganisms reorganize their molecular constituents when they are transferred to a new medium. Depending on the composition of nutrients, new enzymes are synthesized, the synthesis of some other enzymes is repressed, and the internal machinery of cells is adapted to the new environmental conditions. These changes reflect the intracellular mechanisms for the regulation of the metabolic processes discussed in Chapter 4. During this phase, cell mass may increase a little, without an increase in cell number density. When the inoculum is small and has a low fraction of cells that are viable, there may be a pseudolag phase, which is a result, not of adaptation, but of small inoculum size or poor condition of the inoculum.

Low concentration of some nutrients and growth factors may also cause a long lag phase. For example, the lag phase of *Enterobacter aerogenes* (formerly *Aerobacter aerogenes*) grown in glucose and phosphate buffer medium increases as the concentration of Mg^{2+}, which is an activator of the enzyme phosphatase, is decreased. As another example, even heterotrophic cells require CO_2 fixation (to supplement intermediates removed from key energy-producing metabolic cycles during rapid biosynthesis), and excessive sparging can remove metabolically generated CO_2 too rapidly for cellular restructuring to be accomplished efficiently, particularly with a small inoculum.

The age of the inoculum culture has a strong effect on the length of lag phase. The age refers to how long a culture has been maintained in a batch culture. Usually, the lag period increases with the age of the inoculum. In some cases, there is an optimal inoculum age resulting in minimum lag period. To minimize the duration of the lag phase, cells should be adapted to the growth medium and conditions before inoculation, and cells should be young (or exponential phase cells) and active, and the inoculum size should be large (5% to 10% by volume). The nutrient medium may need to be optimized and certain growth factors included to minimize the lag phase. Figure 6.4 shows an example of variation of the lag phase with $MgSO_4$ concentration. Many commercial fermentation plants rely on batch culture; to obtain high productivity from a fixed plant size, the lag phase must be as short as possible.

Multiple lag phases may be observed when the medium contains more than one carbon source. This phenomenon, known as *diauxic growth,* is caused by a shift in metabolic pathways in the middle of a growth cycle (see Example 4.1). After one carbon source is exhausted, the cells adapt their metabolic activities to utilize the second carbon source. The first carbon source is more readily utilizable than the second, and the presence of more readily available carbon source represses the synthesis of the enzymes required for the metabolism of the second substrate.

The *exponential growth phase* is also known as the *logarithmic growth phase*. In this phase, the cells have adjusted to their new environment. After this adaptation period, cells can multiply rapidly, and cell mass and cell number density increase exponentially with time. This is a period of *balanced growth,* in which all components of a cell grow at the same rate. That is, the average composition of a single cell remains approximately constant during this phase of growth. During balanced growth, the net specific growth rate

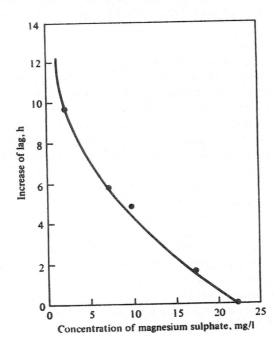

Figure 6.4. Influence of Mg^{2+} concentration on the lag phase in *E. aerogenes* culture. (With permission, from A. C. R. Dean and C. Hinshelwood, *Growth, Function, and Regulation in Bacterial Cells.* Oxford Press, London, 1966, p. 55.)

How Cells Grow   Chap. 6

determined from either cell number or cell mass would be the same. Since the nutrient concentrations are large in this phase, the growth rate is independent of nutrient concentration. The exponential growth rate is first order:

$$\frac{dX}{dt} = \mu_{net} X, \quad X = X_0 \quad \text{at} \quad t = 0 \tag{6.5}$$

Integration of eq. 6.5 yields

$$\ln \frac{X}{X_0} = \mu_{net} t, \quad \text{or} \quad X = X_0 e^{\mu_{net} t} \tag{6.6}$$

where X and X_0 are cell concentrations at time t and $t = 0$.

The time required to double the microbial mass is given by eq. 6.6. The exponential growth is characterized by a straight line on a semilogarithm plot of $\ln X$ versus time:

$$\tau_d = \frac{\ln 2}{\mu_{net}} = \frac{0.693}{\mu_{net}} \tag{6.7}$$

where τ_d is the doubling time of cell mass.

Similarly, we can calculate a doubling time based on cell numbers and the net specific rate of replication. Thus

$$\tau'_d = \frac{\ln 2}{\mu_R} \tag{6.8}$$

where τ'_d is the doubling time based on the replication rate. During balanced growth τ_d will equal τ'_d, since the average cell composition and size will not change with time.

The *deceleration growth phase* follows the exponential phase. In this phase, growth decelerates due to either depletion of one or more essential nutrients or the accumulation of toxic by-products of growth. For a typical bacterial culture, these changes occur over a very short period of time. The rapidly changing environment results in *unbalanced growth*. During unbalanced growth, cell composition and size will change and τ_d and τ'_d will *not* be equal. In the exponential phase, the cellular metabolic control system is set to achieve maximum rates of reproduction. In the deceleration phase, the stresses induced by nutrient depletion or waste accumulation cause a restructuring of the cell to increase the prospects of cellular survival in a hostile environment. These observable changes are the result of the molecular mechanisms of repression and induction that we discussed in Chapter 4. Because of the rapidity of these changes, cell physiology under conditions of nutrient limitation is more easily studied in continuous culture, as discussed later in this chapter.

The *stationary phase* starts at the end of the deceleration phase, when the net growth rate is zero (no cell division) or when the growth rate is equal to the death rate. Even though the net growth rate is zero during the stationary phase, cells are still metabolically active and produce secondary metabolites. *Primary metabolites* are growth-related products and *secondary metabolites* are nongrowth-related. In fact, the production of certain metabolites is enhanced during the stationary phase (e.g., antibiotics, some hormones) due to metabolite deregulation. During the course of stationary phase, one or more of the following phenomena may take place:

1. Total cell mass concentration may stay constant, but the number of viable cells may decrease.
2. Cell lysis may occur and viable cell mass may drop. A second growth phase may occur and cells may grow on lysis products of lysed cells (cryptic growth).
3. Cells may not be growing but may have active metabolism to produce secondary metabolites. Cellular regulation changes when concentrations of certain metabolites (carbon, nitrogen, phosphate) are low. Secondary metabolites are produced as a result of metabolite deregulation.

During the stationary phase, the cell catabolizes cellular reserves for new building blocks and for energy-producing monomers. This is called *endogenous metabolism*. The cell must always expend energy to maintain an energized membrane (i.e., proton-motive force) and transport of nutrients and for essential metabolic functions such as motility and repair of damage to cellular structures. This energy expenditure is called *maintenance energy*. The appropriate equation to describe the conversion of cell mass into maintenance energy or the loss of cell mass due to cell lysis during the stationary phase is

$$\frac{dX}{dt} = -k_d X \quad \text{or} \quad X = X_{so}e^{-k_d t} \tag{6.9}$$

where k_d is a first-order rate constant for endogeneous metabolism, and X_{so} is the cell mass concentration at the beginning of the stationary phase. Because S is zero, μ_g is zero in the stationary phase.

The reason for termination of growth may be either exhaustion of an essential nutrient or accumulation of toxic products. If an inhibitory product is produced and accumulates in the medium, the growth rate will slow down, depending on inhibitor production, and at a certain level of inhibitor concentration, growth will stop. Ethanol production by yeast is an example of a fermentation in which the product is inhibitory to growth. Dilution of toxified medium, addition of an unmetabolizable chemical compound complexing with the toxin, or simultaneous removal of the toxin would alleviate the adverse effects of the toxin and yield further growth.

The *death phase* (or decline phase) follows the stationary phase. However, some cell death may start during the stationary phase, and a clear demarcation between these two phases is not always possible. Often, dead cells lyse, and intracellular nutrients released into the medium are used by the living organisms during stationary phase. At the end of the stationary phase, because of either nutrient depletion or toxic product accumulation, the death phase begins.

The rate of death usually follows first-order kinetics:

$$\frac{dN}{dt} = -k_d' N \quad \text{or} \quad N = N_s e^{-k_d' t} \tag{6.10}$$

where N_s is the concentration of cells at the end of the stationary phase and k_d' is the first-order death-rate constant. A plot of ln N versus t yields a line of slope $-k_d'$. During the death phase, cells may or may not lyse, and the reestablishment of the culture may be possible in the early death phase if cells are transferred into a nutrient-rich medium. In both the death and stationary phases, it is important to recognize that there is a distribution of properties among individuals in a population. With a narrow distribution, cell death will occur nearly simultaneously; with a broad distribution, a subfraction of the population may survive for an extended period. It is this subfraction that would dominate the reestablishment of a culture from inoculum derived from stationary or death-phase cultures. Thus, using an old inoculum may select for variants of the original strain having altered metabolic capabilities.

To better describe growth kinetics, we define some stoichiometrically related parameters. Yield coefficients are defined based on the amount of consumption of another material. For example, the growth yield in a fermentation is

$$Y_{X/S} \equiv -\frac{\Delta X}{\Delta S} \tag{6.11}$$

At the end of the batch growth period, we have an *apparent growth yield* (or observed growth yield). Because culture conditions can alter patterns of substrate utilization, the apparent growth yield is not a true constant. For example, with a compound (such as glucose) that is both a carbon and energy source, substrate may be consumed as:

$$\Delta S = \Delta S_{\substack{\text{assimilation} \\ \text{into biomass}}} + \Delta S_{\substack{\text{assimilated} \\ \text{into an} \\ \text{extracellular} \\ \text{product}}} + \Delta S_{\text{growth energy}} + \Delta S_{\substack{\text{maintenance} \\ \text{energy}}} \tag{6.12}$$

In the section on continuous culture, we will differentiate between the true growth yield (which is constant) and the apparent yield. Yield coefficients based on other substrates or product formation may be defined; for example,

$$Y_{X/O_2} = -\frac{\Delta X}{\Delta O_2} \tag{6.13}$$

$$Y_{P/S} = -\frac{\Delta P}{\Delta S} \tag{6.14}$$

For organisms growing aerobically on glucose, $Y_{X/S}$ is typically 0.4 to 0.6 g/g for most yeast and bacteria, while Y_{X/O_2} is 0.9 to 1.4 g/g. Anaerobic growth is less efficient, and the yield coefficient is reduced substantially (see Fig. 6.5). With substrates that are more or less reduced than glucose, the value of the apparent yield coefficient will change. For methane, $Y_{X/S}$ would assume values of 0.6 to 1.0 g/g, with the corresponding Y_{X/O_2} decreasing to about 0.2 g/g. In most cases the yield of biomass on a carbon-energy source is 1.0 ± 0.4 g biomass per g of carbon consumed. Table 6.1 lists some examples of $Y_{X/S}$ and Y_{X/O_2} for a variety of substrates and organisms.

A *maintenance coefficient* is used to describe the specific rate of substrate uptake for cellular maintenance, or

$$m \equiv -\frac{[dS/dt]_m}{X} \tag{6.15}$$

However, during the stationary phase where little external substrate is available, endogeneous metabolism of biomass components is used for maintenance energy.

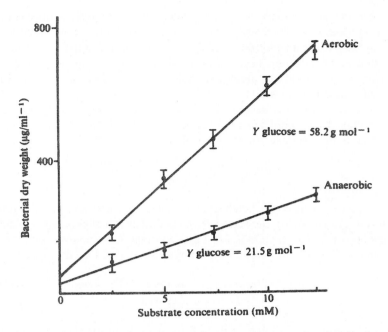

Figure 6.5. Aerobic and anaerobic growth yields of *Streptococcus faecalis* with glucose as substrate. (With permission, from B. Atkinson and F. Mavituna, *Biochemical Engineering and Biotechnology Handbook*, Macmillan, Inc., New York, 1983.)

Cellular maintenance represents energy expenditures to repair damaged cellular components, to transfer some nutrients and products in and out of cells, for motility, and to adjust the osmolarity of the cells' interior volume. Microbial growth, product formation, and substrate utilization rates are usually expressed in the form of specific rates (e.g., normalized with respect to X), since bioreactions are autocatalytic. The specific rates are used to compare the effectiveness of various fermentation schemes and biocatalysts.

Microbial products can be classified in three major categories (see Fig. 6.6):

1. Growth-associated products are produced simultaneously with microbial growth. The specific rate of product formation is proportional to the specific rate of growth, μ_g. Note that μ_g differs from μ_{net}, the net specific growth rate, when endogeneous metabolism is nonzero.

$$q_P = \frac{1}{X}\frac{dP}{dt} = Y_{P/X}\mu_g \qquad (6.16)$$

The production of a constitutive enzyme is an example of a growth-associated product.

2. Nongrowth-associated product formation takes place during the stationary phase when the growth rate is zero. The specific rate of product formation is constant.

TABLE 6.1 Summary of Yield Factors for Aerobic Growth of Different Microorganisms on Various Carbon Sources

Organism	Substrate	$Y_{X/S}$ g/g	$Y_{X/S}$ g/mol	$Y_{X/S}$ g/g-C	Y_{X/O_2}[a] g/g
Enterobacter aerogenes	Maltose	0.46	149.2	1.03	1.50
	Mannitol	0.52	95.2	1.32	1.18
	Fructose	0.42	76.1	1.05	1.46
	Glucose	0.40	72.7	1.01	1.11
Candida utilis	Glucose	0.51	91.8	1.28	1.32
Penicillium chrysogenum	Glucose	0.43	77.4	1.08	1.35
Pseudomonas fluorescens	Glucose	0.38	68.4	0.95	0.85
Rhodopseudomonas spheroides	Glucose	0.45	81.0	1.12	1.46
Saccharomyces cerevisiae	Glucose	0.50	90.0	1.25	0.97
Enterobacter aerogenes	Ribose	0.35	53.2	0.88	0.98
	Succinate	0.25	29.7	0.62	0.62
	Glycerol	0.45	41.8	1.16	0.97
	Lactate	0.18	16.6	0.46	0.37
	Pyruvate	0.20	17.9	0.49	0.48
	Acetate	0.18	10.5	0.43	0.31
Candida utilis	Acetate	0.36	21.0	0.90	0.70
Pseudomonas fluorescens	Acetate	0.28	16.8	0.70	0.46
Candida utilis	Ethanol	0.68	31.2	1.30	0.61
Pseudomonas fluorescens	Ethanol	0.49	22.5	0.93	0.42
Klebsiella sp.	Methanol	0.38	12.2	1.01	0.56
Methylomonas sp.	Methanol	0.48	15.4	1.28	0.53
Pseudomonas sp.	Methanol	0.41	13.1	1.09	0.44
Methylococcus sp.	Methane	1.01	16.2	1.34	0.29
Pseudomonas sp.	Methane	0.80	12.8	1.06	0.20
Pseudomonas sp.	Methane	0.60	9.6	0.80	0.19
Pseudomonas methanica	Methane	0.56	9.0	0.75	0.17

[a] Y_{X/O_2} is the yield factor relating grams of cells formed per gram of O_2 consumed.

With permission, from S. Nagai in *Advances in Biochemical Engineering*, vol. 11, T. K. Ghose, A. Fiechter, and N. Blakebrough, eds., Springer-Verlag, New York, p. 53, 1979.

$$q_P = \beta = \text{constant} \tag{6.17}$$

Many secondary metabolites, such as antibiotics (for example, penicillin), are non-growth-associated products.

3. Mixed-growth-associated product formation takes place during the slow growth and stationary phases. In this case, the specific rate of product formation is given by the following equation:

$$q_P = \alpha\mu_g + \beta \tag{6.18}$$

Lactic acid fermentation, xanthan gum, and some secondary metabolites from cell culture are examples of mixed-growth-associated products. Equation 6.18 is a

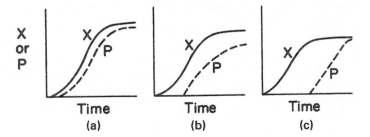

Figure 6.6. Kinetic patterns of growth and product formation in batch fermentations: (a) growth-associated product formation, (b) mixed-growth-associated product formation, and (c) nongrowth-associated product formation.

Luedeking–Piret equation. If $\alpha = 0$, the product is only non-growth associated, and if $\beta = 0$, the product would be only growth associated, and consequently α would then be equal to $Y_{P/X}$.

Some of the concepts concerning growth rate and yield are illustrated in Example 6.1.

Example 6.1.

A strain of mold was grown in a batch culture on glucose and the following data were obtained.

Time (h)	Cell concentration (g/l)	Glucose concentration (g/l)
0	1.25	100
9	2.45	97
16	5.1	90.4
23	10.5	76.9
30	22	48.1
34	33	20.6
36	37.5	9.38
40	41	0.63

    a. Calculate the maximum net specific growth rate.
    b. Calculate the apparent growth yield.
    c. What maximum cell concentration could one expect if 150 g of glucose were used with the same size inoculum?
Solution   A plot of $\ln X$ versus t yields a slope of 0.1 h.

a) $\mu_{\text{net}} = \dfrac{\ln X_2 - \ln X_1}{t_2 - t_1} = \dfrac{\ln 37.5 - \ln 5.1}{36 - 16} \cong 0.1\ \text{h}^{-1}$

b) $Y = -\dfrac{\Delta X}{\Delta S} = -\dfrac{41 - 1.25}{0.625 - 100} \cong 0.4$ g cells/g substrate

c) $X_{\text{max}} = X_0 + YS_0 = 1.25 + 0.4(150) = 60.25$ g cells/l

6.2.3. How Environmental Conditions Affect Growth Kinetics

The patterns of microbial growth and product formation we have just discussed are influenced by environmental conditions such as temperature, pH, and dissolved-oxygen concentration.

Temperature is an important factor affecting the performance of cells. According to their temperature optima, organisms can be classified in three groups: (1) psychrophiles ($T_{opt} < 20°C$), (2) mesophiles (T_{opt} = from 20° to 50°C), and (3) thermophiles ($T_{opt} > 50°$ C). As the temperature is increased toward optimal growth temperature, the growth rate approximately doubles for every 10°C increase in temperature. Above the optimal temperature range, the growth rate decreases and thermal death may occur. The net specific replication rate can be expressed by the following equation for temperature above optimal level:

$$\frac{dN}{dt} = (\mu'_R - k'_d)N \tag{6.19}$$

At high temperatures, the thermal death rate exceeds the growth rate, which causes a net decrease in the concentration of viable cells.

Both μ'_R and k'_d vary with temperature according to the Arrhenius equation:

$$\mu'_R = Ae^{-E_a/RT}, \quad k'_d = A'e^{-E_d/RT} \tag{6.20}$$

where E_a and E_d are activation energies for growth and thermal death. The activation energy for growth is typically 10 to 20 kcal/mol, and for thermal death 60 to 80 kcal/mol. That is, thermal death is more sensitive to temperature changes than microbial growth.

Temperature also affects product formation. However, the temperature optimum for growth and product formation may be different. The yield coefficient is also affected by temperature. In some cases, such as single-cell protein production, temperature optimization to maximize the yield coefficient ($Y_{X/S}$) is critical. When temperature is increased above the optimum temperature, the maintenance requirements of cells increase. That is, the maintenance coefficient (see eq. 6.15) increases with increasing temperature with an activation energy of 15 to 20 kcal/mol, resulting in a decrease in the yield coefficient.

Temperature also may affect the rate-limiting step in a fermentation process. At high temperatures, the rate of bioreaction might become higher than the diffusion rate, and diffusion would then become the rate-limiting step (for example, in an immobilized cell system). The activation energy of molecular diffusion is about 6 kcal/mol. The activation energy for most bioreactions is more than 10 kcal/mol, so diffusional limitations must be carefully considered at high temperatures. Figure 6.7 depicts a typical variation of growth rate with temperature.

Hydrogen-ion concentration (pH) affects the activity of enzymes and therefore the microbial growth rate. The optimal pH for growth may be different from that for product formation. Generally, the acceptable pH range varies about the optimum by ±1 to 2 pH units. Different organisms have different pH optima: the pH optimum for many bacteria ranges from pH = 3 to 8; for yeast, pH = 3 to 6; for molds, pH = 3 to 7; for plant cells, pH = 5 to 6; and for animal cells, pH = 6.5 to 7.5. Many organisms have mechanisms to

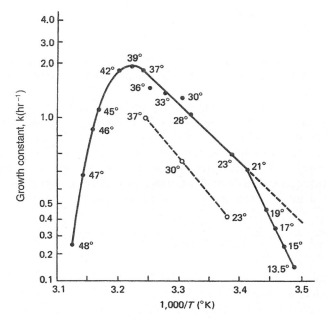

Figure 6.7. Arrhenius plot of growth rate of *E. coli* B/r. Individual data points are marked with corresponding degrees Celsius. *E. coli* B/r was grown in a rich complex medium (●) and a glucose-mineral salts medium (○). (With permission, after S. L. Herendeen, R. A. VanBogelen, and F. C. Neidhardt, "Levels of Major Protein of *Escherichia coli* during Growth at Different Temperatures," *J. Bacteriol. 139*:195, 1979, as drawn in R. Y. Stanier and others, *The Microbial World*, 5th ed., Pearson Education, Upper Saddle River, NJ, 1986, 207.)

maintain intracellular pH at a relatively constant level in the presence of fluctuations in environmental pH. When pH differs from the optimal value, the maintenance-energy requirements increase. One consequence of different pH optima is that the pH of the medium can be used to select one organism over another.

In most fermentations, pH can vary substantially. Often the nature of the nitrogen source can be important. If ammonium is the sole nitrogen source, hydrogen ions are released into the medium as a result of the microbial utilization of ammonia, resulting in a decrease in pH. If nitrate is the sole nitrogen source, hydrogen ions are removed from the medium to reduce nitrate to ammonia, resulting in an increase in pH. Also, pH can change because of the production of organic acids, the utilization of acids (particularly amino acids), or the production of bases. The evolution or supply of CO_2 can alter pH greatly in some systems (e.g., seawater or animal cell culture). Thus, pH control by means of a buffer or an active pH control system is important. Variation of specific growth rate with pH is depicted in Fig. 6.8, indicating a pH optimum.

Dissolved oxygen (DO) is an important substrate in aerobic fermentations and may be a limiting substrate, since oxygen gas is sparingly soluble in water. At high cell concentrations, the rate of oxygen consumption may exceed the rate of oxygen supply, leading to oxygen limitations. When oxygen is the rate-limiting factor, specific growth rate varies with dissolved-oxygen concentration according to saturation kinetics; below a critical concentration, growth or respiration approaches a first-order rate dependence on the dissolved-oxygen concentration.

Above a *critical oxygen concentration*, the growth rate becomes independent of the dissolved-oxygen concentration. Figure 6.9 depicts the variation of specific growth rate with dissolved-oxygen concentration. Oxygen is a growth-rate-limiting factor when the

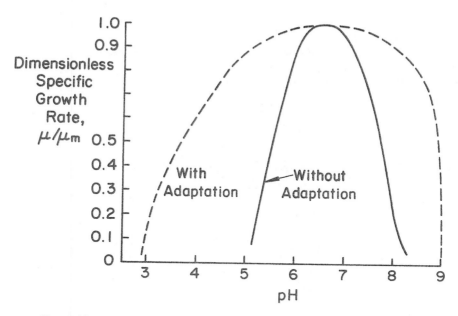

Figure 6.8. Typical variation of specific growth rate with pH. The units are arbitrary. With some microbial cultures, it is possible to adapt cultures to a wider range of pH values if pH changes are made in small increments from culture transfer to transfer.

DO level is below the critical DO concentration. In this case, another medium component (e.g., glucose, ammonium) becomes growth-extent limiting. For example, with *Azotobacter vinelandii* at a DO = 0.05 mg/l, the growth rate is about 50% of maximum even if a large amount of glucose is present. However, the maximum amount of cells formed is not determined by the DO, as oxygen is continually resupplied. If glucose were totally consumed, growth would cease even if DO = 0.05 mg/l. Thus, the extent of growth (mass of cells formed) would depend on glucose, while the growth rate for most of the culture period would depend on the value of DO.

The critical oxygen concentration is about 5% to 10% of the saturated DO concentration for bacteria and yeast and about 10% to 50% of the saturated DO concentration for mold cultures, depending on the pellet size of molds. Saturated DO concentration in water at 25°C and 1 atm pressure is about 7 ppm. The presence of dissolved salts and organics can alter the saturation value, while increasingly high temperatures decrease the saturation value.

Oxygen is usually introduced to fermentation broth by sparging air through the broth. Oxygen transfer from gas bubbles to cells is usually limited by oxygen transfer through the liquid film surrounding the gas bubbles. The rate of oxygen transfer from the gas to liquid phase is given by

$$N_{O_2} = k_L a(C^* - C_L) = \text{OTR}$$

(6.21)

where k_L is the oxygen transfer coefficient (cm/h), a is the gas–liquid interfacial area (cm^2/cm^3), $k_L a$ is the volumetric oxygen transfer coefficient (h^{-1}), C^* is saturated DO

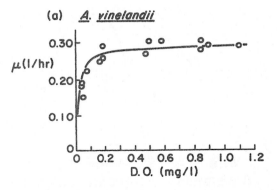

(a) *A. vinelandii*

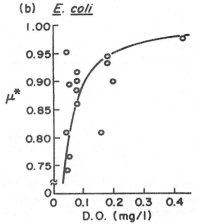

(b) *E. coli*

Figure 6.9. Growth-rate dependence on DO for (a) *Azotobacter vinelandii*, a strictly aerobic organism, and (b) *E. coli*, which is faculative. *E. coli* grows anaerobically at a rate of about 70% of its aerobic growth in minimal medium.

$$\mu^* = \frac{\mu - \mu_m^{anaerobic}}{\mu_m^{aerobic} - \mu_m^{anaerobic}}$$

(With permission, from J. Chen, A. L. Tannahill, and M. L. Shuler, *Biotechnol. Bioeng.* 27: 151, 1985, and John Wiley & Sons, Inc., New York.)

concentration (mg/l), C_L is the actual DO concentration in the broth (mg/l), and the N_{O_2} is the rate of oxygen transfer (mg O_2/l·h). Also, the term *oxygen transfer rate* (OTR) is used.

The rate of oxygen uptake is denoted as *OUR* (*oxygen uptake rate*) and

$$OUR = q_{O_2}X = \frac{\mu_g X}{Y_{X/O_2}} \tag{6.22}$$

where q_{O_2} is the specific rate of oxygen consumption (mg O_2/g dw cells·h), Y_{X/O_2} is the yield coefficient on oxygen (g dw cells/g O_2), and X is cell concentration (g dw cells/l).

When oxygen transfer is the rate-limiting step, the rate of oxygen consumption is equal to the rate of oxygen transfer. If the maintenance requirement of O_2 is negligible compared to growth, then

$$\frac{\mu_g X}{Y_{X/O_2}} = k_L a(C^* - C_L) \tag{6.23}$$

or

$$\frac{dX}{dt} = Y_{X/O_2} k_L a(C^* - C_L) \tag{6.24}$$

Growth rate varies nearly linearly with the oxygen transfer rate under oxygen-transfer limitations. Among the various methods used to overcome DO limitations are the use of oxygen-enriched air or pure oxygen and operation under high atmospheric pressure (2 to 3 atm). Oxygen transfer has a big impact on reactor design (see Chapter 10).

The redox potential is an important parameter that affects the rate and extent of many oxidative–reductive reactions. In a fermentation medium, the redox potential is a complex function of DO, pH, and other ion concentrations, such as reducing and oxidizing agents. The electrochemical potential of a fermentation medium can be expressed by the following equation:

$$E_h = E_0' + \frac{2.3RT}{4F} \log P_{O_2} + 2.3 \frac{RT}{F} \log \ (H^+) \tag{6.25}$$

where the electrochemical potential is measured in millivolts by a pH/voltmeter and P_{O_2} is in atmospheres.

The redox potential of a fermentation media can be reduced by passing nitrogen gas or by the addition of reducing agents such as cysteine HCl or Na_2S. Oxygen gas can be passed or some oxidizing agents can be added to the fermentation media to increase the redox potential.

Dissolved carbon dioxide (DCO_2) concentration may have a profound effect on performance of organisms. Very high DCO_2 concentrations may be toxic to some cells. On the other hand, cells require a certain DCO_2 level for proper metabolic functions. The dissolved carbon dioxide concentration can be controlled by changing the CO_2 content of the air supply and the agitation speed.

The ionic strength of the fermentation media affects the transport of certain nutrients in and out of cells, the metabolic functions of cells, and the solubility of certain nutrients, such as dissolved oxygen. The ionic strength is given by the following equation:

$$I = \frac{1}{2} \Sigma C_i Z_i^2 \tag{6.26}$$

where C is the concentration of an ion, Z_i is its charge, and I is the ionic strength of the medium.

High substrate concentrations that are significantly above stoichiometric requirements are inhibitory to cellular functions. Inhibitory levels of substrates vary depending on the type of cells and substrate. Glucose may be inhibitory at concentrations above 200 g/l (e.g., ethanol fermentation by yeast), probably due to a reduction in water activity. Certain salts such as NaCl may be inhibitory at concentrations above 40 g/l due to high osmotic pressure. Some refractory compounds, such as phenol, toluene, and methanol, are inhibitory at much lower concentrations (e.g., 1 g/l). Typical maximum noninhibitory concentrations of some nutrients are glucose, 100 g/l; ethanol, 50 g/l for yeast, much less for most organisms; ammonium, 5 g/l; phosphate, 10 g/l; and nitrate, 5 g/l. Substrate inhibition can be overcome by intermittent addition of the substrate to the medium.

6.2.4. Heat Generation by Microbial Growth

About 40% to 50% of the energy stored in a carbon and energy source is converted to biological energy (ATP) during aerobic metabolism, and the rest of the energy is released as heat. For actively growing cells, the maintenance requirement is low, and heat evolution is directly related to growth.

The heat generated during microbial growth can be calculated using the heat of combustion of the substrate and of cellular material. A schematic of an enthalpy balance for microbial utilization of substrate is presented in Fig. 6.10. The heat of combustion of the substrate is equal to the sum of the metabolic heat and the heat of combustion of the cellular material.

$$\frac{\Delta H_s}{Y_{X/S}} = \Delta H_c + \frac{1}{Y_H} \qquad (6.27a)$$

where ΔH_s is the heat of combustion of the substrate (kJ/g substrate), $Y_{X/S}$ is the substrate yield coefficient (g cell/g substrate), ΔH_c is the heat of combustion of cells (kJ/g cells), and $1/Y_H$ is the metabolic heat evolved per gram of cell mass produced (kJ/g cells).

Equation 6.27a can be rearranged to yield

$$Y_H = \frac{Y_{X/S}}{\Delta H_s - Y_{X/S}\,\Delta H_c} \qquad (6.27b)$$

ΔH_s and ΔH_c can be determined from the combustion of substrate and cells. Typical ΔH_c values for bacterial cells are 20 to 25 kJ/g cells. Typical values of Y_H are glucose, 0.42 g/kcal; malate, 0.30 g/kcal; acetate, 0.21 g/kcal; ethanol, 0.18 g/kcal; methanol, 0.12 g/kcal; and methane, 0.061 g/kcal. Clearly, the degree of oxidation of the substrate has a strong effect on the amount of heat released.

The total rate of heat evolution in a batch fermentation is

$$Q_{GR} = V_L \mu_{net} X \frac{1}{Y_H} \qquad (6.28)$$

where V_L is the liquid volume (l) and X is the cell concentration (g/l).

In aerobic fermentations, the rate of metabolic heat evolution can roughly be correlated to the rate of oxygen uptake, since oxygen is the final electron acceptor.

$$Q_{GR} \cong 0.12 Q_{O_2} \qquad (6.29)$$

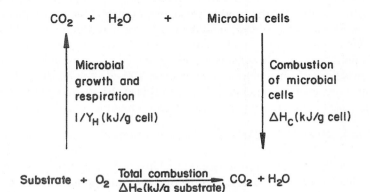

Figure 6.10. Heat balance on microbial utilization of substrate.

Here Q_{GR} is in units of kcal/h, while Q_{O_2} is in millimoles of O_2/h.

Metabolic heat released during fermentation can be removed by circulating cooling water through a cooling coil or cooling jacket in the fermenter. Often, temperature control (adequate heat removal) is an important limitation on reactor design (see Chapter 10). The ability to estimate heat-removal requirements is essential to proper reactor design.

6.3. QUANTIFYING GROWTH KINETICS

6.3.1. Introduction

In the previous section we described some key concepts in the growth of cultures. Clearly, we can think of the growth dynamics in terms of kinetic descriptions. It is essential to recall that cellular composition and biosynthetic capabilities change in response to new growth conditions (*unbalanced growth*), although a constant cellular composition and balanced growth can predominate in the exponential growth phase. If the decelerating growth phase is due to substrate depletion rather than inhibition by toxins, the growth rate decreases in relation to decreasing substrate concentrations. In the stationary and death phases, the distribution of properties among individuals is important (e.g., cryptic death). Although these kinetic ideas are evident in batch culture, they are equally evident and important in other modes of culture (e.g., continuous culture).

Clearly, the complete description of the growth kinetics of a culture would involve recognition of the *structured* nature of each cell and the *segregation* of the culture into individual units (cells) that may differ from each other. Models can have these same attributes. A chemically structured model divides the cell mass into components. If the ratio of these components can change in response to perturbations in the extracellular environment, then the model is behaving analogously to a cell changing its composition in response to environmental changes. Consider in Chapter 4 our discussion of cellular regulation, particularly the induction of whole pathways. Any of these metabolic responses results in changes in intracellular structure. Furthermore, if a model of a culture is constructed from discrete units, it begins to mimic the segregation observed in real cultures. Models may be structured and segregated, structured and nonsegregated, unstructured and segregated, and unstructured and nonsegregated. Models containing both structure and segregation are the most realistic, but they are also computationally complex.

The degree of realism and complexity required in a model depends on what is being described; the modeler should always choose the simplest model that can adequately describe the desired system. An unstructured model assumes fixed cell composition, which is equivalent to assuming *balanced growth*. The balanced-growth assumption is valid primarily in single-stage, steady-state continuous culture and the exponential phase of batch culture; it fails during any transient condition. How fast the cell responds to perturbations in its environment and how fast these perturbations occur determine whether pseudobalanced growth can be assumed. If cell response is fast compared to external changes and if the magnitude of these changes is not too large (e.g., a 10% or 20% variation from initial conditions), then the use of unstructured models can be justified, since the deviation from balanced growth may be small. Culture response to large or rapid perturbations cannot be described satisfactorily by unstructured models.

For many systems, segregation is not a critical component of culture response, so nonsegregated models will be satisfactory under many circumstances. An important exception is the prediction of the growth responses of plasmid-containing cultures (see Chapter 14).

Because of the introductory nature of this book, we will concentrate our discussion on unstructured and nonsegregated models. The reader must be aware of the limitations on these models. Nonetheless, such models are simple and applicable to some situations of practical interest.

6.3.2. Using Unstructured Nonsegregated Models to Predict Specific Growth Rate

6.3.2.1. Substrate-limited growth.
As shown in Fig. 6.11, the relationship of specific growth rate to substrate concentration often assumes the form of saturation kinetics. Here we assume that a single chemical species, S, is growth-rate limiting (i.e., an increase in S influences growth rate, while changes in other nutrient concentrations have no effect). These kinetics are similar to the Langmuir–Hinshelwood (or Hougen–Watson) kinetics in traditional chemical kinetics or Michaelis–Menten kinetics for enzyme reactions. When applied to cellular systems, these kinetics can be described by the *Monod equation:*

$$\mu_g = \frac{\mu_m S}{K_s + S} \tag{6.30}$$

where μ_m is the maximum specific growth rate when $S \gg K_s$. If endogeneous metabolism is unimportant, then $\mu_{net} = \mu_g$. The constant K_s is known as the *saturation constant* or *half-velocity constant* and is equal to the concentration of the rate-limiting substrate when the specific rate of growth is equal to one-half of the maximum. That is, $K_s = S$ when $\mu_g = \frac{1}{2}\mu_{max}$. In general, $\mu_g = \mu_m$ for $S \gg K_s$ and $\mu_g = (\mu_m/K_s)S$ for $S \ll K_s$. The Monod equation is semiempirical; it derives from the premise that a single enzyme system with Michaelis–Menten kinetics is responsible for uptake of S, and the amount of that enzyme or its catalytic activity is sufficiently low to be growth-rate limiting.

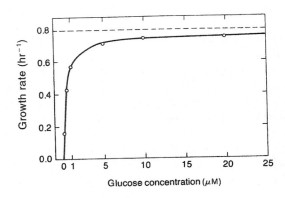

Figure 6.11. Effect of nutrient concentration on the specific growth rate of *E. coli.* (With permission, from R. Y. Stanier, M. Doudoroff, and E. A. Adelberg, *The Microbial World*, 5th ed., Pearson Education, Upper Saddle River, NJ, 1986, p. 192.)

This simple premise is rarely, if ever, true; however, the Monod equation empirically fits a wide range of data satisfactorily and is the most commonly applied unstructured, nonsegregated model of microbial growth.

The Monod equation describes substrate-limited growth only when growth is slow and population density is low. Under these circumstances, environmental conditions can be related simply to S. If the consumption of a carbon–energy substrate is rapid, then the release of toxic waste products is more likely (due to energy-spilling reactions). At high population levels, the buildup of toxic metabolic by-products becomes more important. The following rate expressions have been proposed for rapidly growing dense cultures:

$$\mu_g = \frac{\mu_m S}{K_{s0} S_0 + S} \tag{6.31}$$

or

$$\mu_g = \frac{\mu_m S}{K_{s1} + K_{s0} S_0 + S} \tag{6.32}$$

where S_0 is the initial concentration of the substrate and K_{s0} is dimensionless.

Other equations have been proposed to describe the substrate-limited growth phase. Depending on the shape of μ–S curve, one of these equations may be more plausible than the others. The following equations are alternatives to the Monod equation:

Blackman equation: $\mu_g = \mu_m,$ iff $S \geq 2K_s$

$\mu_g = \dfrac{\mu_m}{2K_s} S,$ iff $S < 2K_s$ (6.33)

Tessier equation: $\mu_g = \mu_m (1 - e^{-KS})$ (6.34)

Moser equation: $\mu_g = \dfrac{\mu_m S^n}{K_s + S^n} = \mu_m (1 + K_s S^{-n})^{-1}$ (6.35)

Contois equation: $\mu_g = \dfrac{\mu_m S}{K_{sx} X + S}$ (6.36)

Although the Blackman equation often fits the data better than the Monod equation, the discontinuity in the Blackman equation is troublesome in many applications. The Tessier equation has two constants (μ_m, K), and the Moser equation has three constants (μ_m, K_s, n). The Moser equation is the most general form of these equations, and it is equivalent to the Monod equation when $n = 1$. The Contois equation has a saturation constant proportional to cell concentration that describes substrate-limited growth at high cell densities. According to this equation, the specific growth rate decreases with decreasing substrate concentrations and eventually becomes inversely proportional to the cell concentration in the medium (i.e., $\mu_g \propto X^{-1}$).

These equations can be described by a single differential equation as

$$\frac{d\upsilon}{dS} = K\upsilon^a (1-\upsilon)^b \tag{6.37}$$

TABLE 6.2 Constants of the Generalized Differential Specific Growth Rate Equation 6.34 for Different Models

	a	b	K
Monod	0	2	$1/K_s$
Tessier	0	1	$1/K$
Moser	$1 - 1/n$	$1 + 1/n$	$n/K_s^{1/n}$
Contois	0	2	$1/K_{sx}$

where $\upsilon = \mu_g/\mu_m$, S is the rate-limiting substrate concentration, and K, a, and b are constants. The values of these constants are different for each equation and are listed in Table 6.2.

The correct rate form to use in the case where more than one substrate is potentially growth-rate limiting is an unresolved question. However, under most circumstances the noninteractive approach works best:

$$\mu_g = \mu_g(S_1) \quad \text{or} \quad \mu_g(S_2) \quad \text{or} \quad \cdots \quad \mu_g(S_n) \tag{6.38}$$

where the lowest value of $\mu_g(S_i)$ is used.

6.3.2.2. Models with growth inhibitors.

At high concentrations of substrate or product and in the presence of inhibitory substances in the medium, growth becomes inhibited, and growth rate depends on inhibitor concentration. The inhibition pattern of microbial growth is analogous to enzyme inhibition. If a single-substrate enzyme-catalyzed reaction is the rate-limiting step in microbial growth, then kinetic constants in the rate expression are biologically meaningful. Often, the underlying mechanism is complicated, and kinetic constants do not have biological meanings and are obtained from experimental data by curve fitting.

1. *Substrate inhibition:* At high substrate concentrations, microbial growth rate is inhibited by the substrate. As in enzyme kinetics, substrate inhibition of growth may be competitive or noncompetitive. If a single-substrate enzyme-catalyzed reaction is the rate-limiting step in microbial growth, then inhibition of enzyme activity results in inhibition of microbial growth by the same pattern.

The major substrate-inhibition patterns and expressions are as follows:

Noncompetitive substrate inhibition:
$$\mu_g = \frac{\mu_m}{\left(1 + \dfrac{K_s}{S}\right)\left(1 + \dfrac{S}{K_I}\right)} \tag{6.39}$$

Or if $K_I \gg K_s$, then:
$$\mu_g = \frac{\mu_m S}{K_s + S + S^2/K_I} \tag{6.40}$$

For competitive substrate inhibition:
$$\mu_g = \frac{\mu_m S}{K_s\left(1 + \dfrac{S}{K_I}\right) + S} \qquad (6.41)$$

Note that eq. 6.41 differs from 6.39 and 6.40, and K_I in 6.40 and 6.41 differ. Substrate inhibition may be alleviated by slow, intermittent addition of the substrate to the growth medium.

2. *Product inhibition:* High concentrations of product can be inhibitory for microbial growth. Product inhibition may be competitive or noncompetitive, and in some cases when the underlying mechanism is not known, the inhibited growth rate is approximated to exponential or linear decay expressions.

Important examples of the product inhibition rate expression are as follows:

Competitive product inhibition:
$$\mu_g = \frac{\mu_m S}{K_s\left(1 + \dfrac{P}{K_p}\right) + S} \qquad (6.42)$$

Noncompetitive product inhibition:
$$\mu_g = \frac{\mu_m}{\left(1 + \dfrac{K_s}{S}\right)\left(1 + \dfrac{P}{K_p}\right)} \qquad (6.43)$$

Ethanol fermentation from glucose by yeasts is a good example of noncompetitive product inhibition, and ethanol is the inhibitor at concentrations above about 5%. Other rate expressions used for ethanol inhibition are

$$\mu_g = \frac{\mu_m}{\left(1 + \dfrac{K_s}{S}\right)}\left(1 - \frac{P}{P_m}\right)^n \qquad (6.44)$$

where P_m is the product concentration at which growth stops, or

$$\mu_g = \frac{\mu_m}{\left(1 + \dfrac{K_s}{S}\right)} e^{-P/K_p} \qquad (6.45)$$

where K_p is the product inhibition constant.

3. *Inhibition by toxic compounds:* The following rate expressions are used for competitive, noncompetitive, and uncompetitive inhibition of growth in analogy to enzyme inhibition.

Competitive inhibition:
$$\mu_g = \frac{\mu_m S}{K_s\left(1 + \dfrac{I}{K_I}\right) + S} \qquad (6.46)$$

Noncompetitive inhibition:
$$\mu_g = \frac{\mu_m}{\left(1 + \dfrac{K_s}{S}\right)\left(1 + \dfrac{I}{K_I}\right)} \qquad (6.47)$$

$$\text{Uncompetitive inhibition:} \quad \mu_g = \frac{\mu_m S}{\left(\dfrac{K_s}{\left(1+\dfrac{I}{K_I}\right)}+S\right)\left(1+\dfrac{I}{K_I}\right)} \tag{6.48}$$

In some cases, the presence of toxic compounds in the medium results in the inactivation of cells or death. The net specific rate expression in the presence of death has the following form:

$$\mu_g = \frac{\mu_m S}{K_s + S} - k_d' \tag{6.49}$$

where k_d' is the death-rate constant (h^{-1}).

6.3.2.3. The logistic equation.

When plotted on arithmetic paper, the batch growth curve assumes a sigmoidal shape (see Fig. 6.3). This shape can be predicted by combining the Monod equation (6.30) with the growth equation (6.2) and an equation for the yield of cell mass based on substrate consumption. Combining eqs. 6.30 and 6.2a and assuming no endogenous metabolism yields

$$\frac{dX}{dt} = \frac{\mu_m S}{K_s + S} X \tag{6.50}$$

The relationship between microbial growth yield and substrate consumption is

$$X - X_0 = Y_{X/S}(S_0 - S) \tag{6.51}$$

where X_0 and S_0 are initial values and $Y_{X/S}$ is the cell mass yield based on the limiting nutrient. Substituting for S in eq. 6.50 yields the following rate expression:

$$\frac{dX}{dt} = \frac{\mu_m (Y_{X/S}S_0 + X_0 - X)}{(K_S Y_{X/S} + Y_{X/S}S_0 + X_0 - X)} X \tag{6.52}$$

The integrated form of the rate expression in this phase is

$$\frac{(K_S Y_{X/S} + S_0 Y_{X/S} + X_0)}{(Y_{X/S}S_0 + X_0)} \ln\left(\frac{X}{X_0}\right) - \frac{K_S Y_{X/S}}{(Y_{X/S}S_0 + X_0)} \ln\{(Y_{X/S}S_0 + X_0 - X)/Y_{X/S}S_0\} = \mu_m t \tag{6.53}$$

This equation describes the sigmoidal-shaped batch growth curve, and the value of X asymptotically reaches to the value of $Y_{X/S}S_0 + X_0$.

Equation 6.52 requires a predetermined knowledge of the maximum cell mass in a particular environment. This maximum cell mass we will denote as X_∞; it is identical to the ecological concept of *carrying capacity*. Equation 6.53 is implicit in its dependence on S.

Logistic equations are a set of equations that characterize growth in terms of carrying capacity. The usual approach is based on a formulation in which the specific growth rate is related to the amount of unused carrying capacity:

$$\mu_g = k\left(1 - \frac{X}{X_\infty}\right)$$

(6.54)

Thus,

$$\frac{dX}{dt} = kX\left(1 - \frac{X}{X_\infty}\right)$$

(6.55)

The integration of eq. 6.55 with the boundary condition $X(0) = X_0$ yields the logistic curve.

$$X = \frac{X_0 e^{kt}}{1 - \frac{X_0}{X_\infty}(1 - e^{kt})}$$

(6.56)

Equation 6.56 is represented by the growth curve in Fig. 6.12.

Equations of the form of 6.56 can also be generated by assuming that a toxin generated as a by-product of growth limits X_∞ (the carrying capacity). Example 6.2 illustrates the use of the logistic approach.

Example 6.2.   Logistic Equation

Ethanol formation from glucose is accomplished in a batch culture of *Saccharomyces cerevisiae*, and the following data were obtained.

Time (h)	Glucose (S), g/L	Biomass (X), g/L	Ethanol (P), g/L
0	100	0.5	0.0
2	95	1.0	2.5
5	85	2.1	7.5
10	58	4.8	20.0
15	30	7.7	34.0
20	12	9.6	43.0
25	5	10.4	47.5
30	2	10.7	49.0

a. By fitting the biomass data to the logistic equation, determine the carrying-capacity coefficient k.

b. Determine yield coefficients $Y_{P/S}$ and $Y_{X/S}$.

Solution

a) Equation 6.55 can be rewritten as:

$$\frac{1}{X}\frac{dX}{dt} = k\left(1 - \frac{\overline{X}}{X_\infty}\right)$$

or

$$k = \frac{1}{\overline{X}}\frac{\Delta X}{\Delta t} \div \left(1 - \frac{\overline{X}}{X_\infty}\right)$$

where $\overline{X}$ is average biomass concentration during Δt, and X_∞ is about 10.8 g/L, since growth is almost complete at 30 h. Thus:

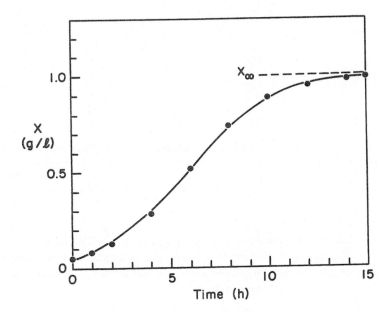

Figure 6.12. Logistic growth curve.

Δt (h)	$\bar{X}$(g/L)	$\dfrac{1}{\bar{X}}\Delta X / \Delta t$ (h^{-1})	$\left(1-\dfrac{\bar{X}}{X_\infty}\right)$	k (h^{-1})
2	0.75	0.333	0.931	0.36
3	1.55	0.236	0.856	0.28
5	3.45	0.156	0.681	0.23
5	6.25	0.093	0.416	0.22
5	8.65	0.044	0.200	0.22
5	10.00	0.016	0.074	0.22
5	10.55	0.0057	0.023	0.25

A value of $k = 0.24$ h^{-1} would describe most of the data, although it would slightly underestimate the initial growth rate. Another approach would be to take the log of the above equation to give:

$$\log \frac{1}{X}\frac{dX}{dt} = \log k + \log\left(1-\frac{\bar{X}_0}{X_\infty}\right)$$

and to fit the data to this equation and estimate k from the intercept. In this case k would be about 0.25 h^{-1}.

b) The yields are estimated directly from the data as:

$$Y_{P/S} = \frac{-\Delta P}{\Delta S} = \frac{-(49-0)}{(2-100)} = 0.5\frac{gP}{gS}$$

$$Y_{X/S} = \frac{-\Delta X}{\Delta S} = \frac{-(10.7-0.5)}{(2-100)} = 0.104\frac{gX}{gS}$$

The above estimate of $Y_{X/S}$ is only approximate, as maintenance effects and endogenous metabolism have been neglected.

6.3.2.4. Growth models for filamentous organisms.

Filamentous organisms such as molds often form microbial pellets at high cell densities in suspension culture. Cells growing inside pellets may be subject to diffusion limitations. The growth models of molds should include the simultaneous diffusion and consumption of nutrients within the pellet at large pellet sizes. This problem is the same one we face in modeling the behavior of bacteria or yeasts entrapped in spherical gel particles (see Chapter 9).

Alternatively, filamentous cells can grow on the surface of a moist solid. Such growth is usually a complicated process, involving not only growth kinetics but the diffusion of nutrients and toxic metabolic by-products. However, for an isolated colony growing on a rich medium, we can ignore some of these complications.

In the absence of mass-transfer limitations, it has been observed that the radius of a microbial pellet in a submerged culture or of a mold colony growing on an agar surface increases linearly with time.

$$\frac{dR}{dt} = k_P = \text{constant} \tag{6.57}$$

In terms of growth rate of a mold colony, eq. 6.57 can be expressed as

$$\frac{dM}{dt} = \rho 4\pi R^2 \frac{dR}{dt} = k_P 4\pi R^2 \rho \tag{6.58a}$$

or

$$\frac{dM}{dt} = \gamma M^{2/3} \tag{6.58b}$$

where $\gamma = k_P(36\pi\rho)^{1/3}$. Integration of eq. 6.58b yields

$$M = \left(M_0^{1/3} + \frac{\gamma t}{3} \right)^3 \approx \left(\frac{\gamma t}{3} \right)^3 \tag{6.59}$$

The initial biomass, M_0, is usually very small compared to M, and therefore M varies with t^3. This behavior has been supported by experimental data.

6.3.3. Models for Transient Behavior

In most practical applications of microbial cultures, the environmental or culture conditions can shift, dramatically leading to changes in cellular composition and biosynthetic capabilities. These cellular changes are not instantaneous but occur over an observable period of time. In this section we examine models that can describe or predict such time-dependent (or transient) changes.

6.3.3.1. Models with time delays.

The unstructured growth models we have described so far are limited to balanced or pseudobalanced growth conditions. These unstructured models can be improved for use in dynamic situations through the addition of time delays. The use of time delays incorporates structure implicitly. It is built on the premise that the dynamic response of a cell is dominated by an internal process with a time delay on the order of the response time under observation. Other internal processes are assumed to be too fast (essentially always at a pseudoequilibrium) or too slow to influence greatly the observed response. By using black-box techniques equivalent to the traditional approach to the control of chemical processes, it is possible to generate transfer functions that can represent the dynamic response of a culture. An example of the results from such an approach are given in Fig. 6.13. However, it should be recognized that such approaches are limited to cultures with similar growth histories and subjected to qualitatively similar perturbations.

6.3.3.2. Chemically structured models.

A much more general approach with much greater a priori predictive power is a model capturing the important kinetic interactions among cellular subcomponents. Initially, chemically structured models were based on two components, but at least three components appear necessary to give good results. More sophisticated models with 20 to 40 components are being used in many laboratories. A schematic of one such model is given in Fig. 6.14.

Writing such models requires that the modeler understand the physical system at a level of greater detail than that at which the model is written, so that the appropriate assumptions can be made. A detailed discussion of such models is appropriate for more advanced texts. However, two important guidelines in writing such models should be understood by even the beginning student. The first is that all reactions should be expressed in terms of *intrinsic* concentrations. An *intrinsic* concentration is the amount of a compound per unit cell mass or cell volume. *Extrinsic* concentrations, the amount of a compound per unit reactor volume, cannot be used in kinetic expressions. Although this may seem self-evident, all the early structured models were flawed by the use of *extrinsic*

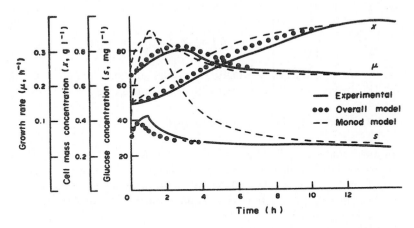

Figure 6.13. Comparison of predictions from a model derived from a system-analysis perspective, predictions from a Monod model, and experiment. The experimental system was a chemostat for a glucose-limited culture of *Saccharomyces cerevisiae* operating at a dilution rate of 0.20 h^{-1}. In this particular experiment, the system was perturbed with a stepwise increase in feed glucose concentration from 1.0 and 2.0 g l^{-1}. X is biomass concentration, μ is growth rate, and S is substrate concentration. (With permission, from T. B. Young III and H. R. Bungay, "Dynamic Analysis of a Microbial Process: A Systems Engineering Approach," *Biotechnol. Bioeng.* 15:377, 1973, and John Wiley & Sons, Inc., New York.)

How Cells Grow   Chap. 6

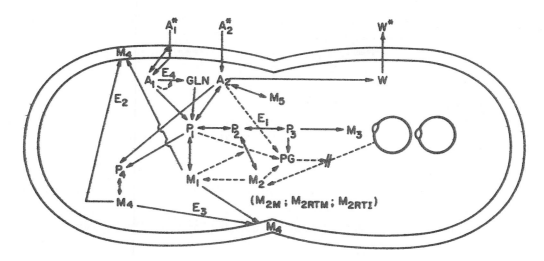

Figure 6.14. An idealized sketch of the model *E. coli* B/rA growing in a glucose–ammonium salts medium with glucose or ammonia as the limiting nutrient. At the time shown the cell has just completed a round of DNA replication and initiated cross-wall formation and a new round of DNA replication. Solid lines indicate the flow of material, while dashed lines indicate flow of information. The symbols are: A_1, ammonium ion; A_2, glucose (and associated compounds in the cell); W, waste products (CO_2, H_2O, and acetate) formed from energy metabolism during aerobic growth; P_1, amino acids; P_2, ribonucleotides; P_3, deoxyribonucleotides; P_4, cell envelope precursors; M_1, protein (both cytoplasmic and envelope); M_{2RTI}, immature "stable" RNA; M_{2RTM}, mature "stable" RNA (*r*-RNA and *t*-RNA—assume 85% *r*-RNA throughout); M_{2M}, messenger RNA; M_3, DNA; M_4, nonprotein part of cell envelope (assume 16.7% peptidoglycan, 47.6% lipid, and 35.7% polysaccharide); M_5, glycogen; PG, ppGpp; E_1, enzymes in the conversion of P_2 to P_3; E_2, E_3, molecules involved in directing cross-wall formation and cell envelope synthesis; GLN, glutamine; E_4, glutamine synthetase; * indicates that the material is present in the external environment. (With permission, from M. L. Shuler and M. M. Domach, in *Foundations of Biochemical Engineering*, H. W. Blanch, E. T. Papoutsakis, and G. Stephanopoulos, ed., ACS Symposium Series 207, American Chemical Society, Washington, DC, 1983, p. 93.)

concentrations. A second consideration, closely related to the first, is that the dilution of *intrinsic* concentration by growth must be considered.

The appropriate equation to use in a nonflow reactor is

$$\underset{\substack{\text{rate of change} \\ \text{in amount of } i \\ \text{in the reactor}}}{\frac{d[V_R C_i]}{dt}} = \underset{\substack{\text{total} \\ \text{biomass} \\ \text{in reactor}}}{V_R X} \times \underset{\substack{\text{rate of} \\ \text{formation} \\ \text{of } i \text{ per} \\ \text{unit biomass} \\ \text{based on intrinsic} \\ \text{concentrations}}}{r_{fi}} \qquad (6.60)$$

where V_R is the total volume in the reactor, X is the extrinsic biomass concentration, and C_i is the extrinsic concentration of component i.

Equation 6.60 can also be rewritten in terms of intrinsic concentrations. For simplicity we will use mass fractions (e.g., C_i/X). Note that

$$\frac{d(C_i/X)}{dt} = \left(\frac{1}{X}\frac{dC_i}{dt}\right) - \left(C_i/X\frac{dX/dt}{X}\right) \tag{6.61}$$

Recalling

$$\mu = \frac{1}{X}\frac{dX}{dt} \tag{6.2a}$$

we have

$$\frac{d(C_i/X)}{dt} = \left(\frac{1}{X}\frac{dC_i}{dt}\right) - \mu C_i/X \tag{6.62}$$

and, substituting eq. 6.60 for $(1/X)(dC_i/dt)$ after dividing eq. 6.60 by $V_R X$ and assuming V_R is a constant,

$$\frac{d(C_i/X)}{dt} = r_{fi} - \frac{\mu_{net}C_i}{X} \tag{6.63}$$

In eq. 6.63, the r_{fi} term must be in terms of intrinsic concentrations and the term $\mu_{net}C_i/X$ represents dilution by growth. These concepts are illustrated in Example 6.3.

The model indicated in Fig. 6.14 is that for a single cell. The single-cell model response can be directly related to culture response if all cells are assumed to behave identically. In this case each cell has the same division cycle. A population will have, at steady state, twice as many cells at "birth" as at division. The average concentrations in the culture will be at the geometric mean (a time equal to $\sqrt{2}$ multiplied by the division time) for each cell component. Used in this way, the model is a structured, nonsegregated model. If, however, a cellular population is divided into subpopulations, with each subpopulation represented by a separate single-cell model, then a population model containing a high level of structure, as well as aspects of segregation, can be built. Such a finite-representation technique has been used and is capable of making good a priori predictions of dynamic response in cultures (see Fig. 6.15). When used in this context, at least one random element in the cell cycle must be included to lead to realistic prediction of distributions. Some examples of such randomness include placement of the cell cross wall, timing of the initiation of chromosome synthesis or cross-wall synthesis, and distribution of plasmids at division.

With this section, the reader should have a good overview of the basic concepts in modeling and some simple tools to describe microbial growth. We need these tools to adequately discuss the cultivation of cells in continuous culture. Example 6.3 illustrates how a structured model might be developed.

Example 6.3.

Write the equations describing the following system. An organism consists of active biomass and a storage component. The storage component is made when the internal concentration of the carbon–energy sources is high and is degraded when it is low. Use the symbols A = active biomass, P = polymeric storage compound, S^* = external concentration of S, and S = internal concentration of S. Assume that S is the growth-rate-limiting nutrient.

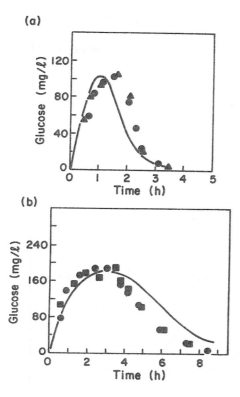

(a)

(b)

Figure 6.15. Prediction of transient response in a continuous-flow system to perturbations in limiting substrate concentration or flow rate. Note that the predictions are not fitted to the data; predictions are completely a priori. (a) Transient glucose concentration in response to step change in glucose feed concentration from 1.0 to 1.88 g/l in anaerobic continuous culture of *E. coli* B/r at dilution rate of 0.38 h^{-1}; ● and ▲, results of two duplicate experimental runs; solid line, computer prediction. (b) Transient glucose concentration in response to step change in dilution rate from $D = 0.16$ to $D = 0.55$ h^{-1}; ● and ■, results of two duplicate experimental runs; solid line, computer prediction. (With permission, from M. M. Ataai and M. L. Shuler, *Biotechnol. Bioeng. 27*:1051, 1985, and John Wiley & Sons, Inc., New York.)

Solution When measuring the concentration of *A, P,* and *S**, the natural units are grams per liter of reactor volume. These are extrinsic concentrations. The total biomass concentration will be

$$X = A + P + S \approx A + P \tag{a}$$

since the internal concentration of *S* will be low [$< 0.001(A + P)$]. By considering *S** and *S* separately, transients in *S* can be captured. However, these transients will be fast in comparison to changes in *A* and *P*. The Monod and other unstructured models assume an instantaneous equilibrium between *S** and *S*.

A number of kinetic descriptions would be defensible. However, let us assume the following pseudochemical equations:

$$A + \alpha_1 S + \cdots \rightarrow 2A + \cdots \tag{b}$$

where α_1 g of *S* is consumed to make 1 g of *A* plus by-products; the reaction requires both *A* as a "catalyst" and *S* for building blocks. Similarly, for *P*,

$$A + \alpha_2 S + \cdots \rightleftharpoons P + A + \cdots \tag{c}$$

where *P* is formed reversibly in this parallel reaction requiring *A* as the catalyst, and α_2 is a stoichiometric coefficient between the monomer *S* and the polymer *P* (g · S consumed/g · P formed).

The actual reaction mechanisms implied in these reactions are likely very complex. Most reactions in a cell are saturable and subject to feedback control. For these reactions, the intracellular level of S is the main control variable. Thus, a reasonable set of equations based on eq. 6.60 for constant V_R is

$$\frac{dS^*}{dt} = X \frac{k_1 S^*}{K_{XS^*} + S^*} \tag{d}$$

where k_1 is rate of uptake per unit cell mass (g S transported/g X-h) and K_{XS^*} is the saturation parameter for uptake. Since S^* is not a cellular component, extrinsic concentrations can be used in the rate expression. For S we write

$$\frac{dS}{dt} = X \left\{ \underset{\substack{\text{rate of} \\ \text{uptake}}}{\frac{k_1 S^*}{K_{XS^*} + S^*}} - \alpha_1 \underset{\substack{\text{rate of } S \text{ used} \\ \text{to form } A}}{\frac{k_2 S/X}{K_{AS} + S/X}(A/X)} \right.$$

$$\left. - \alpha_2 \underset{\substack{\text{net rate of } S \\ \text{consumption to form } P}}{\left[\frac{k_3 S/X}{K_{PSf} + S/X}(A/X) - \frac{k_4 K_{PSd}}{K_{PSd} + S/X}(P/X) \right]} \right\} \tag{e}$$

where k_2 is g A formed/g $\cdot$ A present-h, k_3 is g P formed/g $\cdot$ A present-h, and k_4 is a first-order degradation rate for P (h^{-1}). The parameters α_1 and α_2 are defined in eqs. (b) and (c), while K_{AS}, K_{PSf}, and K_{PSd} are saturation parameters in units of mass fraction. Similarly,

$$\frac{dA}{dt} = X \left[\underset{\substack{\text{rate of} \\ \text{formation} \\ \text{of } A}}{\frac{k_2 S/X}{K_{AS} + S/X}(A/X)} - \underset{\substack{\text{decomposition} \\ \text{of } A \text{ to provide} \\ \text{maintenance energy}}}{k_5(A/X)} \right] \tag{f}$$

where k_5 is a first-order coefficient (h^{-1}) and

$$\frac{dP}{dt} = X \left[\frac{k_3 S/X}{K_{PSf} + S/X}(A/X) - \frac{k_4 K_{PSd}}{K_{PSd} + S/X}(P/X) \right] \tag{g}$$

Equations f and g describe the rate of formation of A and P.

The value of k_1 could be determined by the uptake of a nonmetabolizable analog of the substrate, use of vesicles, or material balance information on exponentially growing cells. Values of k_2 and k_3 can be estimated from measurements on total biomass and polymer under conditions where S/X should be large (e.g., exponential growth), while k_4 and k_5 can be estimated from time-dependent measurements of cell composition in the stationary phase, where S/X should be low. If experimental values of S/X are available at various times, then more refined estimates of k_2, k_3, k_4, and k_5 can be made, as well as estimates of the saturation parameters. The rate form chosen in eq. (g) requires that $K_{PSd} << K_{PSf}$, so that P is made only when S/X is in excess and degraded only when S/X is low. The values for α_1 and α_2 are essentially the reciprocal of the yield coefficients and can be readily estimated from growth experiments and compositional data on P.

The rate forms used here are by no means the only correct solution. A more extensive base of experimental observations would be needed to eliminate other possible formulations.

6.3.4. Cybernetic Models

Another modeling approach has been developed primarily to predict growth under conditions when several substrates are available. These substrates may be complementary (e.g., carbon or nitrogen) or substitutable. For example, glucose and lactose would be substitutable, as these compounds both supply carbon and energy. As the reader will recall, we discussed the diauxic phenomenon for sequential use of glucose and lactose in Chapter 4. That experimental observation led us to an understanding of regulation of the *lac* operon and catabolite repression. This metabolic regulation was necessary for the transition from one primary pathway to another. The reader might infer that the culture had as its objective function the maximization of its growth rate.

One approach to modeling growth on multiple substrates is a *cybernetic* approach. Cybernetic means that a process is goal seeking (e.g., maximization of growth rate). While this approach was initially motivated by a desire to predict the response of a microbial culture to growth on a set of substitutable carbon sources, it has been expanded to provide an alternative method of identifying the regulatory structure of a complex biochemical reaction network (such as cellular metabolism) in a simple manner. Typically a single objective, such as maximum growth rate, is chosen and an objective-oriented mathematical analysis is employed. This analysis is similiar to many economic analyses for resource distribution. For many practical situations this approach describes satisfactorily growth of a culture on a complex medium. However, the potential power of this approach is now being realized in efforts in metabolic engineering and in relating information on DNA sequences in an organism to physiologic function (see Chapter 8).

This approach has limitations, as the objective function for any organism is maximizing its long-term survival as a species. Maximization of growth rate or of growth yield are really subobjectives which can dominate under some environmental conditions; these conditions are often of great interest to the bioprocess engineer. Consequently, the cybernetic approach is often a valuable tool. It is too complex for us to describe in detail in this book; the interested reader may consult the references at the end of this chapter.

6.4. HOW CELLS GROW IN CONTINUOUS CULTURE

6.4.1. Introduction

The culture environment changes continually in a batch culture. Growth, product formation, and substrate utilization terminate after a certain time interval, whereas, in continuous culture, fresh nutrient medium is continually supplied to a well-stirred culture, and products and cells are simultaneously withdrawn. Growth and product formation can be maintained for prolonged periods in continuous culture. After a certain period of time, the system usually reaches a steady state where cell, product, and substrate concentrations remain constant. Continuous culture provides constant environmental conditions for growth and product formation and supplies uniform-quality product. Continuous culture is an important tool to determine the response of microorganisms to their environment and to produce the desired products under optimal environmental conditions. In Chapter 9 we will compare batch and continuous culture in terms of their suitability for large-scale operation.

6.4.2. Some Specific Devices for Continuous Culture

The primary types of continuous cultivation devices are the *chemostat* and *turbidostat*, although plug flow reactors (PFR) are used. In some cases these units are modified by recycle of cells.

Figure 6.16 is a schematic of a continuous culture device (chemostat). Cellular growth is usually limited by one essential nutrient, and other nutrients are in excess. As we will show, when a chemostat is at steady state, the nutrient, product, and cell concentrations are constant. For this reason, the name *chemostat* refers to constant chemical environment.

Figure 6.17 is a schematic of a turbidostat in which the cell concentration in the culture vessel is maintained constant by monitoring the optical density of the culture and controlling the feed flow rate. When the turbidity of the medium exceeds the set point, a

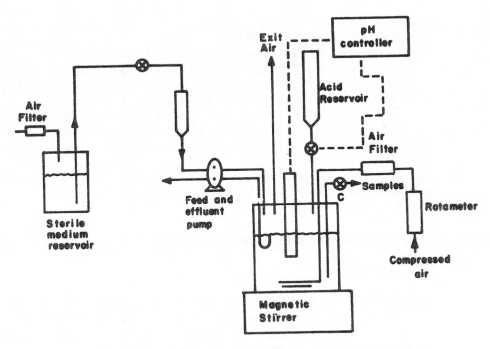

Figure 6.16. A continuous-culture laboratory setup (chemostat). (With permission, from D. I. C. Wang and others, *Fermentation and Enzyme Technology*, John Wiley & Sons, Inc., New York, 1979, p. 99.)

How Cells Grow   Chap. 6

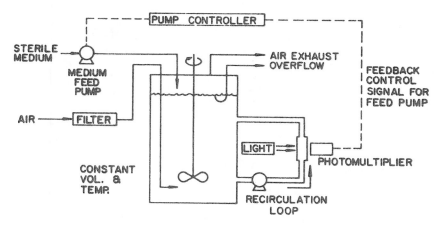

Figure 6.17. Typical laboratory setup for a turbidostat. (With permission, from D. I. C. Wang and others, *Fermentation and Enzyme Technology*, John Wiley & Sons, Inc., New York, 1979, p. 100.)

pump is activated and fresh medium is added. The culture volume is kept constant by removing an equal amount of culture fluid. The turbidostat is less used than the chemostat, since it is more elaborate than a chemostat and because the environment is dynamic. Turbidostats can be very useful in selecting subpopulations able to withstand a desired environmental stress (for example, high ethanol concentrations), because the cell concentration is maintained constant. The selection of variants or mutants with desirable properties is very important (see Chapter 8).

A plug flow reactor (PFR) can also be used for continuous cultivation purposes. Since there is no backmixing in an ideal PFR, fluid elements containing active cells cannot inoculate other fluid elements at different axial positions. Liquid recycle is required for continuous inoculation of nutrient media. In a PFR, substrate and cell concentrations vary with axial position in the vessel. An ideal PFR resembles a batch reactor in which distance along the fermenter replaces incubation time in a batch reactor. In waste treatment, some units approach PFR behavior, and multistage chemostats tend to approach PFR dynamics if the number of stages is large (five or more).

6.4.3. The Ideal Chemostat

An ideal chemostat is the same as a perfectly mixed continuous-flow, stirred-tank reactor (CFSTR). Most chemostats require some control elements, such as pH and dissolved-oxygen control units, to be useful. Fresh sterile medium is fed to the completely mixed and aerated (if required) reactor, and cell suspension is removed at the same rate. Liquid volume in the reactor is kept constant.

Figure 6.18 is a schematic of a simplified chemostat. A material balance on the cell concentration around the chemostat yields

$$FX_0 - FX + V_R\mu_g X - V_R k_d X = V_R \frac{dX}{dt} \tag{6.64}$$

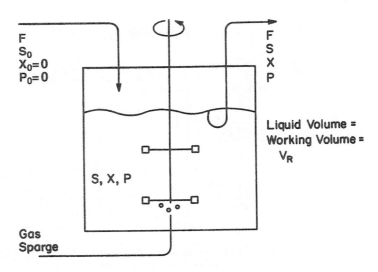

Figure 6.18. Simplified schematic of a chemostat.

where F is the flow rate of nutrient solution (l/h), V_R is the culture volume (l) (assumed constant), X is the cell concentration (g/l), and μ_g and k_d are growth and endogenous (or death) rate constants, respectively (h^{-1}). The reader should note that if cell mass is the primary parameter, it is difficult to differentiate cell death from endogenous metabolism. When we use k_d, we imply that endogenous metabolism is the primary mechanism for cell mass decrease. With k'_d, we imply that cell death and lysis are the primary mechanisms of decrease in mass. The reader should also note that, if eq. 6.64 had been written in terms of cell number, k_d could only be a cell death rate. When balances are written in terms of cell number, the influence of endogenous metabolism can appear only in the substrate balance equation. Since most experiments are done by measuring total cell mass rather than number, we write our examples based on X. However, the reader should be aware of the ambiguity introduced when equations are written in terms of X.

Equation 6.64 can be rearranged as

$$\frac{dX}{dt} = DX_0 + (\mu_g - k_d - D)X \tag{6.65}$$

where D is *dilution rate* and $D = F/V_R$. D is the reciprocal of residence time.

Usually, the feed media are sterile, $X_0 = 0$, and if the endogenous metabolism or death rate is negligible compared to the growth rate ($k_d \ll \mu_g$) and if the system is at steady state ($dX/dt = 0$), then

$$\mu_g = D \quad (\text{if } k_d = 0) \tag{6.66}$$

In a chemostat, cells are removed at a rate equal to their growth rate, and the growth rate of cells is equal to the dilution rate. *This property allows the investigator to manipulate growth rate as an independent parameter* and makes the chemostat a powerful experimental tool.

Since growth rate is limited by at least one substrate in a chemostat, a simple description of chemostat performance can be made by substituting the Monod equation (eq. 6.30) for μ_g in eq. 6.66:

$$\mu_g = D = \frac{\mu_m S}{K_s + S} \tag{6.67}$$

where S is the steady-state limiting substrate concentration (g/l). If D is set at a value greater than μ_m, the culture cannot reproduce quickly enough to maintain itself and is *washed out*. Equation 6.67 is identical to that for Michaelis–Menten kinetics and, as we discussed in Chapter 3, a plot of $1/\mu_g$ versus $1/S$ can be used to estimate values for μ_m and K_s.

Using eq. 6.67, we can relate effluent substrate concentration to dilution rate for $D < \mu_m$.

$$S = \frac{K_s D}{\mu_m - D} \tag{6.68}$$

A material balance on the limiting substrate in the absence of endogenous metabolism yields

$$FS_0 - FS - V_R \mu_g X \frac{1}{Y_{X/S}^M} - V_R q_P X \frac{1}{Y_{P/S}} = V_R \frac{dS}{dt} \tag{6.69}$$

where S_0 and S are feed and effluent substrate concentrations (g/l), q_P is the specific rate of extracellular product formation (g P/μ_g cells h), and $Y_{X/S}^M$ and $Y_{P/S}$ are yield coefficients (g cell/g S and g P/g S). The use of the superscript M on $Y_{X/S}$ denotes a maximum value of the yield coefficient; such a superscript will be important in discussing the effects of maintenance energy.

When extracellular product formation is negligible and the system is at steady state ($dS/dt = 0$),

$$D(S_0 - S) = \frac{\mu_g X}{Y_{Y/S}^M} \tag{6.70}$$

Since $\mu_g = D$ at steady state if $k_d = 0$,

$$X = Y_{X/S}^M (S_0 - S) \tag{6.71}$$

Using eq. 6.68, the steady-state cell concentration can be expressed as

$$X = Y_{Y/S}^M \left(S_0 - \frac{K_s D}{\mu_m - D} \right) \tag{6.72}$$

In eqs. 6.71 and 6.72, the yield coefficient ($Y_{X/S}$) is assumed to be constant, which is an approximation, since endogenous metabolism has been neglected. Usually, $Y_{X/S}$ varies with the limiting nutrient and growth rate. Consider the effect the inclusion of endogenous metabolism will have; eq. 6.66 becomes

$$D = \mu_g - k_d = \mu_{net} \tag{6.73a}$$

or

$$\mu_g = D + k_d \tag{6.73b}$$

By substituting eq. 6.73b into the steady-state substrate balance, assuming no extracellular product formation (eq. 6.70), we find

$$D(S_0 - S) - 1/Y^M_{X/S}(D + k_d)X = 0 \qquad (6.73c)$$

where $Y^M_{X/S}$ denotes the maximum yield coefficient (no endogenous metabolism or maintenance energy). $Y^M_{X/S}$ has a single constant value independent of growth rate. This approach does not allow the direct conversion of S into maintenance energy. Rather, S must be first incorporated into the cell mass, where it is degraded by endogenous metabolism. This viewpoint is the proper one when working with an unstructured model. With a structured model, which explicitly recognizes intracellular S as a subcomponent of the biomass, the direct consumption of S for maintenance functions could be modeled. However, with unstructured models, substrate that is no longer extracellular becomes part of the biomass.

Equation 6.73c can be rearranged to

$$D\left(\frac{S_0 - S}{X}\right) - 1/Y^M_{X/S}(D + k_d) = 0 \qquad (6.74)$$

or

$$D\left(\frac{1}{Y^{AP}_{X/S}}\right) - \frac{D}{Y^M_{X/S}} - \frac{k_d}{Y^M_{X/S}} = 0 \qquad (6.75a)$$

$$\frac{1}{Y^M_{X/S}} + \frac{k_d}{Y^M_{X/S} \cdot D} = \frac{1}{Y^{AP}_{X/S}} \qquad (6.75b)$$

$$\frac{1}{Y^{AP}_{X/S}} = \frac{1}{Y^M_{X/S}} + \frac{m_s}{D} \qquad (6.76)$$

where

$$m_s = \frac{k_d}{Y^M_{X/S}} \qquad (6.77)$$

where m_s is the *maintenance coefficient* based on substrate S. $Y^{AP}_{X/S}$ is the apparent yield. When $Y_{X/S}$ is written, it should be interpreted as $Y^{AP}_{X/S}$. While $Y^M_{X/S}$ is a constant, $Y^{AP}_{X/S}$ varies with growth conditions if $k_d > 0$.

Values of $Y^M_{X/S}$ and m_s can be obtained from chemostat experiments by plotting $1/Y^{AP}_{X/S}$ against $1/D$. The slope is m_s and the intercept is $1/Y^M_{X/S}$ (see Fig. 6.19).

In the presence of endogenous metabolism ($\mu_{net} = \mu_m S/(K_s + S) - k_d$), we can also show that

$$S = \frac{K_s(D + k_d)}{\mu_m - D - k_d} \qquad (6.78)$$

and

$$X = Y^M_{X/S}[S_0 - S] \cdot \frac{D}{D + k_d} \qquad (6.79)$$

In addition to the effects of maintenance, it is often appropriate to consider the conversion of extracellular substrate into extracellular product. With an unstructured

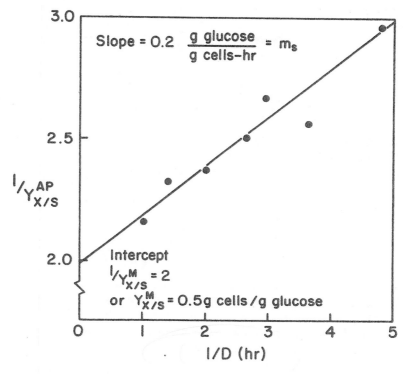

Figure 6.19. Graphical approach to estimating $Y_{X/S}^M$ and m_s for chemostat data for *E. coli* growing on glucose as the limiting nutrient.

model, we must view this conversion as instantaneous. Otherwise, a perceptible period of substrate uptake, prior to conversion to product, would cause a change in the amount of X. This is unlike maintenance, since the period between substrate uptake and use for maintenance functions may be long, so we allow S to become incorporated into X and X is then degraded when needed; the following equations assume instantaneous conversion of substrate to extracellular product, with the cell acting as a catalyst. The balance on product formation is

$$DP = q_P X \tag{6.80}$$

where q_P is described by eqs. 6.16, 6.17, or 6.18. For nongrowth-associated product formation, q_P is a constant (β), while for growth-associated products it is a function of μ_g. For substrate balance, eq. 6.69 becomes

$$D(S_0 - S) = \frac{1}{Y_{X/S}^M}(D + k_d)X + \frac{1}{Y_{P/S}}q_P X \tag{6.81}$$

The biomass balance is unchanged from the case with endogenous metabolism and yields

$$S = \frac{K_s(D + k_d)}{\mu_m - D - k_d} \tag{6.78}$$

Equation 6.81 can be solved for X:

$$X = Y_{X/S}^M(S_0 - S)\left(\frac{D}{D + k_d + q_P\dfrac{Y_{X/S}^M}{Y_{P/S}}}\right) \tag{6.82}$$

The productivity of a chemostat for product and biomass (Pr_x) can be found from DP and DX, respectively. The dilution rate that maximizes productivity is found by differentiating DP or DX with respect to D and setting the derivative equal to zero. The optimal value of D (D_{opt}) will depend on whether endogenous metabolism and/or product formation are considered. When $k_d = 0$ and $q_P = 0$, eqs. 6.71 and 6.68 apply. Then D_{opt} for biomass production (DX) becomes

$$D_{opt} = \mu_m\left(1 - \sqrt{\frac{K_s}{K_s + S_0}}\right) \tag{6.83}$$

Since S_0 is usually much greater than K_s, D_{opt} will approach $D = \mu_m$ or the washout point. Stable chemostat operation with $D \approx \mu_m$ is very difficult, unless the flow rate and liquid volume can be maintained exactly constant. Consequently, a value of D slightly less than D_{opt} may be a good compromise between stability and biomass productivity. It should also be apparent that D_{opt} for biomass formation will not necessarily be optimal for product formation.

Examples 6.4 and 6.5 illustrate the use of these equations to characterize the performance of chemostats.

Example 6.4.

A new strain of yeast is being considered for biomass production. The following data were obtained using a chemostat. An influent substrate concentration of 800 mg/l and an excess of oxygen were used at a pH of 5.5 and $T = 35°C$. Using the following data, calculate μ_m, K_s, $Y_{X/S}^M$, k_d, and m_s, assuming $\mu_{net} = \mu_m S/(K_s + S) - k_d$.

Dilution rate (h^{-1})	Carbon substrate concentration (mg/l)	Cell concentration (mg/l)
0.1	16.7	366
0.2	33.5	407
0.3	59.4	408
0.4	101	404
0.5	169	371
0.6	298	299
0.7	702	59

Solution The first step is to plot $1/Y_{X/S}^{AP}$ versus $1/D$. $Y_{X/S}^{AP}$ is calculated from $X/(S_0 - S)$. The intercept is $1/Y_{X/S}^M = 1.58$ or $Y_{X/S}^M = 0.633$ g X/g S. The slope is 0.06 g S/g X-h, which is the value for m_s. Recall $m_s = k_d/Y_{X/S}^M$; then $k_d = m_s Y_{X/S}^M = 0.06$ g S/g X-h $\cdot$ 0.633 g X/g $S = 0.038$ h^{-1}.

For the second step, recall that

$$D = \mu_g - k_d = \frac{\mu_m S}{K_s + S} - k_d$$

(a)

Then

$$\frac{1}{D + k_d} = \frac{1}{\mu_m} + \frac{K_s}{\mu_m}\frac{1}{S}$$

(b)

We now plot $1/(D + k_d)$ versus $1/S$. The intercept is 1.25 h or $\mu_m = 0.8$ h^{-1}. The slope is 100 h/(mg/l). Thus, $K_s/\mu_m = 100$ or $K_s = 80$ mg/l.

Example 6.5.

The specific growth rate for inhibited growth in a chemostat is given by the following equation:

$$\mu_g = \frac{\mu_m S}{K_s + S + I K_s / K_I}$$

(a)

where

$S_0 = 10$ g/l $K_s = 1$ g/l $I = 0.05$ g/l $Y^M_{X/S} = 0.1\dfrac{\text{g cells}}{\text{g subs}}$

$X_0 = 0$ $K_I = 0.01$ g/l $\mu_m = 0.5$ h^{-1} $k_d = 0$

a. Determine X and S as a function of D when $I = 0$.
b. With inhibitor added to a chemostat, determine the effluent substrate concentration and X as a function of D.
c. Determine the cell productivity, DX, as a function of dilution rate.

Solution

a)

$$S = \frac{K_s D}{\mu_m - D} = \frac{D}{0.5 - D}$$

$$X = Y^M_{X/S}(S_0 - S) = 0.1\left(10 - \frac{D}{0.5 - D}\right)$$

b) In the presence of inhibitor

$$\mu_g = \frac{\mu_m S}{K_s\left(1 + \dfrac{I}{K_I}\right) + S} = D$$

$$S = \frac{K_s\left(1 + \dfrac{I}{K_I}\right)D}{\mu_m - D} = \frac{I\left(1 + \dfrac{0.05}{0.01}\right)D}{0.5 - D}$$

$$S = \frac{6D}{0.5 - D}$$

$$X = Y^M_{X/S}(S_0 - S) = 0.1\left(10 - \frac{6D}{0.5 - D}\right)$$

c) $\mathrm{Pr}_x = DX = DY_{X/S}(S_0 - S) = 0.1D\left(10 - \dfrac{6D}{0.5 - D}\right)$

6.4.4. The Chemostat as a Tool

The chemostat can be used as a tool to study the mutation and selection of cultures and also to study the effect of changes in the environment on cell physiology. The molecular aspects of mutation and selection will be discussed in Chapter 8.

Natural or induced *mutations* can take place in a chemostat culture. Errors in DNA replication take place with an average frequency of about 10^{-6} to 10^{-8} gene per generation. With a cell concentration of 10^9 cells/ml in culture, the probability is high in a chemostat that a wide variety of mutant cells will be formed. The vast majority of natural mutations in a chemostat are of little significance, unless the mutation alters the function of a protein involved in growth in the chemostat environment. If the specific growth rate of the mutant is larger than that of the wild type, then the mutant outgrows the wild type in a chemostat. This selection for a variant cell type can be accomplished by creating a more favorable environment for growth of the mutant organism.

A chemostat culture can be used for the selection of special organisms. Selection or enrichment nutrient media need to be used for this purpose. For example, if it is desired to select an organism growing on ethanol, a nutrient medium containing ethanol and mineral salts is used as a feed to a chemostat culture. An organism capable of oxidizing some toxic refractory compounds can be selected from a mixed culture by slowly feeding this compound to a chemostat. A thermophilic organism can be selected from a natural population by operating a chemostat at an elevated temperature (e.g., 50° to 60°C). Selection in chemostats also presents significant problems in the culture of cells containing recombinant DNA. The most productive cells often grow more slowly and are displaced by less productive cells. We will discuss this problem in more detail in Chapter 14.

6.4.5. Deviations from Ideality

In fermentations such as the utilization or production of polysaccharides or mycelial fermentations, the fermentation broth may be highly viscous and it may be difficult to maintain complete mixing. Also, certain cells tend to grow in the form of a film on solid surfaces in fermenters, such as on fermenter walls or probe surfaces. Incomplete mixing is the rule in industrial-scale fermenters. The presence of incompletely mixed regions in a "chemostat" culture is common. The term "chemostat" is actually applicable only to perfectly mixed vessels. Here we denote nonideality by putting chemostat in quotation marks. A segregated reactor model can be used to analyze an incompletely mixed continuous-flow fermenter.

A simple, two-compartmental model of a "chemostat" culture is depicted in Fig. 6.20, where the reactor is divided into two regions: a well-mixed region and a stagnant region. Feed and effluent streams pass through the well-mixed region 1, and mass exchange takes place between the two regions. The biomass and substrate balances for both regions at steady state are as follows:

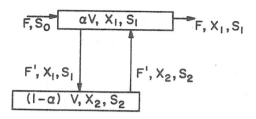

Figure 6.20. Simple two-compartmental model for an incompletely mixed "chemostat."

	Substrate balance	Biomass balance
Region 1	$X_1 = Y(S_0 - S_1)$	$X_2 + \alpha D' \mu_m \dfrac{S_1}{K_s + S_1} X_1 = (1 + DD')X_1$
Region 2	$X_2 - X_1 = Y(S_1 - S_2)$	$X_1 + (1-\alpha)D'\mu_m \dfrac{S_2}{K_s + S_2} X_2 = X_2$

In an imperfectly mixed chemostat with a stagnant biotic phase, the dilution rate can exceed μ_m without washout. The washout dilution rate is obtained by setting $S_0 = S_1 = S_2$ and is

$$D_{wo} = \frac{\mu_m S_0}{K_m + S_0}\left[\frac{(1-\alpha)^2 \mu_m S_0}{K_s + S_0 - (1-\alpha)D'\mu_m S_0} + 1 \right]$$

where D_{wo} is the washout dilution rate.

6.5. SUMMARY

During batch cultivation, a population of cells typically exhibits several different growth phases. During the lag phase, the cell builds the biosynthetic pathways necessary for maximal growth rates in the fresh medium. During *exponential growth*, cell replication is maximal, and the chemical composition of the cell population is nearly constant (e.g., balanced growth). When substrate is nearly exhausted or when toxic metabolic by-products have built to a critical level, the growth rate begins to drop rapidly, causing significant changes in biosynthetic pathways. In the *stationary phase*, there is no net growth; cells now reorient their metabolic machinery to increase the probability of long-term survival. At some point, some cells can no longer obtain enough energy from their reserves or enough of another critical resource, and the culture enters the *death phase*. Dead cells do not have an energized membrane and often lyse (or break apart). Nutrients released by lysed cells can be utilized by survivors, allowing cryptic growth.

Products formed by cells can be related to this batch-culture growth cycle. Primary products are growth associated. Secondary products are nongrowth associated and are made in the stationary phase. Some products have both growth-associated and nongrowth-associated components.

Cell-growth kinetics can be modeled. Models that are *structured* break the population into distinct subcomponents. *Unstructured* models quantify cell mass as a single component. Unstructured models cannot describe transient behavior very well. Some models are *segregated*; they recognize that not all members of the population are identical, and the distribution of properties among individual cells is important. *Nonsegregated models* ignore these distributions and assume that population-average parameters are adequate.

These models apply not only to batch culture but also to continuous culture. The primary form of continuous culture is a steady-state CFSTR or *chemostat*. A chemostat ensures a time-invariant growth environment. The net growth rate is equal to the dilution rate, which is determined by the flow rate to the chemostat. Thus, the growth rate can be manipulated by the investigator. A *turbidostat* adjusts flow rate to maintain a constant cell density. A turbidostat operates well at high flow rates (near the washout point) and is useful in selecting cellular subpopulations that have adapted to a particular stress.

SUGGESTIONS FOR FURTHER READING

AIBA, S., A. E. HUMPHREY, AND N. F. MILLIS, *Biochemical Engineering*, 2d ed., Academic Press, New York, 1973.

BAILEY, J. E., Mathematical Modeling and Analysis in Biochemical Engineering: Past Accomplishments and Future Opportunities, *Biotechnol. Prog. 14*:8, 1998.

BAILEY, J. E., AND D. F. OLLIS, *Biochemical Engineering Fundamentals*, 2d ed., Mc-Graw-Hill Book Co., New York, 1986.

BLANCH, H. W., AND D. S. CLARK. *Biochemical Engineering*. Marcel Dekker, Inc., New York, 1996.

FREDERICKSON, A. G., R. D. MEGEE III, AND H. M. TSUCHIYA, Mathematical Models for Fermentation Processes, *Adv. Appl. Microbiol. 23*:419, 1970.

GADEN, E. L., JR., Fermentation Kinetics and Productivity, *Chem. Ind. Rev.* (London), 1955, p. 154.

HERBERT, D., R. ELLSWORTH, AND R. C. TELLING, The Continuous Culture of Bacteria: A Theoretical and Experimental Study, *J. Gen. Microbiol. 14*:601, 1956.

SHULER, M. L., On the Use of Chemically Structured Models for Bioreactors, *Chem. Eng. Commun. 36*:161–189, 1985.

STRAIGHT, J. V., AND D. RAMKRISHNA, Cybernetic Modeling and Regulation of Metabolic Pathways, Growth on Complementary Nutrients, *Biotechnol. Prog. 10*: 574, 1994.

PROBLEMS

6.1. A simple, batch fermentation of an aerobic bacterium growing on methanol gave the results shown in the table. Calculate:

    a. Maximum growth rate (μ_{max})

    b. Yield on substrate ($Y_{X/S}$)

    c. Mass doubling time (t_d)

    d. Saturation constant (K_s)

    e. Specific growth rate (μ_{net}) at $t = 10$ h

Time (h)	X (g/l)	S (g/l)
0	0.2	9.23
2	0.211	9.21
4	0.305	9.07
8	0.98	8.03
10	1.77	6.8
12	3.2	4.6
14	5.6	0.92
16	6.15	0.077
18	6.2	0

6.2. The growth of a microbial population is a function of pH and is given by the following equation:

$$\mu_g = \frac{1}{X}\frac{dX}{dt} = \frac{\mu_m S}{K_s\left(1+\dfrac{H^+}{k_1}\right)+S}$$

   a. With a given set of experimental data (X and S versus t), describe how you would determine the constants μ_m, K_s, and k_1.

   b. How would the double-reciprocal plot $1/\mu_g$ versus $1/S$ change with pH (or H^+) concentration?

6.3. The following data were obtained for the effect of temperature on the fermentative production of lactic acid by a strain of *Lactobacillus delbrueckii*. From these data, calculate the value of the activation energy for this process. Is the value of the activation energy typical of this sort of biological conversion? (See Chapter 3.)

Temperature (°C)	Rate constant (mol/l-h)
40.4	0.0140
36.8	0.0112
33.1	0.0074
30.0	0.0051
25.1	0.0036

[Courtesy of A. E. Humphrey from "Collected Coursework Problems in Biochemical Engineering," compiled by H. W. Blanch for 1977 Am. Soc. Eng. Educ. Summer School.]

6.4. It is desired to model the growth of an *individual* bacterium. The cell transports S_1 into the cell enzymatically, and the permease is subject to product inhibition. S_1 is converted into precursors, P, that are converted finally into the macromolecular portion of the cell, M. The catalyst of all reactions is M.

   (1) $S_1^* \xrightarrow[M]{} S_1$ (per unit surface area)
   where S^* = outside concentration of S

   (2) $S_1 \xrightarrow[M]{} P$

Problems

$$\left.\begin{array}{l} \text{Energy} + P\xrightarrow{\ M\ } M \\ S_1 \xrightarrow{\ M\ } \text{energy} \end{array}\right\} \text{coupled reaction}$$

(3a)

or

(3b) $\quad S_1 + P\xrightarrow{\ M\ } M$

The dry weight of the cell is T and is equal to

(4) $\quad T = S_1 + P + M_1 = \rho V$

where ρ = cell density and V = cell volume

Write the equations and define all symbols necessary to describe the changes in S_1, P, M, and T within the cell. Remember that the cell volume is always changing.

6.5. A biochemical engineer has determined in her lab that the optimal productivity of a valuable antibiotic is achieved when the carbon nutrient, in this case molasses, is metered into the fermenter at a rate proportional to the growth rate. However, she cannot implement her discovery in the antibiotic plant, since there is no reliable way to measure the growth rate (dX/dt) or biomass concentration (X) during the course of the fermentation. It is suggested that an oxygen analyzer be installed on the plant fermenters so that the OUR (oxygen uptake rate, g/l-h) may be measured.

  a. Derive expressions that may be used to estimate X and dX/dt from OUR and time data, assuming that a simple yield and maintenance model may be used to describe the rate of oxygen consumption by the culture.

  b. Calculate values for the yield (Y_{X/O_2}) and maintenance (m_{O_2}) parameters from the following data:

Time	OUR (g/h)	X (g/l)
0	0.011	0.60
1	0.008	0.63
2	0.084	0.63
3	0.153	0.76
4	0.198	1.06
5	0.273	1.56
6	0.393	2.23
7	0.493	2.85
8	0.642	4.15
9	0.915	5.37
10	1.031	7.59
11	1.12	9.40
12	1.37	11.40
13	1.58	12.22
14	1.26	13.00
15	1.58	13.37
16	1.26	14.47
17	1.12	15.37
18	1.20	16.12
19	0.99	16.18
20	0.86	16.67
21	0.90	17.01

[Courtesy of D. Zabriskie from "Collected Coursework Problems in Biochemical Engineering," compiled by H. W. Blanch for 1977 Am. Soc. Eng. Educ. Summer School.]

6.6. *Pseudomonas sp.* has a mass doubling time of 2.4 h when grown on acetate. The saturation constant using this substrate is 1.3 g/l (which is unusually high), and cell yield on acetate is 0.46 g cell/g acetate. If we operate a chemostat on a feed stream containing 38 g/l acetate, find the following:

  a. Cell concentration when the dilution rate is one-half of the maximum
  b. Substrate concentration when the dilution rate is $0.8 \, D_{max}$
  c. Maximum dilution rate
  d. Cell productivity at $0.8 \, D_{max}$

  [Courtesy of E. Dunlop from "Collected Coursework Problems in Biochemical Engineering," compiled by H. W. Blanch for 1977 Am. Soc. Eng. Educ. Summer School.]

6.7. The following data were obtained in a chemostat for the growth of *E. aerogenes* on a glycerol-limited growth medium.

D, h^{-1} dilution rate	$1/D$	S, mg/ml, glycerol	$1/S$	X, mg/ml cell conc.	ΔS	$\Delta S/X$	$\Delta S/X \cdot D$
0.05	20	0.012	83.3	3.2	9.988	3.12	0.156
0.10	10	0.028	35.7	3.7	9.972	2.7	0.270
0.20	5.0	0.05	20	4.0	9.95	2.49	0.498
0.40	2.5	0.10	10	4.4	9.90	2.25	0.90
0.60	1.67	0.15	6.67	4.75	9.85	2.075	1.245
0.70	1.43	0.176	5.68	4.9	9.824	2.005	1.405
0.80	1.25	0.80	1.25	4.5	9.20	2.045	1.635
0.84	1.19	9.00	0.11	0.5	—	—	—

Note: $S_0 = 10$ mg/ml.

For this system, estimate the values of:

  a. K_s, mg glycerol/ml
  b. μ_m, h^{-1}
  c. $Y_{X/S}$, mg cells/mg glycerol
  d. m_s, mg glycerol/mg cell-h

  [Courtesy of A. E. Humphrey from "Collected Coursework Problems in Biochemical Engineering," compiled by H. W. Blanch for 1977 Am. Soc. Eng. Educ. Summer School.]

6.8. The kinetics of microbial growth, substrate consumption, and mixed-growth-associated product formation for a chemostat culture are given by the following equations:

$$\frac{dX}{dt} = \frac{\mu_m S}{K_s + S} X$$

$$\frac{dS}{dt} = \frac{\mu_m S}{(K_s + S)Y_{X/S}} X$$

$$\frac{dP}{dt} = \alpha \frac{dX}{dt} + \beta X = (\alpha \mu_g + \beta)X$$

The kinetic parameter values are $\mu_m = 0.7$ h^{-1}, $K_s = 20$ mg/l, $Y_{X/S} = 0.5$ g dw/g substrate, $Y_{P/X} = 0.15$ gP/g·dw, $\alpha = 0.1$, $\beta = 0.02$ h^{-1}, and $S_0 = 1$ g/l.

a. Determine the optimal dilution rate maximizing the productivity of product formation (*PD*).

b. Determine the optimal dilution rate maximizing the productivity of cell (biomass) formation (*DS*).

[Problem adapted from one suggested by L. Erickson.]

6.9. Ethanol is to be used as a substrate for single-cell protein production in a chemostat. The available equipment can achieve an oxygen transfer rate of 10 g O_2/l of liquid per hour. Assume the kinetics of cell growth on ethanol is of the Monod type, with $\mu_m = 0.5$ h^{-1}, $K_s = 30$ mg/l, $Y_{X/S} = 0.5$ cells/g ethanol, and $Y_{O_2/S} = 2$ g O_2/g EtOH. We wish to operate the chemostat with an ethanol concentration in the feed of 22 g/L. We also wish to maximize the biomass productivity and minimize the loss of unused ethanol in the effluent. Determine the required dilution rate and whether sufficient oxygen can be provided.

6.10. Plot the response of a culture to diauxic growth on glucose and lactose based on the following: $\mu_{glucose} = 1.0$ h^{-1}; $\mu_{lactose} = 0.6$ h^{-1}; $Y_{glucose} = Y_{lactose} = 0.5$; enzyme induction requires 30 min to complete. Plot cell mass, glucose, and lactose concentrations, assuming initial values of 2 g/l glucose, 3 g/l lactose, and 0.10 g/l cells.

6.11. The following data are obtained in oxidation of pesticides present in wastewater by a mixed culture of microorganisms in a continuously operating aeration tank.

D (h^{-1})	S (Pesticides), mg/l	X (mg/l)
0.05	15	162
0.11	25	210
0.24	50	250
0.39	100	235
0.52	140	220
0.7	180	205
0.82	240	170

Assuming the pesticide concentration in the feed wastewater stream as $S_0 = 500$ mg/l, determine $Y_{X/S}^M$, k_d, μ_m, and K_s.

6.12. In a chemostat you know that if a culture obeys the Monod equation, the residual substrate is independent of the feed substrate concentration. You observe that in your chemostat an increase in S_0 causes an increase in the residual substrate concentration. Your friend suggests that you consider whether the Contois equation may describe the situation better. The Contois equation (eq. 6.36) is:

$$\mu = \frac{\mu_m S}{K_{sx} X + S}$$

a. Derive an expression for S in terms of D, μ_m, K_{sx}, and X for a steady-state CFSTR (chemostat).

b. Derive an equation for S as a function of S_0, D, K_{sx}, $Y_{X/S}^M$, and μ_m.

c. If S_0 increases twofold, by how much will S increase?

6.13. *Pseudomonas putida* with $\mu_m = 0.5$ h^{-1} is cultivated in a continuous culture under aerobic conditions where $D = 0.28$ h^{-1}. The carbon and energy source in the feed is lactose with a concentration of $S_0 = 2$ g/l. The effluent lactose concentration is desired to be $S = 0.1$ g/l. If the growth rate is limited by oxygen transfer, by using the following information:

$$Y_{X/S}^M = 0.45gX/gS, \quad Y_{X/O_2}^M = 0.25gX/gO_2 \text{ and } C^* = 8mg/l$$

   a. Determine the steady-state biomass concentration (X) and the specific rate of oxygen consumption (q_{O_2}).
   b. What should be the oxygen-transfer coefficient (k_La) in order to overcome oxygen-transfer limitation (i.e., $C_L = 2$ mg/l)?

6.14. The maximum growth yield coefficient for *Bacillus subtilis* growing on methanol is 0.4 g X/g S. The heat of combustion of cells is 21 kJ/g cells and for substrate it is 7.3 kcal/g. Determine the metabolic heat generated by the cells per unit mass of methanol consumption.

6.15. Calculate the productivity (i.e., DP) of a chemostat under the following conditions:
   1. Assume Monod kinetics apply. Assume that negligible amounts of biomass must be converted to product ($< 1\%$).
   2. Assume the Luedeking–Piert equation for product formation (eq. 6.18) applies.
   3. Assume steady state:

   $D = 0.8\mu_m$ $Y_{X/S}^M = 0.5$ g X/g S
   $\mu_m = 1.0$ h^{-1} $S_0 = 1000$ mg/l
   $K_s = 10$ mg/l $\beta = 0.5$ h^{-1} mg P/g X
   $\alpha = 0.4$ mg P/g X

6.16. Consider a chemostat. You wish to know the *number* of cells in the reactor and the *fraction* of the cells that are viable (i.e., alive as determined by ability to divide).
   a. Write an equation for viable cell number (n_v). Assume that

   $$\mu_{net,rep.} = \frac{\mu_{m,rep}S}{K_{s,rep} + S} - k_d'$$

   where $\mu_{net,rep}$ = net specific replication rate, $\mu_{m,rep}$ = maximum specific replication rate, and k_d' = death rate. $K_{s,rep}$ is the saturation parameter.
   b. Derive an expression for the value of S at steady state.
   c. Write the number balance in the chemostat on dead cells (n_d).
   d. Derive an expression for the fraction of the total population which are dead cells.

6.17. *E. coli* is cultivated in continuous culture under aerobic conditions with a glucose limitation. When the system is operated at $D = 0.2$ h^{-1}, determine the effluent glucose and biomass concentrations by using the following equations ($S_0 = 5$ g/l):
   a. Monod equation: $\mu_m = 0.25$ h^{-1}, $K_s = 100$ mg/l.
   b. Tessier equation: $\mu_m = 0.25$ h^{-1}, $K = 0.005$ (mg/l)$^{-1}$.
   c. Moser equation: $\mu_m = 0.25$ h^{-1}, $K_s = 100$ mg/l, $n = 1.5$.
   d. Contois equation: $\mu_m = 0.25$ h^{-1}, $K_{sx} = 0.04$, $Y_{X/S}^M = 0.4$ g X/g S.

   $$S_0 = 5 \text{ g/l}$$

   Compare and comment on the results.

6.18. Consider steady-state operation of a chemostat. Assume that growth is substrate inhibited and that endogeneous metabolism can be ignored such that:

   $$\mu_{net} = \frac{\mu_m S}{K_s + S + S^2 / K_I}$$

Problems

a. Derive an expression for the residual substrate concentration (i.e., S) as a function of dilution rate and the kinetic parameters (μ_m, K_S, K_I).

b. What are the implications for operation of a chemostat when the organism is subjected to substrate inhibition?

6.19. Formation of lactic acid from glucose is realized in a continuous culture by *Streptococcus lactis*. The following information was obtained from experimental studies.

$S_0 = 5$ g/l, $\mu_m = 0.2$ h^{-1}, $K_S = 200$ mg/l, $k_d = 0.002$ h^{-1}, $Y_{X/S}^M = 0.4$ g X/g S, $Y_{P/S} = 0.2$ g P/g S, $q_P = 0.1$ g P/g X-h.

a. Plot the variations of S, X, P, DX, and DP with dilution rate.

b. Determine (graphically) the optimum dilution rate maximizing the productivities of biomass (DX) and the product (DP).

7

Stoichiometry
of Microbial Growth
and Product Formation

7.1. INTRODUCTION

Cell growth and product formation are complex processes reflecting the overall kinetics and stoichiometry of the thousands of intracellular reactions that can be observed within a cell. For many process calculations, we wish to compare potential substrates in terms of cell mass yield, or product yield, or evolution of heat. Also, we may need to know how close to its thermodynamic limit a system is operating. (That is, is product yield constrained by kinetic or thermodynamic considerations?) If a system is close to its thermodynamic limit, it would be unwise to try to improve production through mutation or genetic engineering.

Although the cell is complex, the stoichiometry of conversion of substrates into products and cellular materials is often represented by a simple pseudochemical equation. In this chapter we will discuss how these equations can be written and how useful estimates of key yield coefficients can be made.

7.2. SOME OTHER DEFINITIONS

In Chapter 6 we discussed the definitions of yield and maintenance coefficients, and we learned how to estimate their values using chemostat culture. In particular, we discussed the overall growth yield coefficient $Y_{X/S}^M$, which is the maximum yield of cell mass per unit mass of substrate consumed when no maintenance is considered.

TABLE 7.1 Growth Parameters of Some Organisms Growing Anaerobically in a Chemostat

Organism	Growth-limiting Factor	m_{ATP}	$Y^M_{X/ATP}$
Lactobacillus casei	Glucose	1.5	24.3
Enterobacter aerogenes	Glucose[a]	6.8	14.0
	Glucose[b]	2.3	17.6
	Tryptophan	38.7	25.4
	Citrate	2.2	9.0
Escherichia coli	Glucose	18.9	10.3
		6.9	8.5
Saccharomyces cerevisiae	Glucose	0.5	11.0
		0.25	13.0
Saccharomyces cerevisiae (petite)	Glucose	0.7	11.3
Candida parapsilosis	Glucose	0.2	12.5
Clostridium acetobutylicum	Glucose	—	23.8
Streptococcus cremoris	Lactose[c]	2.3	12.6

[a]Minimal medium. [b] Complex medium. [c] In the presence of a high extracellular lactate concentration.
(With permission, from B. Atkinson and F. Mavituna, *Biochemical Engineering and Biotechnology Handbook*, Macmillan, Inc., New York, 1983.)

Two other yield and maintenance coefficients of importance are related to ATP consumption and oxygen. The ATP yield coefficient, $Y_{X/ATP}$, represents the amount of biomass synthesized per mole of ATP generated. Surprisingly, it has been observed that for many substrates and organisms $Y^M_{X/ATP}$ is nearly constant at 10 to 11 g dry weight/mol ATP for heterotrophic growth under anaerobic conditions. The ATP yield for many autotrophic organisms (recall that autotrophic organisms fix CO_2) is $Y^M_{X/ATP} \approx 6.5$ g/mol ATP. Under aerobic conditions, the values for $Y^M_{X/ATP}$ are usually greater than 10.5 (see Table 7.1). Table 7.2 shows calculated ATP yields (maximum theoretical values) for a variety of media. A maintenance coefficient can also be estimated using an equation analogous to the one we developed for substrate yield coefficient in a chemostat:

$$\frac{1}{Y^{AP}_{X/ATP}} = \frac{1}{Y^M_{X/ATP}} + \frac{m_{ATP}}{D} \tag{7.1}$$

where $Y^{AP}_{X/ATP}$ is the "apparent" yield of biomass and m_{ATP} is the rate of ATP consumption for maintenance energy.

Similarly, yields based on oxygen consumption can be defined and calculated.

$$\frac{1}{Y^{AP}_{X/O_2}} = \frac{1}{Y^M_{X/O_2}} + \frac{m_{O_2}}{D} \tag{7.2}$$

TABLE 7.2 ATP Yields in Various Growth Media

Growth Medium	$Y_{X/ATP}$ (g cells/mol ATP)
Glucose + amino acids + nucleic acids	31.9
Glucose + inorganic salts	28.8
Pyruvate + amino acids + nucleic acids	21
Pyruvate + inorganic salts	13.5
CO_2 + inorganic salts (autotrophic growth)	6.5

As indicated in Table 6.1, values of Y_{X/O_2} can vary from 0.17 to 1.5 g biomass/g O_2, depending on substrate and organism.

Information from some measurements can be usefully combined. A particularly important derived parameter is the respiratory quotient (RQ), which is defined as the moles of CO_2 produced per mole of oxygen consumed. The RQ value provides an indication of metabolic state (for example, aerobic growth versus ethanol fermentation in baker's yeast) and can be used in process control.

We have already discussed (Chapter 5) the P/O ratio, which is the ratio of phosphate bonds formed per unit of oxygen consumed (g mole P/g atom O). The P/O ratio indicates the efficiency of conversion of reducing power into high-energy phosphate bonds in the respiratory chain. For eucaryotes, the P/O ratio approaches 3 when glucose is the substrate, while it is significantly less in procaryotes. A closely related parameter is the proton/oxygen ratio (H/O). This ratio is the number of H^+ ions released per unit of oxygen consumed. Electron generation is directly related to proton release. Usually 4 mol of electrons are consumed per mole of oxygen consumed. The generation of electrons results in the expulsion of H^+ that can be used directly to drive the transport of some substrates or to generate ATP.

The complexity of mass and energy balances for cellular growth can be decreased greatly through the recognition that some parameters are nearly the same irrespective of the species or substrate involved. These parameters can be referred to as *regularities*. For example, we have shown in Table 7.1 that $Y_{X/ATP}^M \geq 10.5$ g dry wt/mol ATP. Three important regularities (identified first by I. G. Minkevich and V. K. Eroshin) are 26.95 kcal/g equivalent of available electrons transferred to oxygen (coefficient of variation of 4%), 4.291 g equivalent of available electrons per quantity of biomass containing 1 g atom carbon, and 0.462 g carbon in biomass per gram of dry biomass. It has also been observed that $Y_{X/e^-} = 3.14 \pm 0.11$ g dry wt/g equivalent of electrons. These observed average values of cell composition and yields facilitate estimates of other growth-related parameters.

7.3. STOICHIOMETRIC CALCULATIONS

7.3.1. Elemental Balances

A material balance on biological reactions can easily be written when the compositions of substrates, products, and cellular material are known. Usually, electron–proton balances are required in addition to elemental balances to determine the stoichiometric coefficients in bioreactions. Accurate determination of the composition of cellular material is a major problem. Variations in cellular composition with different types of organisms are shown in Table 7.3. A typical cellular composition can be represented as $CH_{1.8}O_{0.5}N_{0.2}$. One mole of biological material is defined as the amount containing 1 gram atom of carbon, such as $CH_\alpha O_\beta N_\delta$.

Consider the following simplified biological conversion, in which no extracellular products other than H_2O and CO_2 are produced.

$$CH_m O_n + a\,O_2 + b\,NH_3 \longrightarrow c\,CH_\alpha O_\beta N_\delta + d\,H_2O + e\,CO_2 \qquad (7.3)$$

TABLE 7.3 Date on Elemental Composition of Several Microorganisms

Microorganism	Limiting Nutrient	μ (h⁻¹)	Composition (% by wt)							Empirical Chemical Formula	Formula "Molecular" Weight
			C	H	N	O	P	S	Ash		
Bacteria			53.0	7.3	12.0	19.0			8	$CH_{1.666}N_{0.20}O_{0.27}$	20.7
Bacteria			47.1	7.8	13.7	31.3				$CH_2N_{0.25}O_{0.5}$	25.5
Aerobacter aerogenes			48.7	7.3	13.9	21.1			8.9	$CH_{1.78}N_{0.24}O_{0.33}$	22.5
Klebsiella aerogenes	Glycerol	0.1	50.6	7.3	13.0	29.0				$CH_{1.74}N_{0.22}O_{0.43}$	23.7
K aerogenes	Glycerol	0.85	50.1	7.3	14.0	28.7				$CH_{1.73}N_{0.24}O_{0.43}$	24.0
Yeast			47.0	6.5	7.5	31.0			8	$CH_{1.66}N_{0.13}O_{0.40}$	23.5
Yeast			50.3	7.4	8.8	33.5				$CH_{1.75}N_{0.15}O_{0.5}$	23.9
Yeast			44.7	6.2	8.5	31.2	1.08	0.6		$CH_{1.64}N_{0.16}O_{0.52}P_{0.01}S_{0.005}$	26.9
Candida utilis	Glucose	0.08	50.0	7.6	11.1	31.3				$CH_{1.82}N_{0.19}O_{0.47}$	24.0
C. utilis	Glucose	0.45	46.9	7.2	10.9	35.0				$CH_{1.84}N_{0.2}O_{0.56}$	25.6
C. utilis	Ethanol	0.06	50.3	7.7	11.0	30.8				$CH_{1.82}N_{0.19}O_{0.46}$	23.9
C. utilis	Ethanol	0.43	47.2	7.3	11.0	34.6				$CH_{1.84}N_{0.2}O_{0.55}$	25.5

With permission, from B. Atkinson and F. Mavituna, *Biochemical Engineering and Biotechnology Handbook*, Macmillan, Inc., New York, 1983.

where CH_mO_n represents 1 mole of carbohydrate and $CH_\alpha O_\beta N_\delta$ stands for 1 mole of cellular material. Simple elemental balances on C, H, O, and N yield the following equations:

$$C: \quad 1 = c + e$$

$$H: \quad m + 3b = c\alpha + 2d$$

$$O: \quad n + 2a = c\beta + d + 2e \qquad (7.4)$$

$$N: \quad b = c\delta$$

The respiratory quotient (RQ) is

$$RQ = \frac{e}{a} \qquad (7.5)$$

Equations 7.4 and 7.5 constitute five equations for five unknowns a, b, c, d, and e. With a measured value of RQ, these equations can be solved to determine the stoichiometric coefficients.

7.3.2. Degree of Reduction

In more complex reactions, as in the formation of extracellular products, an additional stoichiometric coefficient is added, requiring more information. Also, elemental balances provide no insight into the energetics of a reaction. Consequently, the concept of *degree of reduction* has been developed and used for proton–electron balances in bioreactions. The degree of reduction, γ, for organic compounds may be defined as the number of equivalents of available electrons per gram atom C. The available electrons are those that would be transferred to oxygen upon oxidation of a compound to CO_2, H_2O, and NH_3. The degrees of reduction for some key elements are C = 4, H = 1, N = –3, O = –2, P = 5, and S = 6. The degree of reduction of any element in a compound is equal to the valence of this element. For example, 4 is the valence of carbon in CO_2 and –3 is the valence of N in NH_3. Degrees of reduction for various organic compounds are given in Table 7.4. The following are examples of how to calculate the degree of reduction for substrates.

$$\text{Methane } (CH_4): \qquad 1(4) + 4(1) = 8, \quad \gamma = 8/1 = 8$$

$$\text{Glucose } (C_6H_{12}O_6): \qquad 6(4) + 12(1) + 6(-2) = 24, \quad \gamma = 24/6 = 4$$

$$\text{Ethanol } (C_2H_5OH): \qquad 2(4) + 6(1) + 1(-2) = 12, \quad \gamma = 12/2 = 6$$

A high degree of reduction indicates a low degree of oxidation. That is, $\gamma_{CH_4} > \gamma_{EtOH} > \gamma_{glucose}$.

Consider the aerobic production of a single extracellular product.

$$\underset{\text{substrate}}{CH_mO_n} + aO_2 + bNH_3 \longrightarrow \underset{\text{biomass}}{cCH_\alpha O_\beta N_\delta} + \underset{\text{product}}{dCH_xO_yN_z} + eH_2O + fCO_2 \qquad (7.6)$$

The degrees of reduction of substrate, biomass, and product are

$$\gamma_s = 4 + m - 2n \qquad (7.7)$$

TABLE 7.4 Degree of Reduction and Weight of One Carbon Equivalent of One Mole of Some Substrates and Biomass

Compound	Molecular Formula	Degree of Reduction, γ	Weight, m
Biomass	$CH_{1.64}N_{0.16}O_{0.52}$ $P_{0.0054}S_{0.005}$[a]	4.17 (NH_3) 4.65 (N_2) 5.45 (HNO_3)	24.5
Methane	CH_4	8	16.0
n-Alkane	$C_{15}H_{32}$	6.13	14.1
Methanol	CH_4O	6.0	32.0
Ethanol	C_2H_6O	6.0	23.0
Glycerol	$C_3H_8O_3$	4.67	30.7
Mannitol	$C_6H_{14}O_6$	4.33	30.3
Acetic acid	$C_2H_4O_2$	4.0	30.0
Lactic acid	$C_3H_6O_3$	4.0	30.0
Glucose	$C_6H_{12}O_6$	4.0	30.0
Formaldehyde	CH_2O	4.0	30.0
Gluconic acid	$C_6H_{12}O_7$	3.67	32.7
Succinic acid	$C_4H_6O_4$	3.50	29.5
Citric acid	$C_6H_8O_7$	3.0	33.5
Formic acid	CH_2O_2	2.0	46.0
Oxalic acid	$C_2H_2O_4$	1.0	45.0

With permission, from B. Atkinson and F. Mavituna, *Biochemical Engineering and Biotechnology Handbook,* Macmillan, Inc., New York, 1983.

$$\gamma_b = 4 + \alpha - 2\beta - 3\delta \tag{7.8}$$

$$\gamma_p = 4 + x - 2y - 3z \tag{7.9}$$

Note that for CO_2, H_2O, and NH_3 the degree of reduction is zero.

Equation 7.6 can lead to elemental balances on C, H, O, and N, an available electron balance, an energy balance, and a total mass balance. Of the equations, only five will be independent. If all the equations are written, then the extra equations can be used to check the consistency of an experimental data set. Because the amount of water formed or used in such reactions is difficult to determine and water is present in great excess, the hydrogen and oxygen balances are difficult to use. For such a data set, we would typically choose a carbon, a nitrogen, and an available-electron balance. Thus,

$$c + d + f = 1 \tag{7.10}$$

$$c\delta + dz = b \tag{7.11}$$

$$c\gamma_b + d\gamma_p = \gamma_s - 4a \tag{7.12}$$

With partial experimental data, it is possible to solve this set of equations. Measurements of RQ and a yield coefficient would, for example, allow the calculation of the remaining coefficients. It should be noted that the coefficient, c, is $Y_{X/S}$ (on a molar basis) and d is $Y_{P/S}$ (also on a molar basis).

An energy balance for aerobic growth is

$$Q_0 c \gamma_b + Q_0 d \gamma_p = Q_0 \gamma_s - Q_0 4a \tag{7.13}$$

If Q_0, the heat evolved per equivalent of available electrons transferred to oxygen, is constant, eq. 7.13 is *not* independent of eq. 7.12. Recall that an observed regularity is 26.95 kcal/g equivalent of available electrons transferred to oxygen, which allows the prediction of heat evolution based on estimates of oxygen consumption.

Equations 7.12 and 7.13 also allow estimates of the fractional allocation of available electrons or energy for an organic substrate. Equation 7.12 can be rewritten as

$$1 = \frac{c \gamma_b}{\gamma_s} + \frac{d \gamma_p}{\gamma_s} + \frac{4a}{\gamma_s} \tag{7.14a}$$

$$1 = \xi_b + \xi_p + \varepsilon \tag{7.14b}$$

where ε is the fraction of available electrons in the organic substrate that is transferred to oxygen, ξ_b is the fraction of available electrons that is incorporated into biomass, and ξ_p is the fraction of available electrons that is incorporated into extracellular products.

Example 7.1

Assume that experimental measurements for a certain organism have shown that cells can convert two-thirds (wt/wt) of the substrate carbon (alkane or glucose) to biomass.

a. Calculate the stoichiometric coefficients for the following biological reactions:

Hexadecane: $C_{16}H_{34} + a\,O_2 + b\,NH_3 \longrightarrow c(C_{4.4}H_{7.3}N_{0.86}O_{1.2}) + d\,H_2O + e\,CO_2$

Glucose: $C_6H_{12}O_6 + a\,O_2 + b\,NH_3 \longrightarrow c(C_{4.4}H_{7.3}N_{0.86}O_{1.2}) + d\,H_2O + e\,CO_2$

b. Calculate the yield coefficients $Y_{X/S}$ (g dw cell/g substrate), Y_{X/O_2} (g dw cell/g O_2) for both reactions. Comment on the differences.

Solution

a. For hexadecane,

$$\text{amount of carbon in 1 mole of substrate} = 16(12) = 192 \text{ g}$$

$$\text{amount of carbon converted to biomass} = 192(2/3) = 128 \text{ g}$$

Then, $128 = c(4.4)(12)$; $c = 2.42$.

$$\text{amount of carbon converted to } CO_2 = 192 - 128 = 64 \text{ g}$$

$$64 = e\,(12), \qquad e = 5.33$$

The nitrogen balance yields

$$14b = c(0.86)(14)$$

$$b = (2.42)(0.86) = 2.085$$

The hydrogen balance is

$$34(1) + 3b = 7.3c + 2d$$

$$d = 12.43$$

The oxygen balance yields

$$2a(16) = 1.2c(16) + 2e(16) + d(16)$$

$$a = 12.427$$

For glucose,

amount of carbon in 1 mole of substrate = 72 g

amount of carbon converted to biomass = 72(2/3) = 48 g

Then, $48 = 4.4c(12)$; $c = 0.909$.

amount of carbon converted to $CO_2 = 72 - 48 = 24$ g

$$24 = 12e; \qquad e = 2$$

The nitrogen balance yields

$$14b = 0.86c(14)$$

$$b = 0.782$$

The hydrogen balance is

$$12 + 3b = 7.3c + 2d$$

$$d = 3.854$$

The oxygen balance yields

$$6(16) + 2(16)a = 1.2(16)c + 2(16)e + 16d$$

$$a = 1.473$$

b. For hexadecane,

$$Y_{X/S} = \frac{2.42(MW)_{biomass}}{(MW)_{substrate}}$$

$$Y_{X/S} = \frac{2.42(91.34)}{226} = 0.98 \text{ gdw cells/g substrate}$$

$$Y_{X/O_2} = \frac{2.42(MW)_{biomass}}{12.43(MW)_{O_2}}$$

$$Y_{X/O_2} = \frac{2.42(91.34)}{(12.43)(32)} = 0.557 \text{ gdw cells/g } O_2$$

For glucose,

$$Y_{X/S} = \frac{(0.909)(91.34)}{180} = 0.461 \text{ gdw cells/g substrate}$$

$$Y_{X/O_2} = \frac{(0.909)(91.34)}{(1.473)(32)} = 1.76 \text{ gdw cells/g } O_2$$

The growth yield on more reduced substrate (hexadecane) is higher than that on partially oxidized substrate (glucose), assuming that two-thirds of all the entering carbon is incorporated in cellular structures. However, the oxygen yield on glucose is higher than that on the hexadecane, since glucose is partially oxidized.

7.4. THEORETICAL PREDICTIONS OF YIELD COEFFICIENTS

In aerobic fermentations, the growth yield per available electron in oxygen molecules is approximately 3.14 ± 0.11 gdw cells/electron when ammonia is used as the nitrogen source. The number of available electrons per oxygen molecule (O_2) is four. When the number of oxygen molecules per mole of substrate consumed is known, the growth yield coefficient, $Y_{X/S}$, can easily be calculated. Consider the aerobic catabolism of glucose.

$$C_6H_{12}O_6 + 6\,O_2 \longrightarrow 6\,CO_2 + 6\,H_2O$$

The total number of available electrons in 1 mole of glucose is 24. The cellular yield per available electron is $Y_{X/S} = 24(3.14) = 76$ gdw cells/mol.

The predicted growth yield coefficient is $Y_{X/S} = 76/180 = 0.4$ gdw cells/g glucose. Most measured values of $Y_{X/S}$ for aerobic growth on glucose are 0.38 to 0.51 g/g (see Table 6.1).

The ATP yield ($Y_{X/ATP}$) in many anaerobic fermentations is approximately 10.5 ± 2 gdw cells/mol ATP. In aerobic fermentations, this yield varies between 6 and 29. When the energy yield of a metabolic pathway is known (N moles of ATP produced per gram of substrate consumed), the growth yield $Y_{X/S}$ can be calculated using the following equation:

$$Y_{X/S} = Y_{X/ATP}\, N$$

Example 7.2

Estimate the theoretical growth and product yield coefficients for ethanol fermentation by *S. cerevisiae* as described by the following overall reaction:

$$C_6H_{12}O_6 \longrightarrow 2\,C_2H_5OH + 2\,CO_2$$

Solution   Since $Y_{X/ATP} \approx 10.5$ gdw/mol ATP and since glycolysis yields 2 ATP/mol of glucose in yeast,

$$Y_{X/S} \approx 10.5 \text{ gdw/mol ATP} \cdot 2\frac{\text{moles ATP}}{180 \text{ g glucose}}$$

or

$$Y_{X/S} \approx 0.117 \text{ gdw/g glucose}$$

For complete conversion of glucose to ethanol by the yeast pathway, the maximal yield would be

$$Y_{P/S} = \frac{2(46)}{180} = 0.51 \text{ g ethanol/g glucose}$$

while for CO_2 the maximum yield is

$$Y_{CO_2/S} = \frac{2(44)}{180} = 0.49 \text{ g ethanol}/\text{g glucose}$$

In practice, these maximal yields are not obtained. The product yields are about 90% to 95% of the maximal values, because the glucose is converted into biomass and other metabolic by-products (e.g., glycerol or acetate).

7.5. SUMMARY

Simple methods to determine the reaction stoichiometry for bioreactors are reviewed. These methods lead to the possibility of predicting yield coefficients for various fermentations using a variety of substrates. By coupling these equations to experimentally measurable parameters, such as the respiratory quotient, we can infer a great deal about the progress of a fermentation. Such calculations can also assist in initial process design equations by allowing the prediction of the amount of oxygen required (and consequently heat generated) for a certain conversion of a particular substrate. The prediction of yield coefficients is not exact, because unknown or unaccounted for metabolic pathways and products are present. Nonetheless, such calculations provide useful first estimates of such parameters.

SUGGESTIONS FOR FURTHER READING

ATKINSON, B., AND F. MAVITUNA, *Biochemical Engineering and Biotechnology Handbook*, Macmillan, Inc., New York, 1983.

BAILEY, J. E., AND D. F. OLLIS, *Biochemical Engineering Fundamentals*, 2d ed., McGraw-Hill Book Co., New York, 1986.

ERICKSON, L. E., AND D. Y.-C. FUNG, *Handbook on Anerobic Fermentations*, Marcel Dekker, Inc., New York, 1988. (Five chapters deal with bioenergetics, stoichiometry, and yields.)

————, I. G. MINKEVICH, AND V. K. EROSHIN, Application of Mass and Energy Balance Regularities in Fermentation, *Biotechnol. Bioeng. 20*:1595, 1978.

MINKEVICH, I. G., Mass and Energy Balance for Microbial Product Synthesis: Biochemical and Cultural Aspects, *Biotechnol. Bioeng. 25*:1267, 1983.

ROELS, J. A., *Energetics and Kinetics in Biotechnology*, Elsevier Science Publishing, New York, 1983.

PROBLEMS

7.1. Determine the amount of $(NH_4)_2SO_4$ to be supplied in a fermentation medium where the final cell concentration is 30 g/l in a 10^3 l culture volume. Assume that the cells are 12% nitrogen by weight and $(NH_4)_2SO_4$ is the only nitrogen source.

7.2. The growth of baker's yeast (*S. cerevisiae*) on glucose may be simply described by the following equation:

$$C_6H_{12}O_6 + 3\ O_2 + 0.48\ NH_3 \xrightarrow{\text{yeast}} 0.48\ C_6H_{10}NO_3 + 4.32\ H_2O + 3.12\ CO_2$$

In a batch reactor of volume 10^5 l, the final desired yeast concentration is 50 gdw/l.

Using the above reaction stoichiometry:

a. Determine the concentration and total amount of glucose and $(NH_4)_2SO_4$ in the nutrient medium.

b. Determine the yield coefficients $Y_{X/S}$ (biomass/glucose) and Y_{X/O_2} (biomass/oxygen).

c. Determine the total amount of oxygen required.

d. If the rate of growth at exponential phase is $r_x = 0.7$ gdw/l-h, determine the rate of oxygen consumption (g O_2/l-h).

e. Calculate the heat-removal requirements for the reactor (recall eq. 6.26).

7.3. The growth of *S. cerevisiae* on glucose under anaerobic conditions can be described by the following overall reaction:

$$C_6H_{12}O_6 + \beta\ NH_3 \longrightarrow 0.59\ CH_{1.74}N_{0.2}O_{0.45}\ \text{(biomass)} + 0.43\ C_3H_8O_3 + 1.54$$
$$CO_2 + 1.3\ C_2H_5OH + 0.036\ H_2O$$

a. Determine the biomass yield coefficient $Y_{X/S}$.

b. Determine the product yield coefficients $Y_{EtOH/S}$, $Y_{CO_2/S}$, $Y_{C_3H_2O/S}$.

c. Determine the coefficient β.

7.4. Aerobic growth of *S. cerevisiae* on ethanol is simply described by the following overall reaction:

$$C_2H_5OH + a\ O_2 + b\ NH_3 \longrightarrow c\ CH_{1.704}N_{0.149}O_{0.408} + d\ CO_2 + e\ H_2O$$

a. Determine the coefficients a, b, c, and d, where RQ = 0.66.

b. Determine the biomass yield coefficient, $Y_{X/S}$, and oxygen yield coefficient, Y_{X/O_2} (gdw/g O_2).

7.5. Aerobic degradation of benzoic acid by a mixed culture of microorganisms can be represented by the following reaction.

$$\underset{\text{(substrate)}}{C_6H_5COOH} + a\ O_2 + b\ NH_3 \longrightarrow \underset{\text{(bacteria)}}{c\ C_5H_7NO_2} + d\ H_2O + e\ CO_2$$

a. Determine a, b, c, d, and e if RQ = 0.9.

b. Determine the yield coefficients, $Y_{X/S}$ and Y_{X/O_2}.

c. Determine degree of reduction for the substrate and bacteria.

7.6. Aerobic degradation of an organic compound by a mixed culture of organisms in waste water can be represented by the following reaction.

$$C_3H_6O_3 + a\ O_2 + b\ NH_3 \rightarrow c\ C_5H_7NO_2 + d\ H_2O + e\ CO_2$$

a. Determine a, b, c, d, and e, if $Y_{X/S} = 0.4$ g X/g S.

b. Determine the yield coefficients Y_{X/O_2} and Y_{X/NH_3}.

c. Determine the degree of reductions for the substrate, bacteria, and RQ for the organisms.

7.7. Biological denitrification of nitrate-containing waste waters can be described by the following overall reaction.

$$NO_3^{-1} + a\,CH_3OH + H^+ \longrightarrow b\,C_5H_7NO_2 + c\,N_2 + d\,CO_2 + e\,H_2O$$

a. Determine a, b, c, d, and e, if $Y_{X/S} = 0.5$ g X/g N.
b. Determine the degree of reduction of bacteria and methanol.

How Cellular Information Is Altered

8.1. INTRODUCTION

We have already discussed some aspects of how cells inherit information and how a chemostat (Chapter 6) can be used as a tool to select for individual cells with different or augmented metabolism. The mechanism for DNA replication in procaryotes has been summarized in Chapter 4. This process is a good example of the exchange of genetic information from one generation to another. However, some individuals can receive additional genetic information through natural or artificial means. The initial genetic information within a cell may also undergo rearrangements or alterations. In this chapter we will discuss some mechanisms causing alterations in a cell's content of genetic information and ways that we can manipulate those mechanisms to improve bioprocesses.

8.2. EVOLVING DESIRABLE BIOCHEMICAL ACTIVITIES THROUGH MUTATION AND SELECTION

Although the cell has a well-developed system to prevent errors in DNA replication and an active repair system to correct damage to a DNA molecule, mistakes occur. These mistakes are called *mutations*. Before we discuss mutations, we need to establish the working vocabulary of microbial genetics.

219

The sum of the genetic construction of an organism constitutes its *genotype*. The characteristics expressed by a cell constitute its *phenotype*. The phenotypic response of a culture may change reversibly with alterations in environmental conditions, whereas the genotype is constant irrespective of the environment. A *mutation* is a genotypic change and is irreversible. A whole culture undergoes a phenotypic response, whereas only a rare individual will undergo a genotypic change. For example, if a culture changes in color from white to green when oxygen levels fall and then changes from green to white upon an increase in dissolved oxygen, the change would be phenotypic. Now consider an alternative experiment where white cells were removed from a culture and placed on a plate (a small circular dish filled with nutrients solidified with agar) and allowed to grow into separate colonies. If one colony, but not the others, turned green and if cells obtained from the green colony remain green when cultured under the original conditions, it would be evidence for a genotypic change. In this case, the white cells would be the *wild type* and the green cells the *mutants*.

Let us consider what mechanisms may lead to genotypic change.

8.2.1. How Mutations Occur

Most mutations occur due to mistakes in DNA synthesis. Some examples are shown in Fig. 8.1. One common form is a *point mutation*. A point mutation results from the change of a single base (for example, cytosine instead of thymine). Some point mutations are *silent mutations* because the altered codon still codes for the same amino acid (e.g., UCU and UCA both code for serine). Even if the point mutation causes the substitution of a different amino acid, it may or may not alter protein activity substantially. A change of amino acid near the active site might alter protein activity greatly, whereas the same substitution at another site might not be very critical.

One type of point mutation that usually has a profound effect results in a nonsense or stop codon (e.g., CAA for glutamine to UAA for stop on the *m*-RNA from the altered DNA). This results in the premature termination of translation and an incompletely formed protein.

Generally, *deletion mutations* have profound effects on cellular metabolism. By deleting or adding one or more bases, we can alter the whole composition of a protein, not just a single amino acid. A deletion can shift the *reading frame* when translating the resulting *m*-RNA. This is illustrated in Fig. 8.2.

Additions often take place through *insertion elements* (IS). These elements are about 700 to 1400 base pairs in length; in *E. coli* about five different IS sequences are known and are present on the chromosome. These elements can move on the chromosome from essentially any one site to another. Often they will insert in the middle of a gene, totally destroying its function.

Back mutations or *reversions* are possible. Revertants are cells for which the original wild-type phenotype has been restored. Restoration of a function can occur due to a direct change at the original mutation (e.g., if the original mutation was CAA to UAA, then a second mutation for UAA to CAA restores the original genotype and phenotype). Second-site revertants can occur that restore phenotype (*suppressor mutations*), but not genotype (e.g., a second deletion mutation that restores the gene to the normal reading frame or a mutation in another gene that restores the wild-type phenotype).

Wild type

```
G ‥ C
C ‥ G
A ‥ T
T ‥ A
C ‥ G
G ‥ C
A ‥ T
G ‥ C
T ‥ A
A ‥ T
C ‥ G
```

```
G ‥ C
C ‥ G
A ‥ T
T ‥ A
C ‥ G
G ‥ C
```
Point mutation———T ‥ T
```
G ‥ C
T ‥ A
A ‥ T
C ‥ G
```

Deletion

```
G ‥ C
C ‥ G
A ‥ T      A  G
         ▓▓▓        C
T ‥ A
A ‥ T    C  G
C ‥ G
```

Figure 8.1. DNA base changes from wild type, involving point mutation and deletion. (With permission, from T. D. Brock, D. W. Smith, and M. T. Madigan, *Biology of Microorganisms*, 4th ed., Pearson Education, Upper Saddle River, NJ, 1984, p. 307.)

8.2.2. Selecting for Desirable Mutants

Mutants can serve as powerful tools to better understand cell physiology; they are also valuable as industrial organisms, because mutation can be used to alter metabolic regulation and to cause overproduction of a desired compound. Methods to induce mutations and then select for mutants are important tools for catalyst development in bioprocessing.

Natural (*spontaneous*) rates of mutation vary greatly from gene to gene (10^{-3} to 10^{-9} per cell division), with 10^{-6} mutations in a gene per cell division being typical. Chemical agents (*mutagens*) or radiation are often used in the laboratory to increase mutation rates. Mutagens are nonspecific and may affect any gene.

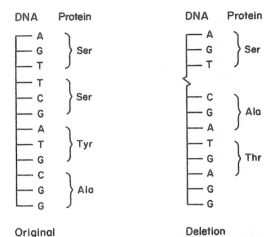

DNA	Protein		DNA	Protein
A			A	
G	Ser		G	Ser
T			T	
T			C	
C	Ser		G	Ala
G			A	
A			T	
T	Tyr		G	Thr
G			A	
C			G	
G	Ala		G	
G				

Original   Deletion

Figure 8.2. Effect of deletion of a base on the reading frame and the protein encoded.

   The selection of a mutant with desirable properties is no easy task. Mutations are classified as *selectable* and *unselectable*. A selectable mutation confers upon the mutant an advantage for growth or survival under a specific set of environmental conditions; thus, the mutant can grow and the wild type will die. An unselectable mutant requires a cell-by-cell examination to find a mutant with the desired characteristics (e.g., green pigment). Even with mutagens, the frequency of mutation is sufficiently low to make prohibitive a brute-force screening effort for most unselectable mutants.

   Selection can be direct or indirect. An example of direct selection would be to find a mutant resistant to an antibiotic or toxic compound. A culture fluid containing 10^8 to 10^{10} cells/ml is subjected to a mutagenic agent. A few drops of culture fluid are spread evenly on a plate, with the antibiotic incorporated into the gelled medium. Only antibiotic-resistant cells can grow, so any colonies that form must arise from antibiotic-resistant mutants. If one in a million cells has this particular mutation, we would expect to find about 10 to 100 colonies per plate if 0.1 ml of culture fluid was tested.

   Indirect selection is used for isolating mutants that are deficient in their capacity to produce a necessary growth factor (e.g., an amino acid or a vitamin). Wild-type *E. coli* grow on glucose and mineral salts. *Auxotrophic* mutants would not grow on such a simple medium unless it were supplemented with the growth factor that the cell could no longer make (e.g., a lysine auxotroph has lost the capacity to make lysine, so lysine must be added to the glucose and salts to enable the cell to grow). The wild-type cell that needs no supplements to a minimal medium is called a *prototroph*. Consider the selection of a rare mutant cell that is auxotrophic for lysine from a population of wild-type cells. This cannot be done directly, since both cell types would grow in the minimal medium supplemented with lysine. A method that facilitates selection greatly is called *replicate plating* (see Fig. 8.3). A master plate using lysine-supplemented medium will grow both the auxotroph and wild-type cells. Once colonies are well formed on the master plate, an imprint is made on sterile velveteen. The bristles on the velveteen capture some of the cells from each colony. The orientation of the master plate is carefully noted. Then a test plate with minimal medium is pressed against the velveteen; some cells, at the point of each previous colony, then serve to inoculate

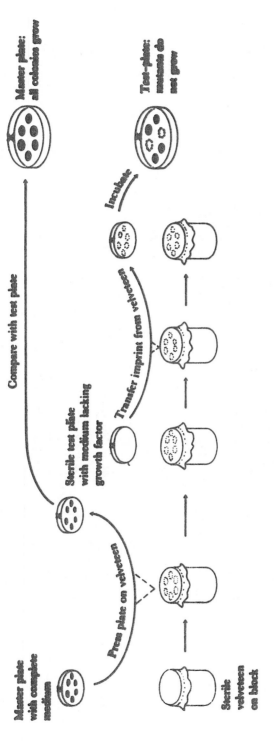

Figure 8.3. Replica-plating method for detecting nutritional mutants. (With permission, from T. D. Brock, D. W. Smith, and M. T. Madigan, *Biology of Microorganisms*, 4th ed., Pearson Education, Upper Saddle River, NJ, 1984. p. 306.)

TABLE 8.1 Kinds of Mutants

Description	Nature of Change	Detection of Mutant
Nonmotile	Loss of flagella; nonfunctional flagella	Compact colonies instead of flat, spreading colonies
Noncapsulated	Loss or modification of surface capsule	Small, rough colonies instead of larger, smooth colonies
Rough colony	Loss or change in lipopolysaccharide outer layer	Granular, irregular colonies instead of smooth, glistening colonies
Nutritional	Loss of enzyme in biosynthetic pathway	Inability to grow on medium lacking the nutrient
Sugar fermentation	Loss of enzyme in degradative pathway	Do not produce color change on agar containing sugar and a pH indicator
Drug resistant	Impermeability to drug or drug target is altered or drug is detoxified	Growth on medium containing a growth-inhibitory concentration of the drug
Virus resistant	Loss of virus receptor	Growth in the presence of large amounts of virus
Temperature sensitive	Alteration of any essential protein so that it is more heat sensitive	Inability to grow at a temperature normally supporting growth (e.g., 37°C) but still growing at a lower temperature (e.g., 25°C)
Cold sensitive	Alteration in an essential protein so that it is inactivated at low temperature	Inability to grow at a low temperature (e.g., 20°C) that normally supports growth

With permission, from T. D. Brock, D. W. Smith, and M. T. Madigan, *Biology of Microorganisms*, 4th ed., Pearson Education, Upper Saddle River, NJ, 1984, p. 306.

the test plate at positions identical to those on the master plate. After incubation (approximately 24 hours for *E. coli*), the test plate is compared to the master plate. Colonies that appear at the same positions on both plates arise from wild-type cells, while colonies that exist only on the master plate must arise from the auxotrophic mutants.

Another class of mutants is called *conditional* mutants. Mutations that would normally be lethal to the cell could not be detected by methods we have described so far. However, mutated proteins are often more temperature sensitive than normal proteins. Thus, temperature sensitivity can often be used to select for conditionally lethal mutations. For example, the mutant may be unable to grow at the normal growth temperature (e.g., 37°C for *E. coli*) but will grow satisfactorily at a lower temperature (e.g., 25°C). Table 8.1 summarizes a variety of mutants and how they may be detected.

Mutation and selection have been used to tremendous advantage to probe the basic features of cell physiology and regulation. They also have been the mainstay of industrial programs for the improvement of production strains. Mutation and selection programs have been primarily responsible for increasing the yield of penicillin from 0.001 g/l in 1939 to current values of about 50 g/l of fermentation broth.

8.3. NATURAL MECHANISMS FOR GENE TRANSFER AND REARRANGEMENT

Bacteria can gain and express wholly different biochemical capabilities (e.g., the ability to degrade an antibiotic or detoxify a hazardous chemical in their environment) literally overnight. These alterations cannot be explained through inheritance and small evolutionary changes in the chromosome. Rather, they arise from gene transfer from one organism to another and/or large rearrangements in chromosomal DNA. In this section we will discuss genetic recombination, gene transfer, and genetic rearrangements—all mechanisms that can be exploited to genetically engineer cells. (See Table 8.1.)

8.3.1. Genetic Recombination

Genetic recombination is a process that brings genetic elements from two different genomes into one unit, resulting in new genotypes in the absence of mutations. Genetic recombination in procaryotes is a rare event, but sufficiently frequent to be important industrially and ecologically. The three main mechanisms for gene transfer are *transformation, transduction,* and *conjugation.* Transformation is a process in which free DNA is taken up by a cell. Transduction is a process in which DNA is transferred by a bacteriophage, and conjugation is DNA transfer between intact cells that are in direct contact with one another.

Once *donor DNA* is inside a cell, the mechanism for recombination is essentially independent of how the donor DNA was inserted. Figure 8.4 summarizes the molecular-level events in general recombination. The donor DNA must be homologous, or nearly so, to a segment of DNA on the recipient DNA.[†] Under the right conditions, cellular enzymes

---

[†]*Illegitimate recombination* between nonhomologous regions of DNA is possible, but rare. See later discussion on transposons.

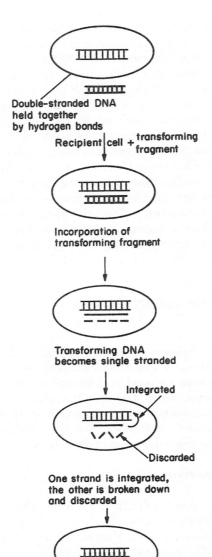

Figure 8.4. Integration of transforming DNA into a recipient cell. (With permission, from T. D. Brock, D. W. Smith, and M. T. Madigan, *Biology of Microorganisms*, 4th ed., Pearson Education, Upper Saddle River, NJ, 1984, p. 353.)

cut out the homologous section of recipient DNA, allow insertion of the donor DNA, and then ligate or join the ends of the donor DNA to the recipient DNA. Pieces of donor DNA that a cell recognizes as foreign are usually degraded by enzymes called *restriction endonucleases* (these enzymes are essential in genetic engineering). A cell marks its own DNA (e.g., through methylation of certain purine or pyrimidine bases) to distinguish it from foreign DNA. These modifications block the action of a cell's own restriction endonucleases on its own DNA. Under natural conditions, gene transfer is effective only if the donor DNA is from the same or closely related species.

Let us now consider some details of how donor DNA can enter a cell.

8.3.2. Transformation

The uptake of naked DNA cannot be done by all genera of bacteria. Even within transformable genera, only certain strains are transformable (*competent*). Competent cells have a much higher capacity for binding DNA to the cell surface than do noncompetent cells. Competency can depend on the physiological state of the cell (current and previous growth conditions). Even in a competent population not all cells are transformable. Typically, about 0.1% to 1.0% are transformable.

E. coli are not normally competent, but their importance to microbial genetics has led to the development of empirical procedures to induce competency. This procedure involves treating *E. coli* with high concentrations of calcium ions, coupled with temperature manipulation. The competency of treated cells varies among strains of *E. coli* but is typically rather low (about one in a million cells becomes successfully transformed). With the use of selective markers, this frequency is still high enough to be quite useful.

Transformation is useful only when the information that enters the cell can be propagated. When doing transformation, we typically use a vector called a *plasmid*. This element forms the basis for most industrially important fermentations with recombinant DNA. A plasmid is an autonomous, self-replicating, double-strand piece of DNA that is normally extrachromosomal. Some plasmids are maintained as low-copy-number plasmids (only a few copies per cell), and others have a high copy number (20 to 100 copies per cell). These plasmids differ in their mechanisms for partitioning at cell division and in the control of their replication. Plasmids encode genes typically for proteins that are nonessential for growth, but that can confer important advantages to their host cells under some environmental circumstances. For example, most plasmids encode proteins that confer resistance to specific antibiotics. Such antibiotic resistance is very helpful in selecting for cells that contain a desired plasmid.

8.3.3. Transduction

DNA transfer from one cell to another can be mediated by viruses and certainly plays an important role in nature. In the most common type of transduction, *generalized transduction*, infection of a recipient cell results in fragmentation of the bacterial DNA into 100 or so pieces. One of these fragments can be packaged accidentally into a phage particle. The defective phage particle then injects bacterial DNA into another cell, where it can recombine with that cell's DNA. With generalized transduction, any bacterial gene may be transferred.

Another method of transfer, which is far more specific with respect to the genes that are transferred, is *specialized transduction*. Here the phage incorporates into specific sites in the chromosome, and the frequency of transduction of a gene is related to its distance away from the site of incorporation. This process is summarized in Fig. 8.5. A *lysogenic cell* is one carrying a prophage or phage DNA incorporated into chromosomal DNA. Phage lambda is an example of such a *temperate phage* (a phage that can either lyse a cell or become incorporated into the chromosome). Such phages almost invariably insert at a specific site in the chromosome. The conversion of a *prophage* (the phage DNA in the chromosome) into the lytic cycle is normally a rare event (10^{-4} per cell division), but it can be induced in almost the whole culture upon exposure to UV light or other agents that interfere with DNA replication.

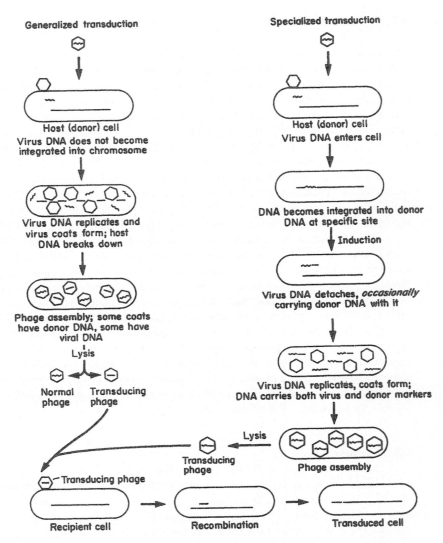

Generalized transduction

Host (donor) cell
**Virus DNA does not become
integrated into chromosome**

**Virus DNA replicates and
virus coats form; host
DNA breaks down**

**Phage assembly; some coats
have donor DNA, some have
viral DNA**

Lysis

Normal   Transducing
phage   phage

Transducing phage

Recipient cell → Recombination → Transduced cell

Specialized transduction

Host (donor) cell
Virus DNA enters cell

**DNA becomes integrated into donor
DNA at specific site**

Induction

**Virus DNA detaches, *occasionally*
carrying donor DNA with it**

**Virus DNA replicates, coats form;
DNA carries both virus and donor markers**

Lysis

Transducing
phage

Phage assembly

Figure 8.5. Transduction, the transfer of genetic material from donor to recipient via virus particles. (With permission, from T. D. Brock, K. M. Brock, and D. M. Ward, *Basic Microbiology with Applications*, 3d ed., Pearson Education, Upper Saddle River, NJ, 1986, p. 161.)

8.3.4. Episomes and Conjugation

A third type of gene transfer involves another genetic element. This element is called an *episome*. It is a DNA molecule that may exist either integrated into the chromosome or separate from it. When it exists separately from the chromosome (extrachromosomally), it is essentially a plasmid. A well-known episome is the F or fertility factor. Such factors are responsible for the process known as *conjugation*.

Most experiments with conjugation are done with the F factor, which is present in low copy number. Direct cell-to-cell contact is required. This DNA molecule encodes at least 13 genes involved in its self-transfer from one cell to another.

In a population of *E. coli* there are frequently some cells with the F plasmid, which are termed F^+ (male). Other cells are F^- (female). F^+ cells encode proteins to make a *sex pilus*. When F^+ and F^- cells are mixed together, the sex pilus connects an F^+ to an F^- cell (see Fig. 8.6). The sex pilus may act as a conduit for the transfer of a copy of the F plasmid to the F^- cell. The actual process of transfer involves replication of the F plasmid.

This process is normal and does not involve transfer of chromosomal genes or recombination. A more rare event is when the F plasmid has been integrated into the chromosome itself to form a single, large, circular molecule. Thirteen sites for integration are known. Such cells are termed Hfr (for high-frequency recombination).

When transfer is initiated, the F plasmid moves not only itself, but also the attached chromosome, to the recipient cell. The time required to transfer a whole *E. coli* chromosome is 100 min. If contact between the two cells is broken during the transfer process, only a proportional amount of the chromosome will have been transferred (that is, at 50 min, about 50%). Since the transfer begins at a known point, Hfr cells can be used to map the location of genes on the chromosome. This technique for gene mapping is being replaced by methods for directly sequencing nucleotide sequences in DNA. If F^+ and F^- cells differ in properties (e.g., the ability to make lysine), conjugation can be used to alter the properties of the F^- cell.

Conjugation, transduction, and transformation all represent forms of gene transfer from one cell to another. However, gene transfer can occur within a cell.

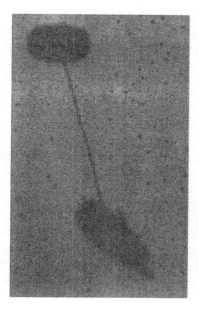

Figure 8.6. Direct contact between two conjugating bacteria is first made via a pilus. The cells are then drawn together for the actual transfer of DNA. (With permission, from T. D. Brock, K. M. Brock, and D. M. Ward, *Basic Microbiology with Applications*, 3d ed., Pearson Education, Upper Saddle River, NJ, 1986, p. 161.)

8.3.5. Transposons: Internal Gene Transfer

Previously, we discussed the presence of *insertion elements* on the chromosome. A closely related phenomenon is a *transposon*, which refers to a gene or genes that have the ability to "jump" from one piece of DNA to another, or to another position on the original piece of DNA. The transposon integrates itself into the new position independently of any homology with the recipient piece of DNA. Transposons differ from insertion sequences in that they code for proteins. Transposons appear to arise when a gene becomes bounded on both sides by insertion sequences. Many of the transposons encode antibiotic resistance.

Transposons are important because (1) they can induce mutations when they insert into the middle of a gene, (2) they can bring once-separate genes together, and (3) in combination with plasmid- or viral-mediated gene transfer, they can mediate the movement of genes between unrelated bacteria (e.g., multiple antibiotic resistance on newly formed plasmids). Transposon mutagensis can be a very powerful tool in altering cellular properties.

8.4. GENETICALLY ENGINEERING CELLS

Our description of DNA replication, mutation, and selection and the natural mechanisms for gene transfer provide the reader with a knowledge of all the tools necessary to genetically engineer a cell. The purposeful, predetermined manipulation of cells at the genetic level, an idea that was farfetched before 1970, is easily within the grasp of beginning college students.

Genetic engineering is a set of tools and not a scientific discipline. Although difficult to define precisely, it involves the manipulation of DNA outside the cell to create artificial genes or novel combinations of genes with predesigned control elements. Because many of these manipulations can be done outside the cell, *we can circumvent species limitations* that limit the age-old techniques of mutation and selection and breeding (e.g., we can express a human protein in *E. coli*). Learning how to use these tools is the basis of modern biotechnology.

8.4.1. Basic Elements of Genetic Engineering

An overview of the strategy typically employed in genetic engineering is given in Fig. 8.7. The strategy makes use of *recombinant DNA* techniques, the ability to isolate genes from one organism and recombine the isolated gene with other DNA that can be propagated in a similar or unrelated host. Most of our discussion will be drawn from approaches for genetically engineering bacteria.

The first step is obtaining the gene of interest. A simple, brute-force approach is *shotgun cloning*. Here the DNA from the donor organism is cut into fragments using *restriction enzymes*. If an efficient screening procedure is available, large numbers of host cells with random fragments of DNA can be screened for those with a property related to the desired gene.

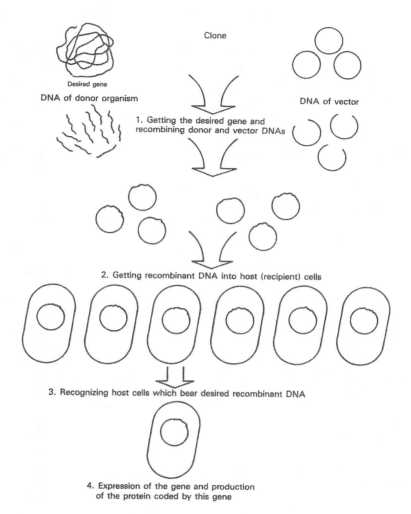

Clone

DNA of donor organism

Desired gene

DNA of vector

1. Getting the desired gene and recombining donor and vector DNAs

2. Getting recombinant DNA into host (recipient) cells

3. Recognizing host cells which bear desired recombinant DNA

4. Expression of the gene and production of the protein coded by this gene

Figure 8.7. Overview of the essential steps in genetic engineering: moving a gene from one organism to another. (With permission, from T. D. Brock, K. M. Brock, and D. M. Ward, *Basic Microbiology with Applications*, 3d ed., Pearson Education, Upper Saddle River, NJ, 1986, p. 169.)

Most often, such a shotgun approach is very inefficient. More specific approaches use *hybridization*. A *probe* can be synthesized chemically to be complementary to a portion of the gene. The probe is usually much shorter than the gene, but sufficiently long that it is unlikely for other genes to have the same complementary DNA sequence. The construction of the probe requires some knowledge of either the nucleotide sequence of the desired gene or a partial amino acid sequence for the desired gene. Since the genetic code is degenerate, the deduction of the actual nucleotide sequence is ambiguous. This ambiguity requires that a variety of probes be generated. Hybridization reactions require the donor DNA to be both fragmented and converted into single strands that can react with the single-stranded probes.

An alternative to hybridization is total *chemical synthesis* of a gene that corresponds to the desired protein. This alternative requires knowledge of the amino acid sequence of the desired protein. An artificial gene may code for exactly the same protein as in nature, even if the sequence of nucleotides on the artificial gene is not identical to the natural gene. Chemical synthesis also allows us to produce specifically modified natural proteins or potentially totally human-designed proteins.

Another method to obtain the desired gene is particularly useful for genes with *introns*. Since bacteria lack the cellular machinery to cutout introns and to do *m*-RNA splicing, eucaryotic genes with introns cannot be directly placed in bacteria to make a desired protein. Often we wish to make these proteins in bacteria, since the bacteria grow much more rapidly and are much easier and cheaper to culture. Often the processed *m*-RNA for the desired gene can be isolated directly from the donor organism's cytoplasm (using hybridization probes). Once the *m*-RNA is isolated, the enzyme *reverse transcriptase* (see Chapter 4) can be used to synthesize a DNA molecule with the corresponding nucleotide sequence; this molecule is called *complementary DNA* or *c DNA*.

Once the desired gene is isolated or made, it can be inserted into a small piece of carrier DNA called a *vector*. Typically, vectors are plasmids, although temperate viruses can be used. The process for preparing the donor DNA and vector for recombination and the actual joining of the DNA segments requires special enzymes (see Fig. 8.8); we have discussed these enzymes in our previous consideration of DNA replication and genetic recombination. A wide variety of *restriction enzymes* exist that will cut DNA at a different prespecified site. Most vectors have maps showing the various restriction sites; important examples are EcoR1 from *E. coli* and Bam H1 from *Bacillus amylofacieus*. Many restriction enzymes leave "sticky ends," a few nucleotides of single-stranded DNA projecting from the cut site. Pieces of DNA with complementary "sticky ends" naturally associate, and in a mixture of cut vector and donor DNA, some pieces of donor DNA will associate with vector DNA. *DNA ligase* can permanently join these ends.

The mixture bearing the desired vector–donor combinations is then moved into the recipient or host cell. In most cases this is done by transformation, although other techniques can be used if transformation of the host is difficult. Note that the construction of the desired vector–donor DNA usually results in a mixture (e.g., some vector molecules may be opened and rejoined without donor DNA being inserted, or multiple copies of donor DNA may be inserted, or DNA contaminants of the donor DNA mixture may become inserted into the vector). Consequently, an efficient method to screen transformants for those with the desired vector–donor DNA combination is important.

Most vectors contain selectable markers such as antibiotic resistance or the genes to make essential growth factors that have been removed by mutation from the host cell. In the latter case, growth in minimal medium is possible only in the presence of the plasmid. These selectable markers allow the isolation of genetic *clones* that have been successfully transformed. A further screening step is then necessary to ensure that the donor DNA is present and being *expressed* (i.e., a functional protein is being made from the donor DNA).

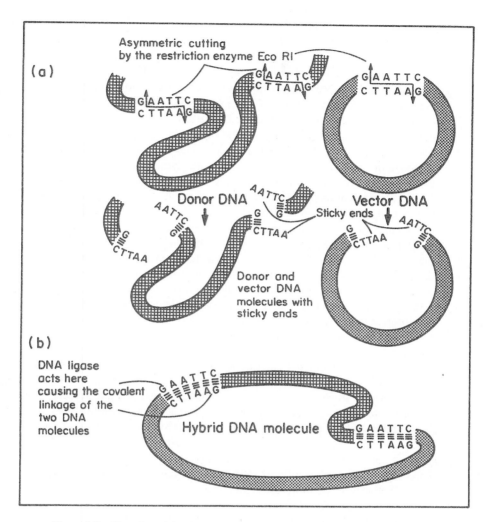

Figure 8.8. Use of special enzymes in genetic engineering. (a) The specific cutting of DNA by a restriction enzyme results in the formation of ends that contain small single-stranded complementary sequences ("sticky ends"). (b) The enzyme DNA ligase links pieces of DNA that have become associated by their sticky ends. (With permission, from T. D. Brock, K. M. Brock, and D. M. Ward, *Basic Microbiology with Applications*, 3d ed., Pearson Education, Upper Saddle River, NJ, 1986, p. 171.)

Obtaining good expression from the donor DNA is often a difficult challenge. Careful selection of stably propagating vectors and of promoters, checks to ensure that the correct reading frame is being used, and the selection of host cell backgrounds that do not interact unfavorably with the "foreign" protein are all important considerations. Discussions of how to obtain and maintain high levels of expression will occupy much of Chapter 14. We can screen for the expression of donor DNA, for example, if the product confers a selectable marker itself (e.g., the ability to grow on a substrate not normally utilized by the host). Also, antibodies to the target protein when tagged with a radiolabel or fluorescent label can be used to identify colonies expressing the target protein.

An important tool in working with both proteins and DNA fragments or plasmids is *electrophoresis*. In protein electrophoresis an electric field is applied to a solution containing proteins placed at the top of a gel (typically made of polyacrylamide). The proteins migrate through the porous structure of the gel in a direction and at a speed that reflect both the size and the net change of the molecule. The gel reduces the effects of convection, although thermally induced convection can be problematic in large gels. For proteins SDS-PAGE (sodium dodecyl sulfate polyacrylamide gel electrophoresis) is commonly used. In this technique the proteins are denatured by heating with SDS and mercaptoethanol. Mercaptoethanol is a reducing agent to break disulfide bands. Individual polypeptide chains form a complex with SDS, which is negatively charged. The negatively charged complex then migrates through the gel at a rate that reflects the molecular weight of the polypeptide. In a typical SDS-PAGE gel there is a lane with polymers of known molecular weight. Other lanes will have samples with unknown proteins. After a defined period of time (typically a few hours) the process is stopped and the gel examined. Each protein forms a band. With some types of electrophoresis the band is highlighted by use of stains. Smaller molecules travel a greater distance. The molecular weight of a protein band in the unknown sample can be estimated by comparing to the lane with the molecular weight standard. The relative amount of a protein can be determined by the intensity of the band.

Various forms of gel electrophoresis can be used to determine if a protein from a clone is being produced and at what relative level. A particularly useful procedure is an *immunoblot* or *Western* blot. Here a nondenaturing gel is run to separate proteins, they are blotted onto nitrocellulose paper, and the protein identified by binding to a specific antibody followed by a radioactive marker that binds antigen-antibody complexes. The band is visualized by exposure of the nitrocellulose paper to x-ray film. Because of the high degree of specificity of antibodies, a positive band is good confirmation that the protein band represents a target product rather than another protein of similar molecular weight.

Similarly DNA molecules can be separated by gel electrophoresis using agarose or polyacrylamide. DNA is negatively charged. DNA fragments (e.g., from digestion with restriction enzymes) can be separated by molecular weight as well as plasmids from larger DNA elements. The DNA separates based on molecular weight. The DNA can be recovered by simply cutting out the part of the gel corresponding to the band for the desired DNA and eluting. The DNA is invisible unless stained. One approach is to expose DNA to a dye that fluoresces when under ultraviolet light.

With shotgun cloning (which produces a large *gene library* or *gene bank*), a desired colony can be isolated by using radiolabeled RNA or DNA probes complementary to the cloned gene. Such a procedure involves transferring colonies to nitrocellulose filter paper, where they are lysed. After lysis, the DNA spills out to bind on the filter paper. The paper is flooded with the probe; the probe only binds to the DNA with a complementary sequence (hybridization) and excess probe is washed away. If the filter is covered with x-ray film, radioactivity will expose the film (visible as a black spot), identifying which colonies had the donor DNA. This procedure identifies colonies that have been transformed with the desired DNA; expression has to be established in separate experiments.

Another approach to using a gene library to find genes that express proteins with certain functions is the use of display technologies. Most common are *phage displays* and *bacterial displays*. For example, a bacteriophage or bacterium may display on its surfaces proteins encoded from genes derived from the library. Each cell may display at most only a few of these gene products. An example of the use of such a system is the isolation of a cell producing a gene product that will bind to a particular molecule (*ligand*). The ligand can be bound to a surface, and only those cells expressing a protein on their surface that binds to that ligand can "stick" to the surface. These cells that stick can be recovered and propagated to make more copies of the gene and protein.

Although our focus is on amplifying the number of genes as a basis for producing target proteins or altering pathways, gene cloning is often done to obtain many copies of a particular gene. The amplification of the gene number facilitates gene sequencing and analysis. This amplification is particularly important for mapping genomes (e.g., the human genome project), for diagnosis of disease-causing organisms (both microbial and viral), for biologists studying evolution, and for forensic scientists.

An alternative technique to traditional gene cloning is a technique called the *polymerase chain reaction*, or *PCR*. PCR is the preferred method to amplify DNA. In this technique a target sequence of interest is a gene on double-stranded DNA.[‡] The technique requires that two short primer sequences (< 20 nucleotides) on either side of the target be known. If heat is applied, the complementary strands of double-stranded DNA separate. While separated, two pieces of chemically synthesized DNA (the primers) are added. Each primer binds to complementary sequences. A heat-stable DNA polymerase from a bacterium that grows in hot springs (the Taq polymerase) is added and quickly synthesizes from the primers the complementary DNA strands using the four nucleotides (A, G, T, and C) added to the reaction mixture. At the end of the cycle, there are two copies of the original gene. If the cycle is repeated, those two copies become four. Thirty cycles can be done in less than a day. Thus, from a single gene copy, 2^{30} copies of the gene can be generated, more than a billionfold increase. Thermal cyclers and PCR kits are commercially available.

This simple summary does not cover all the intricacies investigators often face in obtaining industrially useful clones. However, the procedures discussed here are applicable in most cases where plasmids are used as vectors to transform bacteria such as *E. coli*.

8.4.2. Genetic Engineering of Higher Organisms

The direct genetic engineering of higher organisms can be a great deal more difficult because of a lack of good effective techniques to introduce foreign DNA and of an understanding of host cell genetics. The reader should be aware of the techniques being developed to work with some of these host systems. The introduction of foreign DNA into higher organisms is usually termed *transfection*.

Some plants are subject to infection with the bacterium *Agrobacterium tumefaciens*. *A. tumefaciens* contains a plasmid that contains a section known as T-DNA. This T-DNA can integrate into the plant chromosome. If genes are inserted into the T-DNA region, they

---

[‡]For forensic studies or phylogenic studies on relationships of organisms a non-protein coding sequence may be the target.

can be incorporated ultimately in the plant chromosome. Unfortunately, most cereal plants are not readily susceptible to *Agrobacterium* infection.

The *biolistic process* for obtaining plant transformation is to coat small (1-μm diameter) projectiles (e.g., of tungsten) and shoot these into cells at high velocity. Results with this approach have been remarkably successful, and this technique is fairly general.

Another fairly general approach is *electroporation,* which involves a brief high-voltage electric discharge that renders cells permeable to DNA. Electroporation can be used with animal, plant, fungal, and bacterial systems. The formation of *protoplasts* can enhance transfection but is not essential. A protoplast is a cell in which the outer cell envelope has been removed so that only the cytoplasmic membrane remains.

Chemically or electrically mediated *protoplast fusion* is another technique for transferring genetic information from one cell to another. Such approaches have been particularly useful with some fungi for which few or no plasmids have been identified. Protoplast fusion can be interspecies and can result in stable hybrids with desired properties due to recombination events between the two genomes or extrachromosomal pieces of DNA.

For most animal cells, genetic manipulation can be accomplished by modifying viruses to become vectors. For example, in the insect cell system a *baculovirus* can be modified so as to place a desired gene under the control of a very strong promoter at the expense of a gene product that is unessential for viral replication in cell culture.

Although the basic conceptual approach to genetic engineering is rather straightforward, its implementation can vary widely in difficulty. The level of difficulty depends on the nature of the gene product and its corresponding gene, as well as the character of the desired host cell. Techniques to improve the ease of obtaining desirable genetic modification will undoubtedly continue to be developed. The ultimate limitation will be human imagination and wisdom.

8.5. GENOMICS

Genomics is the set of experimental and computational tools which allow the genetic blueprints of life to be read. A *genome* is an organism's total inheritable DNA. We now have complete genomic information on about 50 microbes, as well as representative animals and plants. Most importantly, we have the genomic sequence for humans. This sequence information is simply a string of nucleotide letters. *Functional genomics* is the process of relating genetic blueprints to the structure and behavior of an organism. To completely relate physiological behavior to this sequence of nucleotides is an extremely challenging problem. It is a problem to which bioengineers can make a significant contribution.

Much of the progress in molecular biology has been due to a reductionist approach in which a subcomponent has been isolated and studied in detail. This approach has been very fruitful in learning about the detailed mechanisms at the heart of living systems. The detailed sequence information now available is the ultimate limit in reduction in biology. There is increasing interest in asking how the individual subcellular components work together. Function arises from the complex interactions of the components. A systems engineering approach allows one to integrate component parts into a functional whole.

While genetic information is linear and static, cellular systems are highly nonlinear, dynamic systems that respond to their environment and regulate gene expression. Over the next decades a focus of biochemical engineering, in conjunction with other disciplines, will be relating this linear sequence information to those nonlinear dynamical systems.

The role of the bioengineer is twofold. One role is as an enabler, by making better tools for rapid analysis of DNA sequences, of expression of *m*-RNAs, and of a cell's total proteins (*proteomics*). A second role is as an interpreter and organizer of genetic information; this role usually involves mathematical modeling.

8.5.1. Experimental Techniques

The primary tools of genomics are used for DNA sequencing, detecting which *m*-RNAs are expressed, and determining which proteins are present in a cell or tissue. The nucleotide sequence of DNA fragments can be determined on a sequencing gel. The key to this method is the use of dideoxyribonucleoside triphosphates, which are derivatives of the natural deoxyribonucleoside triphosphates and lack the hydroxyl (OH) group at the 3′ position. For example, ddATP is the derivative dATP. If a strand of DNA is being replicated and if ddATP is inserted into the position normally occupied by dATP, replication is stopped. The OH group at the 3′ position is essential for continued replication. The basic process for sequencing is shown in Fig. 8.9. It is easiest to imagine reactions in four separate tubes labelled A, T, C, and G. To each tube is added the DNA fragment to be sequenced, DNA polymerase, a stoichiometeric excess of dATP, dTTP, dCTP, and dGTP, and an oligonucleotide primer for the DNA polymerase to use. The reaction is analogous to the first step in the PCR reaction. Only one strand of the DNA fragment to be sequenced is read. In addition to the above reactants, in each tube a small amount of ddNTP is added. For example, in tube A ddATP is added, in tube T, ddTP, and so on. Since the amount of ddNTP added is small in tube N, several reaction products are formed. In tube N, the first time dNTP or ddNTP must be added to the growing copy of the original DNA fragment, there is a high probability that dNTP will be added and the chain can be extended. However, there is also a finite probability (determined by the ratio of ddNTP to dNTP) that ddNTP will be added and the chain extension will be terminated. At the second position where N is required, either ddNTP or dNTP will be added. As above, some chains will be terminated, and others will continue to extend. This reaction continues, and tube N will generate fragments of different sizes, all of which end in the letter N. These fragments are separated by gel electrophoresis, and the sequence of the DNA fragment can be read directly from the gel (as shown in Fig. 8.9). This technique is limited to relatively short DNA fragments (a few hundred nucleotides).

For sequencing genomes this technique can be modified and automated. The basic approach to sequence a large genome is to cut it into millions of overlapping fragments of 2,000 to 10,000 base pairs in length. Each fragment is ligated into a plasmid, which is transformed into *E. coli*. This approach is termed a shotgun approach, and the *E. coli* form a living genomic library. These colonies are robotically picked, identified with a barcode, and placed in a 384-well plate. The amount of the cloned DNA fragment is amplified using PCR. From each end of the fragment 500 letters are replicated using the ddNTP

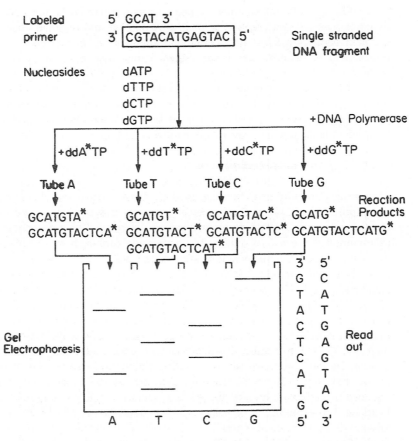

Figure 8.9. Example of a sequencing gel to obtain the nucleotide sequence of a DNA fragment. See text for details.

approach described above. For the purposes of automation a fluorescent dye is used for each ddNTP. Four different dyes are used to distinguish A, T, C, and G. These labelled fragments are then read in automated sequencing machines. These machines rely on 104 glass capillaries; capillary electrophoresis is used to separate the fragments by size. As the fragments exit the capillary, a laser beam detects the color of the dye at the end of the fragment. The information is led to a computer and the 500-letter sequence is determined. One company, Celera, expects to use such machines to read 100 million letters of DNA sequence per day. These sequences are stored in a computer; computer alogrithms are then used to align the overlapping sequences.

Because of the large number of fragments, this information processing is very challenging. For the human genome 70 million separate sequences are necessary to achieve sufficient overlap to reconstruct the whole human genome (about 3 billion letters). This technique cannot do a perfect reconstruction, but one that is effectively complete. One problem is that in much of the human DNA there are long repeated sequences which complicate alignment. Only about 5% of human DNA encodes genes. The coding fraction is much higher in bacteria. Because of the smaller genome size and fewer complications, such as repeated sequences, bacterial genomes are easier to sequence.

DNA sequencing is essentially a problem in information technology. Many feel that the current sequencing technology is comparable to computer technology in 1970. There are tremendous opportunities for engineers to design and construct devices to read and analyze genomic information much more rapidly and cheaply. The intrinsic scale of genomic information is the size of a nucleotide (about 0.34 nm). Recent advances in nanotechnology may allow us to generate such instruments. Indeed, engineers have contributed significantly to a "lab-on-a-chip" device to do DNA sequencing. This device begins to marry microtechnology and biotechnology.

Another technology that is having a large impact on genomics is *microarrays* for measuring which genes are being expressed by measuring the corresponding *m*-RNA levels. These microarrays are high-density oligonucleotide arrays. These oligonucleotides hybridize with the corresponding *m*-RNA. For a known gene or protein an oligonucleotide can be synthesized that binds the corresponding *m*-RNA. Using photolithography, the manufacturing of arrays containing 280,000 individual oligonucleotides on glass substrates of 1.64 cm^2 is now routine. Such arrays can simultaneously analyze expression from 6,000 to 10,000 genes. For bioprocesses the array can be used to determine which genes are up regulated or down regulated in response to a process change (e.g., temperature).

An increase in *m*-RNA levels does not immediately correspond to a change in protein level. Because different *m*-RNAs have different rates of degradation, efficiency of translation, and so on, changes in *m*-RNA expression may not translate directly into changes in the protein content of a cell or tissue. Since the proteins are the primary components responsible for biological activity, a knowledge of the protein content of a cell is highly valuable. Because proteins cannot be easily amplified like nucleic acids, measurements of all of the proteins in a cell (i.e., proteomics) is very difficult. Also proteins vary greatly in properties such as hydrophobicity. Currently there is no method that can measure all proteins in a cell. The best available technique is two-dimensional gel electrophoresis. This technique is derived from the SDS-PAGE technique described earlier in this chapter. For 2D gel electrophoresis two methods are combined. The first is to separate proteins in one direction (say the *x*-axis) according to their *isoelectric point*—the pH at which the protein has no net charge. Using a special set of buffers, a pH gradient can be established in a polyacrylamide gel. When the proteins are subjected to an electric field, they will move to the pH equal to their isoelectric point and remain there. After equilibration with an anionic surfactant, the proteins are then subjected to an electric field in the other direction (say the *y*-axis). The proteins will remain at the same pH but move down the *y*-axis according to their molecular weight. By separating protein mixtures using these two parameters (isoelectric point and molecular weight), a complex protein mixture can be resolved into separate spots. Often those "spots" can be removed and analyzed by tandem mass spectrometry to identify the protein's sequence.

The technique of 2-D gel electrophoresis is time consuming and expensive. Significant improvements will be necessary for routine analysis of cell and tissue protein content. Better gels or methods of protein separation are needed. Promising approaches combine microtechnology, liquid chromatographies, and mass spectrometry techniques. The engineering challenges in making a practical device are significant.

8.5.2. Computational Techniques

The wealth of data emerging from these new experimental techniques is overwhelming. Generally, DNA sequence information could be used to identify which sequences are genes that encode for proteins, what these proteins are, what their biological function is, and ultimately how these genes are regulated.

Deciding what nucleotide sequence corresponds to a gene is not always a clear-cut process. Investigators look for "open reading frames" or ORFs. An ORF is a nucleotide sequence without a "stop" signal that encodes some minimal number of amino acids (about 100). In prokaryotes, identifying ORFs is fairly straightforward. In eukaryotes, because of introns and exons, assignment of ORFs is complicated. Some computer programs can recognize probable (consensus) intron/exon boundaries.

When a prospective gene is identified, we need to know the function of the corresponding protein. In some cases the function can be deduced by comparison to databases with amino acid sequence information on known proteins. In some cases the amino acid sequence and function of a protein is conserved across species.

With a highly conserved protein we might find similiar amino acid sequences—for example, in the fruit fly and humans. In this case we would determine how *homologous* the two amino acid sequences were. We then might infer that the human gene encodes a protein with similiar function. Conservation of amino acids near a catalytic or binding site is particularly critical. Efficient computer algorithms to do such searches are under development.

Ideally one would like to predict protein structure and function solely from the amino acid sequence. Heroic efforts have been made to accomplish this goal, but there is no good general solution. Understanding the folding of proteins into their three-dimensional configuration is a computationally difficult problem.

Even with well-studied organisms, such as *E. coli,* we generally can guess the function of only 50% of the genes. Clearly, knowing the full genome sequence has told us how little we really understand. The process of identifying single genes and the function of the corresponding genes has been the focus of *bioinformatics.*

Even if we knew the identity of every gene in a cell and the function of each corresponding protein, we would have an incomplete understanding of function and cellular physiology. A list of the proteins and their functions needs to be supplemented by an understanding of cellular structure and regulation. A combination of proteins can form metabolic pathways, and there will be a corresponding genetic circuit. Relating the cell's "parts list" to its dynamic, physiological state is an unmet challenge that provides exciting, long-term opportunities for bioengineers.

In particular, models of cells and metabolic circuits as discussed in Chapter 6 provide tools with which to organize data and to understand biological function. Simple stoichiometric models of central metabolism have been used with good success to identify which genes are essential to a cell in a particular environment. Models that also incorporate kinetics and regulatory structure are in development. Such models will be key to relating cellular physiology to the genome. Such an understanding will provide important guidance to development of future bioprocesses.

8.6. SUMMARY

A cell's *genotype* represents the cell's genetic potential, whereas its *phenotype* represents the expression of a culture's potential. The genotype of a cell can be altered by *mutations*. Examples of mutations are *point mutations*, *deletions*, and *additions*. Additions are usually the result of *insertion sequences* that "jump" from one position to another.

Mutations may be *selectable* or *unselectable*. The rate of mutation can be enhanced by the addition of chemicals called *mutagens* or by radiation. *Auxotrophs* are of particular use in genetic analysis and as a basis for some bioprocesses. Another useful class of mutants is *conditional* mutants.

Gene transfer from one cell to another augments genetic information in ways that are not possible through mutation only. *Genetic recombination* of different DNA molecules occurs within most cells. Thus, genetic information transferred from another organism may become a permanent part of the recipient cell. The three primary modes of gene transfer in bacteria are *transformation*, *transduction*, and *conjugation*. Self-replicating, autonomous, extrachromosomal pieces of DNA called *plasmids* play important roles in transformation. *Episomes*, which are closely related to plasmids, are the key elements in conjugation. Bacteriophages are critical to *generalized transduction,* while *temperate phages* are key to *specialized* transduction. Internal gene transfer can occur due to the presence of *transposons*, which probably also play a role in the assembly of new plasmids.

We can use gene transfer in conjunction with *restriction enzymes* and *ligases* to genetically engineer cells. In-vitro procedures to recombine isolated donor DNA genes with *vector* DNA (for example, plasmids, temperate phages, or modified viruses) are called *recombinant DNA techniques*. Once the vector with the DNA donor insert has been constructed, it can be moved to a recipient cell through any natural or artificial method of gene transfer. Although transformation is the most common technique in bacteria, a large variety of artificial methods have been developed (e.g., *biolistic process, electroportation,* modification of infective agents, and *protoplast fusion*) to insert foreign DNA into a host cell.

Genomics is the set of experimental and computational tools which allows the genetic blueprints of a whole organism to be read. *Functional genomics* is the process of relating genetic blueprints to the behavior and structure of an organism.

SUGGESTIONS FOR FURTHER READING

Many of the references cited at the end of Chapter 4 have selections dealing with mutation and selection, gene transfer, and genetic engineering.

The following book explores these same topics, but more from the perspective of the industrial microbiologist:

CRUEGER, W., AND A. CRUEGER, *Biotechnology: A Textbook of Industrial Microbiology*, 2d ed. (T. D. Brock, ed., English edition), Sinauer Associates, Inc., Sunderland, MA, 1990.

The biolistic process and electroporation applied to bacteria are described in:

CHASSY, B. M., A. MERCENIER, AND J. FLICKINGER, Transformation of Bacteria by Electroporation, *Trends Biotechnol. 6:*303–309, 1988.

SANFORD, J. C., The Biolistic Process, *Trends Biotechnol. 6:*299–302, 1988.

A good description of the basic techniques of molecular biology can be found in:

ALBERTS, B., and others. *Essential Cell Biology*, Garland Publishing, Inc., New York, 1998.

DRLICA, K., *Understanding DNA and Gene Cloning*, John Wiley & Sons, New York, 1997.

SAMBROOK, J., AND D. W. RUSSELL, *Molecular Cloning: A Laboratory Manual*, 3d ed., Cold Spring Harbor Laboratory Press, Cold Spring Harbor, NY (to be issued in 2001; updates second edition published in 1989).

Details of our current knowledge of the gene map for *E. coli* can be found in:

BLATTNER, F. R., and others. The Complete Genome Sequence of *Escherichia coli* K–12, *Science* 277:1453–1462, 1997.

For information on genomics and related topics the following may be helpful.

BORMAN, S., Proteomics: Taking Over Where Genomics Leaves Off, *Chem. Eng. News* (July 31):31–37, 2000.

BURNS, M. A., and others, An Integrated Nanoliter DNA Analysis Device, *Science 282:*484–487, 1998.

DUTT, M. J., and K. H. LEE, Proteomic Analysis, *Curr. Opinion Biotechnol. 11:*176–179, 2000.

GINGERAS, T. R., and C. ROSENOW, Studying Microbial Genomes with High-Density Oligonucleotide Arrays, *ASM News 66:*463–469, 2000.

JEGALIAN, K., The Gene Factory, *Technol. Rev* (March/April): 64–68, 1999.

YIN, J., Bio-informatics—A Chemical Engineering Frontier, *Chem. Eng. Prog.* (Nov.) 65–74, 1999.

PROBLEMS

8.1. Would a cell with a point mutation or a deletion be more likely to revert back to the original phenotype? Why?

8.2. Consider the metabolic pathway based on aspartic acid shown in Section 4.9. Describe the procedure you would use to obtain a methionine overproducer. Use mutation-selection procedures, detailing the experiments to be done and their sequence.

8.3. You wish to isolate temperature-sensitive mutants (e.g., those able to grow at 30°C but not at 37°C). Describe experiments to isolate such a cell.

8.4. An important method for screening for carcinogens is called the *Ames test*. The test is based on the potential for mutant cells of a microorganism to revert to a phenotype similar to the nonmutant. The rate of reversion increases in the presence of a mutagen. Many compounds that are mutagens are also carcinogens, and vice versa. Describe how you would set up an experiment and analyze the data to determine if nicotine is mutagenic.

8.5. How many different hybridization probes must you make to ensure that at least one corresponds to a set of four codons encoding the amino acid sequence val-leu-trp-lys?

8.6. You wish to develop a genetically engineered *E. coli* producing a peptide hormone. You know the amino acid sequence of the peptide. Describe the sequence of steps you would use to obtain a culture expressing the gene as a peptide hormone.

8.7. You wish to produce a small protein using *E. coli*. You know the amino acid sequence of the protein. The protein converts a colorless substrate into a blue product. You have access to a high-copy-number plasmid with a penicillin-resistant gene and normal reagents for genetic engineering. Describe how you would engineer *E. coli* to produce this protein. Consider: source of donor DNA; regulatory elements that need to be included; how the donor and vector DNA are combined; how the vector DNA is inserted; and how you would select for a genetically engineered cell to use in production.

8.8. You wish to express a particular peptide in *E. coli* using a high-copy-number plasmid. You have the amino acid sequence for the peptide.

  a. Explain the experimental process for generating and selecting the genetically engineered *E. coli* using restriction enzymes, ligase, *E. coli,* plasmid with neomycin resistance, and the known amino acid sequence.

  b. What control elements would you place on the plasmid to regulate expression and to prevent read-through?

8.9. **a.** There are three primary methods for obtaining donor DNA when doing genetic engineering. What are those methods (two- to six-word descriptions of each are acceptable)?

  b. You need to produce a protein from humans in *E. coli*. You do not know the primary amino acid sequence. You suspect that introns are present. Which method will you use to obtain the donor DNA?

8.10. What is the difference between "transduction" and "transformation" when discussing genes transfer to bacteria?

8.11. You wish to isolate a thymidine auxotrophic mutant of *E. coli*. Describe briefly what experiments you would do to accomplish this.

8.12. For the DNA sequence, TAGGATCATAAGCCA, and using a primer, "ATCC," sketch what the corresponding sequencing gel should look like.

PART 3

Engineering Principles for Bioprocesses

9

Operating Considerations for Bioreactors for Suspension and Immobilized Cultures

9.1. INTRODUCTION

So far we have discussed what cells are, how they work, and how to describe their growth in simple reactors. We now begin our discussion of how to use these cells in processes. We will explore some more complicated reactor strategies and why they might be considered for use in real processes. Chapter 10 will give more details on reactor design, and Chapter 11 will detail how to recover products from these reactors. These chapters should give the reader an understanding of how real bioprocesses can be assembled.

An important decision for constructing any process concerns the configuration the reactor system should take. The choice of reactor and operating strategy determines product concentration, number and types of impurities, degree of substrate conversion, yields, and whether sustainable, reliable performance can be achieved. Unlike many traditional chemical processes, the reactor section represents a very major component (usually > 50%) of the total capital expenditures. Choices at the reactor level and of the biocatalyst determine the difficulty of the separation. Thus, our choice of reactor must be made in the context of the total process: biocatalyst, reactor, and separation and purification train.

9.2. CHOOSING THE CULTIVATION METHOD

One of the first decisions is whether to use a batch or continuous cultivation scheme. Although a simple batch and continuous-flow stirred-tank reactor (CFSTR) represent extremes (we will soon learn about other reactors with intermediate characteristics), consideration of these two extreme alternatives will clarify some important issues in reactor selection.

First, we can consider productivity. The simplest case is for the production of cell mass or a primary product. For a batch reactor, four distinct phases are present: lag phase, exponential growth phase, harvesting, and preparation for a new batch (e.g., cleaning, sterilizing, and filling). Let us define t_l as the sum of the times required for the lag phase, harvesting, and preparation. The value for t_l will vary with size of the equipment and the nature of the fermentation but is normally in the range of several hours (3 to 10 h). Thus, the total time to complete a batch cycle (t_c) is

$$t_c = \frac{1}{\mu_m} \ln \frac{X_m}{X_0} + t_l \tag{9.1}$$

where X_m is the maximal attainable cell concentration and X_0 is the cell concentration at inoculation.

The total amount of cell mass produced comes from knowing the total amount of growth-extent-limiting nutrient present and its yield coefficient:

$$X_m - X_0 = Y_{X/S} S_0 \tag{9.2}$$

The rate of cell mass production in one batch cycle (r_b) is

$$r_b = \frac{Y_{X/S} S_0}{(1/\mu_m)\ln(X_m/X_0) + t_l} \tag{9.3}$$

As discussed in Chapter 6, the maximum productivity of a chemostat is found by differentiating DX with respect to D and setting dDX/dD to zero. The value for D optimal when simple Monod kinetics apply is given by eq. 6.83, and the corresponding X can be determined to be

$$X_{opt} = Y_{X/S}\left\{ S_0 + K_s - \sqrt{K_s(S_0 + K_s)} \right\} \tag{9.4}$$

Thus, the best productivity that could be expected from a chemostat where Monod kinetics apply is $D_{opt} \cdot X_{opt}$, or

$$D_{opt}X_{opt} = Y_{X/S}\,\mu_m\left[1 - \sqrt{\frac{K_s}{K_s + S_0}} \right]\left[S_0 + K_s - \sqrt{K_s(S_0 + K_s)} \right] \tag{9.5}$$

Under normal circumstances $S_0 \gg K_s$, so the rate of chemostat biomass production, r_c, is approximately

$$r_{c,opt} = D_{opt}X_{opt} = \mu_m Y_{X/S} S_0 \tag{9.6}$$

The ratio for rates of biomass formation is

$$\frac{r_{c,\text{opt}}}{r_b} = \ln\frac{X_m}{X_0} + \mu_m t_l \tag{9.7}$$

Most commercial fermentations operate with $X_m/X_0 \approx 10$ to 20. Thus, we would expect continuous systems to always have a significant productivity advantage for primary products. For example, an *E. coli* fermentation with $X_m/X_0 = 20$, $t_l = 5$ h, and $\mu_m = 1.0$ h^{-1} would yield $r_{c,\text{opt}}/r_b = 8$.

Based on this productivity advantage we might be surprised to learn that most commercial bioprocesses are batch systems. Why? There are several answers.

The first is that eq. 9.7 applies only to growth-associated products. Many secondary products are not made by growing cells; growth represses product formation. Under such circumstances, product is made only at very low dilution rates, far below those values optimal for biomass formation. For secondary products, the productivity in a batch reactor may significantly exceed that in a simple chemostat.

Another primary reason for the choice of batch systems over chemostats is *genetic instability*. The biocatalyst in most bioprocesses has undergone extensive selection. These highly "bred" organisms often grow less well than the parental strain. A chemostat imposes strong selection pressure for the most rapidly growing cell. Back-mutation from the productive specialized strain to one similar to the less productive parental strain (i.e., a revertant) is always present. In the chemostat the less productive variant will become dominant, decreasing productivity. In the batch culture the number of generations available (< 25 from slant cultures to a commercial-scale fermenter) for the revertant cell to outgrow the more productive strain is limited. Cells at the end of the batch are not reused. These considerations of genetic stability are very important for cells with recombinant DNA and are discussed in detail in Chapter 14.

Another consideration is operability and reliability. Batch cultures can suffer great variability from one run to another. Variations in product quality and concentration create problems in downstream processing and are undesirable. However, long-term continuous culture can be problematic; pumps may break, controllers may fail, and so on. Maintenance of sterility (absence of detectable foreign organisms) can be very difficult to achieve for periods of months, and the consequences of a loss of sterility are more severe than with batch culture.

One other factor determining reactor choice is market economics. A continuous system forms the basis of a dedicated processing system—dedicated to a single product. Many fermentation products are required in small amounts, and demand is difficult to project. Batch systems provide much greater flexibility. The same reactor can be used for two months to make product A and then for the next three for product B and the rest of the year for product C.

Most bioprocesses are based on batch reactors. Continuous systems are used to make single-cell protein (SCP), and modified forms of continuous culture are used in waste treatment, in ethanol production, and for some other large-volume, growth-associated products such as latic acid.

Let us consider some modifications to these reactor modes.

9.3. MODIFYING BATCH AND CONTINUOUS REACTORS

9.3.1. Chemostat with Recycle

Microbial conversions are autocatalytic, and the rate of conversion increases with cell concentration. To keep the cell concentration higher than the normal steady-state level in a chemostat, cells in the effluent can be recycled back to the reactor. Cell recycle increases the rate of conversion (or productivity) and also increases the stability of some systems (e.g., waste-water treatment) by minimizing the effects of process perturbation. Cells in the effluent stream are either centrifuged, filtered, or settled in a conical tank for recycling.

Consider the chemostat system with cell recycle as depicted in Fig. 9.1. A material balance on cell (biomass) concentration around the fermenter yields the following equation:

$$FX_0 + \alpha FCX_1 - (1+\alpha)FX_1 + V\mu_{net}X_1 = V\frac{dX_1}{dt} \tag{9.8}$$

where α is the recycle ratio based on volumetric flow rates, C is the concentration factor or ratio of cell concentration in the cell recycle stream to the cell concentration in the reactor effluent, F is nutrient flow rate, V is culture volume, X_0 and X_1 are cell concentrations in feed and recycle streams, and X_2 is cell concentration in effluent from the cell separator.

At steady state, and if $dX_1/dt = 0$ and $X_0 = 0$ (that is, sterile feed); then eq. 9.8 becomes

$$\mu_{net} = (1+\alpha-\alpha C)D = [1+\alpha(1-C)]D \tag{9.9}$$

Since $C > 1$ and $\alpha(1-C) < 0$, then $\mu_{net} < D$. That is, *a chemostat can be operated at dilution rates higher than the specific growth rate when cell recycle is used.*

A material balance for growth-limiting substrate around the fermenter yields

$$FS_0 + \alpha FS - V\frac{\mu_g X_1}{Y_{X/S}^M} - (1+\alpha)FS = V\frac{dS}{dt} \tag{9.10}$$

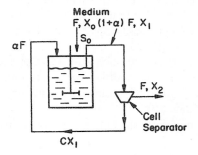

Figure 9.1. Chemostat with cell recycle. The cell separator could be a sedimentation tank, a centrifuge, or a microfiltration device.

At steady state, $dS/dt = 0$ and

$$X_1 = \frac{D}{\mu_g} Y_{X/S}^M (S_0 - S)$$

(9.11)

Substitution of eq. 9.9 when $k_d = 0$ into eq. 9.11 yields

$$X_1 = \frac{Y_{X/S}^M (S_0 - S)}{(1 + \alpha - \alpha C)}$$

(9.12)

Therefore, the steady-state cell concentration in a chemostat is increased by a factor of $1/(1 + \alpha - \alpha C)$ by cell recycle. The substrate concentration in the effluent is determined from eq. 9.9 and the Monod eq. 6.30, where endogenous metabolism is neglected, and is

$$S = \frac{K_s D (1 + \alpha - \alpha C)}{\mu_m - D(1 + \alpha - \alpha C)}$$

(9.13)

Then eq. 9.12 becomes

$$X_1 = \frac{Y_{X/S}^M}{(1 + \alpha - \alpha C)} \left[S_0 - \frac{K_s D (1 + \alpha - \alpha C)}{\mu_m - D(1 + \alpha - \alpha C)} \right]$$

(9.14)

Effluent cell concentrations and productivities in a chemostat with and without cell recycle are compared in Fig. 9.2. Cell concentrations and productivities are higher with cell recycle, resulting in higher rates of substrate consumption. Systems with cell recycle are used extensively in waste treatment and are finding increasing use in ethanol production. The application of cell recycle reactors in waste treatment is detailed in Chapter 16. The equations differ from the case above due to the inclusion of a term for endogenous metabolism (i.e., k_d). The basic concept of operation at flows above the "washout" rate applies when $k_d \neq 0$.

Example 9.1

In a chemostat with cell recycle, as shown in Fig. 9.1, the feed flow rate and culture volumes are $F = 100$ ml/h and $V = 1000$ ml, respectively. The system is operated under glucose limitation, and the yield coefficient, $Y_{X/S}^M$, is 0.5 gdw cells/g substrate. Glucose concentration in the

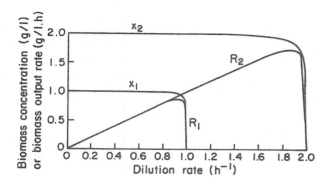

Figure 9.2. Comparison of biomass concentrations and output rates in steady states of chemostat cultures with and without recycle. Symbols: X_1 = biomass concentration in chemostat without recycle; X_2 = biomass concentration in chemostat culture with recycle; R_1 = biomass output rate per unit volume without recycle; R_2 = biomass output rate of chemostat with recycle; $\mu_m = 1.00$ h⁻¹; $S_r = 2.0$ g/l; $K_S = 0.010$ g/l; $Y_{X/S} = 0.5$ g/g; concentration factor, $C = 2.0$; and recycle rate, $\alpha = 0.5$.

feed is $S_0 = 10$ g glucose/l. The kinetic constants of the organisms are $\mu_m = 0.2$ h^{-1}, $K_s = 1$ g glucose/l. The value of C is 1.5, and the recycle ratio is $\alpha = 0.7$. The system is at steady state.

    a. Find the substrate concentration in the recycle stream (S).

    b. Find the specific growth rate (μ_{net}) of the organisms.

    c. Find the cell (biomass) concentration in the recycle stream.

    d. Find the cell concentration in the centrifuge effluent (X_2).

Solution   Using eq. 9.9, we determine μ_{net}.

$$\mu_{net} = [1 + \alpha(1 - C)]D = [1 + (1 - 1.5)0.7](0.1) = \mu_g$$

$$= 0.065 \text{ h}^{-1}$$

Then

$$S = \frac{K_s \mu_{net}}{\mu_m - \mu_{net}} = \frac{(1)(0.065)}{0.2 - 0.065} = 0.48 \text{ g/l}$$

$$X_1 = \frac{D(S_0 - S)Y_{X/S}^M}{\mu_g} = \frac{(0.1)(10 - 0.48)0.5}{0.065}$$

$$= 7.3 \text{ g/l}$$

A biomass balance around the concentrator yields

$$(1 + \alpha)X_1 = \alpha C X_1 + X_2$$

$$X_2 = (1 + \alpha)X_1 - \alpha C X_1$$

$$= (1.7)(7.3) - (0.7)(1.5)(7.3)$$

$$= 4.8 \text{ g/l}$$

9.3.2. Multistage Chemostat Systems

In some fermentations, particularly for secondary metabolite production, the growth and product-formation steps need to be separated, since optimal conditions for each step are different. Conditions such as temperature, pH, and limiting nutrients may be varied in each stage, resulting in different cell physiology and cellular products in multistage systems.

    An example of a multistage system that may be beneficial is in the culture of genetically engineered cells. To improve genetic stability, a plasmid-carrying recombinant DNA usually uses an inducible promoter to control production of the target protein (see Chapter 8). In the uninduced state, the plasmid-containing cell grows at nearly the same rate as the cell that loses the plasmid (a revertant), so the plasmid-free cell holds little growth advantage over the plasmid-containing cell. However, if the inducer is added, the plasmid-containing cells will make large quantities of the desired protein product but will have greatly reduced growth rates. Thus, a single-stage chemostat would not be suitable for the production of the target protein because of resulting problems in genetic stability. A multistage system can circumvent this problem. In the first stage, no inducer is added and the plasmid-containing cell can be maintained easily (usually an antibiotic is added to kill plasmid-free cells; see Chapter 14 for a more complete discussion). In the second stage, the inducer is added and large quantities of product are made. Cells defective in product synthesis should not overtake the culture (at least not completely), because fresh genetically unaltered cells are being continuously fed to the reactor. Thus, the two-stage system can allow the stable continuous production of the target protein when it would be impossible in a simple chemostat.

Perhaps an easier situation to consider is the production of a secondary product (e.g., ethanol or an antibiotic). Here we worry not so much about a mixture of subpopulations, but that conditions that promote growth completely repress product formation. A very large scale multistage system for ethanol production is currently in use. A multistage system of CFSTR approaches PFR behavior. A PFR mimics the batch system, where space time (the time it takes the culture fluid to reach a specific location in the PFR) replaces culture time. A multistage system is much like taking the batch growth curve and dividing it into sections, with each section being "frozen" in a corresponding stage of the multistage system. As in the batch reactor, the culture's physiological state progresses from one stage to the next.

The mathematical analysis of the multistage system that we present here is imperfect. Growth in the second and subsequent stages is intrinsically unbalanced growth, even though it is steady-state growth. New cells entering the second or subsequent stage are continuously adapting to the new conditions in that stage. Consequently, unstructured models are not expected to give completely accurate predictions. However, we use unstructured models here due to their simplicity and to illustrate at least some aspects of multistage systems.

A two-stage chemostat system is depicted in Fig. 9.3. Biomass and substrate balances on the first stage yield the following equations (ignoring endogeneous metabolism):

$$S_1 = \frac{K_s D_1}{\mu_m - D_1}$$

(9.15)

$$X_1 = Y_{X/S}^M (S_0 - S_1)$$

(9.16)

The biomass balance for the second stage yields

$$FX_1 - FX_2 + \mu_2 V_2 X_2 = V_2 \frac{dX_2}{dt}$$

(9.17)

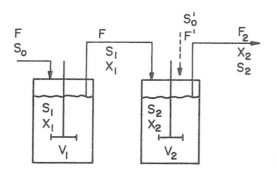

Figure 9.3. Two-stage chemostat system.

At steady state, eq. 9.17 becomes

$$\mu_2 = D_2\left(1 - \frac{X_1}{X_2}\right) \tag{9.18}$$

where $X_1/X_2 < 1$ and $\mu_2 < D_2$.

The substrate balance for the limiting substrate in the second stage is

$$FS_1 - FS_2 - \frac{\mu_2 X_2}{Y_{X/S}^M} V_2 = V_2 \frac{dS_2}{dt} \tag{9.19}$$

At steady state, eq. 9.19 becomes

$$S_2 = S_1 - \frac{\mu_2}{D_2} \frac{X_2}{Y_{X/S}^M} \tag{9.20}$$

where

$$D_2 = F/V_2 \quad \text{and} \quad \mu_2 = \frac{\mu_m S_2}{K_s + S_2}$$

Equations 9.18 and 9.20 can be solved simultaneously for X_2 and S_2 by substituting $\mu_2 = \mu_m S_2/(K_s + S_2)$ in both equations or any other functional form that describes μ_2.

When a feed stream is added to the second stage, then the design equations change. The second feed stream may contain additional nutrients, inducers, hormones, or inhibitors. Biomass balance for the second stage in this case is

$$F_1 X_1 + F'X' - (F_1 + F')X_2 + V_2\mu_2 X_2 = V_2 \frac{dX_2}{dt} \tag{9.21}$$

At steady state when $X' = 0$, eq. 9.21 becomes

$$\mu_2 = D_2' - \frac{F_1}{V_2} \frac{X_1}{X_2} \tag{9.22}$$

where

$$D_2' = \frac{F_1 + F'}{V_2} \quad \text{and} \quad \mu_2 = \frac{\mu_m S_2}{K_s + S_2}$$

Substrate balance for the second stage yields

$$F_1 S_1 + F'S_0' - (F_1 + F')S_2 - \frac{V_2\mu_2 X_2}{Y_{X/S}^M} = V_2 \frac{dS_2}{dt} \tag{9.23}$$

Equations 9.22 and 9.23 need to be solved simultaneously for X_2 and S_2.

We can generalize these equations for a system with no additional streams added to second or subsequent units. If we do a balance around the nth stage on biomass, substrate, and product, we find

$$r_{x,n}(X_x, S_n) = D_n(X_n - X_{n-1}) \tag{9.24a}$$

or

$$\theta_n = \frac{1}{D_n} = \frac{X_n - X_{n-1}}{r_{x,n}(X_n, S_n)} \tag{9.24b}$$

$$r_{s,n} = \frac{1}{Y_{X/S}} r_{x,n}(X_n, S_n) = D_n(S_n - S_{n-1}) \tag{9.25a}$$

or

$$\theta_n = \frac{1}{D_n} = \frac{S_n - S_{n-1}}{r_{s,n}} \tag{9.25b}$$

$$r_{p,n}(X_n, S_n, \ldots) = D_n(P_n - P_{n-1}) \tag{9.26a}$$

or

$$\theta_n = \frac{1}{D_n} = \frac{P_n - P_{n-1}}{r_{p,n}(X_n, S_R, \ldots)} \tag{9.26b}$$

where $D_n = F/V_n$, θ_n is the mean residence time in the nth stage, and $r_{x,n}$, $r_{s,n}$, and $r_{p,n}$ all represent rates of reaction in the nth stage.

The preceding set of equations lends itself to machine calculations. However, graphical approaches to multistage design can also be used and have the advantage that the functional form of the growth or production rate need not be known. All that is required is a batch growth curve. However, the transfer of the information from batch growth curve to predictions of the multistage system still requires the assumption of balanced growth. Hence, the analysis must be used with caution. In at least one case (the production of spores from *Bacillus*), this approach has made experimentally verifiable predictions of the performance of a six-stage system.

The graphical approaches make use of eqs. 9.24 to 9.26. One approach is to use a plot of $1/(dX/dt)$ versus X or $1/(dP/dt)$ versus P derived from batch growth curves. This corresponds to using eqs. 9.24b and 9.26b. The size of the required reactor is determined by the area of the rectangle described with sides $X_n - X_{n-1}$ and $1/(dX/dt)$ or $P_n - P_{n-1}$ and $1/(dP/dt)$. The area of the rectangle is θ, and if F is known, V can be calculated. An alternative approach avoids some trial-and-error solutions that are necessary with the first approach. This second approach requires plots of dX/dt versus X and dP/dt versus P. The intersection of the reaction curve with a line from the mass balance equation (e.g., eq. 9.24a) determines the exit concentration of X or P, while the slope of the line determines D, and if F is known, V can be found. We illustrate the use of these approaches in Example 9.2.

Example 9.2

Data for the production of a secondary metabolite from a small-scale batch reactor are shown in Fig. 9.4. Assume that two reactors, each with 700-l working volume, are available. You will use exactly the same culture conditions (medium, pH, temperature, and so on) as in the batch reactor. If the flow rate is 100 l/h, predict the outlet concentration of the product. Compare that to the value predicted if a single 1400-l reactor were used. Use both graphical approaches.

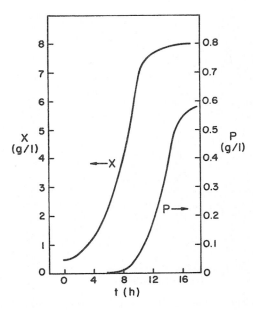

Figure 9.4. Data for Example 9.2. Data are for the production of a secondary product in batch culture.

Solution The first step in using either graphical approach is to differentiate the data in the batch growth curve to yield dX/dt and dP/dt. The differentiation of experimental data can magnify errors present in the original data, so the values of dX/dt and dP/dt must be interpreted cautiously.

For the graphical approach illustrated in Fig. 9.5, we have plotted $1/(dX/dt)$ versus X and $1/(dP/dt)$ versus P, which corresponds to eqs. 9.24b and 9.26b. For $\theta_1 = 7$ h (that is, 700 1/100 1/h), we must determine what value of X_1 will satisfy

$$\theta_1 = (X_1 - X_0)\left(\frac{1}{dX/dt}\right)\Bigg|_{x_1}$$

or

$$7 \text{ h} = X_1\left(\frac{1}{dX/dt}\right)\Bigg|_{x_1}$$

Since a sterile feed is to be used, $X_0 = 0$.

By trial and error, we find on the graph that $X_1 = 7.2$ g/l corresponds to $1/(dX_1/dt) = 0.95$ h/g/l or 7.2 g/l $\cdot$ 0.95 h/g/l = 6.84 h.

Given the accuracy with which Fig. 9.5 can be read, this is an acceptable solution. The product concentration that corresponds to $X_1 = 7.2$ g/l is determined from the batch growth curve. As illustrated, $X_1 = 7.2$ g/l is achieved at 9.4 h after inoculation; at the same time, the value for P_1 is 0.08 g/l.

The effect of the second stage on the process is determined by using eq. 9.26b and noting that again $\theta_2 = 7$ h. Thus,

Bioreactors for Suspension and Immobilized Cultures   Chap. 9

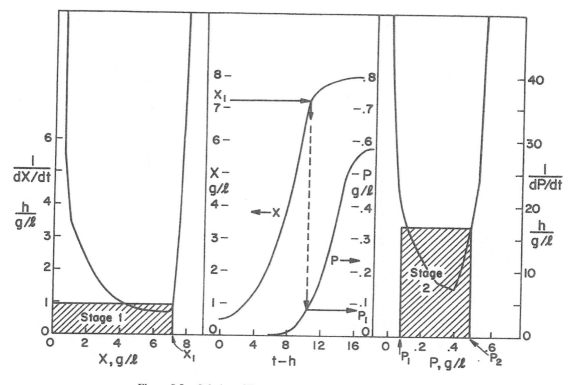

Figure 9.5. Solution of Example 9.2 for two-stage system, each with $\theta = 7$ h.

$$\theta_2 = 7 \ h = (P_2 - P_1)\frac{1}{dP_1/dt}$$

By trial and error, we find that at $P_2 = 0.49$ g/l

$$\theta_2 = (0.49 \ g/l - 0.08 \ g/l)(17 \ h/g/l)$$
$$= 6.97 \ h$$

which corresponds reasonably closely to 7 h.

In this solution the reader should note that for the first stage, only solutions that exist for X_1 greater than the value of X for which $1/(dX/dt)$ is a minimum are practically obtainable. Washout occurs if θ_1 is too small.

We can compare the result to a single-stage system with the same total volume as the two-stage system (Fig. 9.6). Here the trial-and-error approach indicates for $X_1 = 7.35$ g/l that

$$7.35 \ g/l \cdot 1.9 \ h/g/l = 13.97 \ h \approx 14 \ h$$

The value of P_1 that corresponds to $X_1 = 7.35$ g/l is 0.10 g/l. Thus, the use of the two-stage system in this case increased product concentration from 0.10 to 0.49 g/l.

An alternative graphical approach that eliminates the trial-and-error aspect of the first approach is shown in Fig. 9.7. Here eqs. 9.24a and 9.26a have been used. $D_1 = 1/\theta_1 =$

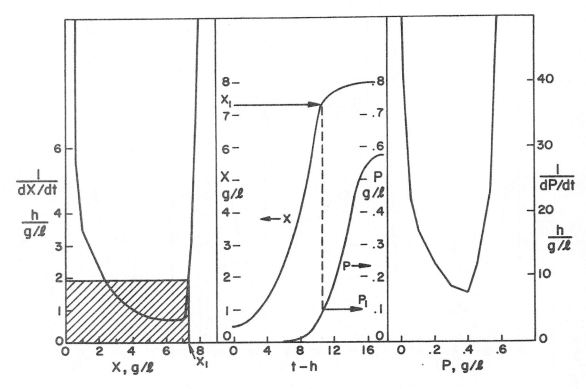

Figure 9.6. Solution of Example 9.2 with a single stage, where $\theta = 14$ h.

0.143 h^{-1}. The intersection of the reaction curve with the straight line determined by $D_1(X_1 - X_0) = D_1X_1$ is the solution to eq. 9.24a. For the second stage, we consider the production phase and use eq. 9.26a. The predicted values of X_1 and P_2 are the same as in the first approach. Note that the dP/dt-versus-P curve is displaced in time from the dX/dt-versus-X curve. Consequently, we use the dX/dt plot before using the dP/dt plot.

9.3.3. Fed-batch Operation

In fed-batch culture, nutrients are continuously or semicontinuously fed, while effluent is removed discontinuously (Fig. 9.8). Such a system is called a *repeated fed-batch culture*. Fed-batch culture is usually used to overcome substrate inhibition or catabolite repression by intermittent feeding of the substrate. If the substrate is inhibitory, intermittent addition of the substrate improves the productivity of the fermentation by maintaining the substrate concentration low. Fed-batch operation is also called the semicontinuous system or variable-volume continuous culture. Consider a batch culture where the concentration of biomass at a certain time is given by

$$X = X_0 + Y_{X/S}^M(S_0 - S) \tag{9.27}$$

Bioreactors for Suspension and Immobilized Cultures Chap. 9

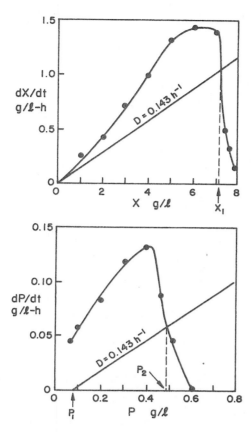

Figure 9.7. Solution to Example 9.2 using alternative graphical approach.

where S_0 is the initial substrate concentration, $Y_{X/S}^M$ is the yield coefficient, and X_0 is the initial biomass concentration. When biomass concentration reaches its maximum value (X_m), the substrate concentration is very low, $S \ll S_0$, and also $X_0 \ll X$. That is, $X_m \approx Y_{X/S}^M S_0$. Suppose that at $X_m \cong Y_{X/S}^M S_0$, a nutrient feed is started at a flow rate F, with the substrate concentration S_0. The total amount of biomass in the vessel is $X^t = VX$, where V is the culture volume at time t. The rate of increase in culture volume is

$$\frac{dV}{dt} = F \tag{9.28}$$

Integration of eq. 9.28 yields

$$V = V_0 + Ft \tag{9.29}$$

where V_0 is the initial culture volume (l).

The biomass concentration in the vessel at any time t is

$$X = X^t/V \tag{9.30}$$

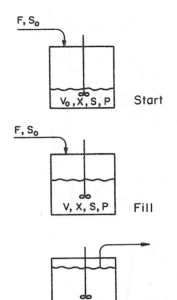

Figure 9.8. Schematic of a fed-batch culture.

The rate of change in biomass concentration is

$$\frac{dX}{dt} = \frac{V(dX^t/dt) - X^t(dV/dt)}{V^2} \tag{9.31}$$

Since $dX^t/dt = \mu_{net}X^t$, $dV/dt = F$, and $F/V = D$, eq. 9.31 becomes

$$\frac{dX}{dt} = (\mu_{net} - D)X \tag{9.32}$$

When the substrate is totally consumed, $S \approx 0$ and $X = X_m = Y_{X/S}^M S_0$. Furthermore, since nearly all the substrate in a unit volume is consumed, then $dX/dt = 0$. This is an example of a quasi-steady state. A fed-batch system operates at quasi-steady state when nutrient consumption rate is nearly equal to nutrient feed rate. Since $dX/dt = 0$ at quasi-steady state, then

$$\mu_{net} = D \tag{9.33}$$

If maintenance energy can be neglected,

$$\mu_{net} = \mu_m \frac{S}{K_s + S}$$

then

$$S \cong \frac{K_s D}{\mu_m - D} \tag{9.34}$$

The balance on the rate-limiting substrate without maintenance energy is

$$\frac{dS^t}{dt} = FS_0 - \frac{\mu_{net}X^t}{Y_{X/S}^M} \tag{9.35}$$

where S^t is the total amount of the rate-limiting substrate in the culture and S_0 is the concentration of substrate in the feed stream.

At quasi-steady state, $X^t = VX_m$ and essentially all the substrate is consumed, so no significant level of substrate can accumulate. Therefore,

$$FS_0 = \frac{\mu_{net}X^t}{Y_{X/S}^M} \tag{9.36}$$

Equation 9.31 at quasi-steady state with $S \approx 0$ yields

$$\frac{dX^t}{dt} = X_m \left(\frac{dV}{dt}\right) = X_m F = FY_{X/S}^M S_0 \tag{9.37}$$

Integration of eq. 9.37 from $t = 0$ to t with the initial amount of biomass in the reactor being X_0^t yields

$$X^t = X_o^t + FY_{X/S}^M S_0 t \tag{9.38}$$

That is, the total amount of cell in the culture increases linearly with time (which is experimentally observed) in a fed-batch culture. Dilution rate and therefore μ_{net} decrease with time in a fed-batch culture. Since $\mu_{net} = D$ at quasi-steady state, the growth rate is controlled by the dilution rate. The use of unstructured models is an approximation, since μ_{net} is a function of time.

Product profiles in a fed-batch culture can be obtained by using the definitions of $Y_{P/S}$ or q_P. When the product yield coefficient $Y_{P/S}$ is constant, at quasi-steady state with $S << S_0$

$$P \cong Y_{P/S} S_0 \tag{9.39}$$

or the potential product output is

$$FP \approx Y_{P/S}S_0F \tag{9.40}$$

When the specific rate of product formation q_P is constant,

$$\frac{dP^t}{dt} = q_P X^t \tag{9.41}$$

where P^t is the total amount of product in culture.

Substituting $X^t = (V_0 + Ft)X_m$ into eq. 9.41 yields

$$\frac{dP^t}{dt} = q_P X_m (V_0 + Ft) \tag{9.42}$$

Integration of eq. 9.42 yields

$$P^t = P_0^t + q_P X_m \left(V_0 + \frac{Ft}{2}\right)t \tag{9.43}$$

In terms of product concentration, eq. 9.43 can be written as

$$P = P_0 \frac{V_0}{V} + q_P X_m \left(\frac{V_0}{V} + \frac{Dt}{2} \right) t \tag{9.44}$$

Figure 9.9 depicts the variation of V, μ $(= D)$, X, S, and P with time at quasi-steady state in a single cycle of a fed-batch culture.

In some fed-batch operations, part of the culture volume is removed at certain intervals, since the reactor volume is limited. This operation is called the *repeated fed-batch culture*. The culture volume and dilution rate $(= \mu_{net})$ undergo cyclical variations in this operation.

If the cycle time t_w is constant and the system is always at quasi-steady state, then the product concentration at the end of each cycle is given by

$$P_w = \gamma P_0 + q_P X_m \left(\gamma + \frac{D_w t_w}{2} \right) t_w \tag{9.45}$$

where $D_w = F/V_w$, V_w is the culture volume at the end of each cycle, V_0 is the residual culture volume after removal, γ is the fraction of culture volume remaining at each cycle $(= V_0/V_w)$, and t_w is the cycle time.

The cycle time is defined as

$$t_w = \frac{V_w - V_0}{F} = \frac{V_w - \gamma V_w}{F} = \frac{1 - \gamma}{D_w} \tag{9.46}$$

Substitution of eq. 9.46 into eq. 9.45 yields

$$P_w = \gamma P_0 + \frac{q_P X_m}{2 D_w} (1 - \gamma^2) \tag{9.47}$$

An example of fed-batch culture is its use in some antibiotic fermentations, where a glucose solution is intermittently added to the fermentation broth due to the repression of pathways for the production of secondary metabolites caused by high initial glucose concentrations. The fed-batch method can be applied to other secondary metabolite fermenta-

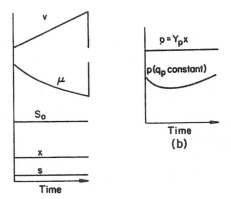

Figure 9.9. (a) Variation of culture volume (V), specific growth rate (μ), cell (X), and substrate (S) concentration with time at quasi-steady state. (b) Variation of product (P) concentration with time at quasi-steady state in a single cycle of a fed-batch culture.

tions such as lactic acid and other plant cell and mammalian cell fermentations, where the rate of product formation is maximal at low nutrient concentrations.

Fed-batch culture is important for *E. coli* fermentations to make proteins from recombinant DNA technology. To make a high concentration of product, it is desirable to grow the culture to very high cell density before inducing production of the target protein. If *E. coli* has an unlimited supply of glucose it will grow at a maximal rate, but produce organic acids (e.g., acetic acid) as by-products. The accumulation of these by-products inhibits growth. If glucose is fed at a rate that substains the growth rate at slightly less than maximal, *E. coli* uses the glucose more efficiently, making less by-product. Very high cell densities (50 to 100 g/l) can be achieved. Fed-batch culture may benefit from active process control. For example, the feed rate of glucose could be controlled by measuring glucose concentration in the medium or the CO_2 evolution rate using a feedback controller.

Example 9.3

In a fed-batch culture operating with intermittent addition of glucose solution, values of the following parameters are given at time $t = 2$ h, when the system is at quasi-steady state.

$$V = 1000 \text{ ml} \qquad\qquad F = \frac{dV}{dt} = 200 \text{ ml/h}$$

$$S_0 = 100 \text{ g glucose/l} \qquad\qquad \mu_m = 0.3 \text{ h}^{-1}$$

$$K_s = 0.1 \text{ g glucose/l} \qquad\qquad Y_{X/S}^M = 0.5 \text{ gdw cells/g glucose}$$

$$X_0^t = 30 \text{ g}$$

a. Find V_0 (the initial volume of the culture).
b. Determine the concentration of growth-limiting substrate in the vessel at quasi-steady state.
c. Determine the concentration and total amount of biomass in the vessel at $t = 2$ h (at quasi-steady state).
d. If $q_P = 0.2$ g product/g cells, $P_0 = 0$, determine the concentration of product in the vessel at $t = 2$ h.

Solution

a. $V = V_0 + Ft$

$V_0 = 1000 - 200(2) = 600$ ml

b. $D = F/V = 0.2 \text{ h}^{-1}$

$$S \approx \frac{K_s D}{\mu_m - D} = \frac{(0.1)(0.2)}{0.3 - 0.2} = 0.2 \text{ g glucose/l}$$

c. $X^t = X_0^t + FY_{X/S}^M S_0 t$

$$= 30 + (0.2)(0.5)(100)(2) = 50 \text{ g}$$

d. $P = P_0 \dfrac{V_0}{V} + q_P X_m \left(\dfrac{V_0}{V} + \dfrac{Dt}{2} \right) t$

$$= 0 + (0.2)(50)\left(\frac{600}{1000} + \frac{(0.2)(2)}{2} \right)(2)$$

$$= 16 \text{ g/l}$$

9.3.4. Perfusion Systems

An alternative to fed-batch culture is a perfusion system. Such systems are used most often with animal cell cultures (see Chapter 12). The basic characteristic is constant medium flow, cell retention, and in some cases selective removal of dead cells. High cell density can be achieved. Cell retention is usually achieved by membranes or screens or by a centrifuge capable of selective cell removal. When a membrane is used, the system has characteristics of an immobilized cell system (see Section 9.4) except the cells are usually maintained in suspension and mixed. With a selective removal/recycle the system approaches the cell recycle reactor discussed earlier in this chapter. Figure 9.10 depicts one type of perfusion system.

The potential advantages of a perfusion system is the potential removal of cell debris and inhibitory by-products, removal of enzymes released by dead cells that may destroy product, shorter exposure time of product to potentially harsh production conditions (compared to batch or fed-batch operation), high per-unit volumetric productivity (due to high cell density and metabolism), and a rather constant environment.

The primary disadvantage is that a large amount of medium is typically used and the nutrients in the medium are less completely utilized than in batch or fed-batch systems. High medium usage is expensive, owing not only to the high cost of raw materials but also to the costs to prepare and sterilize the medium. Additionally, costs for waste treatment increase. Typically the bioprocess engineer must consider the trade-off of improved product quality and reactor productivity with the extra costs associated with a more complex reactor system (membranes, pumps, centrifugal separator, etc.) and increased medium usage. The best choice depends on the specific situation.

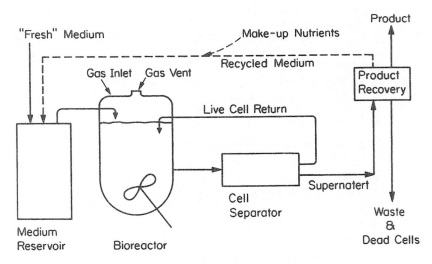

Fig. 9.10. Schematic of a perfusion system with external centrifugation and return of cells. Internal retention of cells is also possible. Return of spent medium is optional.

9.4. IMMOBILIZED CELL SYSTEMS

9.4.1. Introduction

Immobilization of cells as biocatalysts is almost as common as enzyme immobilization. Immobilization is the restriction of cell mobility within a defined space. Immobilized cell cultures have the following potential advantages over suspension cultures.

1. Immobilization provides high cell concentrations.
2. Immobilization provides cell reuse and eliminates the costly processes of cell recovery and cell recycle.
3. Immobilization eliminates cell washout problems at high dilution rates.
4. The combination of high cell concentrations and high flow rates (no washout restrictions) allows high volumetric productivities.
5. Immobilization may also provide favorable microenvironmental conditions (i.e., cell–cell contact, nutrient–product gradients, pH gradients) for cells, resulting in better performance of the biocatalysts (e.g., higher product yields and rates).
6. In some cases, immobilization improves genetic stability.
7. For some cells, protection against shear damage is important.

The major limitation on immobilization is that the product of interest should be excreted by the cells. A further complication is that immobilization often leads to systems for which diffusional limitations are important. In such cases the control of microenvironmental conditions is difficult, owing to the resulting heterogeneity in the system. With living cells, growth and gas evolution present significant problems in some systems and can lead to significant mechanical disruption of the immobilizing matrix.

In Chapter 3 we discussed enzyme immobilization. Figure 3.16 provides a useful summary of immobilization strategies. Many of the ideas in enzyme immobilization have a direct counterpart in whole cells. However, the maintenance of a living cell in such a system is more complex than maintaining enzymatic activity. The primary advantage of immobilized cells over immobilized enzymes is that immobilized cells can perform multistep, cofactor-requiring, biosynthetic reactions that are not practical using purified enzyme preparations.

9.4.2. Active Immobilization of Cells

Active immobilization is entrapment or binding of cells by physical or chemical forces. The two major methods of active immobilization are entrapment and binding.

Physical entrapment within porous matrices is the most widely used method of cell immobilization. Various matrices can be used for the immobilization of cells. Among these are porous polymers (agar, alginate, κ-carrageenan, polyacrylamide, chitosan, gelatin, collagen), porous metal screens, polyurethane, silica gel, polystyrene, and cellulose triacetate.

Polymer beads should be porous enough to allow the transport of substrates and products in and out of the bead. They are usually formed in the presence of cells and can be prepared by one of the following methods:

1. *Gelation of polymers:* Gelatin and agar beads may be prepared by mixing the liquid form of these polymers with cell suspensions and using a template to form beads. Reduction of temperature in the templates causes solidification of the polymers with the cells entrapped. Gel beads are usually soft and mechanically fragile. However, we can use a hard core (glass, plastic) and a soft gelatin shell with entrapped cells to overcome some mechanical problems associated with polymer beads. Because of diffusional limitations, the inner core of such beads is often not active, so this approach does not necessarily decrease the amount of product made per bead.

2. *Precipitation of polymers:* Cells are dispersed in a polymer solution, and by changing the pH or the solvent, the polymer can be precipitated. The starting solution of the polymer has to be prepared with an organic solvent or a water-solvent mixture. Ethanol and acetone are examples of water-miscible solvents. Polymers used for this purpose are polystyrene, cellulose triacetate, and collagen. The direct contact of cells with solvents may cause inactivation and even the death of cells.

3. *Ion-exchange gelation:* Ion-exchange gelation takes place when a water-soluble polyelectrolyte is mixed with a salt solution. Solidification occurs when the polyelectrolyte reacts with the salt solution to form a solid gel. The most popular example of this kind of gelation is the formation of Ca-alginate gel by mixing Na-alginate solution with a $CaCl_2$ solution. Some other polymers obtained by ion-exchange gelation are Al-alginate, Ca/Al carboxymethyl cellulose, Mg pectinate, κ-carrageenan, and chitosan polyphosphate. Alginate and κ-carrageenan are the most widely used polymers for cell-immobilization purposes. Ionic gels can be further stabilized by covalent cross-linking.

4. *Polycondensation:* Epoxy resins are prepared by polycondensation and can be used for cell immobilization. Polycondensation produces covalent networks with high chemical and mechanical stability. Usually, liquid precursors are cured with a multifunctional component. Functional groups usually are hydroxy, amino, epoxy, and isocyanate groups. Some examples of polymer networks obtained by polycondensation are epoxy, polyurethane, silica gel, gelatin–glutaraldehyde, albumin–glutaraldehyde, and collagen–glutaraldehyde. Severe reaction conditions (high temperature, low or high pH values) and toxic functional groups may adversely affect the activity of cells.

5. *Polymerization:* Polymeric networks can be prepared by cross-linking copolymers of a vinyl group containing monomers. Polyacrylamide beads are the most widely used polymer beads, prepared by copolymerization of acrylamide and bisacrylamide. Several different monomers can be used for polymer formation; acrylamide, methacrylamide, and 2-hydroxyethyl methacrylate are the most widely used. Cross-linking is usually initiated by copolymerization with a divinyl compound, such as methylenebis-acrylamide.

Immobilization by polymerization is a simple method. The polymerizing solution is mixed with the cell suspension, and polymerization takes place to form a polymeric block, which is pressed through a sieve plate to obtain regular-shaped particles. Suspension or emulsion polymerization can also be used to form polymeric beads for cell entrapment.

Encapsulation is another method of cell entrapment. Microcapsules are hollow, spherical particles bound by semipermeable membranes. Cells are entrapped within the hollow capsule volume. The transport of nutrients and products in and out of the capsule takes place through the capsule membrane. Microcapsules have certain advantages over gel beads. More cells can be packed per unit volume of support material into capsules, and intraparticle diffusion limitations are less severe in capsules due to the presence of liquid cell suspension in the intracapsule space. Various polymers can be used as capsule membranes. Among these are nylon, collodion, polystyrene, acrylate, polylysine–alginate hydrogel, cellulose acetate–ethyl cellulose, and polyester membranes. Different membranes (composition and MW cutoff) may need to be used for different applications in order to retain some high-MW products inside capsules and provide passage to low-MW nutrients and products.

Another form of entrapment is the use of macroscopic membrane-based reactors. The simplest of these is the hollow-fiber reactor. This device is a mass-transfer analog of the shell-and-tube heat exchanger in which the tubes are made of semipermeable membranes. Typically, cells are inoculated on the shell side and are allowed to grow in place. The nutrient solution is pumped through the insides of the tubes. Nutrients diffuse through the membrane and are utilized by the cells, and metabolic products diffuse back into the flowing nutrient stream. Owing to diffusional limitations, the unmodified hollow-fiber unit does not perform well with living cells. Modifications involving multiple membrane types (for example, for gas exchange or extractive product removal) or changes to promote convective flux within the cell layer have been proposed. Several commercial reactors for animal cell cultivation use membrane entrapment.

In addition to entrapment or encapsulation, cells can be bound directly to a support. Immobilization of cells on the surfaces of support materials can be achieved by physical adsorption or covalent binding.

Adsorption of cells on inert support surfaces has been widely used for cell immobilization. The major advantage of immobilization by adsorption is direct contact between nutrient and support materials. High cell loadings can be obtained using microporous support materials. However, porous support materials may cause intraparticle pore diffusion limitations at high cell densities, as is also the case with polymer-entrapped cell systems. Also, the control of microenvironmental conditions is a problem with porous support materials. A ratio of pore to cell diameter of 4 to 5 is recommended for the immobilization of cells onto the inner surface of porous support particles. At small pore sizes, accessibility of the nutrient into inner surfaces of pores may be the limiting factor, whereas at large pore sizes the specific surface area may be the limiting factor. Therefore, there may be an optimal pore size, resulting in the maximum rate of bioconversion.

Adsorption capacity and strength of binding are the two major factors that affect the selection of a suitable support material. Adsorption capacity varies between 2 mg/g (porous silica) and 250 mg/g (wood chips). Porous glass carriers provide adsorption capacities (10^8 to 10^9 cells/g) that are less than or comparable to those of gel-entrapped cell concentrations (10^9 to 10^{11} cells/ml). The binding forces between the cell and support surfaces may vary, depending on the surface properties of the support material and the type of cells. Electrostatic forces are dominant when positively charged support surfaces (ion-exchange resins, gelatin) are used. Cells also adhere on negatively charged surfaces by covalent binding or H bonding. The adsorption of cells on neutral polymer support surfaces may be mediated by chemical bonding, such as covalent bonding, H bonds, or van der Waals forces. Some specific chelating agents may be used to develop stronger cell-surface interactions. Among the support materials used for cell adsorption are porous glass, porous silica, alumina, ceramics, gelatin, chitosan, activated carbon, wood chips, polypropylene ion-exchange resins (DEAE–Sephadex, CMC-), and Sepharose.

Adsorption is a simple, inexpensive method of cell immobilization. However, limited cell loadings and rather weak binding forces reduce the attractiveness of this method. Hydrodynamic shear around adsorbed cells should be very mild to avoid the removal of cells from support surfaces.

Covalent binding is the most widely used method for enzyme immobilization. However, it is not as widely used for cell immobilization. Functional groups on cell and support material surfaces are not usually suitable for covalent binding. Binding surfaces need to be specially treated with coupling agents (e.g., glutaraldehyde or carbodiimide) or reactive groups for covalent binding. These reactive groups may be toxic to cells. A number of inorganic carriers (metal oxides such as titanium and zirconium oxide) have been developed that provide satisfactory functional groups for covalent binding.

Covalent binding forces are stronger than adsorption forces, resulting in more stable binding. However, with growing cells, large numbers of cell progeny must be lost. Support materials with desired functional groups are rather limited. Among the support materials used for covalent binding are CMC plus carbodiimide; carriers with aldehyde, amine, epoxy, or halocarbonyl groups; Zr(IV) oxide; Ti(IV) oxide; and cellulose plus cyanuric chloride. Support materials with —OH groups are treated with CNBr, materials with —NH$_2$ are treated with glutaraldehyde, and supports with COOH groups are treated with carbodiimide for covalent binding with protein groups on cell surfaces.

The direct cross-linking of cells by glutaraldehyde to form an insoluble aggregate is more like cell entrapment than binding. However, some cells may be cross-linked after adsorption onto support surfaces. Cross-linking by glutaraldehyde may adversely affect the cell's metabolic activity and may also cause severe diffusion limitations. Physical cross-linking may also be provided by using polyelectrolytes, polymers such as chitosan, and salts [CaCl$_2$, Al(OH)$_3$, FeCl$_3$]. Direct cross-linking is not widely used because of the aforementioned disadvantages.

A good support material should be rigid and chemically inert, should bind cells firmly, and should have high loading capacity. In the case of gel entrapment, gels should be porous enough and particle size should be small enough to avoid intraparticle diffusion limitations.

Some examples of cell immobilization by entrapment and by surface attachment (binding) are summarized in Tables 9.1 and 9.2, respectively.

9.4.3. Passive Immobilization: Biological Films

Biological films are the multilayer growth of cells on solid support surfaces. The support material can be inert or biologically active. Biofilm formation is common in natural and industrial fermentation systems, such as biological waste-water treatment and mold fermentations. The interaction among cells and the binding forces between the cell and support material may be very complicated.

TABLE 9.1 Examples of Cell Immobilization by Entrapment Using Different Support Materials

Cells	Support matrix	Conversion
S. cerevisiae	κ-Carrageenan or polyacrylamide	Glucose to ethanol
E. aerogenes	κ-Carrageenan	Glucose to 2,3-butanediol
E. coli	κ-Carrageenan	Fumaric acid to aspartic acid
Trichoderma reesei	κ-Carrageenan	Cellulose production
Z. mobilis	Ca-alginate	Glucose to ethanol
Acetobacter sp.	Ca-alginate	Glucose to gluconic acid
Morinda citrifolia	Ca-alginate	Anthraquinone formation
Candida tropicalis	Ca-alginate	Phenol degradation
Nocardia rhodocrous	Polyurethane	Conversion of testosterone
E. coli	Polyurethane	Penicillin G to G-APA
Catharantus roseus	Polyurethane	Isocitrate dehydrogenase activity
Rhodotorula minuta	Polyurethane	Menthyl succinate to menthol

In mixed-culture microbial films, the presence of some polymer-producing organisms facilitates biofilm formation and enhances the stability of the biofilms. Microenvironmental conditions inside a thick biofilm vary with position and affect the physiology of the cells.

In a stagnant biological film, nutrients diffuse into the biofilm and products diffuse out into liquid nutrient medium. Nutrient and product profiles within the biofilm are important factors affecting cellular physiology and metabolism. A schematic of a biofilm is depicted in Fig. 9.11. Biofilm cultures have almost the same advantages as those of the immobilized cell systems over suspension cultures, as listed in the previous section.

The thickness of a biofilm is an important factor affecting the performance of the biotic phase. Thin biofilms will have low rates of conversion due to low biomass concentration, and thick biofilms may experience diffusionally limited growth, which may or may not be beneficial depending on the cellular system and objectives. Nutrient-depleted regions may also develop within the biofilm for thick biofilms. In many cases, an optimal biofilm thickness resulting in the maximum rate of bioconversion exists and can be determined. In some cases, growth under diffusion limitations may result in higher yields of products as a result of changes in cell physiology and cell–cell interactions. In this case,

TABLE 9.2 Examples of Cell Immobilization by Surface Attachment

Cells	Support surface	Conversion
Lactobacillus sp.	Gelatin (adsorption)	Glucose to lactic acid
Clostridium acetobutylicum	Ion-exchange resins	Glucose to acetone, butanol
Streptomyces	Sephadex (adsorption)	Streptomycin
Animal cells	DEAE-sephadex/cytodex (adsorption)	Hormones
E. coli	Ti(IV) oxide (covalent binding)	
B. subtillis	Agarose–carbodiimide (covalent binding)	
Solanum aviculare	Polyphenylene oxide-glutaraldehyde (covalent binding)	Steroid glycoalkaloids formation

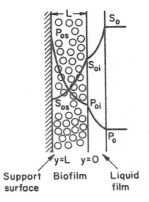

Figure 9.11 Schematic representation of a biofilm.

Support surface   Biofilm   Liquid film

improvement in reaction stoichiometry (e.g., high yield) may overcome the reduction in reaction rate, and it may be more beneficial to operate the system under diffusion limitations. Usually, the most sparingly soluble nutrient, such as dissolved oxygen, is the rate-limiting nutrient within the biofilm. A typical variation of dissolved oxygen concentration within the biofilm is depicted in Fig. 9.12.

9.4.4. Diffusional Limitations in Immobilized Cell Systems

Immobilization of cells may cause extra diffusional limitations as compared to suspension cultures. The presence and significance of diffusional limitations depend on the relative rates of bioconversion and diffusion, which can be described by the Damköhler number (Da) (see eq. 3.52 also).

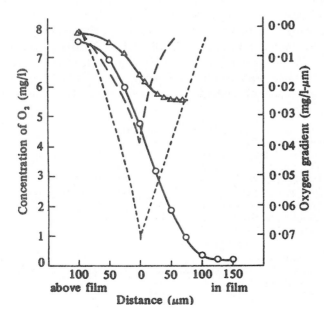

Figure 9.12. Dissolved-oxygen profiles and oxygen gradients in a microbial slime bathed in flowing medium: -Δ-Δ- oxygen profile for 20 ppm nutrient broth, 27.5°C; – –oxygen gradient for this profile; -O-O- oxygen profile for 500 ppm nutrient broth, 26°C; --- oxygen gradient for this profile. (With permission, from H. R. Bungay and others, *Biotechnol. Bioeng. II*:765, 1969, John Wiley & Sons, Inc., New York.)

$$Da = \frac{\text{maximum rate of bioconversion}}{\text{maximum rate of diffusion}} = \frac{r_{max}}{(D_e/\delta)S_0} \qquad (9.48)$$

where r_{max} is the maximum rate of bioconversion (mg S/l h), D_e is the effective diffusivity of the rate-limiting substrate, δ is the thickness of diffusion path (or liquid film), and S_0 is the bulk substrate concentration in liquid phase. When the film-theory model applies, D_e/δ is the mass transfer coefficient (i.e., $k_L = D_e/\delta$).

If Da >> 1, the rate of bioconversion is diffusion limited; for Da << 1, the rate is limited by the rate of bioconversion; and for Da ≈ 1, the diffusion and bioreaction rates are comparable. It is desirable to keep Da < 1 to eliminate diffusion limitations when the productivity of a cell population does not improve upon immobilization due to cell–cell contact and nutrient gradients.

Diffusional limitations may be external (that is, between fluid and support surface in adsorption and covalent binding), intraparticle (i.e., inside particles in entrapment, encapsulation, or immobilization in porous particles), or both. If the external mass transfer is limiting, an increase in liquid-phase turbulence should result in an increase in the reaction rate. In case of intraparticle mass-transfer limitations, a reduction in particle size or an increase in the porous void fraction of the support material should result in an increase in the rate of the bioreaction.

In Chapter 3 we discussed in reasonable detail a mathematical model of the interaction of diffusion and reaction for surface immobilized or entrapped biocatalysts. These models apply directly to immobilized cells when the kinetics of bioconversion are described by a Michaelis–Menten type of kinetic expression. Thus, the reader may wish to consult Chapter 3 again.

Another interesting case is to consider biofilms where we allow cell growth. Models for immobilized enzymes have no terms for biocatalyst replication, so this case presents a new problem.

The thickness of a biofilm or the size of microbial floc increases with time during the growth phase. A *microbial floc* is an aggregation of many cells, and in some processes these aggregates can be more than 1 mm in diameter. However, since the rate of increase in biofilm thickness is much slower than the rate of substrate uptake, the system can be assumed to be at quasi-steady state for relatively short periods. The simplest case is to assume that the system is at quasi-steady state and all the cells inside the biofilm are in the same physiological state. In this situation we write a steady-state substrate balance within the biofilm by using average kinetic constants for the biotic phase (living cells).

A differential material balance for the rate-limiting substrate within the biofilm (see Fig. 9.11) yields at steady state

$$D_e \frac{d^2 S}{dy^2} = \frac{1}{Y_{X/S}} \frac{\mu_m S}{K_s + S} X \qquad (9.49)$$

where D_e is the effective diffusivity (cm²/S) and $Y_{X/S}$ is the growth yield coefficient (g cells/g substrate).

The boundary conditions are

$$S = S_{0i} \quad \text{at } y = 0$$

$$\frac{dS}{dy} = 0 \quad \text{at } y = L$$

where L is the thickness of biofilm.

If it is also assumed that the liquid nutrient phase is vigorously agitated and the liquid film resistance is negligible, then $S_0 \approx S_{0i}$. By defining a maximum rate of substrate utilization as $r_m = \mu_m X / Y_{X/S}$ (g subs/cm³ h), we rewrite eq. 9.49 as

$$D_e \frac{d^2 S}{dy^2} = \frac{r_m S}{K_s + S} \tag{9.50}$$

In dimensionless form, eq. 9.50 can be written as

$$\frac{d^2 \overline{S}}{d\overline{y}^2} = \frac{\phi^2 \overline{S}}{1 + \beta \overline{S}} \tag{9.51}$$

where

$$\overline{S} = \frac{S}{S_0}, \qquad \overline{y} = \frac{y}{L}, \qquad \beta = \frac{S_0}{K_s}$$

and

$$\phi = L \sqrt{\frac{\mu_m X}{Y_{X/S} D_e K_s}} = L \sqrt{\frac{r_m}{D_e K_s}} \tag{9.52}$$

Equation 9.51 can be solved numerically. An analytical solution can be derived for the limiting cases of zero or first-order reaction rates.

The maximum rate of substrate flux in the absence of diffusion limitations is given by the following equation:

$$N_s A_s = -A_s D_e \frac{dS}{dy}\bigg|_{y=0} = \frac{r_m S_0}{K_s + S_0}(L A_s) \tag{9.53}$$

where A_s is a surface area of biofilm available for substrate flux, N_s is the substrate flux, and L is the thickness of the biofilm.

In the presence of diffusion limitation, the rate of substrate consumption or flux is expressed in terms of the effectiveness factor.

$$N_s = -D_e \frac{dS}{dy}\bigg|_{y=0} = \eta \left(\frac{r_m S_0}{K_s + S_0} \right) L \tag{9.54}$$

where η is the effectiveness factor, defined as the ratio of the rate of substrate consumption in the presence of diffusion limitation to the rate of substrate consumption in the absence of diffusion limitation. In the absence of diffusion limitations, $\eta \cong 1$, and in the presence of diffusion limitations, $\eta < 1$. The effectiveness factor is a function of ϕ and β. Figure 9.13 is a plot of η versus β for various values of ϕ. The ϕ value should be low

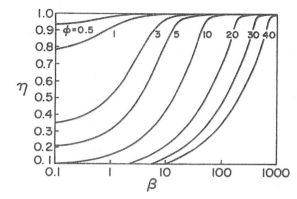

Figure 9.13. Effectiveness factor for a flat biofilm as a function of β, the dimensionless initial substrate concentration, and φ, the Thiele modulus. (With permission, redrawn from B. Atkinson, *Biochemical Reactors,* Pion Ltd., London, 1974, p. 81.)

($\phi < 1$) to eliminate diffusion limitations. As the biofilm grows (slowly), the value of ϕ will gradually increase. If shear forces cause a portion of the film to detach, then ϕ will decrease abruptly.

The effectiveness factor (η) can be calculated as

$$\eta = 1 - \frac{\tanh\phi}{\phi}\left(\frac{\omega}{\tanh\omega} - 1\right), \quad \text{for } \omega \le 1 \tag{9.55}$$

$$\eta = \frac{1}{\omega} - \frac{\tanh\phi}{\phi}\left(\frac{1}{\tanh\phi} - 1\right), \quad \text{for } \omega \ge 1 \tag{9.56}$$

where ω is the modified Thiele modulus and is given by

$$\omega = \frac{\phi(S_0/K_s)}{\sqrt{2}\left(1 + \dfrac{S_0}{K_s}\right)}\left[\frac{S_0}{K_s} - \ln\left(1 + \frac{S_0}{K_s}\right)\right]^{-1/2} \tag{9.57}$$

Some cells such as molds (*A. niger*) form pellets in a fermentation broth, and substrates need to diffuse inside pellets to be available for microbial consumption. Cells may form biofilms on spherical support particles, as depicted in Fig. 9.14. Similar equations need to be solved in spherical geometry in this case to determine the substrate profile within the floc and the substrate consumption rate. The dimensionless substrate transport equation within the microbial floc is

$$\frac{d^2\bar{S}}{d\bar{r}^2} + \frac{2}{\bar{r}}\frac{d\bar{S}}{d\bar{r}} = \frac{\phi^2\bar{S}}{1 + \bar{S}/\beta'} \tag{9.58}$$

where

$$\bar{S} = \frac{S}{S_0}, \quad \bar{r} = \frac{r}{R}, \quad \beta' = \frac{S_0}{K_s}$$

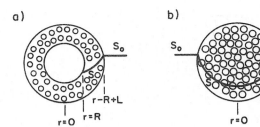

Figure 9.14. (a) Microbial film on inert spherical support particle. (b) Spherical microbial floc.

and

$$\phi = R \sqrt{\frac{\mu_m X}{Y_{X/S} D_e K_s}} = R \sqrt{\frac{r_m}{D_e K_s}} \qquad (9.59)$$

The boundary conditions are

$$\overline{S} = 1, \qquad \text{at } \overline{r} = 1$$

$$\frac{d\overline{S}}{d\overline{r}} = 0, \qquad \text{at } \overline{r} = 0$$

For nonspherical particles, a characteristic length is defined as

$$L = \frac{V_P}{A_P} \qquad (9.60)$$

where V_P and A_P are the volume and surface area of microbial pellet.

The rate of substrate consumption by a single microbial floc is

$$N_S A_P = -A_P D_e \frac{dS}{dr}\bigg|_{r=R} = \eta \frac{r_m S_0}{K_s + S_0} V_P \qquad (9.61)$$

The effectiveness factor (η) is a function of ϕ and β. Variation of η with ϕ and β is similar to that of Fig. 9.13. However, η values for spherical geometry are slightly lower than those of rectangular geometry for intermediate values of ϕ ($1 < \phi < 10$). An analytical solution to eq. 9.58 is possible for first- and zero-order reaction kinetics.

The reaction rate can be approximated to first order at low substrate concentrations.

$$r_s = \frac{\mu_m S}{Y_{X/S} K_s} X = \frac{r_m}{K_s} S \qquad (9.62)$$

where $r_m = (\mu_m/Y_{X/S})X$. The effectiveness factor in this case is given by

$$\eta = \frac{1}{\phi}\left[\frac{1}{\tanh 3\phi} - \frac{1}{3\phi}\right] \qquad (9.63)$$

where

$$\phi = \frac{V_P}{A_P} \sqrt{\frac{r_m/K_s}{D_e}}$$

The rate of bioreaction can be approximated to zero order at values of $S \gg K_S$. Because K_S is often very small, the zero-order limit usefully describes many systems of practical interest.

$$r_s = \frac{\mu_m X}{Y_{X/S}} = r_m$$

(9.64)

The solution to eq. 9.58 in this case is

$$S = S_0 - \frac{r_m}{6D_e}(R^2 - r^2)$$

(9.65)

Substrate concentration may be zero at a certain radial distance from the center of the floc according to eq. 9.65. This distance is called the critical radius (r_{cr}) and is determined by setting $S = 0$ at r_{cr} in eq. 9.65.

$$\left(\frac{r_{cr}}{R}\right)^2 = 1 - \frac{6D_e S_0}{r_m R^2}$$

(9.66)

When $r_{cr} > 0$—that is, $R > (6D_e S_0/r_m)^{1/2}$—then the concentration of the limiting substrate is zero for $0 < r < r_{cr}$. In this case, the limiting substrate is consumed only in the outer shell of the floc, and the effectiveness factor is given by

$$\eta = \frac{r_m \frac{4}{3}\pi(R^3 - r_{cr}^3)}{\frac{4}{3}\pi R^3 \cdot r_m} = 1 - \left(\frac{r_{cr}}{R}\right)^3$$

(9.67)

or

$$\eta = 1 - \left(1 - \frac{6D_e S_0}{r_m R^2}\right)^{3/2}$$

(9.68)

9.4.5. Bioreactor Considerations in Immobilized Cell Systems

Various reactor configurations can be used for immobilized cell systems. Since the support matrices used for cell immobilization are often mechanically fragile, bioreactors with low hydrodynamic shear, such as packed-column, fluidized-bed, or airlift reactors, are preferred. Mechanically agitated fermenters can be used for some immobilized-cell systems if the support matrix is strong and durable. Any of these reactors can usually be operated in a perfusion mode by passing nutrient solution through a column of immobilized cells. Schematic diagrams of immobilized-cell packed-column and fluidized-bed reactors are depicted in Fig. 9.15. These reactors can be operated in batch or continuous mode.

Consider the reactors shown in Fig. 9.15. When the fluid recirculation rate is high, the system approaches CFSTR behavior. One commercial fluidized-bed, immobilized-animal-cell bioreactor system requires high recirculation to maintain uniform conditions in the reactor. The models we have discussed so far can be applied to such systems. The

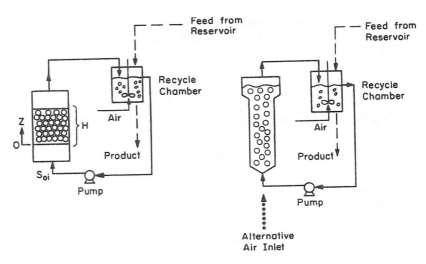

Figure 9.15. Schematics of a packed-bed and a fluidized-bed biofilm or immobilized-cell bioreactors are shown. In batch operation, only the streams with solid lines exist. In continuous operation, the streams shown by dashed lines are added. For the fluidized bed, fluidization can be accomplished by liquid recirculation only or a mixture of liquid and gas flows.

other extreme involves some waste-treatment systems where the rate of fluid recirculation is low or even zero. In the latter case, the system cannot be modeled as a CFSTR but must be treated as a PFR. To analyze such a system, consider a material balance on the rate-limiting substrate over a differential element:

$$-F \, dS_0 = N_S a A \, dz \qquad (9.69)$$

where S_0 is the bulk liquid-phase substrate concentration (mg S/cm^3) and is a function of height, F is the liquid nutrient flow rate (cm^3/h), N_s is flux of substrate into the biofilm (mg S/cm^2 h), a is the biofilm or support particle surface area per unit reactor volume (cm^2/cm^3), A is the cross-sectional area of the bed (cm^2), and dz is the differential height of an element of the column (cm). Substituting eq. 9.54 into eq. 9.69 yields the following equation:

$$-F \frac{dS_0}{dz} = \eta \frac{r_m S_0}{K_s + S_0} LaA \qquad (9.70)$$

Integration of eq. 9.70 yields

$$K_s \ln \frac{S_{0i}}{S_0} + (S_{0i} - S_0) = \frac{\eta r_m LaA}{F} H \qquad (9.71)$$

where S_{0i} is the inlet bulk substrate concentration, L is the biofilm thickness or the characteristic length of the support particle ($L = V_P/A_P$), and H is the total height of the packed bed.

For low substrate concentrations in the feed, the rate of substrate consumption is first order and eq. 9.71 has the following form:

$$\ln\frac{S_0}{S_{0i}} = -\frac{\eta r_m LaA}{FK_s}H \qquad (9.72)$$

Substrate concentration drops exponentially with the height of the column in this case, and a plot of $\ln S_0$ versus H results in a straight line. Equation 9.71 or 9.72 can be used as the design equation for immobilized–biofilm column reactors to determine the height of the column for a desired level of substrate conversion.

Example 9.4

Glucose is converted to ethanol by immobilized *S. cerevisiae* cells entrapped in Ca-alginate beads in a packed column. The specific rate of ethanol production is $q_P = 0.2$ g ethanol/ g cell · h, and the average dry-weight cell concentration in the bed is $\bar{X} = 25$ g/l bed. Assume that growth is negligible (i.e., almost all glucose is converted to ethanol) and the bead size is sufficiently small that $\eta \cong 1$. The feed flow rate is $F = 400$ l/h, and glucose concentration in the feed is $S_{0i} = 100$ g glucose/l. The diameter of the column is 1 m, and the product yield coefficient is $Y_{P/S} \approx 0.49$ g ethanol/g glucose.

a. Write a material balance on the glucose concentration over a differential height of the column and integrate it to determine $S = S(z)$ at steady state.
b. Determine the column height for 98% glucose conversion at the exit of the column.
c. Determine the ethanol concentration in the effluent.

Solution

a. A material balance on the glucose concentration over a differential height of the column yields

$$-F\,dS_0 = \frac{q_P\bar{X}}{Y_{P/S}}dV = \frac{q_P\bar{X}}{Y_{P/S}}A\,dz$$

Integration yields

$$-F\int_{S_{0i}}^{S_n}dS_0 = \frac{q_P\bar{X}}{Y_{P/S}}A\int_0^H dz$$

$$S_{0i} - S_0 = \frac{q_P\bar{X}}{Y_{P/S}}\frac{A}{F}H$$

This equation differs from the form of eq. 9.72 because S_{0i} is high and the reaction rate is effectively zero order.

b. $S_0 = 0.02(100) = 2$ g glucose/l. Substituting the given values into the above equation yields

$$(100-2) = \frac{(0.2)(25)}{0.49}\frac{(\pi/4)(10)^2}{400}H$$

$$H = 49 \text{ dm} = 4.9 \text{ m}$$

c. $P = Y_{P/S}(S_{0i} - S_0) = 0.49(98) = 48$ g/l.

9.5. SOLID-STATE FERMENTATIONS

Solid-state fermentations (SSF) are fermentations of solid substrates at low moisture levels or water activities. The water content of a typical submerged fermentation is more than 95%. The water content of a solid mash in SSF often varies between 40% and 80%. Solid-state fermentations are usually used for the fermentation of agricultural products or foods, such as rice, wheat, barley, corn, and soybeans. The unique characteristic of solid-state fermentations is operation at low moisture levels, which provides a selective environment for the growth of mycelial organisms, such as molds. In fact, most solid-state fermentations are mold fermentations producing extracellular enzymes on moist agricultural substrates. Since bacteria and yeasts cannot tolerate low moisture levels (water activities), the chances of contamination of fermentation media by bacteria or yeast are greatly reduced in SSF. Although most SSFs are mold fermentations, SSFs based on bacteria and yeast operating at relatively high moisture levels (75% to 90%) are also used. Solid-state fermentations are used widely in Asia for food products, such as tempeh, miso, or soy sauce fermentations, and also for enzyme production.

The major advantages of SSFs over submerged fermentation systems are (1) the small volume of fermentation mash or reactor volume, resulting in lower capital and operating costs, (2) a lower chance of contamination due to low moisture levels, (3) easy product separation, (4) energy efficiency, and (5) the allowing of the development of fully differentiated structures, which is critical in some cases to product formation. The major disadvantage of SSFs is the heterogeneous nature of the media due to poor mixing characteristics, which results in control problems (pH, DO, temperature) within the fermentation mash. To eliminate these control problems, fermentation media are usually mixed either continuously or intermittently. For large fermentation mash volumes, the concentration gradients may not be eliminated at low agitation speeds, and mycelial cells may be damaged at high agitation speeds. Usually, a rotating-drum fermenter is used for SSF systems, and the rotational speed needs to be optimized for the best performance.

Solid-substrate fermentations imply a more general method of fermentations in which moisture content may not need to be low, but the substrate is in the form of submerged solid particles in liquid media. Bacterial ore leaching (i.e., growth and microbial oxidation on surfaces of mineral sulfide particles) or fermentation of rice in a packed column with circulating liquid media are examples of solid-substrate fermentations. Solid-state (or solid-phase) fermentations are a special form of solid-substrate fermentations for which the substrate is solid and the moisture level is low.

The *koji process* is an SSF system that employs molds (*Aspergillus, Rhizopus*) growing on grains or foods (wheat, rice, soybean). A typical SSF process involves two stages. The first and the primary stage is an aerobic, fungal, solid-state fermentation of grains called the *koji*. The second stage is an anaerobic submerged fermentation with a mixed bacterial culture called the *moromi*. The products listed in Table 9.3 are the products of aerobic SSF, the koji process. Fermentation in the second stage (moromi) may be realized by using the natural flora, or, usually, with externally added bacteria and yeasts. Some strains of *Saccharomyces, Torulopsis*, and *Pediococcus* are used as flavor producers in soy sauce manufacture. The moromi is usually fermented for 8 to 12 months. However, the processing time can be reduced to 6 months by temperature profiling. The final product is pressed to recover the liquid soy sauce and is pasteurized, filtered, and bottled.

TABLE 9.3 Some Traditional Food Fermentations

| Product | Primary genus | Common substrate | Thermal processing | | Initial moisture (%) | Incubation | | Further processing |
			Temp. (°C)	Time (min)		Time (h)	Temp. (°C)	
Soy sauce	*Aspergillus*	Soybean, wheat	110	30	45	72	30	Yes
Miso	*Aspergillus*	Rice, soybean	100	40	35	44	30	Yes
Tempeh	*Rhizopus*	Soybean	100	30	40	22	32	No
Hamanatto	*Aspergillus*	Soybean, wheat				36		Yes
Sufu	*Actinomucor*	Tofu	100	10	74		15	Yes

With permission, from R. E. Midgett, in A. L. Demain and N. A. Solomon, eds., *Manual of Industrial Microbiology and Biotechnology*, ACS Publications, Washington, D.C., 1986.

The major industrial use of the koji process is for the production of enzymes by fungal species. Fungal amylases are produced by SSF of wheat bran by *A. oryzae* in a rotating-drum fermenter. Wheat bran is pretreated with formaldehyde, and the initial pH of the bran is adjusted to pH = 3.5 to 4.0 to reduce the chance of contamination. Usually, perforated pans, rotating drums, or packed beds with air ventilation are used. A typical rotary-drum type of koji fermenter is depicted in Fig. 9.16. Enzymes other than amylases, such as cellulase, pectinase, protease, and lipases, can also be produced by koji fermentations. *Trichoderma viride* species have been used for the production of cellulases from wheat bran in a rotary-tray fermenter.

Some secondary metabolites, such as antibacterial agents, are produced by *Rhizopus* and *Actinomucor* species in some koji processes. Certain mycotoxins, such as aflatoxins, were produced by SSF of rice (40% moisture) by *A. parasiticus*. Ochratoxins were also produced by *Aspergillus* species on wheat in a rotary-drum koji fermenter. Microbial degradation of lignocellulosics can also be accomplished by solid-state fermentations for waste-treatment purposes or in biopulping of wood chips for use in paper manufacture. Spores from some molds have found use as insecticides. Proper spore formation is difficult to obtain in submerged culture, and SSF must be used.

Major process variables in SSF systems are moisture content (water activity), inoculum density, temperature, pH, particle size, and aeration/agitation. Optimization of these parameters to maximize product yield and rate of product formation is the key in SSF systems and depends on the substrate and organism used. Most natural substrates (e.g., grains) require pretreatment to make the physical structure of substrates more susceptible to mycelial penetration and utilization. Solid substrates are usually treated with antimicrobial agents, such as formaldehyde, and are steamed in an autoclave. Nutrient media addition, pH adjustment, and the adjustment of moisture level are realized before inoculation of the fermentation mash. Koji fermentations are usually realized in a controlled-humidity air environment with air ventilation and agitation. Many solid-state mycelial fermentations are shear sensitive due to disruption of the mycelia at high

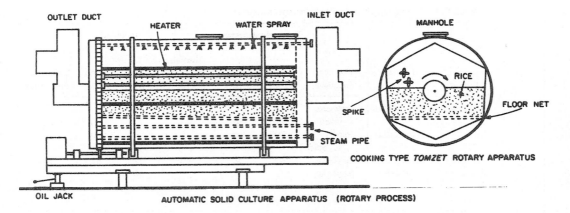

OUTLET DUCT　　HEATER　　WATER SPRAY　　INLET DUCT　　MANHOLE

RICE

SPIKE

STEAM PIPE

FLOOR NET

COOKING TYPE *TOMZET* ROTARY APPARATUS

OIL JACK　　AUTOMATIC SOLID CULTURE APPARATUS (ROTARY PROCESS)

Figure 9.16. Rotary-drum type of koji-making apparatus used for rice solids culture by *A. oryzae*. All operations (washing, cooking, inoculation, loosening of solids, water spraying, cooling, air circulation, filling, and exhausting) can be done in this apparatus. (With permission, from N. Toyama, *Biotechnol. Bioeng. Symp.*, Vol. 6, pp. 207–219, 1976, John Wiley & Sons, Inc., New York.)

agitation/rotation speeds. At low agitation rates, oxygen transfer and CO_2 evolution rates become limiting. Therefore, an optimal range of agitation rate or rotation speed needs to be determined. Similarly, there is a minimum level of moisture content (~30% by weight) below which microbial activity is inhibited. At high moisture levels (>60%), solid substrates become sticky and form large aggregates. Moreover, moisture level affects the metabolic activities of cells. Optimal moisture level needs to be determined experimentally for each cell–substrate system. For most of the koji processes, the optimal moisture level is about 40% ± 5%. Particle size should be small enough to avoid any oxygen–CO_2 exchange or other nutrient transport limitations. Porosity of the particles can be improved by pretreatment to provide a larger intraparticle surface to volume ratio for microbial action.

Most of the SSF processes are realized using a rotary-tray type of reactor in a temperature- and humidity-controlled chamber where controlled-humidity air is circulated through stacked beds of trays containing fermented solids. Figure 9.17 depicts a rotary-tray chamber for koji fermentations. Rotary-drum fermenters are used less frequently because of the shear sensitivity of mycelial cells.

9.6. SUMMARY

Bioreactors using suspended cells can be operated in many modes intermediate between a batch reactor and a single-stage chemostat. Although a chemostat has potential productivity advantages for primary products, considerations of genetic instability, process flexibility, and process reliability have greatly limited the use of chemostat units. The use of cell recycle with a CSTR increases volumetric productivity and has found use in large-

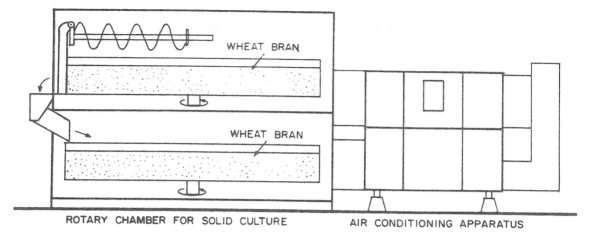

ROTARY CHAMBER FOR SOLID CULTURE   AIR CONDITIONING APPARATUS

Figure 9.17. Rotary, automatic koji-making apparatus. The apparatus has a two-storied chamber. Each chamber has a large rotary tray on which wheat bran is heaped evenly. After inoculated fungus has grown sufficiently, solid culture is transferred by a screw conveyor to the lower rotary-tray hopper. (With permission, from N. Toyama, *Biotechnol. Bioeng. Symp.*, Vol. 6, pp. 207–219, 1976, John Wiley & Sons, Inc., New York.)

volume, low-product-value processes (e.g., waste treatment and fuel-grade ethanol production). Multistage continuous systems improve the potential usefulness of continuous processes for the production of secondary metabolites and for the use of genetically unstable cells. The fed-batch system is widely used in commercial plants and combines the features of continuous culture and batch that allow the manufacturer to maintain flexibility and ease of intervention. The perfusion system is another option that is particularly attractive for animal cells.

Immobilized cell systems offer a number of potential processing advantages, and the commercialization of such systems is proceeding rapidly where cell culture is expensive and difficult (e.g., animal cell tissue culture). Physical entrapment or encapsulation is used in most cases, although adsorption onto surfaces or covalent binding of cells to surfaces is possible.

In some cases, self-immobilization on surfaces is possible and a biofilm is formed. Biofilm reactors can apply to tissue culture, mold, and bacterial systems. Biofilm-based reactors are very important in waste-treatment applications and in natural ecosystems. The analysis of immobilized cell reactors is analogous to that for immobilized enzyme reactors except for the feature of biocatalyst replication.

Solid-state fermentations share some characteristics with immobilized cell systems, but differ in that no discernible liquid is present. SSFs have found important uses in the production of some traditional fermented foods and may have use in upgrading agricultural or forest materials and in the production of mold products requiring full mold differentiation.

Sec. 9.6   Summary

The reactor options described in this chapter are many. The best choice of reactor systems will ultimately be determined by the choice of biocatalyst and the requirements for product recovery and purification.

SUGGESTIONS FOR FURTHER READING

AIBA, S., A. E. HUMPHREY, AND N. F. MILLIS, *Biochemical Engineering*, 2d ed., Academic Press, New York, 1973.

ATKINSON, B., AND F. MAVITUNA, *Biochemical Engineering and Biotechnology Handbook*, 2d ed., Stockton Press, New York, 1991.

BAILEY, J. E., AND D. F. OLLIS, *Biochemical Engineering Fundamentals*, 2d ed., McGraw-Hill Book Co., New York, 1986.

BLANCH, H. W., AND D. S. CLARK, *Biochemical Engineering*, Marcel Dekker, Inc., New York, 1996.

CHARACKLIS, W. G., R. BAKKE, AND M. G. TRULEAR, 1991, "Fundamental Considerations of Fixed Film Systems," in M. Moo-Young, ed., *Comprehensive Biotechnology*, Vol. 4, pp. 945–961, 1985.

CHIBATA, I., T. TOSA, AND T. SATO, "Methods of Cell Immobilization," in *Manual of Industrial Microbiology and Biotechnology*, A. L. Demain and N. A. Solomon, eds., American Society for Microbiology, Washington, DC, pp. 217–229, 1986.

DE GOOIJER, C. D., W. A. M. BAKKER, H. H. BEEFTINK, AND J. TRAMPER, Bioreactors in Series: An Overview of Design Procedures and Practical Applications, *Enzyme Microbiol Technology 18:* 202–219, 1996.

KARGI, F., AND M. MOO-YOUNG, "Transport Phenomena in Bioprocesses," in M. Moo-Young, ed., *Comprehensive Biotechnology*, Vol. 2, Pergamon Press, Elmsford, NY, pp. 5–55, 1985.

KLEIN, J., AND K. D. VORLOP, "Immobilization Techniques: Cells," in M. Moo-Young, ed., *Comprehensive Biotechnology*, Vol. 2, Pergamon Press, Elmsford, NY, pp. 203–334, 1985.

MERCILLE, S., M. JOHNSON, S. LAUTHIER, A. A. KAMEN, AND B. MASSIA, Understanding Factors that Limit the Productivity of Suspension-Based Perfusion Cultures Operated at High Medium Renewal Rates, *Biotechnology Bioengineering, 67:* 435–450, 2000.

MIDGETT, R. E., "Solid State Fermentations," in A. L. Demain and N. A. Solomon, *Manual of Industrial Microbiology and Technology*, American Society for Microbiology, Washington, DC, pp. 66–83, 1986.

MOO-YOUNG, M., *Bioreactor Immobilized Enzymes and Cells: Fundamentals and Applications*, Elsevier Science Publishing, Inc., New York, 1988.

SCHROEDER, E. D., *Water and Wastewater Treatment*, McGraw-Hill Book Co., New York, 1977.

WANG, D. I. C., AND OTHERS, *Fermentation and Enzyme Technology*. John Wiley & Sons, Inc., 1979.

WEBSTER, I. A., M. L. SHULER, AND P. RONY. The Whole Cell Hollow Fiber Reactor: Effectiveness Factors, *Biotech. Bioeng. 21:* 1725–1748, 1979.

PROBLEMS

9.1. Consider a 1000-l CSTR in which biomass is being produced with glucose as the substrate. The microbial system follows a Monod relationship with $\mu_m = 0.4$ h^{-1}, $K_S = 1.5$ g/l (an unusually high value), and the yield factor $Y_{X/S} = 0.5$ g biomass/g substrate consumed. If normal operation is with a sterile feed containing 10 g/l glucose at a rate of 100 l/h:

a. What is the specific biomass production rate (g/l-h) at steady state?

b. If recycle is used with a recycle stream of 10 l/h and a recycle biomass concentration five times as large as that in the reactor exit, what would be the new specific biomass production rate?

c. Explain any difference between the values found in parts a and b.

9.2. In a two-stage chemostat system, the volumes of the first and second reactors are $V_1 = 500$ l and $V_2 = 300$ l, respectively. The first reactor is used for biomass production and the second is for a secondary metabolite formation. The feed flow rate to the first reactor is $F = 100$ l/h, and the glucose concentration in the feed is $S = 5.0$ g/l. Use the following constants for the cells.

$$\mu_m = 0.3 \ h^{-1}, \qquad K_S = 0.1 \ g/l, \qquad Y_{X/S} = 0.4 \ \frac{g \ dw \ cells}{g \ glucose}$$

a. Determine cell and glucose concentrations in the effluent of the first stage.

b. Assume that growth is negligible in the second stage and the specific rate of product formation is $q_P = 0.02$ g P/g cell h, and $Y_{P/S} = 0.6$ g P/g S. Determine the product and substrate concentrations in the effluent of the second reactor.

9.3. Consider the following batch growth data:

Time	X	P	dX/dt	dP/dt
h	g/l	g/l	g/l-h	g/l-h
0	0.3	<0.01	—	—
3	1.0	<0.01	0.30	—
6	2.3	<0.01	0.55	—
8	4.0	0.010	1.0	0.005
9	5.1	0.025	1.3	0.010
10	6.5	0.060	1.4	0.045
10.5	7.0	—	1.4	—
11	7.4	0.10	0.60	0.059
12	7.7	0.17	0.20	0.072
13	7.8	0.26	0.02	0.105
14	—	0.36	—	0.130
15	8.0	0.47	~0	0.087
16	8.0	0.54	~0	0.042
17	—	0.58	—	0.021
18	—	0.60	—	0.005

You have available three tanks of different volumes: 900, 600, and 300 l. Given a flow rate of 100 l/h, what configuration of tanks would maximize product formation?

9.4. Penicillin is produced by *P. chrysogenum* in a fed-batch culture with the intermittent addition of glucose solution to the culture medium. The initial culture volume at quasi-steady state is $V_0 = 500$ l, and glucose-containing nutrient solution is added with a flow rate of $F = 50$ l/h. Glucose concentration in the feed solution and initial cell concentration are $S_0 = 300$ g/l and $X_0 = 20$ g/l, respectively. The kinetic and yield coefficients of the organism are $\mu_m = 0.2 \ h^{-1}$, $K_S = 0.5$ g/l, and $Y_{X/S} = 0.3$ g dw/g glucose.

a. Determine the culture volume at $t = 10$ h.

b. Determine the concentration of glucose at $t = 10$ h at quasi-steady state.

c. Determine the concentration and total amount of cells at quasi-steady state when $t = 10$ h.

d. If $q_P = 0.05$ g product/g cells h and $P_0 = 0.1$ g/l, determine the product concentration in the vessel at $t = 10$ h.

9.5 The bioconversion of glucose to ethanol is carried out in a packed-bed, immobilized-cell bioreactor containing yeast cells entrapped in Ca-alginate beads. The rate-limiting substrate is glucose, and its concentration in the feed bulk liquid phase is $S_{0i} = 5$ g/l. The nutrient flow rate is $F = 2$ l/min. The particle size of Ca-alginate beads is $D_P = 0.5$ cm. The rate constants for this conversion are

$$r_m = 100 \text{ mg } S/cm^3 \cdot h$$

$$K_S = 10 \text{ mg } S/cm^3$$

for the following rate expression:

$$r_s = \frac{r_m S}{K_S + S}$$

The surface area of the alginate beads per unit volume of the reactor is $a = 25$ cm^2/cm^3, and the cross-sectional area of the bed is $A = 100$ cm^2. Assuming a first-order reaction-kinetics (e.g., relatively low substrate concentrations), determine the required bed height for 80% conversion of glucose to ethanol at the exit stream. *Hint:* To calculate the effectiveness factor, we can use the following equations:

$$\eta = \frac{1}{\phi}\left[\frac{1}{\tanh 3\phi} - \frac{1}{3\phi}\right]$$

where

$$\phi = \frac{V_P}{A_P}\sqrt{\frac{k_1}{D_S}}$$

and

$$k_1 = \frac{r_m}{K_S}, \qquad D_S = 10^{-6} \text{ cm}^2/s$$

9.6. A fluidized-bed, immobilized-cell bioreactor is used for the conversion of glucose to ethanol by *Z. mobilis* cells immobilized in κ-carrageenan gel beads. The dimensions of the bed are 10 cm (diameter) by 200 cm. Since the reactor is fed from the bottom of the column and because of CO_2 gas evolution, substrate and cell concentrations decrease with the height of the column. The average cell concentration at the bottom of the column is $X_0 = 45$ g/l, and the average cell concentration decreases with the column height according to the following equation:

$$X = X_0(1 - 0.005Z)$$

where Z is the column height (cm). The specific rate of substrate consumption is $q_S = 2$ g S/g cells $\cdot$ h. The feed flow rate and glucose concentration in the feed are 5 l/h and 160 g glucose/l, respectively.

a. Determine the substrate (glucose) concentration in the effluent.

b. Determine the ethanol concentration in the effluent and ethanol productivity (g/l $\cdot$ h) if $Y_{P/S} = 0.48$ g ethanol/g glucose.

9.7. In a fluidized-bed biofilm reactor, cells are attached on spherical plastic particles to form biofilms of average thickness $L = 0.5$ mm. The bed is used to remove carbon compounds from a waste-water stream. The feed flow rate and concentration of total fermentable carbon compounds in the feed are $F = 2$ l/h and $S = 2000$ mg/l. The diameter of the column is 10 cm. The kinetic constants of the microbial population are $r_m = 50$ mg S/cm$^3 \cdot$ h and $K_S = 25$ mg S/cm^3. The specific surface area of the biofilm in the reactor is 2.5 cm^2/cm^3. Assuming first-order reaction kinetics and an average effectiveness factor of $\eta = 0.7$ throughout the column, determine the required height of the column for effluent total carbon concentration of $S_{0i} = 100$ mg/l.

9.8 Glucose is converted to ethanol by immobilized yeast cells entrapped in gel beads. The specific rate of ethanol production is: $q_P = 0.2$ g ethanol/g-cell-h. The effectiveness factor for an average bead is 0.8. Each bead contains 50 g/L of cells. The voids volume in the column is 40%. Assume growth is negligible (all glucose is converted into ethanol). The feed flow rate is $F = 400$ l/h and glucose concentration in the feed is $S_{0i} = 150$ g glucose/l. The diameter of the column is 1 m and the yield coefficient is about 0.49 g ethanol/g glucose. The column height is 4 m.

   a. What is the glucose conversion at the exit of the column?

   b. What is the ethanol concentration in the exit stream?

9.9 Consider the batch growth curve in Fig. 9.4 and the corresponding plots of dX/dt vs. X and dP/dt vs. P. (Fig. 9.7). You are asked to design a two-stage reactor system with continuous flow that will produce product P at a concentration of 0.55 g/l. You wish to minimize total reactor volume. For a flow rate of 1000 l/h what size reactors (and in what order) would you recommend?

9.10 Consider Fig. 9.9, which applies to a fed-batch system. Assume at $t = 0$, $V = 100$ l, $X = 2$ g/l, $\mu = 1$ h^{-1}, $S_0 = 4$ g/l, and $S = 0.01$ g/l. V is increased at a constant rate such that $dV/dt = 20$ l/h $= F$ (or flow rate) and X is constant at all times.

   a. Derive a formula to relate μ to V and dV/dt.

   b. What is μ at $t = 5$ h?

9.11 An industrial waste-water stream is fed to a stirred-tank reactor continuously and the cells are recycled back to the reactor from the bottom of the sedimentation tank placed after the reactor. The following are given for the system:

$F = 100$ l/h; $S_0 = 5000$ mg/l; $\mu_m = 0.25$ h^{-1}; $K_s = 200$ mg/l; α (recycle ratio) $= 0.6$; C (cell concentration factor) $= 2$; $Y_{X/S}^M = 0.4$. The effluent concentration is desired to be 100 mg/l.

   a. Determine the required reactor volume.

   b. Determine the cell concentration in the reactor and in the recycle stream.

   c. If the residence time is 2 h in the sedimentation tank, determine the volume of the sedimentation tank and cell concentration in the effluent of the sedimentation tank.

9.12 A waste-water stream is treated biologically by using a reactor containing immobilized cells in porous particles. Variation of rate of substrate removal with particle size is given in the following table.

r_s (mg/l-h)	D_P(mm)
300	1
300	2
250	3
200	4
150	5
100	7
50	10

a. What are the effectiveness factors for $D_P = 4$ mm and $D_P = 7$ mm?

b. The following data were obtained for $D_P = 4$ mm at different substrate concentrations. Assuming no liquid film resistance, determine the r_m and K_s for the microbial system.

S_0 (mg/l)	r (mg/l-h)
100	85
250	200
500	360
1000	630
2000	1000

9.13. A waste-water stream of $F = 1$ m³/h with substrate at 2000 mg/l is treated in an upflow packed bed containing immobilized bacteria in form of biofilm on small ceramic particles. The effluent substrate level is desired to be 30 mg/l. The rate of substrate removal is given by the following equation:

$$r_s = \frac{kXS}{K_s + S}$$

By using the following information, determine the required height of the column (H).

$k = 0.5$ h⁻¹, $X = 10$ g/l, $K_s = 200$ mg/l, $L = 0.2$ mm, $a = 100$ m²/m³, $A = 4$ m², $\eta = 0.8$

10

Selection, Scale-up, Operation, and Control of Bioreactors

10.1. INTRODUCTION

In our previous discussion we focused on idealized descriptions of homogeneous bioreactors with regard to operating mode (batch, fed-batch, continuous, and multistage) and generalized discussions of heterogeneous reactors (immobilized-cell systems and SSF). However, real processes demand greater insight into the details of reactor configuration. It would be impossible in a book of this type to provide the reader with all the information necessary to become a design engineer. However, we will introduce some of the concepts that drive the choice of real bioreactors. Our focus will be primarily on reactors for bacteria and fungi; some of the special reactor issues for animal and plant cell tissue culture will be handled in chapters on tissue culture.

10.2. SCALE-UP AND ITS DIFFICULTIES

10.2.1. Introduction

Why would the performance of a fungal culture making an antibiotic be so different at 10,000 l than at 10 l? At first glance the reader may have difficulty understanding why performance should change with scale. The answers lie in the difficulty of maintaining homogeneity in large systems, changes in surface to volume ratios, and changes in the cultures themselves due to the increased length of culture time. To understand the problems in scale-up, we first need to describe what traditional culture vessels are and how they are operated.

10.2.2. Overview of Reactor Types

Rather than catalog the large array of suggested fermenter and bioreactor designs, we will restrict ourselves to considering some basic types:

1. Reactors with internal mechanical agitation.
2. *Bubble columns*, which rely on gas sparging for agitation.
3. *Loop reactors,* in which mixing and liquid circulation are induced by the motion of an injected gas, by a mechanical pump, or by a combination of the two.

All three reactor types (see Fig. 10.1 for schematics) invariably are concerned with three-phase reactions (gas–liquid–solid). Three-phase reactors are difficult to design, because the mass transfer of components among the three phases must be controlled. Inadequate tools for the complex fluid mechanics in such systems coupled with the complex nature of cells as reactive solids make the prediction of system performance from first principles very difficult.

The traditional fermenter is the stirred-tank reactor (Fig. 10.2), the prime example of a reactor with internal mechanical agitation. The main virtues of such systems are that they are highly flexible and can provide high $k_L a$ (volumetric mass-transfer coefficient) values for gas transfer. Stirred reactors of up to 400 m^3 are used in antibiotic production, with stirrer powers of up to 5 kW/m^3. Stirred reactors can be used commercially up to viscosities of about 2000 centipoises (2 Pa sec).

Gas under pressure is supplied to the *sparger* (usually either a ring with holes or a tube with a single orifice). The size of the gas bubbles and their dispersion throughout the tank are critical to reactor performance. Although a sparging ring will initially provide smaller bubble size and better gas distribution, spargers with a single discharge point are often preferred for media with high levels of suspended solids because they are more resistant to plugging.

Gas dispersion is mainly the function not of the sparger but of the *impeller*. The impeller must provide sufficiently rapid agitation to disperse bubbles throughout the tank, to increase their residence time within the liquid, and to shear larger bubbles into small bubbles. Too much stirring can be detrimental, owing to the shear sensitivity exhibited to varying degrees by some cells (e.g., animal cells) and the stratification of reactor contents

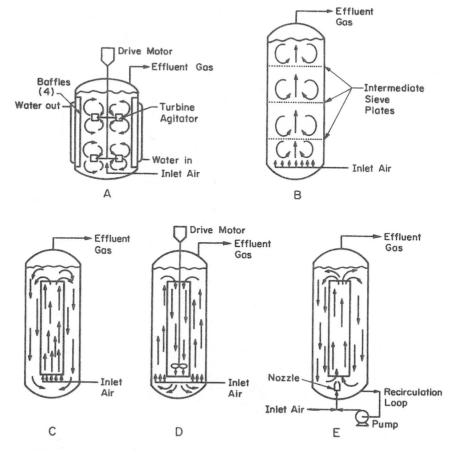

Figure 10.1. Bioreactor types. (a) Stirred-tank reactor, (b) bubble-column reactor, (c) airlift loop reactor with central draft tube, (d) propeller loop reactor, and (e) jet loop reactor. Arrows indicate fluid circulation patterns. (With permission, from D. N. Bull, R. W. Thoma, and T. E. Stinnett, *Adv. Biotechnol. Processes I*, 1, 1985, and Alan R. Liss, Inc., New York.)

with multiple-impeller systems. Although a wide variety of impeller designs have been proposed, the predominant choices are disc- and turbine-type impellers, while marine and paddle impellers are of particular interest for cellular systems with high levels of shear sensitivity.

The Rushton impeller (Fig. 10.3) is a disc with typically 6 to 8 blades designed to pump fluid in a radial direction. This was the predominant impeller design up to the mid-1980s and is commonly found on industrial and laboratory fermenters. The axial flow hydrofoil impellers (Fig. 10.3) have become increasingly popular. These axial flow systems can pump liquid either down or up. They have been shown to give superior performance (compared to Rushton radial flow impellers) with respect to lower energy demands for the same level of oxygen transfer. Further, they show reduced maximum shear rates,

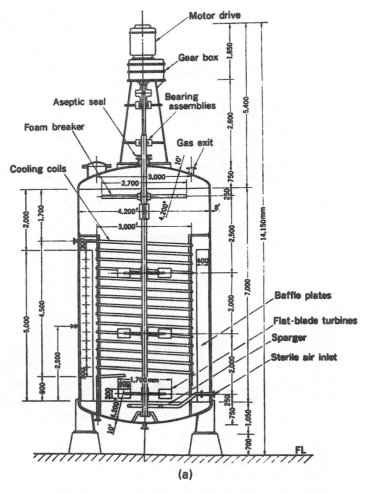

Figure 10.2. (a) Mechanically stirred 100,000-liter fermenter. (With permission, from S. Alba, A. E. Humphrey, and N. F. Millis, *Biochemical Engineering*, 2d ed., University of Tokyo Press, Tokyo, 1973.) (b; see p. 289) Installation of mechanically stirred fermenter: S, steam, C, condensate; W, water, and A, air. The steam lines permit in-place sterilization of valves, pipes, and seals. The input air can be sterilized by both incineration and filtration. (With permission, from W. Crueger and A. Crueger, *Biotechnology: A Textbook of Industrial Microbiology*, R. Oldenbourg Verlag, München, Germany, 1984.)

making them usable with sensitive cultures such as animal cells, while still being capable of giving excellent performance with viscous mycelial fermentations. Axial flow systems break up the compartmentalization often observed with multiple radial flow impellers on the same shaft. Combinations of axial flow and radial flow impeller systems are sometimes used. To augment mixing and gas dispersion, baffles are employed. A typical arrangement includes four baffles, the width of each being about 8% to 10% of the reactor

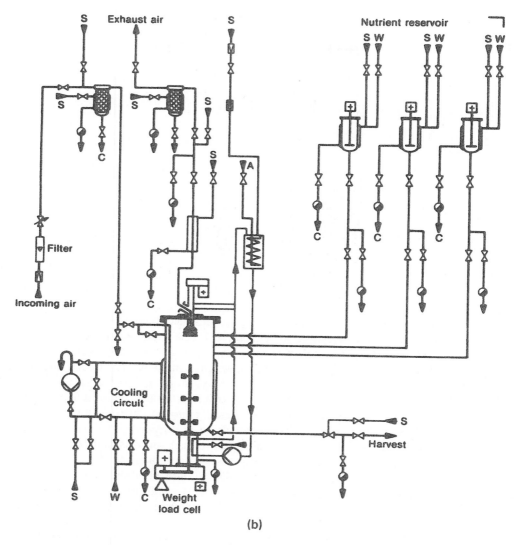

Figure 10.2 (continued)

diameter. With animal cell cultures, baffles cause shear damage; instead of baffles bottom-drive axial impellers slightly offset from center are used.

The vessel itself is almost always stainless steel. Type 316 is used on all wetted parts, type 304 on covers and jackets. With plant and animal cell tissue cultures, a low-carbon version (type 316L) is often used. Many bench-scale fermenters are glass vessels with stainless-steel cover plates. Glass fermenters are rarely used at the 50 l scale, and material considerations limit the maximum size of glass vessels to about 500 l. The use of stainless steel and glass is called for by the need to sterilize the reactor and the corrosive and/or abrasive nature of many fermentation media.

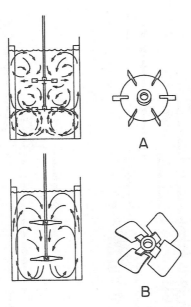

Figure 10.3. Liquid flow in bottled tanks with (A) Rushton radial flow impellers and with (B) axial flow hydrofoil impellers.

Most fermenters are built with a height-to-diameter ratio of 2 to 3, although bioreactors for animal cell cultures tend to be closer to 1. The Rushton impeller diameter is typically 30% to 40% of the tank diameter, while it may reach 50% with the axial flow hydroflow impellers. In large reactors the two main limitations on size are the abilities of the design to provide an adequate supply of oxygen and to remove metabolic heat efficiently. Large reactors usually use either internal coils for heat removal or a jacketed vessel. Although copper coils have better thermal transport characteristics, stainless-steel coils are almost always used, although this choice can depend on the nature of the media and culture. Internal coils provide advantages over jacketed vessels in terms of efficiency of heat removal, owing to the larger surface area for heat transfer. However, in many systems the coils become rapidly fouled by microbial growth, decreasing heat transfer and often adversely affecting mixing. In such cases the jacketed vessel offers advantages.

Another problem often encountered in commercial fermentations is *foaming*. If foam escapes from the fermenter, it can wet filters, increasing pressure drop and decreasing gas flow. Of greater concern is that it provides a pathway for contaminating cells to enter the fermenter. For most products of commercial interest, absolute sterility is required, and contamination may cause the loss of much product, time, and money. Foam can be controlled with a mechanical foam breaker or the addition of surface-active chemical agents. Although such chemicals can be very effective in controlling foam, there are penalties. Foam-breaking chemicals usually lower k_La values, reducing the reactor's capacity to supply oxygen or other gases, and in some cases they can be inhibitory to cell growth. Foam formation is not well understood, but complex media or the formation of high levels of extracellular polymers (e.g., proteins) tend to promote foaming. Foam can limit the ultimate productivity of a fermenter. All stirred-tank fermenters provide head space for the gas to disengage from the liquid. The *working volume* (the amount of culture) in a fermenter is typically about 75% of the total fermenter volume.

Sterility is a prime design consideration for fermenter hardware. Pressurized steam is used for in-place sterilization of the reactor, seals, probes, and valves. The number of openings into the fermenter should be limited to what is essential. A trade-off is necessary between the use of many probes, which improves fermenter control, and of few probes, which improves the chances of maintaining sterility. Small openings are made leakproof with O-rings, while flat gaskets are satisfactory for larger openings. Although small fermenters can use magnetically coupled agitators, industrial-size fermenters have moving shafts that must penetrate into the fermenter. Prevention of contamination due to the entry of foreign organisms through such a shaft is a major challenge in the mechanical design of fermenters. Stuffing-box seals are common on old fermenters, while double mechanical seals are used in newer ones. All surfaces must be smooth. Crevices in the tank surface, pipes, and valves can trap large quantities of particulate organics and contaminating organisms. Such clumps of cells increase the chances for a contaminant to survive the sterilization procedure. Cleanability of all surfaces is important.

Cleaning is generally done "in place," and fermenter design includes spray balls to allow for clean-in-place (CIP) technology. Highly alkaline detergents are often used, and this factor helps dictate the selection of materials. Surfaces, especially in bioreactors for animal or plant tissue cultures, often undergo electropolishing, an electrolytic process that removes the sharp microscopic projections often resulting from mechanical polishing. All ports and valves involved in sampling and injection should be protected with steam-sterilizable closures. The application of a continuous flow of live steam to sample valves is one strategy that is often applied in antibiotic plants.

The same concern for sterility applies to bubble columns and loop (e.g., airlift) reactors. Bubble columns offer distinct advantages for some systems. They are suitable for low-viscosity Newtonian broths; satisfactory mixing may not be possible in highly viscous broths. Bubble columns provide a higher energy efficiency than stirred-tank systems, where by *energy efficiency* we mean the amount of oxygen transferred per unit of power input. An additional advantage often mentioned for bubble columns is that they provide a low-shear environment, which may be a critical consideration with some cells. Cells tend to accumulate at the bubble surface, however, and bubble bursting is highly detrimental to cells. The absence of mechanical agitation also reduces cost and eliminates one potential entry point for contaminants.

Besides having less vigorous mixing capabilities than stirred tanks, bubble column operation is often limited by considerations of foaming and bubble coalescence. Because of bubble coalescence, bubble columns work over a rather narrow range of gas flow rates. The range of appropriate gas flows varies with the nature of the broth. Thus, bubble columns are less flexible than stirred tanks. Partial relief from the problem of coalescence can be found by using multistage columns; each stage (perforated plate) acts to redistribute gas flow. This gas redispersion, however, carries an energy penalty.

Loop reactors have intermediate characteristics between bubble columns and stirred tanks. We will consider primarily the *airlift* system (Fig. 10.1). Here the motion of the gas carries fluid and cells up a draft tube. At the top, gas disengages from the liquid, and the degassed liquid (which is denser than the gassed liquid) descends in the annulus outside the tube. At the bottom of the reactor, the descending fluid again encounters the gas stream and is carried back up the draft tube. Airlift systems can generally handle somewhat more viscous fluids than bubble columns, and coalescence is not so much of a problem. The largest agitated fermenter ($1500 \, \text{m}^3$) ever built (by ICI) is an airlift design for the production of single-cell protein. With very large fermenters ($> 200 \, \text{m}^3$) the use of nonmechanically agitated designs is preferred, as high oxygen transfer rates and better cooling can be attained. With an airlift design the interchange of material between fluid elements is small, so the transient time to circulate from the bottom of the draft tube to the top and back again is important. In the ICI design, multiple injection points for the substrate were used to prevent the cells from becoming substrate starved during circulation. The addition of mechanical stirring to a loop reactor increases flexibility (the operating range).

Although we focus our attention on agitated tanks, the reader should realize that most fermentations (in terms of total volume) are conducted in nonstirred and nonaerated vessels. Such vessels are used often in food fermentations, such as for beer, wine, and dairy products (e.g., cheese). In anaerobic fermentations, gas evolution by the fermenting organisms can provide some mixing, but gases are not normally sparged into the vessel. However, agitated and/or aerated vessels are used on a wider variety of fermentations (in terms of number of processes), are more difficult to scale, and are more likely to be chosen for the production of new products.

10.2.3. Some Considerations in Aeration, Agitation, and Heat Transfer

The basic equations describing oxygen transfer, carbon dioxide evolution, and heat generation by microbial growth are given in Chapter 6. For industrial-scale fermenters, oxygen supply and heat removal are the key design limitations.

The severity of the oxygen requirements depends on the choice of organism. Equation 6.24 can be rewritten as

$$\text{OUR} = X \cdot q_{O_2} = k_L a(C^* - C_L) \tag{10.1}$$

where q_{O_2} is the specific uptake rate of oxygen (mol O_2/g-h). Typical values of q_{O_2} are given in Table 10.1. The value of q_{O_2} (or OUR or oxygen uptake rate) is the demand side of the equation; typical requirements in large-scale systems are 40 to 200 mmol O_2/l-h, with most systems in the 40- to 60-mmol O_2/l-h range.

On the supply side, the critical parameter is $k_L a$ (h^{-1}), the volumetric transfer coefficient. A wide range of equations has been suggested for the estimation of $k_L a$. References at the end of this chapter detail some of these correlations. A typical correlation is of the form

$$k_L a = k(P_g/V_R)^{0.4} (v_S)^{0.5} (N)^{0.5} \tag{10.2a}$$

where k is an empirical constant, P_g is the power requirement in an aerated (gassed) bioreactor, V_R is bioreactor volume, v_S is the superficial gas exit speed, and N is the rotational speed of the agitator. With equations of the form of 10.2a, all the parameters must be in the units used in fitting the equation to a data set. An example of one of the original correlations of this type (when partial pressures are used) is

$$k_L a \, (\text{mmol/l-h-atm}) = 0.6[P_g/V_R(hp/1000 \, \text{l})]^{0.4}[v_S(\text{cm/min})]^{0.5}[N(\text{rpm})]^{0.5} \tag{10.2b}$$

TABLE 10.1 Typical Respiration Rates of Microbes and Cells in Culture

Organism	q_{O_2} (mmol O_2/g dw-h)
Bacteria	
E. coli	10–12
Azotobacter sp.	30–90
Streptomyces sp.	2–4
Yeast	
Saccharomyces cerevisiae	8
Molds	
Penicillium sp.	3–4
Aspergillus niger	ca. 3
Plant cells	
Acer pseudoplatanus (sycamore)	0.2
Saccharum (sugar cane)	1–3
Animal cells	
HeLa	$0.4\dfrac{\text{mmol } O_2/\text{l-h}}{10^6 \text{ cells/ml}}$
Diploid embryo WI-38	$0.15\dfrac{\text{mmol } O_2/\text{l-h}}{10^6 \text{ cells/ml}}$

A value of P_g can be estimated from other correlations, such as

$$P_g = K\left(\frac{P_u^2 \cdot N \cdot D_i^3}{Q^{0.56}}\right)^{0.45} \tag{10.3}$$

where K is a constant based on reactor geometry, P_u is the power required in the ungassed fermenter, D_i is the impeller diameter, and Q is the aeration rate (volume of gas supplied per minute divided by the liquid volume in the reactor).

The preceding correlations are not very good for Newtonian systems and are even worse for non-Newtonian or highly viscous systems. They also neglect the effects of medium components on $k_L a$. The presence of salts and surfactants can significantly alter bubble size and liquid film resistance around the gas bubble. These factors also can affect *oxygen solubility* (C^*). Temperature and pressure also affect $k_L a$ and C^*. Finally, on the supply side, it should be noted that C_L is maintained at a value above the critical oxygen concentration (see Chapter 6) but at a low enough value to provide good oxygen transfer. For many bacterial fermentations, a C_L of about 1 mg/l provides a good margin of safety if mixing is incomplete and yet allows a good rate of oxygen transfer.

Although $k_L a$ is difficult to predict, it is a measurable parameter. Four approaches are commonly used: unsteady state, steady state, dynamic, and sulfite test. The way in which these methods are applied depends on whether the test is being made on the system in the presence or absence of cells.

A new reactor prior to operation can be filled with pure water or a medium in which C^* can be accurately measured. Oxygen is removed from the system by sparging with N_2.

With the *unsteady-state method,* air is then introduced and the change in dissolved oxygen (DO) is monitored until the solution is nearly saturated. In this case,

$$\frac{dC_L}{dt} = k_L a(C^* - C_L) \qquad (10.4a)$$

or

$$\frac{-d(C^* - C_L)}{dt/(C^* - C_L)} = k_L a \qquad (10.4b)$$

or

$$\ln(C^* - C_L) = -k_L at \qquad (10.5)$$

As shown in Fig. 10.4, a plot of log $(C^* - C_L)$ versus time will give an estimate of $k_L a$.

Another approach is the *sulfite method.* In the presence of Cu^{2+}, the sulfur in sulfite (SO_3^{2-}) is oxidized to sulfate (SO_4^{2-}) in a zero-order reaction. This reaction is very rapid, and consequently C_L approaches zero.

The rate of sulfate formation is monitored and is proportional to the rate of oxygen consumption ($\frac{1}{2}$ mol of O_2 is consumed to produce 1 mol of SO_4^{2-}). Thus,

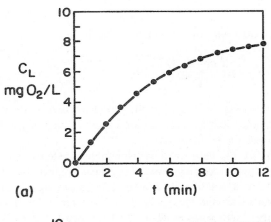

(a)

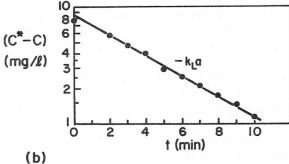

(b)

Figure 10.4. (a) Typical data for unsteady-state accumulation of oxygen in a large stirred tank. (b) Assuming $C^* = 8.65$ mg O_2/l, a value of $k_L a = 11.5$ h^{-1} is determined. (With permission, from D. W. Sundstrom and H. E. Klei, *Wastewater Treatment,* Pearson Education, Upper Saddle River, NJ, 1979, p. 53.)

Selection, Scale-up, Operation, and Control of Bioreactors   Chap. 10

$$\tfrac{1}{2}dC_{SO_4}/dt = k_L a C^* \tag{10.6}$$

where C_{SO_4} is the concentration of SO_4^{2-}. The factor, $\tfrac{1}{2}$, requires that C_{SO_4} and C^* be expressed in terms of moles. Oxygen solubility, C^*, is a constant dependent on medium composition, temperature, and pressure and can be measured separately. Thus,

$$k_L a = \frac{1}{2}\frac{dC_{SO_4}/dt}{C^*} \tag{10.7}$$

The sulfite method probably overestimates $k_L a$. The astute reader may have already noted that if the chemical reaction takes place in the liquid film around the gas bubble, the apparent oxygen transfer will be greater than that for a system with no chemical reaction in the liquid film. (An unresolved question is whether respiring organisms can enhance oxygen transfer by a similar method.)

Perhaps the best way to determine $k_L a$ is the *steady-state* method, in which the whole reactor is used as a respirometer. Such an approach requires the accurate measurement of oxygen concentration in all gas exit streams and a reliable measurement of C_L. A mass balance on O_2 in the gas allows the rate of O_2 uptake, OUR, to be calculated:

$$k_L a = \frac{\text{OUR}}{C^* - C_L} \tag{10.8}$$

OUR could also be estimated with off-line measurements of a sample of the culture in a respirometer, but using information from the actual fermenter is ideal. One complication, however, in this (and other methods) is the value of C^* to use. In a large fermenter, gas is sparged under significant pressure (due at least to the liquid height in an industrial fermenter). C^* is proportional to pO_2, which depends on total pressure as well as the fraction of the gas that is oxygen. At the sparger point, pO_2 will be significantly higher than at the exit, due to both higher pressure and the decrease in the fraction of the gas that is O_2 because of consumption by respiration. In a bubble column, the log mean value of C^* based on pO_2 at the entrance and exit would be a justifiable choice. In a perfectly mixed vessel, the composition of gas in the exit stream should be the same as bubbles dispersed anywhere in the tank, and consequently C^* would be based on pO_2 at the exit. In an actual tank, a knowledge of the residence time distribution of gas bubbles is necessary to estimate a volume averaged value of C^*.

The *dynamic* method shares similarities with the steady-state method in that it uses a fermenter with active cells. It is simpler in that it requires only a dissolved oxygen (DO) probe and a chart recorder rather than off-gas analyzers as well as the DO probe required in the steady-state method. The governing equation for DO levels is:

$$dC_L/dt = \text{OTR} - \text{OUR} \tag{10.9a}$$

or

$$dC_L/dt = k_L a\,(C^* - C_L) - q_{O_2} X \tag{10.9b}$$

The dynamic method requires that the air supply be shut off for a short period (eq < 5 min) and then turned back on. The anticipated result is shown in Fig. 10.5. Since there are no gas bubbles when the gas is off, $k_L a$ will be zero. Hence,

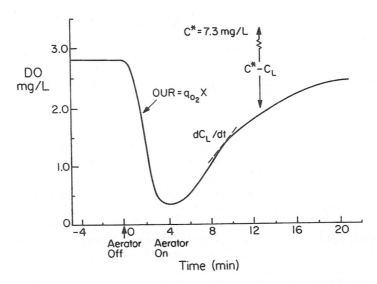

Figure 10.5. Example of response of DO in a fermenter when stopping and then restarting air flow. The dynamic method can be used to estimate OUR and k_La.

$$dC_L/dt = -q_{O_2} X \tag{10.10}$$

and the slope of the descending curve will give the OUR or $-q_{O_2}X$. The lowest value of C_L obtained in the experiment must be above the critical oxygen concentration (recall Chapter 6) so that q_{O_2} is independent of C_L. Further complications can arise due to the measurement lag in the DO probe (about 30 to 45 s) and the dissolution of oxygen from the headspace gas into the liquid (which may be significant primarily in small vessels, since the ratio of surface area for gas transfer is high compared to liquid volume). When air sparging is resumed, the ascending curve can be used to calculate k_La. Values of dC_L/dt can be estimated from the slope of the ascending curve calculated at various time points. A plot of $(dC_L/dt + OUR)$ versus $(C^* - C_L)$ results in a line with a slope of k_La. The primary advantage of the dynamic method is that k_La can be estimated under actual fermentation conditions. Additionally, if q_{O_2} is known, the value of OUR can be used to estimate X.

Variations on these methods to determine k_La exist, but the general principles remain the same. However, once the OTR (oxygen transfer rate) can be estimated from either correlations or experimentally determined values k_La, it is possible to quickly estimate the rate of metabolic heat generation. Equation 6.29 provides such a relationship. Equations 6.27 to 6.29 apply in all cases and lead to an estimate of the rate of heat generation. The total amount of cooling surface (either jacket or coils) required can then be calculated, given the temperature of the cooling water, the maximum flow rates allowable, the desired temperature differential between the exiting coolant and the reactor, and the overall heat transfer coefficient. The latter presents the greatest problem, since it can vary greatly from fermentation to fermentation. Highly viscous fermentations, which decrease fluid mixing, obviously decrease the rate of heat transfer. Many fermentations will cause solids deposition on coils, greatly decreasing the heat transfer coefficient. In extreme cases, solids deposition can plug spaces between coils, greatly decreasing convective flow by coils and consequently the heat transfer coefficient. Cooling coils are usually placed so that high fluid velocities will strike the coils to promote cleaning. If a reactor is to be maintained as a highly flexible piece of equipment adaptable to a wide range of fermentations, very conservative estimates of the overall heat transfer coefficient must be made.

Example 10.1.

Use the data in Fig 10.5 to estimate k_La from the dynamic method. Also, if the cell dry weight has been measured as 2 g/l, evaluate the specific respiration rate of the culture.

Solution   When the aerator is off, we can use the declining part of the curve in Fig 10.5 to estimate OUR. For the straight-line part of the curve (0.5 to 3 min after the aerator is off) we estimate OUR as

$$-dC_L/dt = \text{OUR} = -\frac{2.5 - 0.5 \ \text{mg/l}}{0.5 - 3 \ \text{min}}$$

$$= 0.8 \ \text{mg DO/l-min}$$

To estimate k_La we use the ascending curve formed during reaeration or

$$k_La = \frac{dC_L/dt + \text{OUR}}{C^* - C_L}$$

The best method to estimate k_La is to plot $dC_L/dt + \text{OUR}$ versus $(C^* - C_L)^{-1}$, but you can just make use of single points at individual times. In this problem $k_La = 0.16 \ \text{min}^{-1}$. The specific respiration rate is estimated as

$$\text{OUR} = q_{O_2} X$$

or

$$q_{O_2} = \text{OUR}/X$$

$$q_{O_2} = \frac{0.8 \ \text{mg DO/l-min}}{2 \ \text{g cells}/l}$$

$$q_{O_2} = 0.4 \ \text{mg DO/g cells-min}$$

10.2.4. Scale-up

The discussion in the previous sections should have alerted the reader to the complex nature of industrial fermenters. Now we consider how these complexities affect approaches to scaling fermenters.

Generally, fermenters maintain a height-to-diameter ratio of 2 to 1 or 3 to 1. If the height-to-diameter ratio remains constant, then the surface-to-volume ratio decreases dramatically during scale-up. This change decreases the relative contribution of surface aeration to oxygen supply and dissolved-carbon-dioxide removal in comparison to the contribution from sparging. For traditional bacterial fermentations, surface aeration is unimportant, but for shear-sensitive cultures (e.g., animal cells) it can be critical because of restrictions on stirring and sparging.

More important in bacterial and fungal fermentations is *wall growth*. If cells adhere to surfaces, and if such adherent cells have altered metabolism (e.g., due to mass transfer limitations), then data obtained in a small fermenter may be unreliable in predicting culture response in a larger fermenter. This point is more clearly illustrated in Example 10.2.

Perhaps even more importantly, it can be shown that the physical conditions in a large fermenter can never exactly duplicate those in a smaller fermenter if *geometric similarity* is maintained. In the case described in Table 10.2, a stirred-tank diameter has been increased by a factor of 5, resulting in a 125-fold increase in volume, since the height-to-diameter ratio was maintained constant. Four cases are treated in Table 10.2: scale-up based on constant power input (P_0/V), constant liquid circulation rate inside the vessel (pumping rate of impeller per unit volume, Q/V), constant shear at impeller tip (ND_i), and constant Reynolds number ($ND_i^2\rho/\mu$). Note that $P \propto N^3D_i^5$, $V \propto D_i^3$, $Q \propto ND_i^3$, $P/V \propto N^3D_i^2$, and $Q/V \propto N$. Thus, fixing N and D_i fixes all the quantities in Table 10.2. Since these quantities have different dependencies on N and D_i, a change of scale *must result* in changes in the physical environment that the cells experience. When these changes alter the distribution of chemical species in the reactor, or they destroy or injure cells, the metabolic response of the culture will differ from one scale to another. In some cases cells respond to modest changes in mechanical stress by changing physiological functions even when there is no visible cell injury or cell lysis. Thus, different scale-up rules (constant P/V implies constant OTR, constant Re implies geometrically similar flow patterns, constant N to give constant mixing times, and constant tip speed to give constant shear) can give very different results.

These scale-up problems are all related to transport processes. In particular, the relative time scales for mixing and reaction are important in determining the degree of heterogeneity in a fermenter. As we scale up, we may move from a system where the microkinetics (the cellular reactions) control the system response at small scale to one where transport limitations control the system response at large scale. When a change in the controlling regime takes place, the results of small-scale experiments become unreliable with respect to predicting large-scale performance.

TABLE 10.2 Interdependence of Scale-up Parameters

Scale-up criterion	Designation	Small fermenter, 80 l	Production fermenter, 10,000:1			
			Constant, P_0/V	Constant, N	Constant, $N \cdot D_i$	Constant, Re
Energy input	P_0	1.0	125	3125	25	0.2
Energy input/volume	P_0/V	1.0	1.0	25	0.2	0.0016
Impeller rotation number	N	1.0	0.34	1.0	0.2	0.04
Impeller diameter	D_i	1.0	5.0	5.0	5.0	5.0
Pump rate of impeller	Q	1.0	42.5	125	25	5.0
Pump rate of impeller/volume	Q/V	1.0	0.34	1.0	0.2	0.04
Maximum impeller speed (max. shearing rate)	$N \cdot D_i$	1.0	1.7	5.0	1.0	0.2
Reynolds number	$ND_i^2\rho/\mu$	1.0	8.5	25.0	5.0	1.0

With permission, from J. Y. Oldshue, *Biotechnol. Bioeng. 8:3–24 (1996)* John Wiley & Sons, Inc.

TABLE 10.3 Some Time Constants (Equations)[a]

Transport process	Equation
Flow	L/v or V/Q
Diffusion	L^2/D
Oxygen transfer	$1/k_L a$
Heat transfer	$V\rho C_p/UA$
Mixing	$t_m = 4V/(1.5ND^3)$, stirred vessel
Conversion processes:	
Growth	$1/\mu$
Chemical reaction	C/r
Substrate consumption	C_s/r_{max} ($C_s \gg K_s$)
	K_s/r_{max} ($C_s \ll K_s$)
Heat production	$\rho C_p \Delta_k T/r\Delta H$

[a]With permission, from N. W. F. Kossen, in T. K. Ghose, ed., *Biotechnology and Bioprocess Engineering*, United India Press Link House, New Delhi, 1985, pp. 365–380.

One approach to predicting possible reactor limitations is the use of characteristic time constants for conversion and transport processes. Table 10.3 defines some of the important time constants, and Table 10.4 shows the application of these time constants to a 20 m³ fermenter for the production of gluconic acid.

Processes with time constants that are small compared to the main processes appear to be essentially at equilibrium. If, for example, $1/k_L a \ll t_{O_2}$ conversion (t_{O_2} conversion is the time constant for O_2 consumption), then the broth would be saturated with oxygen, because oxygen supply is much more rapid than conversion. If, on the other hand, consumption is of the same order of magnitude as oxygen supply (e.g., $1/k_L a \approx t_{O_2}$ conversion), the dissolved oxygen concentration may be very low. This is precisely the case in Table 10.4 and Fig. 10.6. The resulting experimental measurements of dissolved oxygen show the

TABLE 10.4 Time Constants, 20-m³ Fermenter[a]

Transport phenomenon	Time constant(s)
Oxygen transfer	5.5 (noncoal.)–11.2 (coal.)[b]
Circulation of the liquid	12.3
Gas residence	20.6
Transfer of oxygen from a gas bubble	290 (noncoal.)–593 (coal.)
Heat transfer	330–650
Conversion	
Oxygen consumption, zero order	16
First order	0.7
Substrate consumption	$5.5 * 10^4$
Growth	$1.2 * 10^4$
Heat production	350

[a]With permission, from N. W. F. Kossen, in T. K. Ghose, ed., *Biotechnology and Bioprocess Engineering*, United Press Link House, New Delhi, 1985, pp. 365–380.

[b]Coal. = coalescing air bubbles.

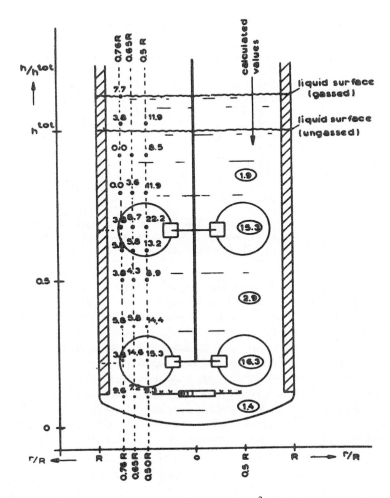

Figure 10.6. Measured oxygen concentrations in a 20-m^3 production fermenter (figures in circle are model estimates). (With permission, from N. W. F. Kossen, in T. K. Ghose, ed., *Biotechnology and Bioprocess Engineering*, United India Press, New Delhi, 1985, pp. 365–380.)

great variability in oxygen concentration in the reactor with some values at zero. This heterogeneity means that cells pass periodically through anaerobic regions. Since many cells have regulatory circuits to respond to changes from aerobic to anaerobic conditions, these circuits may be constantly altering cellular metabolism.

Traditional scale-up is highly empirical and makes sense only if there is no change in the controlling regime during scale-up, particularly if the system is only reaction or only transport controlled. Common scale-up rules are the maintenance of constant power-to-volume ratios, constant $k_L a$, constant tip speed, a combination of mixing time and Reynolds number, and the maintenance of a constant substrate or product level (usually dissolved-oxygen concentration). Each of these rules has resulted in both successful and unsuccessful examples (described in several of the references at the end of the chapter). Results are more favorable with Newtonian broths than with non-Newtonian systems.

The failure of any of these rules is related to changes in the controlling regime upon scale-up. No empirical rule can satisfactorily address such situations. Advances are being made in models that predict flow distribution, mixing times, and gas dispersion in fermenters, as well as models to predict explicitly cellular responses to changes in the local environment. If these models can be integrated, they may provide a much more fundamentally sound basis for scale-up. With the advent of improved supercomputers, the computational demands of such sophisticated models can be met. An approach to estimating mixing times in a stirred fermenter is given in Example 10.4.

It has been shown empirically for fermenter volumes of 0.1 to 100 m^3 that mixing time can be correlated to reactor volume according to an expression like[†]:

$$t_m = T_k \cdot V^{0.3} \tag{10.11}$$

where t_m is mixing time and V is vessel volume. The constant, T_k, is a function of impeller type, placement, and vessel design. Equation 10.11 assumes multiple Rushton-type impellers and is based on data from vessels of different sizes at practical operating conditions.

In summary, scale-up as currently practiced is an empirical, imprecise art. As long as geometric similarity is demanded at various scales, microenvironmental conditions cannot be made scale independent. Models and the use of scale-down procedures are potential approaches to improving scale-up, but these approaches are not yet fully developed.

The nature of the practical operating boundaries for an aerated, agitated fermenter can be summarized as in Fig. 10.7. The exact placement of such boundaries depends on the fermentation and change as the system is scaled up or scaled down. The boundaries are fuzzy rather than sharp. Nonetheless, the bioprocess engineer must appreciate the existence of such constraints.

10.2.5. Scale-down

Although scale-up models and the use of characteristic time analysis are potentially attractive, a more immediate approach to the rational scaling of reactors is *scale-down*. The basic concept is to provide at a smaller scale an experimental system that duplicates exactly the

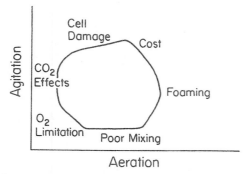

Fig. 10.7. Practical operating boundaries for aerated, agitated fermenters. (After Lilly, M. D., (1983) in *Bioactive Microbial Products 2;* D. J. Winstanley and L. J. Nisbet, eds., Academic Press, London, pp. 79–89.)

[†]B. C. Buckland and M. D. Lilly. In *Biotechnology* 2d ed., H.-J. Rehm and G. Reed. Vol, 3. G. Stephanopoulos, VCH, New York, 1983, pp. 12–13.

same heterogeneity in environment that exists at the larger scale. In many cases, scale-up will require using existing production facilities, so it is important to mimic those production facilities at a smaller scale. At the smaller scale, many parameters can be tested more quickly and inexpensively than at the production scale. Also, such a small-scale system can be used to evaluate proposed process changes for an existing operating process.

Figure 10.8 is a sketch of a smaller-scale apparatus to approximate the types of variations in substrate and dissolved-oxygen concentrations that might be expected from a mixing-time analysis (or from estimates from time constants) or from actual data on residence-time distributions from a large reactor. The difference between this scale-down apparatus and a traditional smaller-scale fermenter in a pilot plant is that the type of apparatus described in Fig. 10.8 is constructed specifically to mimic a known piece of equipment. It should be noted that the apparatus in Fig. 10.8 would not mimic temperature heterogeneity as it might exist in the larger system. Modifications to control temperature separately in each subvessel could be made.

Such a scale-down apparatus could be used to estimate the system's response (e.g., growth rate, product formation, and formation of contaminating by-products) to changes in medium composition (e.g., a new supplier of raw materials), introduction of modified production strains, use of different inoculum preparations, new antifoam agents, and for testing for O_2 and CO_2 tolerance. Corrective protocols can be suggested also for use in the large-scale system by simulating the response to pH or oxygen-probe failure or compressor failure during different phases of the fermentation.

Construction of scale-down apparatus can be a powerful complement to mathematical models, scale-up rules, and traditional pilot-plant operation.

Examples 10.2, 10.3 and 10.4 illustrate scale-up issues.

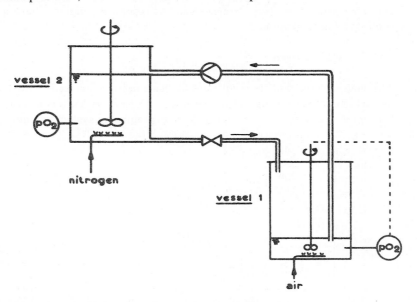

Figure 10.8. Experimental setup for scale-down experiments. (With permission, from N. W. F. Kossen, in T. K. Ghose, ed., *Biotechnology and Bioprocess Engineering*, United India Press, New Delhi, 1985, pp. 365–380.)

Example 10.2.

After a batch fermentation, the system is dismantled and approximately 75% of the cell mass is suspended in the liquid phase (2 l), while 25% is attached to the reactor walls and internals in a thick film (ca. 0.3 cm). Work with radioactive tracers shows that 50% of the target product (intracellular) is associated with each cell fraction. The productivity of this reactor is 2 g product/l at the 2 l scale. What would be the productivity at 20,000 l scale if both reactors had a height-to-diameter ratio of 2 to 1?

Solution Both tanks are geometrically similar, so we can calculate the diameter and the resulting surface area and volume in both tanks.

$$V = \tfrac{1}{4}\pi D^2 \cdot H = \tfrac{1}{4}\pi D^2 \cdot 2D = \tfrac{1}{2}\pi D^3$$

$$S = \pi D \cdot H = \pi D \cdot 2D = 2\pi D^2$$

For the 2 l system

$$D = 10.8 \text{ cm}$$

$$S = 738 \text{ cm}^2$$

$$V = 2000 \text{ cm}^3$$

For the 20,000 l systems,

$$D = 233.5 \text{ cm}$$

$$S = 342,600 \text{ cm}^2$$

$$V = 2 \times 10^7 \text{ cm}^3$$

At the bench scale (2 l) the amount of product made by surface-attached cells is

$$2 \text{ g/l} \cdot 2 \text{ l} \cdot \tfrac{1}{2} = 2 \text{ g}, \quad \text{for 738 cm}^2 \text{ of surface area}$$

The amount of product formed at 20,000 l due to surface-attached growth is

$$342,600 \text{ cm}^2 / 738 \text{ cm}^2 \times 2 \text{ g} = 928 \text{ g}$$

The overall yield in the 2 l system is

$$2 \text{ g/l} \cdot 2 \text{ l} = 4 \text{ g}$$

The overall yield in the 20,000 l system is

$$928 \text{ g} + (2 \text{ g/l} \cdot \tfrac{1}{2} \cdot 20,000 \text{ l}) = 20,928 \text{ g}$$

If no wall growth had been present, the 20,000 l tank would have yielded 40,000 g. Thus, wall growth, if present, can seriously alter the productivity of a large-scale system upon scale-up.

Example 10.3.

Consider the scale-up of a fermentation from a 10 l to 10,000 l vessel. The small fermenter has a height-to-diameter ratio of 3. The impeller diameter is 30% of the tank diameter. Agitator speed is 500 rpm and three Rushton impellers are used. Determine the dimensions of the large fermenter and agitator speed for:

a. Constant P/V

b. Constant impeller tip speed

c. Constant Reynolds number

Assume geometric similarity and refer to Table 10.2

Solution   Assume the vessel is a cylinder.

Thus,

$$V = (\pi/4) \, D_t^2 H.$$

but $H = 3 \, D_t$, so

$$V = (\pi/4) \, 3 \, D_t^3 = 10,000 \text{ cm}^3$$

Solving for D_t, H, and $D_i = 0.3 \, D_t$ gives:

$$D_t = 16.2 \text{ cm}$$

$$H = 48.6 \text{ cm}$$

$$D_i = 4.9 \text{ cm}$$

The scale-up factor is the cube root of the ratio of tank volumes, or

$$(10,000 \, 1 / 10 \, 1)^{\frac{1}{3}} = 10$$

To maintain geometric simularity the larger vessel will have its dimensions increased by a factor of 10. That is:

$$D_t = 1.62 \text{ m}$$

$$H = 4.86 \text{ m}$$

$$D_i = 0.49 \text{ m}$$

a. For constant P/V to apply $N^3 D_t^2$ must be the same in both vessels. Let subscript 1 refer to the small vessel and 2 to the large vessel.

$$N_2 = N_1 \left(\frac{D_{t1}}{D_{t2}} \right)^{2/3} = 500 \text{ rpm } (1/10)^{2/3} = 107 \text{ rpm}$$

b. For constant tip speed to apply ND_t must be the same in both vessels.

$$N_2 = N_1 \left(\frac{D_{t1}}{D_{t2}} \right) = 500 \text{ rpm } (1/10) = 50 \text{ rpm}$$

c. For constant Re to apply ND_t^2 must be the same in both vessels.

$$N_2 = N_1 \left(\frac{D_{t1}}{D_{t2}} \right)^2 = 500 \text{ rpm } (1/10)^2 = 5 \text{ rpm}$$

Scale-up on the basis of constant P/V would be the most likely choice unless the culture was unusually shear sensitive. Scale-up based on constant Re would not be used.

Example 10.4.[†]

As fermenters are scaled up, the mixing time usually increases. Mixing time, t_M^{95}, can be defined as the time it takes for the concentration of a compound to return to 95% of the equilib-

[†]Adapted from J. Jost, Chapter 3 in S. L. Sandler and B. A. Finlayson, eds., *Chemical Engineering Education in a Changing Environment*, Engineering Foundation, New York, 1988.

rium value after a local perturbation in its concentration. Mixing times are experimentally determinable by step addition of an electrolyte. The conductivity can be measured continuously at various locations distant from the injection site.

A production fermenter usually contains multiple impellers. An effective modeling approach is to divide the contents of the large tank into mixing compartments, where each compartment is perfectly mixed. As indicated in Fig. 10.9, a simple model is to consider that a separate compartment is associated with each impeller. In this problem, we let H be the overall mass transfer coefficient between compartments. The transient mass balances and experimental data can be used to estimate a value of H.

Consider the case where H has been determined to be 0.43 s^{-1} for a 10 l vessel and 0.075 s^{-1} for a 10,000 l vessel. With *E. coli* fermentations, glucose feed rates in a fed batch are adjusted to maintain a constant, relatively low concentration of glucose to prevent the formation of toxic metabolites (e.g., acetate) that would limit the ultimate cell concentration. Assume that the desired glucose concentration is 25 mg/l and that the Monod kinetics can be approximated as first order with a rate constant of about 0.05 s^{-1}. Assume the cell concentration changes slowly. Assume that the glucose supplemental feed is sufficiently concentrated that the total fluid volume in the reactor is constant. Also, assume that F, the mass addition rate of glucose per unit reactor volume, changes slowly in comparison to the characteristic mixing and reaction times. Compare the variation in glucose concentrations in the small and large tanks when an ideal probe (no error or lag in measurement) is used to maintain the set-point concentration at 25 mg/l in the middle compartment. Consider the response if the probe is placed in the top compartment instead of the middle compartment.

Solution Note that F and k_1 are closely related. As the cells grow, k_1 increases, which changes the demand for glucose (F). Because F changes more slowly than the characteristic time constants for mixing, we assume a quasi-steady state.

Following Fig. 10.9, we write

$$\frac{dC_1}{dt} = 0 = H(C_2 - C_1) + F - k_1 C_1 \tag{a}$$

$$\frac{dC_2}{dt} = 0 = H(C_1 - C_2) - H(C_2 - C_3) - k_1 C_2 \tag{b}$$

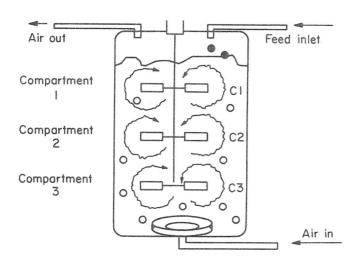

Air out

Feed inlet

Compartment 1

C1

Compartment 2

C2

Compartment 3

C3

Air in

Figure 10.9. Schematic of a simple mixing model for Example 10.4. C_i = concentration of a component in the ith mixing compartment. (With permission, from J. L. Jost, in S. L. Sandler and B. A. Finlayson, eds., *Chemical Engineering Education in a Changing Environment*, Engineering Foundation, New York, 1980, pp. 21–36.)

$$\frac{dC_3}{dt} = 0 = H(C_2 - C_3) - k_1 C_3 \tag{c}$$

where C_i is the glucose concentration in compartment i and k_1 is the first-order reaction constant.

The solution to the preceding set of equations is

$$C_1 = \left(2 + \frac{k_1}{H} - \frac{H}{H + k_1}\right) \frac{F}{3k_1 + \left(k_1^2 / H\right)} \tag{d}$$

$$C_2 = \frac{F}{3k_1 + \left(k_1^2 / H\right)} \tag{e}$$

$$C_3 = \frac{H}{(H + k_1)} \frac{F}{\left(3k_1 + \left(k_1^2 / H\right)\right)} \tag{f}$$

If the probe is placed in the middle compartment, then $C_2 = 25$ mg/l if the control system is perfect. Using eq. e, we calculate $F = 3.9$ mg/l-s in the small tank and $F = 4.6$ mg/l-s in the large tank. Values for C_1 and C_3 are found from eqs. d, e, and f. The process is repeated for the second case where the probe is placed in the upper compartment.

The results are summarized as follows:

A. Probe in the middle compartment
  A1. Small tank (10 l); $F = 3.9$ mg/l-s

$$C_1 = 30.5 \text{ mg/l}; C_2 = 25 \text{ mg/l}; C_3 = 22.4 \text{ mg/l}$$

  A2. Large tank (10,000 l); $F = 4.6$ mg/l-s

$$C_1 = 51.7 \text{ mg/l}; C_2 = 25 \text{ mg/l}; C_3 = 15 \text{ mg/l}$$

B. Probe in the top compartment
  B1. Small tank (10 l); $F = 3.2$ mg/l-s

$$C_1 = 25 \text{ mg/l}; C_2 = 20.5 \text{ mg/l}; C_3 = 18.4 \text{ mg/l}$$

  B2. Large tank (10,000 l); $F = 2.2$ mg/l-s

$$C_1 = 25 \text{ mg/l}; C_2 = 12.1 \text{ mg/l}; C_3 = 7.3 \text{ mg/l}$$

These results illustrate the importance of the alteration of mixing time on nutrient consumption rates (F) and on the concentration gradients in a large vessel. It should also be clear that probe placement in a less than perfectly mixed vessel will have significant effects on reactions within that vessel.

10.3. BIOREACTOR INSTRUMENTATION AND CONTROL

10.3.1. Introduction

The maintenance of optimal conditions for product formation in the complex environment in a bioreactor requires the control and measurement of at least a few parameters. Almost all fermenters have pH, temperature, and dissolved-oxygen control (by control of agitator rotational speed). New probes and techniques for the measurement of various ions, substrates, and products are being developed and have been implemented in particular situations. Actual process-control strategies for bioreactors are very primitive in comparison to the petrochemical industry, because of a lack of sensors for on-line measurements and of reliable, quantitative, dynamically accurate models. In this section we will summarize briefly not only the state of the art but some ideas on opportunities to improve the control of bioreactors.

10.3.2. Instrumentation for Measurements of Active Fermentation

The maintenance of sterility in a fermenter imposes a severe limitation on obtaining on-line measurements of fermentation parameters. Some probes (e.g., pH and dissolved oxygen) enter the reactor through penetrations in the fermenter shell. Each penetration significantly increases the probability of contamination. Thus, the benefit in increased productivity from the use of each probe must outweigh the economic losses that result from an increased number of contamination events. The probes themselves must also be sterilizable, preferably with steam. Thus, probes must ideally be able to withstand moderately high temperatures (121°C) in the presence of 100% humidity. Chemical sterilization, which is less desirable, may allow the use of a temperature-sensitive device if it has sufficient chemical resistance. Any general technique for monitoring fermentations must be compatible with the limitations imposed by sterility requirements.

Techniques to monitor the physical environment are summarized in Table 10.5. Except for viscosity and turbidity, these parameters would be monitored on most pilot-plant fermenters and many industrial fermenters. Each parameter is generally subject to its own closed-loop control system. These individual control loops may be integrated into an overall control package, especially at the pilot-plant scale, but in many cases higher-level control is not practiced, owing to the difficulty of measuring key parameters (e.g., product concentration) and the lack of dynamically accurate process models.

On-line measurements of concentrations beyond pH and dissolved oxygen are difficult, although exciting progress is being made. Table 10.6 summarizes techniques that have been considered for determining the concentrations of key components.

With insertable probes, we need to worry not only about probe performance but also about probe placement. The heterogeneity in a large fermenter means that dissolved oxygen and substrate levels (and to a lesser extent pH) will be position dependent (see Example 10.4). Although placement in the midsection of the vessel often gives the most representative values, mechanical design considerations may dictate placement elsewhere. Because probe fouling is a potential problem, the probe should be inserted in a region

TABLE 10.5 Monitoring and Control of the Physical Environment

Parameter	Measuring devices(s)	Comments
1. Temperature	Resistance thermometer or thermistor	Thermistors are used when small size and rapid response are required
2. Pressure	Diaphragm gages	Pressure is regulated by a simple back-pressure regulator in gas exit line
3. Agitator shaft power	Wattmeter or strain gages	Strain gages are used in bench or pilot-scale equipment
4. Foam	Rubber-sheathed electrode	Mechanical foam breakers are self-regulating; sensor is used to activate solenoid valve to release antifoam agent
5. Flow rate (gas)	Rotameters or thermal mass flow meter	Position of rotameter float is converted to an electrical signal; controller manipulates flow valve
6. Flow rate (liquid)	Magnetic-inductive flow meters or change in weight of additional vessels determined by load cell	Magnetic-inductive meters are good for viscous fluids or fluids with high level of particulates; controller can manipulate flow valves
7. Liquid level	Load cells to measure amount of liquid in vessel. Liquid height, conductivity sensors; capacitance probes; ultrasound	Liquid height is a function of gas sparge rate and gas hold-up; foam can complicate measurement
8. Viscosity	Rotational viscometers	On-line measurement is difficult; broths with high solids content present special problems
9. Turbidity (to indicate cell mass)	Photometer (either as a probe into the reactor or in a slip stream)	Many problems: fouling and interference from gas bubbles and suspended solids

with sufficient turbulence to help keep it clean. Although the use of multiple probes would be desirable, the increased risk of contamination often argues against it. Even with accurate probe response, the interpretation of that response in a large fermenter must be done carefully.

Since many fermentations require extensive periods for completion (2 to 20 days), it is important that probe response be stable for extended periods. Industrial fermentation broths contain many proteins and other organics that have a significant tendency to adsorb onto surfaces. Many microbes also have a strong tendency to adhere to surfaces. Thus, *probe fouling* in an extended fermentation is a constant problem. Drift is a problem in some probes, and recalibration *in situ* is not always possible. If sterility is to be maintained, the removal and replacement of probes is practically impossible. In some cases, special designs allowing some back-flushing are possible. However, the quality of information available tends to decrease as the length of a fermentation cycle increases.

Exit-gas measurements, particularly with mass spectrometers, are of increasing interest. Advances in building robust process instruments at lower cost have made this a more attractive proposition, particularly when such an instrument can be used for several fermenters (e.g., using a computer-controlled switching manifold). The main limitation on exit-gas analysis is that only volatile components can be monitored.

TABLE 10.6 Approaches to Monitoring and Control of the Chemical Environment

Approach	Possible measuring devices and compounds monitored	Comments
1. Insertable probes	pH electrode; H^+	Special design to facilitate repeated autoclaving; protein fouling of diaphragm can be a problem; pH probes used extensively
	Redox electrodes; redox potential	Measurement fairly reliable; interpretation and use of information often difficult
	Ion-sensitive electrodes; NH_3, NH_4, Br, Cd, Ca, Cl, Cu, CN, F, BF_4, I, Pb, NO_3, ClO_4, K, Ag, Na, S, SCN	Cross-sensitivity major problem; drift and calibration can be troublesome; response time can be slow; sterilizability is often poor
	O_2 probes, either galvanic or polarographic; pO_2	Measure partial pressure of O_2, not O_2 concentration; slow response; drift, fouling, and recalibration during extended fermentations can be a problem; O_2 probes widely used and extremely important
	CO_2 probes; activity of dissolved CO_2	CO_2 is known to have important physiological effects, but relating these effects to dissolved CO_2 is difficult
	Fluorescence probes; NADH	Commercially available to monitor intracellular concentrations; fouling is potential problem
	Biosensors (use enzymes, antibodies, organelles, or immobilized whole cells for chemical reaction with physicochemical probes to measure resulting change, e.g., pH); wide range of compounds potentially detectable, (e.g., glucose, glutamic acid, ammonia, acetate, ethanol formate, cephalosporin, penicillin)	Sterilizability, enzyme cost, cofactor regeneration, usable lifetime are all constraints on utilization of such probes; high level of development activity will lead to at least partial solution of these problems
2. Exit gas analyzers	Paramagnetic analyzer (O_2); thermal conductivity or long-path infrared analyzers (CO_2)	These instruments are specialized for measurements of these specific gases; important in fermentation balances; sample conditioning important
	Flame ionization detector; low levels of organically bound carbon, especially useful for volatile organics such as ethanol or methanol	Flame ionization measures total hydrocarbon; automatic process gas chromatography equipment available to separate into individual compounds
	Mass spectrometer; O_2, CO_2, volatile substances (can also be used on liquid streams)	Highly specific, rapid, and accurate; expensive; has not been used for complex molecules; depends on gas phase for volatile compounds

(continued)

Sec. 10.3   Bioreactor Instrumentation and Control

309

TABLE 10.6   (*continued*)

Approach	Possible measuring devices and compounds monitored	Comments
3. Measurements from liquid slip stream	Semiconductor gas sensors; flammable or reducing gases or organic vapors	Has been used in bioreactors primarily for ethanol measurements
	HPLC (high-performance liquid chromatography); dissolved organics; particularly useful for proteins; auto analyzer	Highly specific; compound need not be volatile; response time is long (min); solids must be removed by filtration; membrane fouling can be a problem; guard columns must remove compounds that may foul columns; expensive; main value is to differentiate among closely related proteins or variants of a particular protein
	Mass spectrometers; dissolved compounds that can be volatilized	Same instrument can be used also for exit gas analysis; rapid, specific, but expensive; complex molecules present problems
	Enzymatic methods; potentially wide range, but glucose has received most attention	Slow response (min), not highly flexible, but very discriminating; potential for contamination; has limited acceptance

On-line HPLC (high-performance, or high-pressure, liquid chromatography) is potentially very powerful for measuring the levels of dissolved solutes, particularly proteins made from genetically engineered cells. Although this assay is not routine, many vendors are actively developing systems for this application. However, this method, as well as most others using a liquid slip stream, requires sample preparation and has a significant time delay associated with sampling. Typically, a small side stream is pumped from the reactor, and microfiltration or an ultrafilter is used to remove cells and particulates. The filtrate usually requires further processing (e.g., the removal of compounds that could foul the analytical-grade columns). After injection of the sample, there can be a significant time (min) before a response is obtained. One other problem with slip streams is the potential for contamination; the membrane filtration step is of particular concern. Nonetheless, the use of liquid slip streams allows the relatively rapid determination of important product information.

Other methods not listed in Table 10.6 have the potential to significantly affect the on-line measurement of fermentation parameters. Nuclear magnetic resonance (NMR) has given important information on intracellular metabolism for off-line or small-scale growth experiments. *In situ* NMR has the potential for use with at least bench-scale systems, although significant technical advances must first be made. Selective fluorescence in combination with flow cytometry has given important information, off line, on the distribution of intracellular parameters (e.g., plasmid content) in a population. The adaptation of such techniques to on-line measurement is conceivable, but again a number of technical problems would need to be solved. Another analytical method that offers a good potential for on-line determination of solutes is Fourier transform infrared (FTIR) spectra analysis. Finally, the use of spectra analysis combined with optical fibers may offer a good technique to determine cell mass in the presence of other suspended solids.

In summary, current instrumentation to monitor fermenters on line is very limited. However, many techniques currently under development promise to solve this problem.

10.3.3. Using the Information Obtained

Having information is one thing; using it wisely is another. Current fermenter design and control techniques are rather limited. We do not have a fully satisfactory approach to the effective use of information on the extracellular and intracellular chemical environment.

Computer-controlled fermenters are fairly common, particularly at the pilot-plant scale. Figure 10.10 displays an overview of the software functions for a typical antibiotic fermentation facility, and Fig. 10.11 shows the relationship of primary measurements to secondary parameters.

An important function of such systems is *data logging*. This information is unusually important in the food and pharmaceutical industries. An actual record for the manufacture of each product batch is required by regulatory agencies. If the product is later found to be unsafe, such information can be used to trace the problem. Beyond the regulatory concerns, the lack of good high-level control strategies leads to a control strategy based on exactly duplicating a particular recipe. Consistency from batch to batch is an important consideration. Expert control systems may be useful in such situations.

In many cases, fermentation control schemes have begun to move beyond simple open-loop environmental control strategies. The use of computers allows the rapid manipulation of information from the data-logging operation to yield estimates of secondary parameters. For example, a secondary parameter may be cell concentration estimated from estimates of oxygen consumption or carbon dioxide evolution based on primary measurements of off-gas composition and dissolved-gas concentrations. Essentially, cell mass is estimated from mass balances using information from what have been termed *gateway sensors*. The main limitation to this approach is the accumulation of error. For example, in estimating cell mass X at time t, the value of X at time $t - \Delta t$ must be known. Any systematic errors in cell mass estimates or the inability of an algorithm to respond to an unusual perturbation can lead to significant difficulties over the length of a typical culture cycle. Data-handling procedures to mitigate this problem have been suggested (e.g., Kalman filters).

In some cases our partial knowledge of cellular metabolism is sufficient to solve important bioreactor problems. For many years it was thought impossible to grow *E. coli* to high cell densities (> 15 or 20 g/l). An improved understanding of cell physiology led to the observation that it was the buildup of toxic metabolic by-products (primarily acetate) that inhibited growth. Acetate is formed in *E. coli* when a good carbon source, such as glucose, is available in high concentrations. Control strategies that feed glucose at a rate sufficient to maintain at least moderately good growth without forming acetate were seen as possibilities to improve reactor volumetric productivities.

Thus, several possible control strategies are immediately apparent. The simplest perhaps would be to measure acetate concentration on line and reduce nutrient feed rates

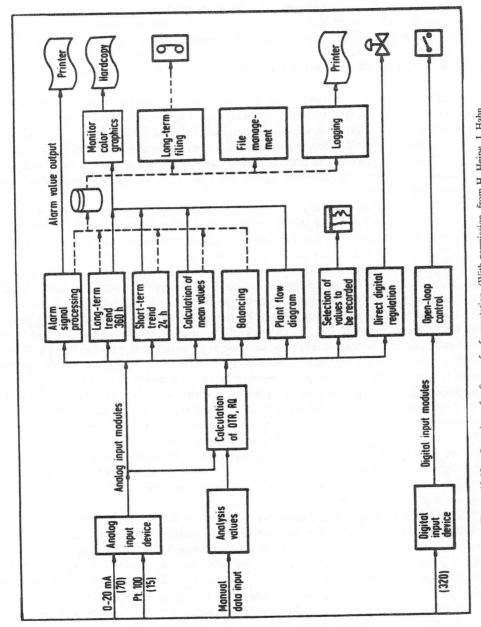

Figure 10.10. Overview of software for fermentation. (With permission, from H. Heine, J. Hahn, and A. Mangold, in *Biotechnology Focus I*, R. K. Finn and P. Prave, eds., Hanser Publishers, New York, 1988, p. 203.)

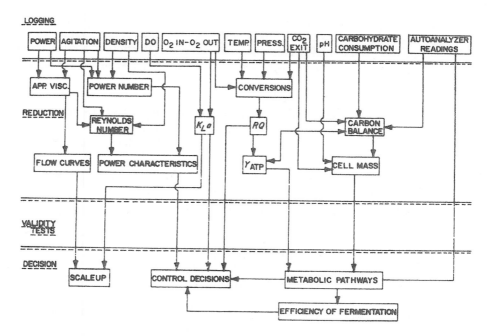

Figure 10.11. Primary measurements, shown on top, may be used to calculate many related process properties and parameters. (With permission, from L. P. Tannen and L. K. Nyiri, "Instrumentation of Fermentation Systems," p. 331, in *Microbial Technology*, 2d ed., Vol. II, H. J. Peppler and D. Perlman, eds., Academic Press, New York, 1979.)

in response to acetate accumulation. On-line measurement of acetate is not now routine, but improvements in sensors certainly make this a tenable approach. A strategy that is easier to implement with current sensors is based on measuring the rates of carbon dioxide evolution. Measurement of carbon dioxide in the off-gas, coupled with information on substrate concentration in the feed and nutrient flow rates, allows the glucose feed rate to be manipulated to maintain glucose concentrations at an optimal level. This indirect strategy depends on the use of mass balances and the gateway sensor concept. Another strategy is to control glucose concentration with a feedback control system, if a glucose sensing system is available. The easiest approach is to use a predetermined glucose feeding schedule, although this usually results in significant periods of sub- or superoptimal glucose concentrations. These strategies have been used successfully. High-density *E. coli* cultures (ca. 100 g dry wt/l) have greatly increased product concentration and productivities in *E. coli*-based fermentations to make proteins from recombinant DNA. At these high densities, special strategies for oxygen supply (e.g., O_2-enriched gas) are often required. This example illustrates the important interaction between process-control strategies and cellular metabolism.

As we develop better sensors and a better understanding of metabolic pathways, it becomes more feasible to combine information from primary sensors with models of metabolism to develop better control strategies. Without direct measurement of the product, simple feedback control on the process itself is impossible. Without extremely good models for dynamic response, feedforward control strategies are ineffective. Because of the highly nonlinear nature of most culture systems, black-box techniques to develop dynamic models are usually ineffective. Good process control for fermentations awaits improved models of cultures, more sophisticated sensors, and advances in nonlinear control theory.

In addition to reactor control, a computer monitored and controlled process involves the control of medium preparation, sterilization, and some downstream recovery processes. An important factor in bioprocesses is the predominance of batch processing. The scheduling of equipment is critical, and these tasks are being done increasingly with the aid of in-house computers.

In summary, exciting progress is being made in instrumentation and modeling, but the possible contributions of process control to bioprocesses have been only marginally realized.

10.4. STERILIZATION OF PROCESS FLUIDS

10.4.1. Introduction and the Kinetics of Death

Modern fermenters consume large amounts of media and gas. The tens of thousands of liters of medium and millions of liters of air used in a typical antibiotic fermentation must be absolutely devoid of any contaminating organism. The economic penalty for contamination is high. With bioprocesses to make proteins from recombinant DNA, the exit streams must also be treated to prevent the discharge of any living cell. The ability to ensure the destruction of viable organisms is critical.

Sterility is an absolute concept; a system is never partially or almost sterile. However, with a 100,000-l fermenter, we cannot sample every drop of fluid for a foreign organism. On a practical basis, sterility means the absence of any detectable viable organism, and a *pure culture* means that only the desired organism is detectably present.

Disinfection differs from sterilization. A disinfecting agent will greatly reduce the number of viable organisms, often a specific type of organism, to a low, but nonzero value.

Fluid streams can be sterilized through the physical removal of cells and viruses or the inactivation of living particles by heat, radiation, or chemicals. If sterilization is accomplished by inactivating living cells, spores, or viruses, we need to understand the kinetics of death. *Death* in this case means the failure of the cell, spore, or virus to reproduce or germinate when placed in a favorable environment. When dealing with sterilization, the probabilistic nature of cell death cannot be ignored.

The simplest case presumes that all the viable cells or spores are identical. The *probability of extinction* of the total population, $P_0(t)$, is

$$P_0(t) = \left[1 - p(t)\right]^{N_0} \tag{10.12}$$

where $p(t)$ is the probability that an individual will still be viable at time t, and N_0 is the number of individuals initially present. The expected value of the number of individuals present at time t, $E[N(t)]$, is

$$E[N(t)] = N_0 p(t) \tag{10.13}$$

while the variance about this number is

$$V[N(t)] = N_0 p(t)[1 - p(t)] \tag{10.14}$$

The specific death rate, k_d, is

$$k_d = \frac{1}{-E[N(t)]} \cdot \frac{d}{dt} E[N(t)] \tag{10.15a}$$

and if $p(t)$ is not a function of $N(t)$, then

$$k_d = -\frac{d}{dt} \ln p(t) \tag{10.15b}$$

The functional form for $p(t)$ depends on the organism and environment. The simplest form is to assume a first-order death model in which k_d is a constant and has units of reciprocal time:

$$p(t) = e^{-k_d t} \tag{10.16}$$

In the microbiological literature, the term *decimal reduction time* (D) is often used. It is the time for the number of viable cells to decrease tenfold. For example, with k_d as a constant, D is

$$E[N(t)] = N_0 e^{-k_d t} \tag{10.17a}$$

$$0.1 = e^{-k_d D} \tag{10.17b}$$

$$D = \frac{2.303}{k_d} \tag{10.17c}$$

A plot of $\ln N(t)/N_0$ versus time is called a *survival curve* (see Fig. 10.12). Such a curve is implicitly deterministic; the deterministic and probabilistic approaches give essentially the same result as long as $N(t)$ is >> 1 (about 10 to 100). Extrapolation of a survival curve to one organism or less is not proper.

Most real populations are not homogeneous. Often a subgroup is more resistant to the sterilization agent. Also, organisms growing in clumps tend to be more resistant to death. These factors are important in processing (see Fig. 10.13 for the effects on survival curves). For example, consider a tragic error in the early development of a polio vaccine. The virus is inactivated ("killed"). The required processing time was estimated from linear extrapolation (on a semilog plot) of a survival curve. The original data for the plot did not extend to sufficiently low numbers to detect the presence of a resistant subpopulation. Consequently, some patients were inoculated with live rather than killed virus.

10.4.2 Sterilization of Liquids

The kinetics of death apply to the sterilization of liquids. Values for k_d can be determined for inactivation by chemical, thermal, or radioactive agents. Thermal inactivation is very

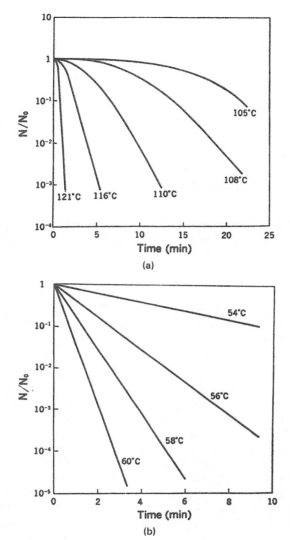

Figure 10.12. (a) Typical death-rate data for spores of *Bacillus stearothermophilus* Fs 7954 in distilled water, where N = number of viable spores at any time, N_0 = original number of viable spores. (b) Typical death rate for *E. coli* in buffer, where N = number of viable cells at any time and N_0 = original number of viable cells. (With permission, from S. Aiba, A. E. Humphrey, and N. F. Millis, *Biochemical Engineering*, 2d ed., University of Tokyo Press, Tokyo, 1973, p. 241.)

much preferred for the economic, large-scale sterilization of equipment and liquids. However, heat-sensitive equipment must be sterilized with chemical agents or radiation.

The use of ultraviolet (UV) radiation is effective to sterilize surfaces, but UV cannot penetrate fluids easily, particularly those with high amounts of suspended material. Although x-rays can penetrate more deeply, cost and safety considerations preclude their use in large-scale systems.

A chemical agent for sterilization must leave no residue that would be toxic to the desired culture. Ethylene oxide, a gas, can be used to sterilize equipment. A 70% ethanol–water mixture acidified to a pH of 2 with HCl kills virtually all vegetative cells

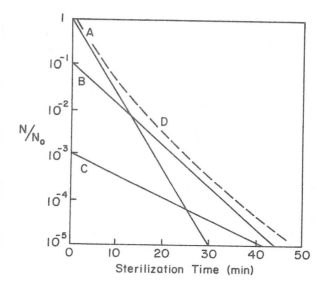

Figure 10.13. Overall survival curve (D) for a system with three distinct subpopulations (A, B, and C). Population A is normally dominant, but it is the one most sensitive to the sterilizing agent, while subpopulation C is far more resistant.

and many spores and can be used to sterilize equipment. A formaldehyde solution is often effective. Sodium hypochlorite solution (3%) has been used in sterilizing (or disinfecting) small-scale, heat-sensitive equipment. Some chemicals (e.g., ozone) cannot usually be used to sterilize fluids without adverse side reactions affecting medium quality.

For most large-scale equipment and liquids, thermal sterilization is used (filter sterilization is the only practical alternative for liquids). Usually, the dependence of the specific death rate on temperature is given by an Arrhenius equation:

$$k_d = \alpha e^{-E_{0d}/RT} \tag{10.18}$$

where R is the gas constant, T is absolute temperature, and E_{0d} is the activation energy for the death of the organism. Values for E_{0d} range from about 50 to 150 kcal/g-mol. For spores of *Bacillus stearothermophilus*, $E_{0d} \approx 70$ kcal/g-mol, and values of 127 kcal/g-mol have been determined for *E. coli*. Most thermal sterilizations take place at 121°C. The values for k_d in such situations are very high for vegetative cells (often $> 10^{10}$ min^{-1}). For spores, the values of k_d typically range from 0.5 to 5.0 min^{-1}. In most cases we are only concerned about spores when steam-sterilizing equipment and media, since the value of k_d is so much lower with spores than vegetative cells.

The E_{0d} for vitamins and growth factors in many media is about 2 to 20 kcal/g-mol. The inactivation of viability is much more sensitive to temperature changes than the degradation of important growth factors in the medium. This factor is important to the design of sterilization equipment and protocols so as to assure complete killing of foreign organisms without the destruction of necessary growth factors in the medium.

The main factors in any sterilization protocol are the temperature, time of exposure, and initial number of organisms that must be killed. The problems of sterilization increase with scale-up. Consider the probability of an unsuccessful fermentation, $[1 - P_0(t)]$, in a reactor,

$$1 - P_0(t) = 1 - \left[1 - p(t)\right]^{N_0} \qquad (10.19a)$$

or with a simple model of killing in homogeneous populations:

$$1 - P_0(t) = 1 - \left[1 - e^{-k_d t}\right]^{N_0} \qquad (10.19b)$$

N_0 in eq. 10.19 corresponds to the number of individuals in the reactor, not the concentration of organisms. Let n_0 be the concentration of particles. Now consider the probability of an unsuccessful sterilization in a 1 l and 10,000 l reactor, where each contains the same identical solution (n_0 is the same in both tanks), and the temperature (k_d) and time of sterilization are identical. Thus, for the 1-l tank

$$1 - P_0(t) = 1 - \left[1 - e^{-k_d t}\right]^{1 \cdot n_0} \qquad (10.20a)$$

and for the 10,000-l tank

$$1 - P_0(t) = 1 - \left[1 - e^{-k_d t}\right]^{10,000 \cdot n_0} \qquad (10.20b)$$

If $k_d t$ is 15 and $n_0 = 10^4$ spores/l in both cases, the probability of an unsuccessful fermentation is 0.003 in the 1-l vessel and about 1 in the 10,000-l tank ($5 \cdot 10^{-14}$ probability of extinction of the spores). Thus, the sterilization protocol that would be acceptable for a laboratory bench-scale experiment would be totally unacceptable for the larger-scale system. The larger tank would require much longer exposure to the same temperature to achieve the same degree of sterility, and the longer exposure to higher temperatures could lead to greater changes in the chemical composition of the medium.

A *sterilization chart* can be constructed from eq. 10.19b. An example is depicted in Fig. 10.14. To use such a chart, you need to specify the probability of failure that is acceptable (for example, 10^{-3}) and the number of particles initially present in the fluid ($n_0 \cdot$ total volume). For 10^{-3} and $N_0 = 10^8$, the corresponding value of $k_d t$ is about 26. If $k_d = 1$ min^{-1} at 121°C, then $t = 26$ min corresponds to the exposure time at 121°C required to ensure that 999 sterilizations out of a 1000 are successful.

Steam sterilizations can be accomplished batchwise, often *in situ* in the fermentation vessel, or in a continuous apparatus (Fig. 10.15). The greatest difficulty with batch sterilization is thermal lags and incomplete mixing. Typically, batch sterilization occurs at 121°C. The time required to heat the fluid to 121°C and to cool it back to growth temperatures (e.g., 37°C) is often much longer than the time of exposure to the desired temperature. For most spores, k_d falls very rapidly with temperatures (e.g., a tenfold decrease for k_d at $T = 110$°C rather than 121°C), so the heat-up and cool-down periods do little to augment spore killing. However, the elevated temperatures during heat-up and cool-down can be very damaging to vitamins and proteins, can lead to carmelization reactions for sugars, and can greatly alter medium quality.

Continuous sterilization, particularly a high-temperature, short-exposure time, can achieve complete sterilization with much less damage to the medium. Both the heat-up

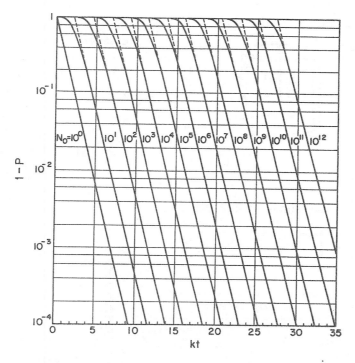

Figure 10.14. Sterilization chart. (With permission, from M. L. Shuler, *Encyclopedia of Physical Science and Technology*, Vol. 2, p. 427, Academic Press, New York, 1987.)

and cool-down periods are very rapid. Continuous sterilization is easier to control and reduces downtime in the fermenters. Two potential disadvantages of the continuous process are dilution of the medium with steam injection and foaming. The flow pattern inside the pipe is critical, since the fluid residence time near the wall can be different from that in the center. The average flow rate and the length of the sterilizing section should be designed to ensure high Peclet numbers (above 500), so that the velocity distribution approaches piston-like flow.

In addition to steam sterilization, process fluids can be filter sterilized. *Filter sterilization* is necessary when the medium contains heat-sensitive materials. An important example of filter sterilization in bioprocesses is the sterilization of medium to support the growth of animal cells. Microporous filters (pore sizes < 0.2 μm) are typically used. The filters should be absolute filters; no pores larger than the nominal pore size must exist. A very narrow pore-size distribution is advantageous. The medium may be first prefiltered to remove large particulates that might plug the microporous filter. The actual microporous filter must be sterilized. The equipment that receives the filtered–sterilized fluid is also sterilized.

Filter sterilization is not as reliable as steam sterilization. Any defect in the membrane can lead to failure. Viruses and mycoplasma (small wall-less bacteria) can pass the filter. Consequently, filtered–sterilized medium is usually quarantined for a period of time

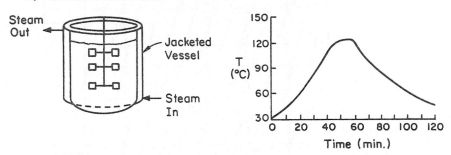

A) Batch Sterilization

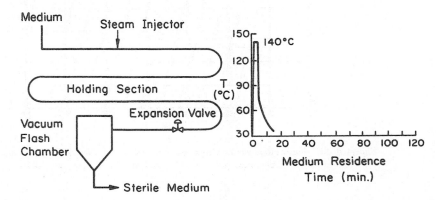

B) Continuous Sterilization

Figure 10.15. Comparison of a batch (A) with a continuous sterilization strategy (B) for the temperature profile of the medium sterilized. The continuous unit allows short-time, high-temperature sterilization.

to allow any contaminants to multiply to a detectable number. Clearly, a period of quarantine will detect only vegetative cells and not host-dependent contaminants (e.g., viruses).

Filter sterilization is used not only to filter medium, but also to sterilize process air.

10.4.3. Sterilization of Gases

Aerobic fermentations require huge volumes of air. At 0.1 to 1 volume of gas per volume of liquid per minute, a 50,000-l fermenter requires $7 \cdot 10^6$ l to $7 \cdot 10^7$ l per day of air. Many fermentations are several days in duration. For a five-day fermentation, as much as $2 \cdot 10^8$ l of air may be required. This volume of air must be absolutely sterile. Typically, the concentration of microbes in air is 10^3 to $10^4/m^3$ or 1 to 10 microbes/l. Fed-batch or continuous culture systems place even more stringent demands on air sterilization, since air filters cannot be replaced during the run period.

An air supply of this magnitude requires sizable air compressors. Adiabatic compression of air can increase the air temperature (typically 150° to 220°C). Dry heat is less effective in killing organisms than moist heat. To kill spores, a temperature typically of 220°C for 30 s is required. Consequently, the adiabatic compression of process air certainly aids air sterilization. Since the exit air cools rapidly and the pipes connecting the compressor to the fermenter are difficult to maintain as absolutely sterile, an air-filtration step is almost always used. (Alternatives such as UV radiation, ozone injection, and scrubbing are not practical.)

The filtration of gases can be accomplished using either depth filters or surface filters. Historically, depth filters using glass wool were used, but *depth filters* have been almost totally replaced with membrane cartridge filters, which are *surface filters*.

Glass wool filters rely on a combination of mechanisms for the capture of particles. Possible mechanisms of removal for particles of about 1-µm diameter are direct interception, electrostatic effects, diffusion (or Brownian motion), and inertial effects. As the flowing gas approaches a fiber, the flow streamlines must curve around the fiber. If a particle had no density, it would follow the streamlines around the fiber. Only particles whose centers are on streamlines less than a particle radius away from the fiber could be intercepted. However, for real particles, inertial effects mean that particles with mass will have a tendency to maintain a straight-line trajectory as they approach the fiber. Thus, such particles deviate from the streamline and crash into the fiber. Both interception and inertial effects are important in removing bacteria.

Diffusional effects may be important for virus removal, but bacteria are sufficiently large that diffusion is relatively unimportant. The removal of a particle by these mechanisms is probabilistic. The deeper the filter is, the smaller the probability of a particle penetrating the depth of the filter.

Depth filters using glass wool can show shrinkage and hardening upon steam sterilization. In such cases, channeling can occur, and the filter becomes far less effective than would be predicted. More recent advances in the design of fiberglass filter cartridges have overcome much of this disadvantage. Another serious problem with such fibrous filters is wetting. If a filter wets, an easy path becomes available for a contaminant to penetrate through it. A wet filter also greatly increases pressure drop. Thus, any condensation within such a filter must be avoided.

Surface filters (membrane cartridges) work using another mechanism for particle removal, a sieving effect. Figure 10.16 depicts a membrane cartridge unit and its housing. Membranes with uniformly small pores prevent the passage of particles with a radius larger than the pore radius. Such filters can be steam-sterilized many times. Also, any condensate formed on the nonsterile side cannot pass into the sterile side.

With both depth filters and membrane cartridges, pressure drop is critical. The energy input for compressed air for a commercial-scale process is significant. Air treatment can account for 25% of total production costs. Thus, the design engineer has to balance the assurance of sterility against pressure drop.

Generally, the bioprocess engineer is not involved in the design of membrane cartridge units (a number of competent vendors offer suitable products), but the testing of such units for effectiveness is important. In the United States, an example is an aerosol test using corn oil nebulized to 1.0 to 1.5 mg oil/l of air. This test is recognized by the FDA. Such an aerosol would contain particles primarily in the range of 0.2- to 1.0-µm diameters. Aerosol in the exit gas from a filter cartridge can be monitored by a photometer. The number of sterilization cycles a cartridge can undergo before failure is an important criterion in the selection of membrane cartridge units.

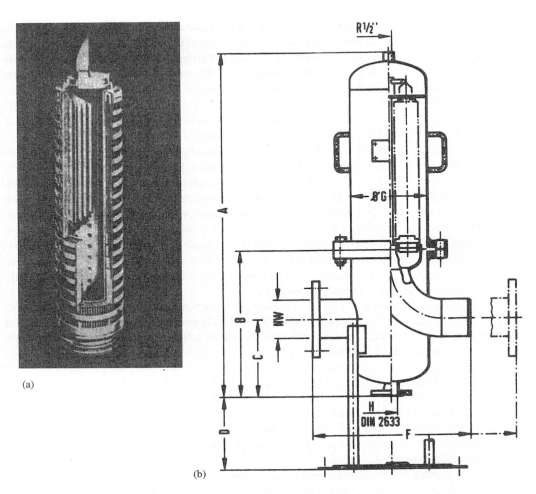

Figure 10.16. (a) A Pull-Emflon membrane cartridge for filter sterilization of air. (b) Housing for air sterilization filter. (With permission, from W. Crueger, "Sterile Techniques in Biotechnology," in R. K. Finn and P. Prave, eds., *Biotechnology Focus 2*, Hanser Publishers, New York, 1990, p. 413.)

Other tests for the integrity of the unit are pressure-drop-versus-flow-rate and bubble-point tests, which detect defects in the membrane and maximum pore size. "Grow-through" tests use a sterile nutrient solution on one side of the membrane and a similar nutrient solution inoculated with a test strain (e.g., *Pseudomonas diminuta*) on the nonsterile side. The greater the integrity of the filter, the longer it will take before growth on the nominally sterile side will occur. A filter should be evaluated with several different types of tests.

Because of the high costs associated with the loss of a batch due to contamination, the choice of air filter to give dependable protection for a fermentation while minimizing pressure drop is critical.

So far we have considered only inlet gas streams. With fermentations involving pathogens (disease-causing organisms) or recombinant DNA, *all organisms* must be removed from the exhaust gas. The concentration of microbes in the exit gas is far higher than in the inlet gas. Catalytically aided combustion (incineration) of the exit gas is an effective, but expensive, solution. Consequently, filtration of the exit air is of increasing importance.

10.5. SUMMARY

Scale-up of reactors is a task primarily for the biochemical engineer. Three basic reactor types for the aerobic cultivation of suspended cells are (1) *systems with internal mechanical agitation*, (2) *bubble columns*, and (3) *loop reactors*. Although the bubble and loop reactors offer advantages in terms of energy efficiency and reduced shear damage to cells, the traditional stirred-tank system is more flexible and can better handle broths that become highly viscous.

The primary limitations on the size of stirred-tank bioreactors are gas supply (e.g., O_2) and heat removal. The value of $k_L a$, the *volumetric transfer coefficient*, is of prime concern. Its value depends not only on the equipment used, gas flow rates, and agitator speed, but also on the nature of the fluid (salt content, presence of surface active agents, and viscosity). The properties of the fluid can change during the fermentation. By directly monitoring oxygen concentration in the gas phase and the dissolved-oxygen level, it is possible to make on-line estimates of $k_L a$. If the rate of oxygen uptake is known, the rate of heat generation can be readily estimated in aerobic fermentations.

Scale-up is difficult, because conditions in a large vessel are much more heterogeneous than in a small vessel. If geometrically similar tanks are used, it is impossible to maintain shear, mixing times, and $k_L a$ simultaneously identical in both the large and small tank. Scale-up would be simplified if good reaction models could be coupled to good transport models; since this is not yet possible, *scale-down techniques* are a good alternative.

Bioreactor instrumentation and control are less advanced than in the petrochemical industry. Improvements in sensor technology and the dynamical models of bioreactors are critical to improvements in control technology.

The large-scale bioreactor places heavy demands on processes to *sterilize* (kill or remove all organisms from) fluids entering the bioreactor. Liquid streams are thermally sterilized or filter sterilized. Steam terilization is preferred, but the sterilization process must not damage the ability of the medium to support growth. Since longer periods of exposure to high temperatures are necessary to assure sterility in larger volumes of liquid, the sterilization process can alter the medium composition more for large-scale than for small-scale systems. This is an additional factor that can lead to differences in productivity upon scale-up. *Continuous sterilization* protects the medium components from degradation better than batch sterilization, because the heat-up and cool-down periods are greatly minimized. *Filter sterilization* of liquids is used when the medium contains particularly heat-sensitive components. Air streams are typically filter sterilized. Surface filters (membrane cartridges) are commonly used in gas sterilization.

SUGGESTIONS FOR FURTHER READING

A number of general references address problems in reactor design, scale-up, control, and sterilization. The following are some examples.

AIBA, S., A. E. HUMPHREY, AND N. F. MILLIS, *Biochemical Engineering*, 2d ed., Academic Press, New York, 1973. (Although this book is somewhat dated, it provides a good practical overview of all the topics in this chapter.)

BAILEY, J. E., AND D. F. OLLIS, *Biochemical Engineering Fundamentals*, 2d ed., McGraw-Hill Book Co., New York, 1986.

BLANCH, H. W., AND D. S. CLARK, *Biochemical Engineering*. Marcel Dekker, Inc., New York, 1996.

DORAN, P. M. *Bioprocess Engineering Principles,* Academic Press, San Diego, 1995.

PRAVE, P., AND OTHERS, *Fundamentals of Biotechnology*, VCH Verlagsgesellschaft, Weinheim, Germany, 1987.

STEPHANOPOULOS, G., ed., *Bioprocessing* (Vol. 3 of *Biotechnology*, 2d ed., H.-J. Rehm and G. Reed, ed.), VCH, New York, 1993.

More specific articles and books are the following:

BULL, D. N., R. W. THOMA, AND T. E. STINNETT, Bioreactors for Submerged Culture, *Adv. Biotechnol. Processes 1:*1–30, 1983.

BYLUND, F., F. GUILLARD, S.-O ENFORS, C. TRÄGARDH, AND G. LARSSON, Scale Down of Recombinant Protein Production: A Comparative Study of Scaling Performance, *Bioprocess Eng. 20:*327–389, 1999.

CHISTI, Y., AND M. MOO-YOUNG, Improve the Performance of Airlift Reactors, *Chem. Eng. Prog. 89*(6):38–45, June, 1993.

CRUEGER, W., "Sterile Techniques in Biotechnology," in R. K. Finn and P. Prave, eds., *Biotechnology Focus 2*, Hanser Publishers, New York, pp. 391–422, 1990.

HEINE, H., J. HAHN, AND A. MANGOLD, "Computers in the Production of Antibiotics," in R. K. Finn and P. Prave, eds., *Biotechnology Focus I*, Hanser Publishers, New York, pp. 193–211, 1988.

JOST, J., "Selected Bioengineering Problems in Stirred-Tank Fermenters," in S. L. Sandler and B. A. Finlayson, eds., *Chemical Engineering Education in a Changing Environment*, Engineering Foundation, New York, 1988.

JUNKER, B. H., M. STANIK, C. BARNA, P. SALMON, E. PAUL, AND B. C. BUCKLAND, Influence of Impeller Type on Power Input in Fermentation Vessels, *Bioprocess Eng. 18:* 401–412, 1998.

KARGI, F., AND M. MOO-YOUNG, "Transport Phenomena in Bioprocesses," in *Comprehensive Biotechnology*, M. Moo-Young, ed., Pergamon Press, Oxford, UK, Vol. 2, pp. 5–55, 1985.

KOSSEN, N. W. F., "Problems in the Design of Large Scale Bioreactors," T. K. Ghose, ed., *Biotechnology and Bioprocess Engineering*, United India Press, New Delhi, pp. 365–385, 1985.

LUONG, J. H. T., A. MULCHANDANI, AND G. G. GUILBAULT, Developments and Applications of Biosensors, *Trends Biotechnol. 6:*310–316, 1988.

McFARLANE, C. M., AND A. W. NIENOW, Studies of High Solidity Ratio Hydrofoil Impellers for Aerated Bioreactors. I. Review, *Biotechnol. Prog. 11:*601–607, 1995.

OLDSHUE, J., *Fluid Mixing Technology,* McGraw-Hill, New York, 1983.

PROBLEMS

10.1. The air supply to a fermenter was turned off for a short period of time and then restarted. A value for C^* of 7.3 mg/l has been determined for the operating conditions. Use the tabulated measurements of dissolved oxygen (DO) values to estimate the oxygen uptake rate and $k_L a$ in this system.

	Time (min)	DO (mg/l)
	−1	3.3
Air off	0	3.3
	1	2.4
	2	1.3
	3	0.3
	4	0.1
	5	0.0
Air on	6	0.0
	7	0.3
	8	1.0
	9	1.6
	10	2.0
	11	2.4
	12	2.7
	13	2.9
	14	3.0
	15	3.1
	16	3.2
	17	3.2

10.2. A value of $k_L a = 30$ h^{-1} has been determined for a fermenter at its maximum practical agitator rotational speed and with air being sparged at 0.5 l gas/l reactor volume-min. *E. coli* with a q_{O_2} of 10 mmol O_2/g-dry wt-h are to be cultured. The critical dissolved oxygen concentration is 0.2 mg/l. The solubility of oxygen from air in the fermentation broth is 7.3 mg/l at 30°C.

    a. What maximum concentration of *E. coli* can be sustained in this fermenter under aerobic conditions?

    b. What concentration could be maintained if pure oxygen was used to sparge the reactor?

10.3. a. Estimate the required cooling-water flow rate for a 100,000-l fermenter with an 80,000-l working volume when the rate of oxygen consumption is 100 mmol O_2/l-h. The desired operating temperature is 35°C. A cooling coil is to be used. The minimum allowable temperature differential between the cooling water and the broth is 5°C. Cooling water is available at 15°C. The heat capacities of the broth and the cooling water are roughly equal.

    b. Estimate the required length of cooling coil if the coil has a 2.5-cm diameter and the overall heat transfer coefficient is 1420 J/s-m²-°C.

10.4. Consider Example 10.4. What would be the substrate concentrations in each compartment in the 10-l and 10,000-l tanks if the probe were placed in the bottom compartment?

10.5. Consider the 10-l and 10,000-l tanks described in Example 10.4. Suppose that fully continuous operation is to be used, and F was fixed at 5 mg/l-s for both tanks, and $D = 0.2$ h^{-1} for each tank with fluid removal from the top. What fraction of the inlet substrate would be consumed in each tank? If the biomass yield coefficient were 0.5 g cells/g substrate and $Y_{P/X} = 0.1$ g product/g cells, what would be the effect on volumetric productivity upon scale-up?

10.6. A continuous culture system is being constructed. The fermentation tank is to be 50,000 l in size and the residence time is to be 2 h. A continuous sterilizer is to be used. The unsterilized medium contains 10^4 spores/l. The value of k_d has been determined to be 1 min^{-1} at 121°C and 61 min^{-1} at 140°C. For each temperature (121°C and 140°C), determine the required residence time in the holding section so as to ensure that 99% of the time four weeks of continuous operation can be obtained without contamination (due to contaminants in the liquid medium).

10.7. Discuss the effects on sterilization of mixing in a batch fermenter.

10.8. A medium containing a vitamin is to be sterilized. Assume that the number of spores initially present is 10^5/l. The values of the pre-Arrhenius constant and E_{0d} for the spores are

$$E_{0d} = 65 \text{ kcal/g-mol}$$
$$\alpha = 1 \cdot 10^{36} \text{ min}^{-1}$$

For the inactivation of the vitamin, the values of E_{0d} and α are

$$E_{0d} = 10 \text{ kcal/g-mol}$$
$$\alpha = 1 \cdot 10^{4} \text{ min}^{-1}$$

The initial concentration of the vitamin is 30 mg/l. Compare the amount of active vitamin in the sterilized medium for 10-l and 10,000-l fermenters when both are sterilized at 121°C when we require in both cases that the probability of an unsuccessful fermentation be 0.001. Ignore the effects of the heat-up and cool-down periods.

10.9. Consider the data given in the table on the temperature changes in a 10,000-l fermenter, which includes the heat-up and cool-down periods. Use the values for the Arrhenius parameters given in Problem 10.8 and assume an initial spore concentration of 10^5/l and a vitamin concentration of 30 mg/l.

Time (min)	Temperature (°C)
0	30
10	40
20	54
30	70
40	95
50	121
55	121
60	121
65	106
70	98
90	75
100	64
120	46
140	32

a. What is the probability of a successful sterilization?

b. What fraction of the vitamin remains undegraded?

c. What fraction of the vitamin is degraded in the sterilization period?

d. What fraction of the vitamin is degraded in the heat-up and cool-down periods?

e. What is the fraction of spores deactivated in the heat-up and cool-down cycles?

10.10. *E. coli* have a maximum respiration rate, q_{O_2max}, of about 240-mg O_2/g-dry wt-h. It is desired to achieve a cell mass of 20 g dry wt/l. The k_La is 120 h^{-1} in a 1000-l reactor (800 l working volume). A gas stream enriched in oxygen is used (i.e., 80% O_2) which gives a value of $C* =$ 28 mg/L. If oxygen becomes limiting, growth and respiration slow; for example,

$$q_{O_2} = \frac{q_{O_2 max} C_L}{0.2 \text{ mg/l} + C_L}$$

where C_L is the dissolved oxygen concentration in the fermenter. What is C_L when the cell mass is at 20 g/l?

10.11. The temperature history of the heating and cooling of a 40,000-l tank during sterilization of medium is: 0 to 15 min, $T = 85°C$; 15 to 40 min, $T = 121°C$; 40 to 50 min, $T = 85°C$; 50 to 60 min, $T = 55°C$; > 60 min, $T = 30°C$. The medium contains vitamins, the most fragile of the vitamins has an activation energy for destruction of 10 kcal/g-mol, and the value of α (see eq. 10.18) is $1 \cdot 10^4$ min^{-1}. Assume vitamin destruction is first order and the initial concentration is 50 mg/l. R is 1.99 cal/g-mol-°K. The medium contains $2.5 \cdot 10^3$ spores/l. The spores have an $E_{0d} = 65$ kcal/g-mol, and k_d at 121°C is 1.02 min^{-1}. *Estimate:* (a) the probability of a successful sterilization, and (b) what fraction of the vitamin remains active?

10.12. Estimate k_La from Fig. 10.5 if $C*$ is 35 mg/l due to the use of nearly pure oxygen rather than air.

10.13. In cultivation of baker's yeast in a stirred and aerated tank, lethal agents are added to the fermentation medium to kill the organisms immediately. Increase in dissolved oxygen (DO) concentration upon addition of lethal agents is followed with the aid of a DO analyzer and a recorder. Using the following data, determine the oxygen transfer coefficient (k_La) for the reactor. Saturation DO concentration is $C* = 9$ mg/l.

Time (min)	DO (mg/l)
1	1
2	3
2.5	4
3	5
4	6.5
5	7.2

10.14. A stirred-tank reactor is to be scaled down from 10 m^3 to 0.1 m^3. The dimensions of the large tank are: $D_t = 2$ m; $D_i = 0.5$ m; $N = 100$ rpm.

a. Determine the dimensions of the small tank (D_t, D_i, H) by using geometric similarity

b. What would be the required rotational speed of the impeller in the small tank if the following criteria were used?

1) Constant tip speed

2) Constant impeller Re number

10.15. An autoclave malfunctions, and the temperature reaches only 119.5°C. The sterilization time at the maximum temperature was 20 min. The jar contains 10 l of complex medium that has 10^5 spores/l. At 121°C $k_d = 1.0$ min^{-1} and $E_{0d} = 90$ kcal/g-mol. What is the probability that the medium was sterile?

11

Recovery and Purification of Products

11.1. STRATEGIES TO RECOVER AND PURIFY PRODUCTS

The recovery and purification of a fermentation product is essential to any commercial process. The difficulty entailed depends heavily on the nature of the product. Products may be the biomass itself, an extracellular component, or an intracellular component. Since the chemical nature of a fermentation broth is quite complex and extremely high purity is required for some products (e.g., some pharmaceuticals), recovery and purification often require many processing steps and in many cases represent a manufacturing cost higher than that involved in producing the product. Since the compounds of interest are often fragile (e.g., heat sensitive) and present in dilute aqueous solution, the traditional separation techniques of the chemical engineer must be augmented with more specialized ones.

Recovery and purification are major expenses in production of most fermentation products; often they comprise more than 50% of the total manufacturing costs, especially for an intracellular product. There is an excellent correlation between the price of a product (on a per-kg basis) and how dilute the product is as it exits the fermentation. For example, citric acid is on the order of 100 g/l when it enters recovery processes and sells for order of 1 to 2 $/kg, while a therapeutic protein might be at the concentration of 0.00001 g/l and sell for 100,000,000 $/kg. The efficiency of the production process and the difficulty and cost of the recovery and purification processes are tightly coupled.

The major unit operations used for product recovery and purification and the major principles of separations are summarized in Fig. 11.1. Separation methods vary with the

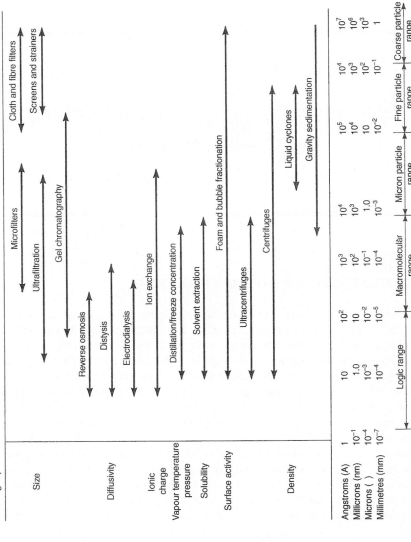

Figure 11.1. Ranges of applications of some standard unit operations. (With permission, from B. Atkinson and F. Mavituna, *Biochemical Engineering and Biotechnology Handbook*, Macmillan, Inc., New York, 1983.)

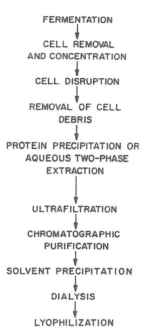

FERMENTATION

CELL REMOVAL
AND CONCENTRATION

CELL DISRUPTION

REMOVAL OF CELL
DEBRIS

PROTEIN PRECIPITATION OR
AQUEOUS TWO-PHASE
EXTRACTION

ULTRAFILTRATION

CHROMATOGRAPHIC
PURIFICATION

SOLVENT PRECIPITATION

DIALYSIS

LYOPHILIZATION

Figure 11.2. Major steps involved in the separation and purification of intracellular enzymes.

size and nature of the product, and a number of different methods may therefore need to be used for a fermentation broth containing soluble products of different molecular size. Figure 11.2 summarizes the major steps involved in the separation and purification of an enzyme from a fermentation broth, and this approach is generally applicable for many protein products.

The strategy in Fig. 11.2 can be generalized to involve four primary functions: (1) separation of insoluble products and other solids, (2) primary isolation or concentration of product and removal of most of the water, (3) purification or removal of contaminating chemicals, and (4) product preparation, such as drying. Process economics dictate that water must be removed very early in the process train so that the size of the equipment in the following steps will be minimized. Because a step designed primarily to concentrate product and the steps designed to remove contaminating chemicals often remove solvent, there is no distinct separation between steps 2 and 3. We will treat these steps by describing in the same section operations to recover soluble compounds. First, however, we consider those steps that remove solids, such as cells or the insoluble material.

11.2. SEPARATION OF INSOLUBLE PRODUCTS

The separation of solids such as biomass, insoluble particles, and macromolecules from the fermentation broth is usually the first step in product recovery. In some cases, the broth may need to be pretreated to facilitate solids separation. Examples of pretreatment are heat treatment, pH and ionic strength adjustment, and the addition of coagulants and flocculants. If the product is biomass, then separation of solids is the major step in product recovery, which results in a significant volume reduction. If the product is a soluble compound, solids need to be separated from liquid before the liquid is further treated to recover and purify the soluble product. For the recovery of intracellular products, the cells need to be disrupted and other cellular products need to be separated from the desired product. The major methods used for the separation of cellular material (biomass) are (1) filtration (both rotary vacuum filtration and micro- or ultrafiltration), (2) centrifugation, and (3) coagulation and flocculation.

11.2.1. Filtration

Filtration is probably the most cost-effective method for the separation of large solid particles and cells from fermentation broth. Fermentation broth is passed through a filter medium, and a filter cake is formed as a result of deposition of solids on the filter surface. Continuous rotary filters or *rotary vacuum precoat filters* are the most widely used types in the fermentation industry. The drum is covered with a layer of precoat, usually of diatomaceous earth, prior to filtration. A small amount of coagulating agent or filter aid is added to the broth before it is pumped into the filter. As the drum rotates under vacuum, a thin layer of cells adhere to the drum. The thickness of the cell layer increases in the section designed for forming the cake. The layer of solids is washed and dewatered during its passage to the discharge point, where a knife blade cuts off the cake. A vacuum maintained in the drum provides the driving force for liquid and air flow. A schematic diagram of a continuous rotary vacuum filter is shown in Fig. 11.3.

Filtration is commonly used for separating mycelium from fermentation broth in antibiotic fermentations. It is also commonly used in waste-water treatment facilities.

The rate of filtration (the flow of filtrate) for a constant-pressure (vacuum) filtration operation is determined primarily by the resistance of the cake and filter medium:

$$\frac{dV}{dt} = \frac{g_c \Delta p A}{(r_m + r_c)\mu} \tag{11.1}$$

where V is the volume of filtrate, A is the surface area of the filter, Δp is the pressure drop through the cake and filter medium, μ is the viscosity of the filtrate, r_m is the resistance of

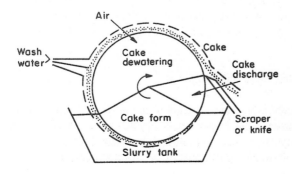

Figure 11.3. Schematic of a continuous rotary vacuum filter. A vacuum (subatmospheric pressure) is maintained within the drum, causing the pressure-driven flow of liquid (during cake formation) and air (during dewatering).

the filter medium, and r_m is the resistance of the cake. The value of r_m is characteristic of the filter medium. However, the cake resistance, r_c, increases during filtration, and after a start-up period, r_c exceeds r_m. The value of r_c is given by

$$r_c = \alpha \frac{W}{A} = \alpha \frac{CV}{A} \tag{11.2}$$

where W is the total weight of the cake on filter, C is the weight of the cake deposited per unit volume of filtrate, and α is the average specific resistance of the cake.

The total weight of cake is related to the total volume of filtrate by

$$W = CV \tag{11.3}$$

Substituting eqs. 11.2 and 11.3 into eq. 11.1 with constant A yields

$$\frac{d(V/A)}{dt} = \frac{g_c \Delta p}{\left(r_m + \alpha \dfrac{CV}{A} \right) \mu} \tag{11.4}$$

Integration of eq. 11.4 from $V = 0$ to $V = V$ and $t = 0$ to $t = t$ yields

$$V^2 + 2VV_0 = Kt \tag{11.5}$$

where

$$V_0 = \frac{r_m}{\alpha C} A \quad \text{and} \quad K = \left(\frac{2A^2}{\alpha C \mu} \right) \Delta p \cdot g_c$$

Equation 11.5 is known as the Ruth equation for constant-pressure filtration and can be rearranged to give

$$\frac{t}{V} = \frac{1}{K}(V + 2V_0) \tag{11.6}$$

A plot of t/V versus V yields a straight line with a slope of $1/K$ and intercept of $2V_0/K$, as depicted in Fig. 11.4. The values for r_m and α are calculated from experimentally determined values of K and V_0.

In a rotating drum filter (Fig. 11.3), the drum rotates at a constant speed (n rps) and only a fraction of drum-surface area is immersed in suspension reservoir (φ). The period of time during which filtration is carried out is φ/n per revolution of the drum. Equation 11.5 can be rewritten in this case as

$$\left(\frac{V'}{n} \right)^2 + 2 \frac{V'}{n} V_0 = K \frac{\varphi}{n} \tag{11.7}$$

where $V' = $ filtrate volume per unit time (volume/time) and V'/n represents the volume of filtrate filtered for one revolution of the drum.

This analysis of filtration is based on several assumptions; the primary one is an *incompressible cake* which results in constant specific cake resistance. Usually, fermentation cakes are compressible, so α varies with Δp. The concentration of filter aid (1% to 5%) also has a significant effect on specific cake resistance. As depicted in Fig. 11.5, the

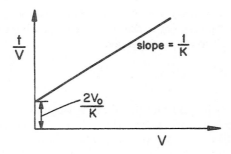

Figure 11.4. A plot of t/V versus V is used in filtration to calculate α and r_m.

specific cake resistance decreases as the concentration of filter aid increases. The filterability of fermentation broth is also affected by fermentation conditions. The pH, viscosity, and composition of the medium affect the cake resistance. Figure 11.6 depicts the effect of pH on cake resistance. Specific cake resistance decreases with decreasing pH. The coagulation of mycelial protein with heat treatment improves the filterability of the

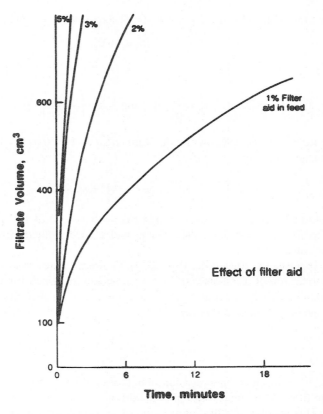

Figure 11.5. Effect of filter aid on the specific resistance to filtration of *Streptomyces griseus* broth (pH = 3.7 to 3.8). (With permission, from P. A. Belter, E. L. Cussler, and W-S. Hu, *Bioseparations: Downstream Processing for Biotechnology*, John Wiley & Sons, New York, 1988, p. 19.)

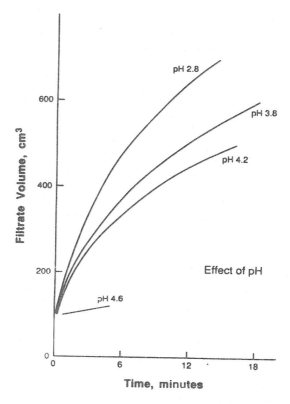

Figure 11.6. Effect of pH on the rate of filtration of *Streptomyces griseus* with 2% filter aid. (With permission, from P. A. Belter, E. L. Cussler, and W-S. Hu, *Bioseparations: Downstream Processing for Biotechnology*, John Wiley & Sons, New York, 1988, p. 18.)

fermentation broth in mycelial fermentations such as penicillin. The duration of fermentation also affects filterability and antibiotic activity in penicillin fermentations and represents an example of the interplay of upstream processing (e.g., fermentation) with downstream processing. Fermentation times of 180 to 200 h are used in penicillin fermentations to minimize the cake resistance and maximize antibiotic activity. Typical filtration conditions for streptomycin fermentations are pH = 3.6, 2% to 3% filter aid, following a heat treatment of 30 to 60 min at $T = 80°$ to $90°C$.

Cell separation can also be accomplished by cross-flow filtration using ultrafiltration or microporous filtration. These systems are described in more detail in Section 11.4.8. This technique has achieved rather rapid acceptance and has replaced centrifugation and vacuum filtration in many applications. Cross-flow filtration is well suited to the removal of dispersed cells such as *E. coli* and yeast. Microporous filters with pores of 0.02 and 0.2 µm are useful for bacterial separations, while larger sizes 0.2 to 2 µm can be used with yeast.

Example 11.1.

The following data were obtained in a constant-pressure filtration unit for filtration of a yeast suspension.

t (min)	4	20	48	76	120
V (l filtrate)	115	365	680	850	1130

Characteristics of the filter are as follows:

$$A = 0.28 \text{ m}^2, \quad C = 1920 \text{ kg/m}^3, \quad \mu = 2.9 \times 10^{-3} \text{ kg/m-s}, \quad \alpha = 4 \text{ m/kg}$$

a. Determine the pressure drop across the filter.
b. Determine the filter medium resistance (r_m).
c. Determine the size of filter for the same pressure drop to process 4000 l of cell suspension in 20 min.

Solution

a.

V(l)	115	365	680	850	1130
t/V (min/l)	0.035	0.055	0.07	0.089	0.106

A plot of t/V versus V results in a straight line with a slope of 0.67×10^{-4} (min/l²) and an intercept of 0.028 (min/l).

$$\text{slope} = \frac{1}{K} = 0.67 \times 10^{-4}, \quad K = 1.5 \times 10^4 \text{ l}^2/\text{min}$$

$$\Delta p = \frac{K\alpha C\mu}{2A^2 g_c}, \quad g_c = 9.8 \frac{\text{kg}_m}{\text{kg}_f \cdot \text{s}^2}$$

$$\Delta p = 2.3 \times 10^{-4} \text{ N/m}^2$$

b. y intercept $= \dfrac{2V_0}{K} = 0.028$

$$V_0 = 2101$$

$$r_m = \frac{\alpha V_0 C}{A} = 5760 \text{ m}^{-1}$$

c. $V^2 + 2VV_0 = Kt$, where $K = (2A^2/\alpha C\mu)\Delta p g_c$ and $V_0 = (r_m A/\alpha C)$. By substituting numerical values for V, Δp, and t, we obtain

$$A^2 - 25A - 66.67 = 0$$

The solution to the quadratic equation yields

$$A = 27.43 \text{ m}^2$$

11.2.2. Centrifugation

Centrifugation is used to separate particles of size between 100 and 0.1 µm from liquid by centrifugal forces. The theory of solid–liquid separations in a gravitational field should be clearly understood before centrifugal and gravity separations such as sedimentation are covered. Particle settling in a high-particle-density suspension is known as *hindered settling*, which resembles solid–liquid separations in a centrifugal field.

The major forces acting on a solid particle settling in a liquid by gravitational forces are gravitational force (F_G), drag force (F_D), and buoyant force (F_B). When the particles reach a terminal settling velocity, forces acting on a particle balance each other, resulting in a zero net force. That is,

$$F_G = F_D + F_B \tag{11.8}$$

where

$$F_G = \frac{\pi}{6} D_p^3 \rho_p \frac{g}{g_c} \tag{11.9}$$

$$F_B = \frac{\pi}{6} D_p^3 \rho_f \frac{g}{g_c} \tag{11.10}$$

and

$$F_D = \frac{C_D}{2g_c} \rho_f U_0^2 A \tag{11.11}$$

F_D is the drag force exerted by the fluid on solid particles, C_D is the drag coefficient, ρ_f is fluid density, U_0 is the relative velocity between the fluid and particle or the terminal velocity of a particle, and A is the cross-sectional area of the particles perpendicular to the direction of fluid flow; for a sphere, $A = (\pi/4)D_p^2$. For spherical particles, when $Re_p < 0.3$, the drag force, F_D, is given by the Stokes equation:

$$F_D = 3\pi\mu D_p U_0 \frac{1}{g_c} \tag{11.12}$$

Substitution of eq. 11.12 into eq. 11.11 results in the following relationship between C_D and Re_p:

$$C_D = \frac{24}{Re_p} \tag{11.13a}$$

When Re_p is between 1 and 10,000, C_D for a sphere is approximated by

$$C_D = \frac{24}{Re_p} + \frac{3}{\sqrt{Re_p}} + 0.34 \tag{11.13b}$$

where $Re_p = D_p U_0 \rho_f / \mu$.

Substitution of eqs. 11.9, 11.10, and 11.12 into eq. 11.8 results in

$$3\pi\mu D_p U_0 = \frac{\pi}{6} D_p^3 (\rho_p - \rho_f) g \tag{11.14}$$

or

$$U_0 = \frac{g D_p^2 (\rho_p - \rho_f)}{18\mu} \tag{11.15}$$

where D_p and ρ_p are particle diameter and density, respectively.

Sec. 11.2   Separation of Insoluble Products

In a centrifugal field, the terminal separation velocity of particles, U_{0c}, is given by the following equation, where the centrifugal acceleration is substituted for the gravitational acceleration:

$$U_{0c} = \frac{r\omega^2 \cdot D_p^2(\rho_p - \rho_f)}{18\mu} \tag{11.16}$$

or

$$U_{0c} = \frac{gZD_p^2(\rho_p - \rho_f)}{18\mu} = ZU_0 \tag{11.17}$$

where $Z = r\omega^2/g$ is the centrifugal factor, r is the radial distance from the central axis of rotation, and ω is angular velocity of rotation ($\omega = 2\pi Nr$).

The analysis presented so far is valid only for dilute cell suspensions where particle–particle interactions are negligible. When particle concentration is high, particles interact to form a swarm, and their terminal velocity decreases from U_0 to U. This is known as *hindered settling*. The separation of cells or particles in a centrifugal field is similar to hindered settling under gravity, since particle concentration is high under centrifugation conditions.

In hindered settling, the drag force on particles is

$$F_D' = \frac{1}{g_c} 3\pi\mu D_p U \left(1 + \beta_0 \frac{D_p}{L}\right) \tag{11.18}$$

where U is the terminal velocity of the particles under hindered-settling conditions, L is the average distance between adjacent particles, and β_0 is the hindered-settling coefficient. β_0 is 1.6 for a rectangular arrangement of particles. In dilute solutions, since $D_p/L \ll 1$, F_D' approaches F_D. Hindered settling becomes important when D_p and L are comparable.

The parameter β_0 is a function of α, the volume fraction of particles, and the shape of the particles. It can be shown that

$$\frac{U_c}{U_{0c}} = \frac{1}{1 + \alpha'\alpha^{1/3}} \tag{11.19}$$

where α' is empirically correlated with α and depends on particle shape also. U_c is the terminal velocity of the particles of a given size and shape in the centrifugal field under conditions of hindered settling.

When designing or sizing a centrifuge, we are concerned primarily with the capacity of the centrifuge to handle a given flow rate of broth. If we know both the velocity of the particle in the centrifugal field and the distance the particle must travel to be captured, then we can proceed with the design.

Examples of two popular centrifuges are shown in Fig. 11.7. Clearly, the geometry of the centrifuge influences the travel distance.

Consider a simple one-dimensional case where the distance of travel of a particle is

$$y = U_{0c}t \tag{11.20}$$

Substituting eq. 11.17 for U_{0c} and noting that the time of interest is $t = V_c/F_c$, we write

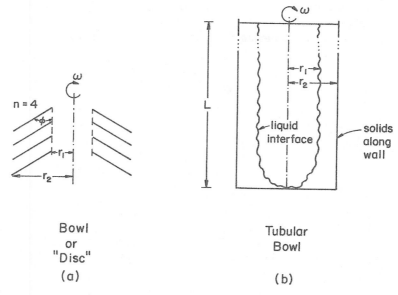

Figure 11.7. Two types of common centrifuges are depicted. (a) A disc or bowl centrifuge with multiple separator bowls or discs. The scale-up factor depends on the two radial distances, r_1 and r_2, the angle ϕ, the rotational speed, ω, and n, the number of discs or effective separator bowls. (b) A tubular bowl centrifuge can operate with continuous liquid feed with the span of continuous operation determined by the capacity of the bowl to collect solids. Key scale-up parameters are the distances r_1 and r_2, the height of the unit, L, and the rotational speed, ω.

$$y = \frac{r\omega^2 D_p^2 (\rho_p - \rho_f)}{18\mu} \cdot \frac{V_c}{F_c} \tag{11.21}$$

where V_c is liquid volume in the centrifuge and F_c is liquid flow rate through the centrifuge.

By substituting $y = L_e/2$ and $r = r_e$ into eq. 11.21, we obtain

$$F_c = 2\left[\frac{gD_p^2(\rho_p - \rho_f)}{18\mu}\right]\frac{r_e\omega^2 V_c}{gL_e} \tag{11.22}$$

or

$$F_c = 2U_0\Sigma \tag{11.23}$$

where $\Sigma = r_e\omega^2 V_c/gL_e$, which is known as the *centrifugation coefficient,* U_0 is the free settling velocity of the particles under gravity, L_e is the effective distance of settling ($= 2y$), and r_e is the effective radius of rotation. Σ is equal to the surface area of the gravity settling basin whose separation performance is equal to that of a particular centrifuge.

For a tubular centrifuge, Σ is

$$\Sigma = \frac{2\pi L\omega^2}{g}\left(\frac{3}{4}r_2^2 + \frac{1}{4}r_1^2\right) \tag{11.24}$$

where L is the length of cylindrical separator, and r_1 and r_2 are inner and outer radii of centrifugation distance.

For a bowl centrifuge, Σ is

$$\Sigma = \frac{2\pi n'\omega^2}{3g\tan\phi'}\,(r_2^3 - r_1^3) \tag{11.25}$$

where n' is the number of separator bowls, and ϕ' is the angle of inclination of the side walls.

Assuming that the free settling velocities of particles are the same, the following equation can be used for the scale-up of centrifuges.

$$\frac{F_{c_2}}{F_{c_1}} = \frac{\Sigma_2}{\Sigma_1} \tag{11.26}$$

where F_{c_1} and F_{c_2} are the flow rates through the small and large centrifuges, respectively.

Cross-flow microfiltration processes (Section 11.4.8) have replaced centrifugation in many bioprocesses, particularly when it is essential to prevent the escape of cells into the atmosphere. This consideration is particularly relevant when the organism is genetically modified.

11.2.3. Coagulation and Flocculation

Coagulation and flocculation are usually used to form cell aggregates before centrifugation, gravity settling, or filtration to improve the performance of these separation processes. Coagulation is the formation of small flocs from dispersed colloids using coagulating agents, which are usually simple electrolytes. Flocculation is the agglomeration of these small flocs into larger settlable particles using flocculating agents, which are usually polyelectrolytes or certain salts, such as $CaCl_2$.

Simple electrolytes used in coagulation are acids, bases, salts, and multivalent ions, which are relatively inexpensive, but less effective than polyelectrolytes. Certain fine solid particles, such as clays, activated carbon, or silica, may serve as nucleation sites for coagulation. Polyelectrolytes used in flocculation are high-molecular-weight, water-soluble organic compounds, which can be either anionic, cationic, or nonionic.

The analysis of gravity sedimentation is based on Stokes's and Newton's laws and is similar to the analysis presented in the centrifugation section. Sedimentation is a critical feature in the activated waste-treatment process and has been applied in some schemes to stabilize a population of genetically engineered cells. A list of widely used flocculants and their dosage ranges is given in Table 11.1. The important parameters to be considered in flocculation are flocculant–coagulant concentration, concentration of colloidal cells,

TABLE 11.1 Flocculant Dosages

| Agent | Flocculant dose [g (100 g dry cell wt)$^{-1}$] | | | | |
	Glucose broth	Hydro-carbon broth	Resuspended cells in buffer	Penicillin broth	Dilute slurries
Polyelectrolytes					0.045–4.5
Anionic polyelectrolytes					
Polystyrene sulphate	0.2	0.1	0.06		
Polyacrylamide	Ineffective	Ineffective	Ineffective		
Cationic polyelectrolytes					
Polyethylene imine	10		7.0		
Calcium chloride				200	
Colloidal clay, bentonite	2.0	20.0	0.6		
Inorganic coagulants					0.045–4.5

With permission, from B. Atkinson and F. Mavituna, *Biochemical Engineering and Biotechnology Handbook*, Macmillan, Inc., New York, 1983.

kinetics of the binding, and settling phenomena. The selection criteria usually are flocculation–sedimentation rate, floc size, and the clarity of the supernatant liquid.

11.3. CELL DISRUPTION

After cells are separated from liquid broth, if the desired product is intracellular, then the cells need to be disrupted to release the intracellular products. The method of disruption varies with the type of cells and the nature of intracellular products. Major methods of cell disruption can be classified as mechanical and nonmechanical. With small bacteria, efficient, large-capacity cell-disruption processes can be difficult to construct.

11.3.1. Mechanical Methods

Mechanical methods can be applied to a liquid or solid medium. First, consider some methods applied to a liquid medium.

Ultrasonic vibrators (sonicators) are used to disrupt the cell wall and membrane of bacterial cells. Wave density is usually around 20 kc/s. Rods are broken more readily than cocci, and gram-negative cells more easily than gram-positive cells. This method is not as effective for molds. An electronic generator is used to generate ultrasonic waves, and a transducer converts these waves into mechanical oscillations by a titanium probe immersed in a cell suspension. Intracellular compounds (enzymes, metabolites) are released into the broth upon cell disruption. Ultrasonic disruption in some cases results in denaturation of sensitive enzymes and fragmentation of cell debris. Heat dissipation is an important problem in cell disintegration, particularly if the volume subjected to sonication is large. Consequently this method is used primarily at the laboratory scale.

The Gaulin–Manton and French presses work well on a laboratory scale. The French press is a hollow cylinder in a stainless-steel block that is filled with cell paste and subjected to a high pressure. The cylinder has a needle valve at the base, and the cells disrupt as they are extruded through the valve to atmospheric pressure. The Ribi fractionator is similar to a French press but is capable of continuous operation. It requires a preconcentrated cell paste or slurry. The Rannie high-pressure homogenizer is advertised to operate up to 1000 bar and to have a capacity of up to 40,000 l/h.

Other mechanical homogenizers make use of small beads (20 to 50 mesh). Some of these can be effectively used at rather large scales. The Dyno-Mill is available in sizes up to 275 l and can process 2000 kg/h of a cell suspension or about 340 kg dw/h of yeast. The Dyno-Mill can work with algae, bacteria, and fungi. The principle of operation is to pump the cell suspension through a horizontal grinding chamber filled with about 80% beads. Within the grinding chamber is a shaft with specially designed discs. When rotated at high speeds, high shearing and impact forces from millions of beads break cell walls. The broken cells are then discharged. A primary advantage of bead over pressure devices is better temperature control. Also this device is fully enclosed, preventing release of organisms in an aerosol, which is a critical consideration with pathogenic or genetically modified organisms.

For a solid medium, such as frozen cell paste or cells attached to or within a solid matrix, different processing options are available. Ball mills can be operated on large-scale processes. More specialized pieces of equipment include the Hughes and X-presses. The Hughes press consists of a split block with a half-cylinder hollowed in each face. The frozen cell paste is placed in the hollow, and pressure is applied to it from the cylinder into channels cut in the block. In the X-press, frozen cells are forced continuously through a small hole in a disc between two cylinders at low temperature and pressure. Cell disruption takes place because of deformation of organisms embedded in the ice. Pressure levels of several hundred bars can be obtained with the X-press.

With any of these approaches there is potential for damage to the product. Clearly heat denaturation is a major potential problem. The release of proteases from cellular compartments can lead to enzymatic degradation of the product. Some techniques, such as bead mills, have comparatively long residence times, so that if 100% cell breakage is desired, some fragile compounds released early in the process may be damaged. Once cell breakage occurs, products are released and encounter an oxidizing environment (due to air) that can cause, for protein products, rapid denaturation and aggregation. Typically this effect can be mitigated by addition of a reducing agent (e.g., β-mercaptoethanol) and a chelating agent (e.g., EDTA) to complex metal ions that may accelerate the oxidation process. Product quality considerations are essential to selection of a method for cell breakage.

11.3.2 Nonmechanical Methods

Osmotic shock and rupture with ice crystals are commonly used methods. By slowly freezing and then thawing a cell paste, the cell wall and membrane may be broken, releasing enzymes into the media. Changes in the osmotic pressure of the medium may result in the release of certain enzymes, particularly periplasmic proteins in gram-negative cells.[†] Cell-wall preparation of certain organisms can be made by heat-shocking. Freeze-drying after treatment with acetone, butanol, and buffers results in cell-wall disruption.

[†]The periplasmic space in a gram-negative cell is the space between the inner and outer membranes.

Enzymes such as lysozyme (a carbohydrase) can be used to lyse cell walls of bacteria. Gram-positive bacteria are far more susceptible to enzymatic lysis than gram-negative bacteria. The cells may be treated with EDTA or by freezing and thawing before treatment with lysozyme. Ethylenediaminetetraacetic acid (EDTA) is a chelating agent; high levels of EDTA will extract divalent ions that are part of the cell envelope. Enzymatic hydrolysis is an expensive method and is not very widely used in industry.

Actively growing cells can be treated with an antibiotic, such as penicillin or cycloserine, that interferes with cell-wall synthesis and, coupled with the correct osmotic conditions, can lead to cell disruption.

After cells have been lysed and the products are released into the medium, cell debris can be separated by ultracentrifugation or ultrafiltration, and soluble products are recovered using the following methods.

11.4. SEPARATION OF SOLUBLE PRODUCTS

Most microbial products, such as antibiotics, organic acids, solvents, amino acids, and extracellular enzymes, are soluble and extracellular. Various methods have been developed to recover such soluble products, including extraction, adsorption, ultrafiltration, and chromatography.

11.4.1. Liquid–Liquid Extraction

Liquid extraction is commonly used to separate inhibitory fermentation products such as ethanol and acetone–butanol from a fermentation broth. Antibiotics are also recovered by liquid extraction (using amylacetate or isoamylacetate). Ideally, the liquid extractant should be nontoxic, selective, inexpensive, and immiscible with fermentation broth and should have a high distribution coefficient for the product.

The extraction of a compound from one phase to the other is based on solubility differences of the compound in one phase relative to the other. When a compound is distributed between two immiscible liquids, the ratio of the concentrations in the two phases is known as the distribution coefficient:

$$K_D = \frac{Y_L}{X_H} \tag{11.27}$$

where Y_L and X_H are concentrations of the solute in light and heavy phases, respectively. In most, but certainly not all cases, the light phase will be the organic solvent and the heavy phase will be the aqueous fermentation broth.

Assuming that K_D is constant and the solvents are totally immiscible (that is, the mass flows of the light and heavy phases are conserved so that $L_0 = L_1 = L$ and $H_0 = H_1 = H$), a mass balance on the extracted solute yields (Fig. 11.8a)

$$H(X_0 - X_1) = LY_1 \tag{11.28}$$

or

$$X_1 = X_0 - \frac{L}{H}Y_1 \tag{11.29}$$

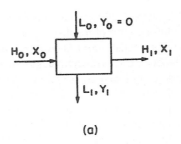

(a)

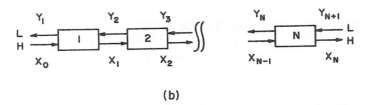

(b)

Figure 11.8. (a) Single-stage and (b) multistage countercurrent liquid extraction systems.

Since $K_D = Y_1/X_1$, eq. 11.29 can be written as

$$X_1 = X_0 - \frac{LK_D}{H} X_1 \tag{11.30}$$

$$\frac{X_1}{X_0} = \frac{1}{1 + (LK_D/H)} = \frac{1}{1+E} \tag{11.31}$$

where $E = LK_D/H$ is the extraction factor.

For a multistage countercurrent operation, as depicted in Fig. 11.8b, a material balance on extracted solute yields

$$R = \frac{E(E^n - 1)}{E - 1} \tag{11.32}$$

where R is the rejection ratio, which is the weight ratio of the solute leaving in the light phase to that leaving in the heavy phase, and n is the number of equilibrium stages.

The fraction of solute extracted is then

$$\% \text{ extraction} = 1 - \frac{1}{R+1} \tag{11.33}$$

When the solute enters the system in the light phase, the rejection ratio is

$$R = \frac{E^n(E-1)}{E^n - 1} \tag{11.34}$$

where n is the number of stages. Figure 11.9 depicts the relationship between E, X_n/X_0, and n, which is used to determine the number of stages (n) required for a certain degree of extraction (X_n/X_0) for a given system (E).

Most antibiotics are extracted from the fermentation broth by using solvents such as amylacetate or isoamylacetate. Usually, continuous *centrifugal Podbielniak extractors* are used for the extraction of antibiotics. Penicillin is more soluble in organic phase at low pHs (pH = 2 to 3) and is highly soluble in aqueous phase at high pHs (pH = 8 to 9). Therefore, penicillin is extracted between organic and aqueous phases several times by shifting the pH to improve the purity of the product. Figure 11.10 depicts the variation of distribution coefficient (solvent–aqueous) with pH in penicillin extraction by amylacetate.

Some products to be recovered from the fermentation broth are weak acids or weak bases. Since compounds that are not ionized are soluble in the organic phase, the pH conditions are selected such that the extracted compound is neutral and is soluble in an organic solvent. Therefore, weak bases are extracted at high pHs and weak acids are extracted at low pH values in the neutral form.

If fermentation broth contains more than one component, then the selectivity (β) of the solvent becomes an important parameter. The selectivity coefficient (β_{ij}) is defined as

$$\beta_{ij} = \frac{K_i}{K_j} \tag{11.35}$$

The ease of separation depends on the β_{ij} value: the higher the β_{ij} value is, the easier the separation of i from j. In some cases, the β_{ij} value changes appreciably with a pH shift and

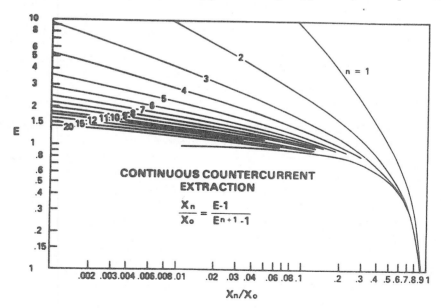

Figure 11.9. Relationship between unextracted solute, extraction factor, and number of stages in continuous countercurrent extraction. (With permission, from L. C. Craig, D. Craig, and E. C. Scheibel in *Technique of Organic Chemistry*, A. Weissberger, ed., 2d ed., Vol. 3, Part 1, Wiley—Interscience, New York, 1956.)

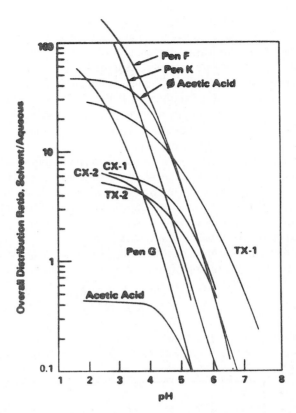

Figure 11.10. Distribution ratios for various penicillins and impurities. (With permission, from M. Sounders, G. J. Pierotti, and C. L. Dunn, *Chem. Eng. Prog. Symp. Ser. 66,* p. 40, 1970, AIChE, New York.)

the separation becomes easier (Fig. 11.10). For example, at low pH values, penicillins can be separated from other impurities by selective extraction into the organic solvent. However, the separation of similar compounds by extraction is difficult. If only nonionized species are soluble in both phases at extreme pH values and the solvents are immiscible, the aforementioned design equations can be used. However, at intermediate pH values, when compounds are partially ionized, the analysis of the system becomes more complicated. The extraction of ionized species (weak acid or bases) is known as *dissociation extraction.*

The apparent distribution coefficients for weak acids and bases are

Weak acids:
$$K_D^{AP} = \frac{K_D^0[H^+]}{[H^+]+K_1} \quad \text{or} \quad pH - pK_1 = \log\left(\frac{K_D^0}{K_D^{AP}} - 1\right) \tag{11.36}$$

Weak bases:
$$K_D^{AP} = \frac{K_D^0 K_1}{K_1 + [H^+]} \quad \text{or} \quad pK_1 - pH = \log\left(\frac{K_D^0}{K_D^{AP}} - 1\right) \tag{11.37}$$

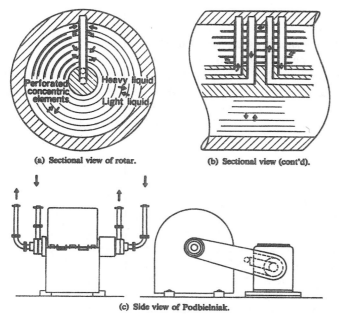

(a) Sectional view of rotar.

(b) Sectional view (cont'd).

(c) Side view of Podbielniak.

Figure 11.11. Liquid-liquid extraction equipment (Podbielniak). Bold arrow indicates heavy liquid flow, while open arrow represents the flow of a mixture or light liquid. (With permission, from S. Aiba, A. E. Humphrey, and N. F. Millis, *Biochemical Engineering*, 2d ed., University of Tokyo Press, Tokyo, 1973.)

where K_D^{AP} is the apparent distribution coefficient, K_D^0 is the distribution coefficient for neutral species, and K_1 is the dissociation equilibrium constant for weak acids or bases.

A particularly important device for liquid–liquid extraction for fermentation products is the Podbielniak centrifugal extractor (see Fig. 11.11). Many fermentation products are unstable (e.g., penicillin). The use of mixer–settlers can be problematic, because the residence time of the product in the pH-adjusted broth is too long. The rapid rotation of the Podbielniak extractor produces a centrifugal field that rapidly drives the two fluids countercurrent to each other, as depicted in Fig. 11.11. A product can be extracted and returned to another aqueous phase (e.g., a phosphate buffer) within minutes.

Example 11.2.

Penicillin is extracted from a fermentation broth using isoamylacetate as the organic solvent in a continuous countercurrent cascade extraction unit. The flow rates of organic (L) and aqueous (H) phases are $L = 10$ l/min and $H = 100$ l/min, respectively. The distribution coefficient of penicillin between organic and aqueous phases at pH = 3 is $K_D = Y_L/X_H = 50$. If the penicillin concentration in the feed stream is 20 g/l, determine the number of stages required to reduce the penicillin concentration to 0.1 g/l in the effluent of the extraction unit.

Solution

$$E = \frac{LK_D}{H} = \frac{(10)(50)}{(100)} = 5$$

$$\frac{X_n}{X_0} = \frac{0.1}{20} = 5 \times 10^{-3}$$

Using Fig. 11.9, we can obtain $n = 4$. If a Podbielniak centrifugal extractor were used, it would have to correspond to four ideal stages.

11.4.2. Aqueous Two-phase Extraction

Aqueous two-phase extraction is an approach under active development for the extraction of soluble proteins such as enzymes between two aqueous phases containing incompatible polymers, such as polyethylene glycol (PEG) and dextran. The phases containing PEG and dextran are more than 75% water and are immiscible. Typical aqueous phases used for this purpose are PEG–water/dextran–water and PEG–water/K-phosphate–water. PEG/dextran and PEG/K phosphate are reasonably immiscible. The partition coefficient, K_p, varies with the molecular weight (MW) of the soluble protein in the form of an exponential function.

$$K_p = e^{AM/T} \tag{11.38}$$

where M is the MW of protein, T is the absolute temperature, and A is a constant.

The partition coefficient, K_p, of many enzymes between the two phases (C_{PEG}–C_{DEX}) varies between 1 and 3.7, resulting in poor separations in a single stage. The partition coefficient can be improved by including ion-exchange resins or certain salts, such as $(NH_4)_2SO_4$ and KH_2PO_4, in one of the phases. Figure 11.12 depicts the variation of K_p of an enzyme with a concentration of KH_2PO_4 in a PEG/$(NH_4)_2SO_4$ system. More than ten-fold increases in K_p values may be achieved by increasing the concentration of KH_2PO_4 from 0.1 to 0.3 M. In PEG–salt systems, salting-out may result in protein precipitation at the interface.

Ion-exchange resins can be used to increase the value of a partition coefficient. PEG can be derivatized to yield cation or anion exchange properties and used in a two-phase

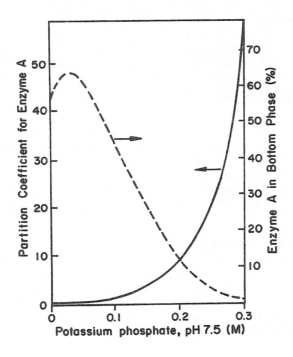

Figure 11.12. The expected behavior of an enzyme in a two-phase solution is depicted. Note the dependence of the partition coefficient of a hypothetical protein on the concentration of potassium phosphate in a polyethylene glycol 4000 ammonium sulphate system. A typical condition would be 14% PEG 4000 and 9.5% $(NH_4)_2SO_4$.

system. Affinity ligands may also be used to increase K_p values; examples are PEG–NADH and PEG–cibacron blue systems. This method is known as *two-aqueous-phase affinity partition extraction*. Also, partial hydrolysis of dextran and PEG may result in an increase in the K_p value, since lower-MW polymers may interact with proteins more effectively. By mixing two kinds of PEGs (PEG_{400} and PEG_{4000}), the K_p value may be increased by a factor of 6 for the partition of fumarase. Two-aqueous-phase extraction can also be used for the recovery of cell debris, polysaccharides, and nucleic acids. The partition coefficient for whole cells and DNA is between 100 and 0.01, for proteins it is between 10 and 0.1, and for small ions it is around 1.

After the extraction step, the phases can be separated by centrifugation or decantation, and PEG can be recovered by ultrafiltration. Figure 11.13 is a block diagram of an enzyme separation unit using the two-phase partitioning method with PEG recovery. This separation method is fast and can be operated under mild conditions of temperature, pressure, and pH. Dextran and PEG are recovered and reused, since the cost of polymers is the major economic factor.

11.4.3. Precipitation

The first step in the purification of intracellular proteins after cell disruption is usually *precipitation*. Proteins in a fermentation broth (before or after cell lysis) can be separated from other components by precipitation using certain salts. Examples include streptomycin sulfate and ammonium sulfate.

.   The two major methods used for protein precipitation are as follows:

1. Salting-out by adding inorganic salts such as $(NH_4)_2SO_4$ at high ionic strength.
2. Solubility reduction at low temperatures by adding organic solvents ($T < -5°$).

Salting-out of proteins is achieved by increasing the ionic strength of a protein-containing solution by adding salts such as $(NH_4)_2SO_4$ or Na_2SO_4. The added ions interact with water more strongly, causing protein molecules to precipitate. The solubility of proteins in a solution as a function of the ionic strength of the solution is given by

$$\log \frac{S}{S_0} = -K'_S(I) \qquad (11.39)$$

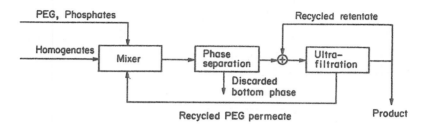

Figure 11.13.   Two-phase extraction process with PEG recovery.

where S is the solubility of protein in solution (g/l), S_0 is the solubility of protein when $I = 0$, I is the ionic strength of solution, and K_s' is the salting-out constant, which is a function of temperature and pH.

The ionic strength of a solution is defined as

$$I = \frac{1}{2} \Sigma C_i Z_i^2 \qquad (11.40)$$

where C_i is molar concentration of the ionic species (mol/l), and Z_i is the charge (valence) on ions.

Figure 11.14 depicts variation of the solubility of hemoglobin with inorganic salt concentrations. At high ionic strengths, the solubility of the protein decreases logarithmically with ionic strength.

Organic solvent addition at low temperatures ($T < -5°C$) causes the precipitation of proteins by reducing the dielectric constant of the solution. The solubility of protein as a function of the dielectric constant of a solution is given by

$$\log \frac{S}{S_0} = -K'/D_s^2 \qquad (11.41)$$

where D_S is the dielectric constant of the water–solvent solution.

A reduction in the dielectric constant of a solution results in stronger electrostatic forces between the protein molecules and facilitates protein precipitation. The addition of solvents also reduces protein–water molecule interactions and therefore decreases protein solubility. Solvents may cause protein denaturation. However, denaturation of proteins in the salting-out method is less likely. *Solvent precipitation* can be used with salt addition, pH adjustment, and low temperature to improve precipitation. Among other protein precipitation methods, the following methods are the most widely used.

Isoelectric precipitation is the precipitation of proteins at their isoelectric point, which is the pH at which proteins have no net charge. The isoelectric point of a protein is defined as $pI = \frac{1}{2}(pK_1 + pK_2)$. When pH = pI, protein becomes free of charge and precipi-

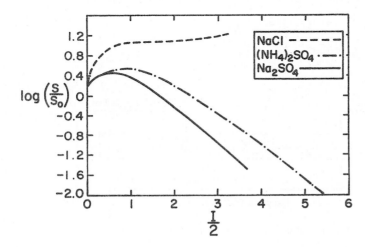

Figure 11.14. Effect of inorganic salts on solubility of a typical protein.

Recovery and Purification of Products   Chap. 11

tates. This method is effective for proteins with high surface hydrophobicity (that is, a nonpolar surface). It is inexpensive, but low pH values may result in the denaturation of proteins.

The use of ionic polyelectrolytes, such as ionic polysaccharides, polyphosphates, and polyacrylic acids, changes the ionic strength of the media and causes protein precipitation. However, polyelectrolytes may cause protein denaturation and structural damage.

The use of nonionic polymers, such as dextrans and polyethylene glycol, reduces the amount of water available to interact with protein molecules and results in protein precipitation. High concentrations of polymers are most effective for low-MW proteins, and nonionic polymers do not interfere with further protein recovery.

11.4.4. Adsorption

Adsorption of solutes from liquid media onto solids is a common practice in separating soluble materials from fermentation broth. Various mechanisms may be involved in adsorption. In physical adsorption, weak forces, such as van der Waals forces, are dominant; however, in ion-exchange adsorption, strong ionic bonds are utilized. Solute is transferred from liquid to solid phase, and an equilibrium is reached after a while in a batch operation. The type of adsorbent used depends on the particular application. The most widely used adsorbent for waste-water treatment applications is activated carbon, since it has large internal surface area per unit weight. Ion-exchange resins and other polymeric adsorbents can be used for protein separations of small organics. For example, a carboxylic acid cation exchange resin is used to recover streptomycin. Adsorption capacity varies depending on adsorbent, adsorbate, physicochemical conditions, and the surface properties of the adsorbent and adsorbate. Usually, the exact mechanism of adsorption is not well understood, and the equilibrium data must be determined experimentally.

One typical equilibrium relationship between solute concentrations in liquid and solid phases is

$$C_S^* = K_F C_L^{*(1/n)}$$
(11.42)

where C_L^* and C_S^* are equilibrium concentrations of the solute in liquid and solid phases. C_S is in units of mass of solute adsorbed per unit volume of resin. This relationship is an example of a type of Freundlich adsorption isotherm, with K_F and n being empirical parameters and n being greater than 1.

Various types of solid–liquid contactors have been developed for the adsorption of solutes. Among these are packed-bed, moving-bed, fluidized-bed, or agitated-vessel contactors. Packed- and moving-bed contactors are the most widely used, since the adsorption area per unit volume of reactor is the largest among the others. Analysis of adsorption phenomena in a *packed-bed* column is based on a differential material balance of the solution in the column.

$$U\frac{\partial C_L}{\partial Z} + \varepsilon\frac{\partial C_L}{\partial t} = -(1-\varepsilon)\frac{\partial C_S}{\partial t}$$
(11.43)

where U is the superficial velocity of the liquid, ε the porosity of the bed, and C_L the concentration of the solute in the liquid phase.

The first term in eq. 11.43 is the convective transfer of solute in the bed, the second term represents the time course of change (transient) of the solute concentration in liquid, and the last term represents the rate of solute transfer from the liquid to the solid phase.

The rate of solute transfer from the liquid into the solid phase is

$$\frac{dC_S}{dt} = K_a(C_L - C_L^*) \tag{11.44}$$

where K_a is the overall mass transfer coefficient describing the internal and external mass transfer resistance, and C_L^* is the concentration of solute in the liquid phase, which is in equilibrium with C_S.

Substituting eq. 11.44 into eq. 11.43 yields

$$U\frac{\partial C_L}{\partial Z} + \varepsilon\frac{\partial C_L}{\partial t} = -(1-\varepsilon)K_a(C_L - C_L^*) \tag{11.45}$$

Equation 11.45, along with the equilibrium relationship (eq. 11.42), can be solved numerically to determine the solute profile, $C_L = C_L(Z, t)$, in the column.

The preceding approach is based on plug flow conditions in the bed. However, in practice, due to back-mixing and dispersion, irregular flow profiles and channeling occur. A dispersion term $[-D_E(\partial^2 C_L/\partial Z^2)]$ is added to the right side of eq. 11.45 to account for dispersion effects. A fixed-bed adsorption unit can be approximated to a large number of well-mixed tanks in series.

For very long *moving-bed columns* to which adsorbent is added continuously, the operation may be considered to be at steady state (see Fig. 11.15). For the analysis, equilibrium is assumed to occur at the top of the column. If the dispersion effects are negligible, then at steady state eq. 11.45 can be written as

$$U\frac{dC_L}{dZ} = -K_a(1-\varepsilon)(C_L - C_L^*) \tag{11.46}$$

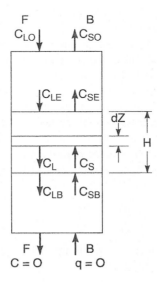

Figure 11.15. Schematic diagram of steady-state adsorption column. (With permission, from D. W. Sundstrom and H. E. Klei, *Wastewater Treatment*, Pearson Education, Upper Saddle River, NJ, 1979, p. 260.)

with the boundary condition $C_L = C_{L0}$ at $Z = H$ for liquid feed into the top of a column.

Equation 11.46 can be integrated to determine the height of the column for a certain degree of separation.

$$-\frac{U}{K_a(1-\varepsilon)}\int_{C_L}^{C_{L0}}\frac{dC_L}{C_L - C_L^*} = \int_0^H dZ = H \tag{11.47}$$

By using an equilibrium relationship, we can integrate eq. 11.47 to determine the height of the column. To complete the analysis, a relationship between the unsaturated resin and the liquid concentration is needed. This relationship is called an *operating line* (Fig. 11.16) and represents a mass balance. Note that the volumetric resin flow is denoted by B and the volumetric liquid flow rate by F. Also note that

$$F = UA \tag{11.48}$$

where A is the cross-sectional area of the column.

From mass balance considerations on the solute,

$$F(C_{L0} - 0) = B(C_{S0}^* - 0) \tag{11.49}$$

or

$$\frac{B}{F} = \frac{C_L}{C_S} = \frac{C_{L0}}{C_{S0}^*} \tag{11.50}$$

where C_{S0}^* is the resin concentration of solute in equilibrium with C_{L0}. As noted in Fig. 11.16, the operating line is straight.

For fixed-bed columns, the absorptive capacity is limited to a maximum level. In this view, a column would consist of three zones: a *saturated zone*, an *adsorption zone*, and a *virgin zone*. The adsorption zone moves down a column at a rate determined by the feed rate of the solute and the mass transfer characteristics of the bed. When the leading edge of the adsorption zone reaches the bottom of the column, there will be a rapid increase in the exit concentration of solute. Normally, the column would be shut down at this point and the resin replaced or regenerated. However, if the column were operated for a period of time equal to the time for the adsorption zone to travel its own height, we

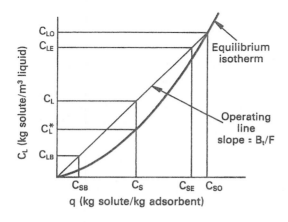

Figure 11.16. Operating line for steady-state adsorption column. (With permission, from D. W. Sundstrom and H. E. Klei, *Wastewater Treatment*, Pearson Education, Upper Saddle River, NJ, 1979, p. 261.)

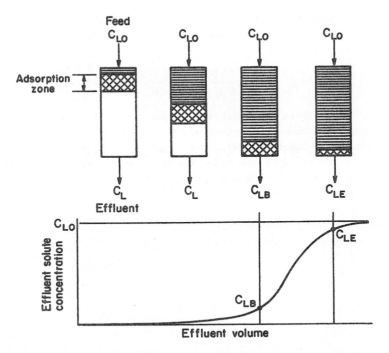

Figure 11.17. Movement of adsorption zone through a fixed-bed downflow adsorber and the corresponding breakthrough curve. (With permission, from D. W. Sundstrom and H. E. Klei, *Wastewater Treatment*, Pearson Education, Upper Saddle River, NJ, 1979, p. 281.)

would observe the *breakthrough* curve of the adsorption zone. This process is depicted in Fig. 11.17.

Under these circumstances, eq. 11.47 would describe the concentration profile in the adsorption zone of a height H. If the column were of total height H_t, then the time to exhaustion of the column would be equal to the ratio of H to H_t multiplied by the velocity of the adsorption zone, plus the time required to form the adsorption zone (usually negligible).

Example 11.3.

Cephalosporin is separated from fermentation broth by adsorption on an ion-exchange resin in a moving-bed column. The bed is 4 cm in diameter and contains 0.8 cm³ resin/cm³ bed. The density of the resin is 1.3 g/cm³, and the feed solution contains 5 g/l of the antibiotic. If the superficial velocity of liquid is 1.5 m/h and the overall mass transfer coefficient is 15 h⁻¹, determine the height of the column for an effluent antibiotic concentration of $C_L = 0.2$ g/l. Assume an equilibrium relationship of

$$C_S = 25(C_L^*)^{1/2}$$

where C_S is g solute/l resin and C_L is g solute/l solution. The ratio of volumetric flow rate of broth to resin is 10.

Solution Equation 11.47 can be written as

$$-\frac{U}{K_a(1-\varepsilon)}\int_{C_L}^{C_{L0}}\frac{dC_L}{C_L-C_L^*}=H$$

Using the equilibrium relationship, we obtain

$$C_L^* = \frac{1}{625}C_S^2$$

and by using the operating-line relationship, which is $F/B = 10$, we note that $C_S = 10C_L$.

$$C_L^* = \frac{1}{625}100C_L^2 = \frac{1}{6.25}C_L^2 = 0.16C_L^2$$

Substituting this equation into eq. 11.47 yields

$$-\frac{U}{K_a(1-\varepsilon)}\int_{C_L}^{C_{L0}}\frac{dC_L}{C_L-0.16C_L^2}=H$$

Integration with the numerical values yields

$$-\frac{1.5 \text{ m/h}}{(15 \text{ h}^{-1})(0.8)}\int_{0.2}^{5}\frac{dC_L}{C_L-0.16C_L^2}=H$$

$H = 0.6$ m.

11.4.5. Dialysis

Dialysis is a membrane separation operation used for the removal of low-MW solutes such as organic acids ($100 < \text{MW} < 500$) and inorganic ions ($10 < \text{MW} < 100$) from a solution. A well-known example is the use of dialysis membranes to remove urea ($\text{MW} = 60$) from urine in artificial kidney (dialysis) devices. In biotechnology dialysis can be used to remove salts from a protein solution, which is often a step in resolubilizing proteins that were initially in inclusion bodies. A schematic of a dialysis membrane is depicted in Fig. 11.18. Note that the membrane is selective. The dialysis membrane separates two phases containing low-MW and high-MW solutions. Since the MW cutoff of a dialysis membrane is very small, low-MW solutes move from a high- to a low-concentration region. At equilibrium, the chemical potentials of diffusing compounds on both sides of a membrane are equal.

$$\mu_1^\alpha = \mu_1^\beta \tag{11.51}$$

where μ is the chemical potential of the diffusing compound. In terms of concentrations, eq. 11.51 can be written as

$$RT\ln C_1^\alpha \; \gamma_1^\alpha = RT\ln C_1^\beta \; \gamma_1^\beta \tag{11.52}$$

or

$$C_1^\alpha \; \gamma_1^\alpha = C_1^\beta \; \gamma_1^\beta \tag{11.53}$$

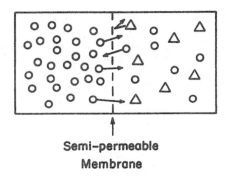

Figure 11.18. A typical dialysis membrane separation. Low-MW component 1 (○) diffuses through membrane from high to low concentration region. High-MW component (△) cannot pass.

Semi-permeable Membrane

where C_1 is the concentration and γ_1 is the activity coefficient of diffusing component 1. For ideal (very dilute) solutions, $\gamma_1 \approx 1$ and

$$C_1^\alpha = C_1^\beta \qquad (11.54)$$

The concentrations represented in eqs. 11.51 through 11.54 are concentrations of dissolved (unbound) component 1.

This dialysis equilibrium is based on the assumption of uncharged solute molecules. If macromolecules are polyelectrolytes, such as proteins and nucleic acids, then the charge equilibrium of the macromolecules needs to be considered. This equilibrium is known as the Donnan equilibrium.

11.4.6. Reverse Osmosis

For fermentation broths, osmosis is the transport of water molecules from a high- to a low-concentration region (i.e., from a pure water phase to a salt-containing aqueous phase) when these two phases are separated by a selective membrane. The water passes the membrane easily, while the salt does not. At equilibrium, the chemical potential of water must be the same on both sides of the membrane. As the water passes into the salt solution, its pressure increases. This *osmotic pressure* can be expressed by

$$\pi = CRT(1 + B_2 C + B_3 C^2 + \cdots) \qquad (11.55)$$

where C is the concentration of the solute, T is temperature, R is the gas constant, and B_2, B_3 are the virial coefficients for the solute. For very dilute, ideal solutions, $B_2 = B_3 = 0$, and

$$\pi = CRT \qquad (11.56)$$

In reverse osmosis (RO), a pressure is applied onto a salt-containing phase, which drives water (solvent) molecules from a low- to a high-concentration region and results in the concentration of solute (salt) molecules on one side of the membrane. The pressure required to move solvent from a low- to high-concentration phase is equal to or slightly larger than the osmotic pressure. When $\Delta p > \pi$, a solvent flux takes place in the direction against the concentration gradient (see Fig. 11.19).

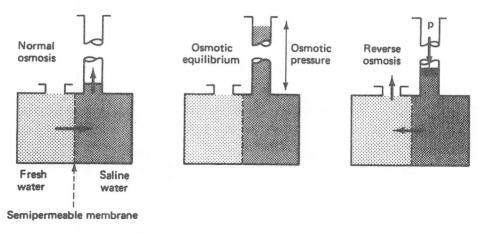

Figure 11.19. Osmotic flows across a membrane separating fresh water and saline water. (With permission, from D. W. Sundstrom and H. E. Klei, *Wastewater Treatment*, Pearson Education, Upper Saddle River, NJ, 1979, p. 281.)

In some reverse osmosis applications, membranes may allow the passage of solute molecules along with solvent. Solute transport can be by diffusion or convection. A *reflection coefficient* (σ) for each solute can be defined as the fraction of solute molecules retained on one side of the membrane in the presence of a solvent flux. Therefore, for σ = 0, complete solute passage is obtained, and for σ = 1, no solute passage is achieved (that is, perfect reflection). Solvent and solute fluxes in reverse osmosis can be expressed as

$$\text{Solvent:} \quad N_1 = K_p(\Delta P - \pi) \tag{11.57}$$

$$\text{Solute:} \quad N_2 = C(1-\sigma)N_1 + K_{p'}\,\Delta C \tag{11.58}$$

where K_p and $K_{p'}$ are permeability coefficients for solvent and solute, respectively, C is the average solute concentration in solution, and ΔC is the solute concentration difference across the membrane.

The magnitude of the required pressure varies with the concentration of solutions. Pressures on the order of 30 to 40 atm are required for a 0.6 M salt solution. The salt level in fermentation fluids may be much higher than 0.6 M, requiring high pressures for RO separations. The applications of RO in bioseparations are limited, since the method requires high pressures and is based on solvent removal. RO membranes are usually used for dewatering and concentration purposes, but not for protein separations.

Another problem with RO membranes is the deposition of solute molecules on membrane surfaces, resulting in large resistances for solvent flow. This phenomenon, known as *concentration polarization,* can be overcome by increasing the degree of turbulence on the membrane surface.

The osmotic pressure for multicomponent systems is equal to the sum of the individual osmotic pressures:

$$\pi = \sum_i \pi_i = \sum_i C_i RT(1 + B_{2i}C_i + B_{3i}C_i^2 + \cdots) \tag{11.59}$$

The solvent and solute fluxes in this case are

$$N_1 = K_{pl}(\Delta P - \pi) \tag{11.60}$$

$$N_i = C_i(1 - \sigma_i)N_1 + K'_{pi}\Delta C_i; \quad i \geq 2 \tag{11.61}$$

11.4.7. Ultrafiltration and Microfiltration

Membranes are widely used to separate solute molecules such as proteins on the basis of their size and to concentrate cells from fermentation broth. Membrane separations may be used to separate proteins after precipitation to remove many of the contaminants. Membranes can serve as a molecular sieve to separate solute molecules of different molecular size. Depending on molecular-size cutoff, different membranes can be used for the separation of different MW proteins. *Microfiltration* or *microporous filtration* (MF) is used to separate species, such as bacteria and yeast, that range from 0.1 to 10 μm in width. Ultrafilters are used for macromolecules with a molecule-weight range of 2000 to 500,000. All these membrane sieving methods (microfiltration, ultrafiltration, reverse osmosis) are based on the same driving force, namely pressure, but have some minor differences. Some *ultrafiltration* (UF) membranes have *anisotropic structure*. In an anisotropic membrane, a thin skin with small pores is formed on top of a thick, highly porous structure. The thin layer provides selectivity, while the thicker layer provides mechanical support. With newer membrane materials available, anisotropic membranes are used less frequently. *Microporous filters* are usually isotropic and may have an open, tortuous path structure, which causes particle entrapment within the filter, or may have well-defined pores of uniform size.

While most membranes in use are made from polymers, ceramic membranes have increased in popularity. Factors that affect choice of membrane materials are interactions with proteins, mechanical stability, chemical stability (especially to cleaning agents), biocompatibility, flux rates, ease of sterilization (e.g., thermal stability), and cost. Typical polymeric materials are cellulose acetate, nylon, polytetrafluoroethylene (PTFE), polyvinyldine difluroide (PVDF), and polysulfone. Ceramic membranes have a high level of chemical resistance, can be steam sterilized, and have a much longer life time (ca. 10 yr) versus many polymeric membranes. Such membranes have high initial cost and, while mechanically strong, are sometimes brittle. The actual choice of membrane depends on the application, the composition of the feed stream, and required product characteristics.

UF and MF membranes are widely used in the pharmaceutical, chemical, and food industries for the separation of vaccines, fermentation products, enzymes, and other proteins. Ultrafiltration is an energy-efficient, economical separation method used to concentrate chemicals and biologicals at a high degree of purity.

Typical UF and MF operations are pressure-driven processes in which low-MW solutes and water pass through the filter and high-MW solutes are retained on the membrane surface. Therefore, a concentration gradient builds up between the surface of the membrane and the bulk fluid. This gradient results in concentration polarization. As a result, solute diffuses back from the membrane surface to the solution (Fig. 11.20). At steady state, the rate of convective transfer of solute toward membrane is equal to the rate

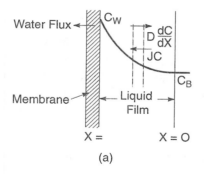

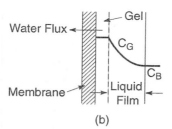

Figure 11.20. Solute transfer in a UF membrane: (a) without gel formation; (b) with gel formation.

of diffusion of solute in the opposite direction because of concentration polarization. That is,

$$D_e \frac{dC}{dX} = JC \tag{11.62}$$

where D_e is the effective diffusivity of solute in liquid film (cm²/s), J is the volumetric filtration flux of liquid (cm³/cm² s), and C is the concentration of solute (mol/cm³ liquid).

Integration of eq. 11.62 with the boundary conditions of $C = C_B$ at $X = 0$ and $C = C_W$ at $X = \delta$ yields

$$J = \frac{D_e}{\delta} \ln \frac{C_w}{C_B} \tag{11.63}$$

or

$$J = k \ln \frac{C_w}{C_B} \tag{11.64}$$

where $k = D_e/\delta$ is the mass transfer coefficient and δ is film thickness.

The mass transfer coefficient is a function of fluid and solute properties and flow conditions and is correlated with Re (Reynolds number) and Sc (Schmidt number).

$$\text{Sh} = \frac{dk}{D_e} = a\,\text{Re}^b\,\text{Sc}^b \tag{11.65}$$

where $\text{Re} = dv\rho/\mu$, $\text{Sc} = \mu/\rho D_e$, and $\text{Sh} = kd/D_e$, where Sh is the Sherwood number. The value of a is $\frac{1}{3}$ according to the boundary-layer theory; b is approximately 0.5 for laminar flow and 1.0 for turbulent flow. That is, $k \propto v^{0.5}$ for laminar and $k \propto v$ for turbulent flow.

A gel layer is formed on the surface of the UF membrane when slowly diffusing large macromolecules accumulate at the surface. A gel layer forms if the protein concentration in the solution exceeds 0.1%. For biological fluids, protein accumulation is usually dominant (Fig. 11.20). The protein concentration in the gel, C_G, is the maximum value of C_W. In this case, the liquid flux through the filter is given by

$$J = k\ln\frac{C_G}{C_B} \tag{11.66}$$

Gel formation depends on the nature and concentration of the solute, pH, and pressure. Once gel is formed, C_G becomes constant, and liquid flux decreases logarithmically with increasing solute concentration in the bulk liquid. The gel layer causes a hydraulic resistance against flow and acts somewhat like a second membrane. Figure 11.21 depicts the variation of liquid flux with the log of solute concentration in the absence and presence of gel formation.

11.4.8 Cross-flow Ultrafiltration and Microfiltration

Gel formation can be partially eliminated by *cross-flow filtration*, where pressure is not applied directly perpendicular to the membrane, but parallel to the membrane surface. This process is also called *tangential flow filtration*. That is, fluid flows parallel to the membrane surface and passes through the membrane, leaving solutes in a liquid phase above the membrane. Mechanical agitation or vibration of the membrane surface can also be used to alleviate gel formation.

A schematic of cross-flow filtration is shown in Fig. 11.22. The pressure drop driving fluid flow is

$$\Delta P = P_i - P_0 \tag{11.67}$$

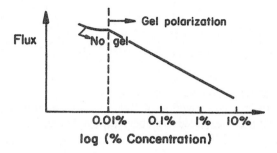

Figure 11.21. Variation of flux with solute concentration.

Recovery and Purification of Products   Chap. 11

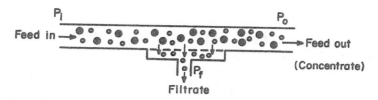

Figure 11.22. Schematic of cross-flow filtration.

For laminar flow, using the Hagen–Poiseuille equation, we can derive the following:

$$P_i - P_0 = \Delta P = \frac{C_1 \mu L V}{d^2} = \frac{C_2 \mu L Q}{d^4} \tag{11.68}$$

where L is the length of the tube, μ is fluid viscosity, Q is the volumetric flow rate of the liquid, and d is diameter of the tube.

For turbulent flow, the following equation is used for ΔP:

$$P_i - P_0 = \Delta P = \frac{C_3 \mu L V^2}{d} = \frac{C_4 f L Q^2}{d^5} \tag{11.69}$$

where f is the Fanning friction factor, which is a function of Re. For cross-flow filtration, turbulent flow is desired.

The average transmembrane pressure drop is

$$\Delta P_M = \frac{P_i + P_0}{2} - P_f \tag{11.70}$$

where P_f is the filtrate pressure, which is usually near atmospheric pressure.

Assuming that $P_f = P_{atm}$ or P_f is zero gauge pressure, we can relate ΔP_M to ΔP.

$$\Delta P_M = P_i - \tfrac{1}{2}\Delta P \tag{11.71}$$

High inlet pressure and low fluid velocities need to be used to obtain high ΔP_M.

The filtration flux (J), as a function of transmembrane pressure drop, is given by

$$J = \frac{\Delta P_M}{R_G + R_M} \tag{11.72}$$

where R_G and R_M are gel and membrane resistances, respectively. R_M is constant and R_G varies with the solute concentration and the tangential velocity across the membrane, which can retard or eliminate gel formation. Also, the filtration flux (J) is a function of fluid velocity, as described by eqs. 11.71 and 11.72. Usually, there is an optimal fluid velocity range maximizing the filtration rate. At low velocities, the mass transfer coefficient (k) is low, resulting in high gel resistance (R_G) and low filtrate flux. At high fluid velocities, ΔP is high, resulting in low ΔP_M and therefore low J values. Also, there is an optimal value of ΔP_M resulting in maximum flux. This is because of a maximum-pressure limitation on P_i. That is, the maximum value of ΔP_M is P_i, and P_i is limited by the physical properties of the membrane. With modern membranes, especially the new ceramic membranes, it may be possible to push P_i to high levels. Therefore, at low ΔP_M values, flux increases with ΔP_M; however, at high ΔP_M values, filtration flux drops as a result of the decrease in velocity. Figure 11.23 describes optimal values of ΔP_M resulting in the maximum UF or MF rate for various solute concentrations.

In the absence of gel formation, the filtration flux increases linearly with ΔP_M. Figure 11.24 depicts the variation of filtration flux (J) with ΔP_M. At low ΔP_M values, J increases linearly with ΔP_M because of the absence of gel polarization. However, at high ΔP_M, gel formation takes place, and gel resistance (R_G) increases with increasing ΔP_M, resulting in a constant filtration flux (J) over a large range of high ΔP_M values. At higher solute concentrations, flux levels off at lower ΔP_M values.

The *rejection coefficient* of an ultrafilter is defined as

$$R = \frac{C_B - C_F}{C_B} = 1 - \frac{C_F}{C_B} \qquad (11.73)$$

where C_F is the concentration of the solute in the filtrate. When $C_F = 0$, only water passes through the filter and $R = 1$, which is complete solute rejection. If $C_F = C_B$, complete solute transfer to the filtrate takes place and $R = 0$ (no rejection). Usually, $0 < R < 1$ and is closer to 1 (i.e., $R \approx 0.95$ or 0.98). The value of R is a measure of the selectivity of the membrane for certain solutes. Selective separations of various compounds can be achieved using membranes with the right molecular-weight cutoff. Figure 11.25 depicts the variation of the rejection coefficient (R) with MW of the solute. For MW $> 10^5$ and MW $< 10^3$, $R = 1$ and $R = 0$, respectively. That is, the membrane allows complete passage

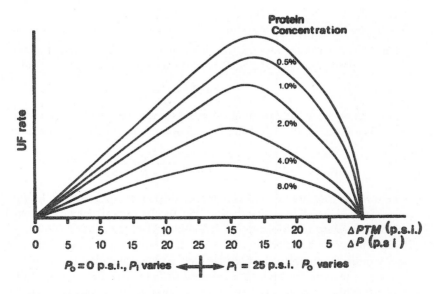

Figure 11.23. Filtration flux optimized as a function of transmembrane pressure at varying solute concentrations and fixed maximum inlet pressure. (With permission, from R. S. Tutunjian, in M. Moo-Young, ed., *Comprehensive Biotechnology*, Vol. 2, Elsevier Science, London, 1985.)

Recovery and Purification of Products   Chap. 11

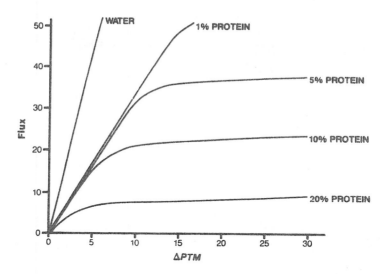

Figure 11.24. Filtration flux as a function of transmembrane pressure at varying solute concentrations. (With permission, from R. S. Tutunjian, in M. Moo-Young, ed., *Comprehensive Biotechnology*, Vol. 2, Elsevier Science, London, 1985.)

of compounds with MW $< 10^3$, retains compounds with MW $> 10^5$, and partially retains compounds of $10^3 <$ MW $< 10^5$.

Three major membrane configurations are depicted in Fig. 11.26: hollow fibers, flat sheets, and spiral-wound cartridges. Fiber cartridges provide a large surface-to-volume ratio. However, they plug more easily than the other two configurations. Flat-plate configurations are easy to construct and allow easy membrane replacement. The channel width can be altered to reduce plugging problems. Membrane support must be added to allow operation at higher pressures. Flat-plate systems have a low membrane surface-to-volume ratio. Spiral cartridges contain rolled membranes and are essentially flat-plate systems configured to increase the surface-to-volume ratio. Most UF membranes operate with $\Delta P < 5$ to 7 bars and MF membranes at a slightly lower ΔP.

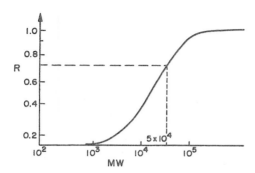

Figure 11.25. Rejection coefficient as a function of the molecular weight of the solute.

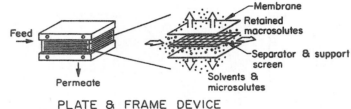

PLATE & FRAME DEVICE

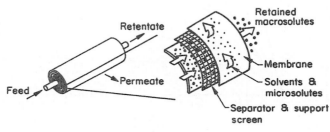

SPIRAL CARTRIDGE

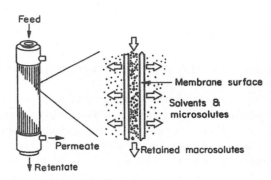

HOLLOW FIBER CARTRIDGE

Figure 11.26. Examples of different membrane configurations. (Adapted with permission, from R. S. Tutunjian in M. Moo-Young, ed., *Comprehensive Biotechnology*, Vol. 2, Elsevier Science, London, 1985.)

Three major applications of ultrafiltration are concentration, purification, and diafiltration. In concentration, water is removed from the aqueous solution of solute using an UF membrane, and solute is concentrated. In purification, solvent and low-MW products are separated from high-MW compounds using a UF membrane. The product contains low-MW compounds and the solvent. In diafiltration, low-MW solutes such as salts, sugars, and alcohols pass through the filter, and the retained stream contains the product. The permeate that leaves the system is replaced with deionized water.

Ultrafiltration membranes are used for the separation of proteins, enzymes, pyrogens (lipopolysaccharides from bacterial cell wall), cell debris, and viruses from fermentation media. Ultrafiltration is usually the second step in protein recovery after protein precipitation. UF separations operate under mild conditions ($T = 20°$ to $30°C$, $P = 5$ to 50 psi) and are energy efficient, inexpensive, nondestructive, and easy to operate.

Microfiltration is used primarily to concentrate bacterial and yeast suspensions and provide a cell-free supernatant. Cross-flow filtration and improved membranes have been key to making this process viable. In many applications it has replaced centrifugation. Initial devices were primarily open plate and frame, owing to possible blockages, although hollow-fiber systems are now available. Application to animal cells, which are mechanically fragile, is possible but difficult, owing to high levels of hydrodynamic sheer that can break cells.

11.4.9. Chromatography

Chromatography separates mixtures into components by passing a fluid mixture through a bed of adsorbent material. We will be interested primarily in *elution chromatography*. Typically a column is packed with adsorbent particles, which may be solid, a porous solid, a gel, or a liquid phase immobilized in or on a solid. A *mobile phase* or fluid phase with a mixture of solutes is injected. This pulse is followed by a solvent or eluent. The pulse enters as a narrow concentrated peak, but exits dispersed and diluted by additional solvent. Different solutes in the mixture interact differently with the adsorbent material (*stationary phase*); some interact weakly and some interact strongly. Solutes that interact weakly with the matrix pass out of the column rapidly (see Fig. 11.27), while those that interact strongly with the matrix exit slowly. These differential rates of migration separate the

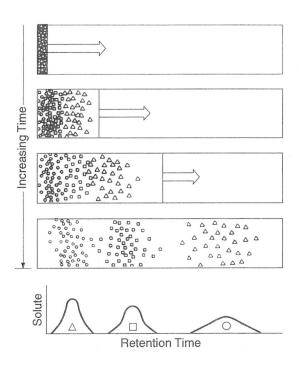

Figure 11.27. Concentrations in elution chromatography. Three solutes, shown schematically as circles, squares, and triangles, are injected into one end of a packed bed. When solvent flows through the bed from left to right, the three solutes move at different rates because of different adsorption. They exit at different times, and hence are separated. (With permission, from P. A. Belter, E. L. Cussler, and W. S. Hu, *Bioseparations: Downstream Processing for Biotechnology*, John Wiley & Sons, New York, 1988, p. 184.)

solutes into different peaks that exit at a characteristic retention time (which will change if operating conditions are altered).

Elution chromatography is similiar to fixed-bed adsorption (Section 11.4.4). However, in fixed-bed adsorption (sometimes called frontal chromatography) the solute is captured in the adsorbent and then eluted as a concentrate. In elution chromatography you purify the product even while it is being diluted.

When large amounts of material are being processed, *displacement chromatography* may be attractive. In this method the column is subjected to sequential step changes in inlet conditions (e.g., nature of solvent). In this method the feed mixture is introduced, followed by a constant infusion of displacer solution. The displacer must have a higher affinity for the stationary phase than any compound in the feed solution. The displacer "pushes" solute off the stationary phase and back into the mobile phase. If conditions are chosen correctly, the feed components are forced into adjacent square-wave-like zones of concentrated, pure solutes. These zones then break through the end of the column with the zone having the solute with the lowest affinity for the stationary phase exiting first. The primary advantage of displacement chromatography over elution chromatography is the potential for higher throughput, but operation is more difficult, and high *resolution* (separation of solutes) can be difficult to obtain in some situations.

These chromatographic processes are almost always run as batch operations, although schemes for continuous or semicontinuous operation have been proposed. For the remainder of our discussion we focus on elution chromatography as a batch process. Some important types of chromatographic methods are:

1. *Adsorption chromatography* (ADC) is based on the adsorption of solute molecules onto solid particles, such as alumina and silica gel, by weak van der Waals forces and steric interactions.

2. *Liquid–liquid partition chromatography* (LLC) is based on the different partition coefficients (solubility) of solute molecules between an adsorbed liquid phase and passing solution. Often the adsorbed liquid is nonpolar.

3. *Ion-exchange chromatography* (IEC) is based on the adsorption of ions (or electrically charged compounds) on ion-exchange resin particles by electrostatic forces.

4. *Gel-filtration (molecular sieving) chromatography* is based on the penetration of solute molecules into small pores of packing particles on the basis of molecular size and the shape of the solute molecules. It is also known as size exclusion chromatography.

5. *Affinity chromatography* (AFC) is based on the specific chemical interactions between solute molecules and ligands (a functional molecule covalently linked to a support particle). Ligand–solute interaction is very specific, such as enzyme–substrate interaction, which may depend on covalent, ionic forces or hydrogen-bond formation. Affinity binding may be molecular size and shape specific.

6. *Hydrophobic chromatography* (HC) is based on hydrophobic interactions between solute molecules (e.g., proteins) and functional groups (e.g., alkyl residues) on support particles. This method is a type of *reverse phase chromatography* which requires that the stationary phase is less polar than the mobile phase.

7. *High-pressure liquid chromatography* (HPLC) is based on the general principles of chromatography, the only difference being high liquid pressure applied to the

packed column. Owing to high-pressure liquid (high liquid flow rate) and dense column packing, HPLC provides fast and high resolution of solute molecules.

The choice of the stationary phase and consequently the type of chromatography depends on the nature of the solutes and process goals. Ion-exchange chromatography is widely used, particularly for recovery of proteins. This method offers good resolution of peaks, high capacity, and good speed. Hydrophobic chromatography and LLC share many of these attributes, but are best applied when the aqueous phase is at high ionic strength. Also organic solvents may be needed, which may denature proteins or create environmental problems. Adsorption chromatography is often relatively inexpensive, but resolution is often not very sharp. Gel-filtration chromatography is good for buffer exchange and desalting and offers decent resolution. Because gels are compressible, capacity and throughput are often low; new rigid packings for size exclusion chromatography allow faster flow rates. Affinity methods offer the possibility of very high selectivity and good capacity and speed. However, such methods can be extremely expensive, especially if based on antibodies. For real bioprocesses (e.g., recovery of proteins) multiple types of chromatography are used, since greater purity is possible if more than one basis of separation is used. Ion-exchange and affinity chromatography are the most widely used methods to recover proteins from bioprocesses.

A single chromatographic separation process may include more than one of these mechanisms. For example, adsorption chromatography using silica gel functions properly when the water content of silica gel is less than 15%. The system may operate as liquid–liquid partition chromatography when the water content exceeds 30%. In some cases, electrostatic and weak forces may be simultaneously functional.

A typical chromatographic column is shown in Fig. 11.28, where a solution is flowing downward in the column and a solute is adsorbed on adsorbent solids by forming a

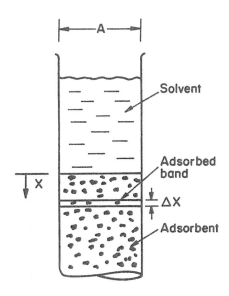

Figure 11.28. Schematic of a chromatography column.

band. The solvent carries the solute downward as it moves along the axis of the column. It is assumed that a rapid equilibrium is established between the solute and adsorbent and that radial gradients are negligible.

Assume that a small volume of solvent, ΔV, is poured through the band, which carries the adsorbed solute downward in the column. A material balance on solute over a differential column height of ΔX yields

$$\begin{pmatrix} \text{total rate of} \\ \text{solute removal} \\ \text{from solvent} \end{pmatrix} = \begin{pmatrix} \text{rate of solute} \\ \text{removal in} \\ \text{void space} \end{pmatrix} + \begin{pmatrix} \text{rate of solute} \\ \text{removal in} \\ \text{solid phase} \end{pmatrix}$$

$$-\left[\left(\frac{\partial C_L}{\partial X}\right)\Delta X\right]\Delta V = \varepsilon A\, \Delta X\left(\frac{\partial C_L}{\partial V}\right)\Delta V + A\,\Delta X\left(\frac{\partial C'_S}{\partial V}\right)\Delta V \qquad (11.74)$$

where C'_S is the amount of adsorbed solute per unit volume of column. Further simplification of eq. 11.74 yields

$$\frac{\partial C_L}{\partial X} + A\left(\varepsilon\frac{\partial C_L}{\partial V} + \frac{\partial C'_S}{\partial V}\right) = 0 \qquad (11.75)$$

In the simplest case, the amount of solute adsorbed per unit volume of column is closely related to the adsorption isotherm of the solute.

$$C'_S = Mf(C_L) \qquad (11.76)$$

where M is amount of adsorbent per unit volume of column (mg/ml) and $f(C_L)$ is the adsorption isotherm or amount of adsorbed solute per unit amount of absorbent, which is a function of C_L. By substituting the derivative of eq. 11.76 into eq. 11.75, we obtain

$$-\frac{\partial C_L}{\partial X} = A[\varepsilon + Mf'(C_L)]\frac{\partial C_L}{\partial V} \qquad (11.77)$$

Substitution of $\partial C_L/\partial X = (\partial C_L/\partial V)(\partial V/\partial X)_c$ into eq. 11.77 yields

$$\left(\frac{\partial V}{\partial X}\right)_c = A[\varepsilon + Mf'(C_L)] \qquad (11.78)$$

Integration of eq. 11.78 from $X = X_0$ to $X = X$ and $V = V_0$ to $V = V$ yields

$$\Delta X = \frac{\Delta V}{A[\varepsilon + Mf'(C_L)]} \qquad (11.79)$$

Since ΔV, A, ε, and M are constants, ΔX is determined by $f'(C_L)$, which is the derivative of $f(C_L)$, or the adsorption isotherm. That is, the location of the solute band formed is mainly determined by the adsorption characteristics of a particular solute on an adsorbent. When the solvent contains two components, I and J, the solutes form two distinct bands at different axial distances, depending on their adsorption characteristics (see Fig. 11.29).

If we define L_I as the distance traveled by solute I, L_S as the distance traveled by solvent above the adsorption column ($\Delta V/A$), and L_C as the penetration distance of the sol-

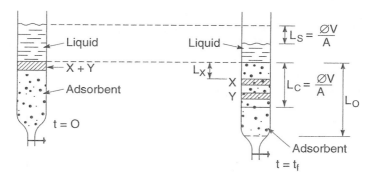

Figure 11.29. Seperation of a binary mixture in a chromatographic column.

vent into the column ($\Delta V/A\varepsilon$), then the resolution (R) or the ratio of the distance traveled by solute I to that of solvent in the space above the column is given by

$$\frac{L_I}{L_S} = R = \frac{1}{\varepsilon + MK} \tag{11.80}$$

where $K = f'(C_L) = df(C_L)/dC_L =$ partition coefficient. Similarly, we can define R_f as

$$R_f = \frac{L_I}{L_C} = \frac{\varepsilon}{\varepsilon + MK} \tag{11.81}$$

where R_f is the ratio of the travel distance of I to that of solvent in the column.

When component I is moved out of the column, then $L_I = L_0$ and $L_S = V_0/A$, where V_0 is the volume of solvent required to displace out of the packed column. Then L_S/L_I becomes

$$V_r = \frac{L_S}{L_I} = \frac{V_0}{L_0 A} = \varepsilon + MK \tag{11.82}$$

where V_r is the ratio of solvent volume to column volume.

This theory of chromatography is applicable to fixed column adsorption, partition, ion-exchange, and affinity chromatography. The theory of *gel-filtration chromatography* is somewhat different. In gel-filtration (molecular sieving) chromatography, solute molecules diffuse into porous structures of support particles, depending on their molecular size and shape. Small solute molecules get into the fine pore structures of the support particles and remain in contact with the solid for a "long" period, while large molecules are adsorbed on the outer surface and remain in contact with the solid for a "short" period. When the column is eluted, the largest molecules appear first in the eluent solvent, and the smallest molecules last. Different bands of solute molecules are obtained on the basis of their size and shape. Consider the case where a buffer with the solutes of interest is added to the column, followed by sufficient washes to elute the original buffer solution.

The total volume of the buffer solution eluted from the column over a certain period of time is

$$V_e = V_0 + K_D V_i \qquad (11.83)$$

where V_e is the total eluent buffer volume, V_0 is the void volume in the column (= $V\varepsilon$), which is the volume of fluid filling the void space outside the gel particles, V_i is the total void volume inside the gel particles, and K_D is the partition coefficient.

For large molecules that do not penetrate inside the gel structure, $K_D = 0$; for small molecules that completely penetrate inside the gel, $K_D = 1$. Equation 11.83 can be rewritten as

$$\frac{V_e}{V_0} = 1 + K_D \frac{V_i}{V_0} \qquad (11.84)$$

For a given solute mixture and gel beads, K_D and the V_i/V_0 ratio are fixed, and the V_e/V_0 ratio is approximately constant, irrespective of column geometry and V_e.

Gel-filtration chromatography can be used to determine the molecular weight of macromolecules. A plot of V_e/V_0 versus log (MW) yields a straight line with a negative slope that is proportional with K_D.

Affinity chromatography is based on the highly specific interaction between solute molecules and ligands attached on polymeric or ceramic beads in a packed column. The concept of affinity chromatography is described in Fig. 11.30. The matrix is usually agarose. However, polyacrylamide, hydroxyethyl methacrylate, cellulose, and porous glass can also be used as the matrix bead. Spacer arms between the matrix and ligand are usually linear aliphatic hydrocarbons. The use of spacer arms may reduce the steric hindrance generated by the matrix. Coupling between the matrix and ligand depends on the functional groups present on the matrix and ligand. Chemically reactive groups on the support matrix usually are —OH,—NH$_2$, or —COOH groups. If the reactive group on the matrix is an —OH group (polysaccharides, glass, hydroxyalkyl methacrylate), then cyanogen bromide (CNBr) is used as a coupling agent. The cyanogen bromide-activated agarose reacts with primary amine groups present in proteins that act as ligands. After the desired solutes are bound to the ligand, elution is achieved by changing the pH or ionic strength in the column. Ligand-solute molecule interactions in affinity chromatography

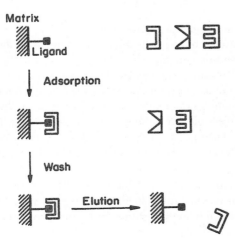

Figure 11.30. Basic principles of affinity chromatographic separations. A suitable ligand is covalently attached to an insoluble matrix. In the adsorption step, only those molecules with a specific binding site for the ligand bind to the adsorbent; molecules without the proper geometric fit pass through unaffected. In the elution step, the bound molecules are disengaged from the column and collected. (With permission, from M. L. Yarmush and C. K. Colton, in M. Moo-Young, ed., *Comprehensive Biotechnology*, Vol. 2, Elsevier Science, London, 1985.)

are very specific, such as enzyme–substrate or antigen–antibody interactions. That is, an enzyme inhibitor or substrate may be used as a ligand in separating a specific enzyme from a mixture. Monoclonal antibodies (MAb) may be used as a ligand to separate specific antigen molecules by affinity chromatography.

Another type of affinity chromatography is *IMAC* (immobilized-metal-affinity chromatography) which exploits the different affinities that solutes (esp. proteins) have for metal ions chelated to a support surface. A very attractive feature is the low cost of metals and the ease of regeneration of the stationary phase, particularly in contrast to affinity chromatography using antibodies. Typically, iminodiacetic acid (IDA) or nitrilotriacetic acid (NTA) linked to an agarose or silica gel support provide attachment sites for a metal ion. A variety of target proteins have been purified using zinc, copper, and nickel chelates. Water molecules will normally solvate metal ions, but in the presence of a strong Lewis base (e.g., histidine residue in a protein) the water molecules can be displaced. The bonds that form between the immobilized metal and the base (protein) result in an adsorbed protein. Factors such as the accessibility, microenvironment of the binding residue (i.e., histidine, cystein, and tryptophan), cooperation between neighboring amino acid side groups, and local conformations play important roles in the strength of binding and thus protein retention. Subsequent destruction of the bond by lowering the pH or adding a competing ligand in the mobile phase results in the elution of the metal binding protein.

Not all proteins naturally bind metals. However, IMAC can still be used if the genetic information for the desired protein is altered to include a *tail* that has high metal-binding affinity. Typically the structural gene encoding the target protein is extended to code for a metal-binding peptide, resulting in production of a fusion protein of the target protein and tail. The amino acid sequence of the tail is based on a sequence of amino acids known to bind metals. A common choice is six adjacent histidines. Peptide libraries have also been used to generate effective affinity tails. The biggest drawback to the method is that for many applications the tail must be removed. The system can be designed with a linker between the tail and target protein that can be cleaved chemically.

Our discussion of the theory of chromatography is based on many simplifying assumptions. Usually the analysis of real columns is more complicated, owing to dispersion, wall effects, and lack of local equilibrium. Wall effects can be reduced by keeping the column-diameter-to-bead-diameter ratio larger than 10 ($D_c/D_b > 10$). Bed compression can occur under some operating conditions and can further complicate analysis of the system by causing flow irregularities such as viscous fingering.

Understanding these hydrodynamical and kinetic effects is also critical to scale-up of chromatographic processes. These effects alter the shape of the peaks exiting the column. In addition to these flow effects, the step controlling the rate of adsorption will alter peak shape. For a porous support internal diffusion, diffusion through the external film, or external dispersion (Taylor dispersion) in laminar flow can control the rate of mass transfer. In the following we consider a more quantitative description of these ideas.

Figure 11.31 depicts the output from a chromatography column. Refering to this figure, we can define some key characteristics of the output. The *resolution* of two adjacent peaks, R_S, is

$$R_S = \frac{t_{\max,j} - t_{\max,i}}{\frac{1}{2}(t_{w,i} + t_{w,j})} \tag{11.85}$$

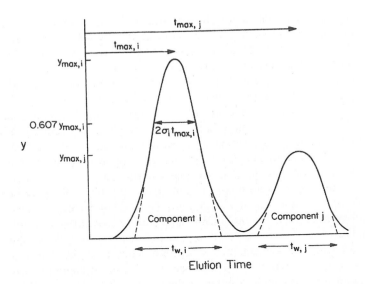

Figure 11.31. Schematic of output from a chromatography column. The relative t_{max} values are used to calculate resolution, while the standard deviation of a peak ($\sigma t_{max,i}$) is used to estimate yields, purity, and capacity and is used in scale-up.

The yield is defined as the amount of a solute, such as i, collected between times t_1 and t_2 divided by the total amount of i injected into the column. Thus,

$$Y_{\Delta t,i} = \frac{\int_{t_1}^{t_2} y_i \, F \, dt}{\int_0^{\infty} y_i \, F \, dt} \tag{11.86}$$

where y_i is the concentration of solute, i, and F is the solvent flow rate.

The value of y_i can be estimated from the peak height and shape if the peak is assumed to be Gaussian. This assumption leads to

$$y_i = y_{max,i} \exp\left[-\frac{(t - t_{max,i})^2}{2(\sigma t_{max,i})^2}\right] \tag{11.87}$$

where $t_{max,i}$ is the time at which the maximum amount of i elutes, t is elution time, and $\sigma t_{max,i}$ is the standard deviation for peak i. Then the yield can be calculated as

$$Y_{\Delta t,i} = \frac{1}{2}\left\{\text{erf}\left[\frac{(t_2 - t_{max,i})}{\sqrt{2}\,\sigma t_{max,i}}\right] - \text{erf}\left[\frac{(t_1 - t_{max,i})}{\sqrt{2}\,\sigma t_{max,i}}\right]\right\} \tag{11.88}$$

where erf is the error function as defined by

Recovery and Purification of Products   Chap. 11

$$\text{erf}(x) = \frac{2}{\sqrt{\pi}} \int_0^x e^{-u^2} du \tag{11.89}$$

The purity is defined as

$$P_{\Delta t,i} = \frac{\int_{t_1}^{t_2} y_i \, F \, dt}{\sum_k \int_{t_1}^{t_2} y_k \, F \, dt} \tag{11.90}$$

or the amount of solute i eluted in Δt divided by the total sum of all solutes eluted in Δt.

The key in the above analysis is σ. The value of σ will depend on dispersion and adsorption kinetics. With a more detailed analysis it can be shown that

$$\sigma^2 = \frac{v}{kal} \tag{11.91}$$

where v is the superficial fluid velocity (i.e., volumetric flow rate divided by column diameter A), ka is the surface "reaction" rate consisting of a reaction rate (k) and available surface area (a), and l is column length. The reader may note that σ^2 in eq. 11.91 equals a flow rate divided by the reaction rate and is effectively a Peclet Number. The value of σ^2 depends on rate controlling step; for example[†]

$$\sigma^2 \propto \frac{vd^2}{l} \qquad \text{(internal diffusion control)} \tag{11.92}$$

$$\sigma^2 \propto \frac{v^{1/2}/d^{3/2}}{l} \qquad \text{(external film control)} \tag{11.93}$$

$$\sigma^2 \propto \frac{vd^2}{Dl} \qquad \text{(Taylor dispersion)} \tag{11.94}$$

In the above equations d is the diameter of the particle used to pack the column and D is the effective diffusion coefficient of the solute.

Consider the problem of scale-up of a chromatography process to handle increasing amount of product. One approach might be to increase solute concentration while using the same column. The danger here is that the packing would become saturated, which would reduce purity and yield. Another approach might be to increase both d and A so as to maintain a constant particle-size-to-bed-diameter ratio. Such an approach maintains flow patterns, but note from eqs. 11.92 to 11.94 that σ would increase if d increased. Higher values of σ indicate broadening of peaks and reduced resolution, yield, and purity. Another approach is to fix d and increase both v and l, but maintain the ratio of v to l constant. While this approach will mathematically keep σ constant, there are practical limitations. Lengthening the column increases pressure drop. Since many packings are soft, they compress further, thereby increasing pressure drop, reducing flow, and altering flow patterns. Some chromatography columns have been made in segments, each less than

[†]See P. A. Belter, E. L. Cussler, and W. S. Hu, *Bioseparations*, John Wiley & Sons, New York, 1988.

30 cm in height; segmentation reduces the bed-compression problem. Another approach is to increase A and volumetric flow to maintain a constant v and hence constant σ. This yields a "pancake" reactor. It is quite workable if the liquid can be applied uniformly over the surface of the column.

Recent advances in the design of packings (particles) for chromatography columns have also assisted in scale-up. Rigid, mechanically strong particles with macropores allow convective flow both around and through particles. The macropores are intimately connected with the micropores, so good mass transfer is maintained. These particles allow fast flow of solvent and solutes and reduce problems of bed compression, allowing column length to be increased.

Example 11.4.

In a chromatographic separation column used for the adsorption of solute A onto an adsorbent solid B, the adsorption isotherm is given by

$$C_S = k_1 C_L^3 = f(C_L)$$

where C_S is mg solute adsorbed/mg adsorbent, C_L is the solute concentration in liquid medium (mg solute/ml liquid), and k_1 is a constant. $k_1 = 0.2$ (mg solute adsorbed/mg adsorbent)/(mg solute/ml liquid)^3. The porosity (void fraction) of the packed column is $\varepsilon = 0.35$. The cross-sectional area of the column is 10 cm^2 and M is 5 g adsorbent per 100 ml column volume. If the volume of the liquid added is $\Delta V = 250$ ml:

a. Determine the position (ΔX) of the solute band in the column when the solute concentration in the liquid phase at equilibrium is $C_L = 5 \times 10^{-2}$ mg/ml.

b. Find the ratio of the travel distance of solute A (L_A) to that of solvent B in the column (R_f) when $C_L = 5 \times 10^{-2}$ mg/ml (i.e., R_f).

Solution

a. $C_L = 5 \times 10^{-2}$ mg/ml

$$f(C_L) = k_1 C^3, \quad f'(C_L) = 3k_1 C_L^2$$

$$f'(C_L) = 3(0.2)(5 \times 10^{-2})^2 = 0.0015$$

$$M = 5\text{g}/100 \text{ ml} = 50 \text{ mg/ml}$$

$$\Delta X = \frac{\Delta V}{A(\varepsilon + Mf'(c))} = \frac{250 \text{ cm}^3}{10 \text{ cm}^2[0.35 + 50(0.0015)]}$$

$$\Delta X = 58.5 \text{ cm}$$

b. $R_f = \dfrac{\varepsilon}{\varepsilon + MK} = \dfrac{0.35}{0.35 + 50(0.0015)}$

$$R_f = 0.824$$

Example 11.5.

Consider an ion-exchange chromatography column used to purify 20 g of a particular protein. At a superficial velocity of 20 cm/h the peak exits the column with y_{max} at 80 min. The standard deviation of the peak is 10 min. Estimate:

a. How long must the column be run to achieve 95% yield?

b. How long must we run if the flow is increased to 40 cm/h and Taylor dispersion controls?

Solution

a. Note that

$$\sigma = \frac{\sigma t_{max}}{t_{max}} = \frac{10 \text{ min}}{80 \text{ min}} = 0.125$$

and from eq. 11.88 the yield where $t_1 = 0^\dagger$ is

$$Y_{\Delta t} = 0.95 = \frac{1}{2}\left[1 + \text{erf}\left(\frac{t - t_{max}}{\sqrt{2}\,\sigma t_{max}}\right)\right]$$

$$= 0.95 = \frac{1}{2}\left[1 + \text{erf}\left(\frac{t - 80}{\sqrt{2}\,10 \text{ min}}\right)\right]$$

or $t = 96.4$ min.

b. With twice the flow t_{max} becomes 40 min. Since for Taylor dispersion $\sigma \propto v^{1/2}$, then

$$\sigma_2 = \sigma_1\left(\frac{v_2}{v_1}\right)^{1/2} = 0.125\left(\frac{40}{20}\right)^{1/2} = 0.177$$

For yield = 0.95 the time required is estimated as

$$0.95 = \frac{1}{2} + \text{erf}\left[\frac{t - 40}{\sqrt{2} \cdot 0.177 \cdot 40}\right]$$

or $t = 51.6$ min.

Notice that in this second case ($v = 40$ cm/h), that the peak is broader and $t_{95\%}/t_{max} = 96.4/80 = 1.205$ for $v = 20$ cm/h and $t_{95\%}/t_{max} = 51.6/40 = 1.29$ for $v = 40$ cm/h.

11.4.10. Electrophoresis

Electrophoresis is used for the separation of charged biomolecules according to their size and charge in an electric field. In an electric field, the drag force on a charged particle is balanced by electrostatic forces when the particle is moving with a constant terminal velocity. A force balance on a charged particle moving with a terminal velocity in an electric field yields the following equation:

$$qE = 3\pi\mu D_P V_t \tag{11.95}$$

or

$$V_t = \frac{qE}{3\pi\mu D_P} \tag{11.96}$$

where q is the charge on the particle, E is electric field intensity, D_P is particle diameter, μ is the viscosity of the fluid, and V_t is the terminal velocity of the particle.

Depending on the pH of the medium, electrostatic charges on protein molecules will be different. When pH > pI, the protein will be negatively charged; and when pH < pI, the

---

†Note that erf $(-x) = -\text{erf } x$ and erf $(-5.9) = -1$.

charge on the protein will be positive. The net charge on the protein will determine the velocity of the protein. When a protein molecule is placed in a pH gradient, the electrophoretic velocity becomes zero when pH = pI, since the net charge on the protein is zero at pH = pI. Precipitation of proteins in a pH gradient at their isoelectric point is known as *isoelectric focusing*. Since the size and charge of protein molecules are different at a given pH, the terminal velocity of the proteins will be different according to eq. 11.96 and, therefore, proteins will be separated from each other in an electric field. Certain gels, such as agar or polyacrylamide, are used for protein separations by gel electrophoresis. Gel electrophoresis is an important analytical separation technique. Scale-up is problematic due to thermal convection resulting from electrical heating.

One analytical and micropreparative version of electrophoresis that has excellent resolution is called two-dimensional protein electrophoresis (2DE). The 2DE procedure is actually the series combination of two electrophoretic separations which resolve protein mixtures based on two independent characteristics—charge and size. Proteins are first separated in a polyacrylamide gel matrix using isoelectric focusing—an equilibrium separation technique that resolves proteins based on their respective isoelectric points. Proteins are subsequently coated with an anionic surfactant and separated by size in another polyacrylamide gel. Finally, proteins are detected using a chemical stain, by autoradiography or by other methods. The result of this two-dimensional separation is a high-resolution fingerprint of protein expression that is characteristic of a particular biological system. This technique has at least two orders of magnitude better resolution than any other analytical tool for protein analysis. Separated protein spots can be cut out of the polyacrylamide gels and subjected to further microchemical analysis to determine the amino acid sequence.

11.4.11. Electrodialysis

Electrodialysis (ED) is a membrane separation method used for the separation of charged molecules from a solution by application of a direct electric current. The membranes contain ion-exchange groups and have a fixed electrical charge. Positively charged membranes (anion membranes) allow the passage of anions and repel cations; negatively charged membranes (cation membranes) allow the passage of cations and repel anions. This method is very effective in the concentration of electrolytes and proteins.

The ED process is driven by electrostatic forces and can be used to transfer salts from low to high concentration. Salt solutions can be concentrated or diluted by this method. The ED method is much faster than dialysis and is a more efficient desalting method.

Ion-exchange membranes (IEM) used in ED units are in essence ion-exchange resins in sheet form. IEMs contain mobile counterions that carry electric current. The cation-exchange membrane consists of polystyrene with negatively charged sulfonate groups bonded to phenyl groups in polystyrene. Anion membranes contain positively charged quaternary ammonium groups ($—N_3R^+$) chemically bonded to the phenyl groups in the polystyrene. Mobile counterions account for the low electrical resistance of the membranes. To control swelling of sulfonated polystyrene, some cross-linking agents such as divinyl benzene are included in the polymer. A typical ED membrane has a pore size of 10 to 20 Å with a capacity of 2 to 3 mEq/g dw resin. Membrane thickness is on the order of 0.1 to 0.6 mm, and electrical resistance is approximately 3 to 30 ohm/cm².

ED units can be used for (1) concentration and dilution of salts, (2) ion substitution, and (3) electrolysis. A schematic of an ED unit used for concentration and dilution of salts is depicted in Fig. 11.32. The unit contains a number of diluting ($D*$) and concentrating ($C*$) compartments separated by alternating anion–cation membranes and two electrodes, the cathode and anode. ED units can also be used for ion-substitution purposes to change the ionic composition of a process stream. If the ionic liquid stream is passed between two membranes of the same charge (two cation or two anion exchange membranes), ions will be transferred from the solution in one direction through one membrane, and equivalent ions of the same charge will be transferred into the solution through the opposite membrane from a makeup solution on the other side of the membrane. In electrolytic ED units, electrodes play an important fundamental role in the process. Two electrode compartments are separated by

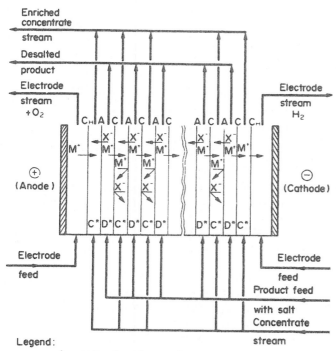

Legend:
 A = anion–transfer membrane
 C = cation–transfer membrane
 C_H = heavy cation transfer membrane
 X^- = anion
 M^+ = cation
 D^* = diluting cell
 C^* = concentrating cell

Figure 11.32. Schematic illustration of concentrating-diluting multicell pair electrodialysis process. (With permission, from S. M. Jain and P. B. Reed, in M. Moo-Young, ed., *Comprehensive Biotechnology*, Vol. 2, Elsevier Science, London, 1985.)

a membrane in electrolytic units. An example of an electrolytic ED unit is a chlor-alkali cell in which an NaCl solution is electrolyzed and converted to NaOH and Cl.

11.5. FINISHING STEPS FOR PURIFICATION

The major finishing steps in fermentation-product purification are crystallization and drying.

11.5.1. Crystallization

Crystallization is usually the last step in producing highly purified products such as antibiotics. Crystallization operates at low temperatures, which minimize thermal degradation of heat-sensitive materials. Operations are conducted at high concentrations and, therefore, unit costs are low and separation factors are high. The determination of optimal crystallization conditions is a matter of empirical experimentation, since phase diagrams and pertinent kinetics for specific systems are usually not available.

High-purity crystals are recovered by using batch Nutsche-type filters or centrifugal filters. The following equation is used for the filtration of crystal slurries, assuming the crystals are noncompressible and the resistance of the filter medium is negligible.

$$\frac{dV}{dt} = K\frac{A\,\Delta P}{\Delta X} \tag{11.97}$$

where ΔX is the thickness of crystal layer, t is time, V is liquid volume, ΔP is pressure drop, A is surface area, and K is the transfer coefficient. Substituting $W \propto V$, where W is the solids concentration in the cake, and integrating eq. 11.97, we obtain

$$t = K\frac{W^2}{\Delta P \cdot A^2} \tag{11.98}$$

A typical centrifugal filter combines a bowl centrifuge with a filter, and the following relationship is used to determine the filtration–crystallization rate for centrifugal filters.

$$\frac{dV}{dt} = \frac{(2\pi N)^2 \rho L(r_0^2 - r_L^2)}{2\mu(\alpha W/A_m^2 + R_M/A_0)} \tag{11.99}$$

where N is the rotational speed, α is the specific cake resistance, W is the solids concentration in the cake, A_m is the log mean area, and R_M is the medium resistance.

The nature and size of crystals affect the centrifugation and washing rates; therefore, crystallization and recovery steps are interrelated and should be considered together in the optimization of crystallization operations. After washing, the crystals are discharged for drying.

11.5.2. Drying

The removal of solvent from purified wet product (crystal or dissolved solute) is usually achieved by drying. In selecting drying conditions, the physical properties of the product, its heat sensitivity, and the desirable final moisture content must be considered. The parameters affecting drying can be classified in four categories: physical properties of the

Recovery and Purification of Products   Chap. 11

solid–liquid system, intrinsic properties of the solute, conditions of the drying environment, and heat-transfer parameters.

The major types of driers used for drying fermentation product are the following:

1. *A vacuum-tray drier* consists of heated shelves in a single chamber and is usually used in pharmaceutical products. This is a good method for small batches of expensive materials, where product loss and heat damage must be minimized.

2. *Freeze drying (lyophilization)* is a method where water is removed by sublimation (from solid ice to vapor) from the frozen solution. The freezing can be accomplished either outside or inside the vacuum chamber prior to drying. This method is used for antibiotics, enzyme solutions, and bacterial suspensions.

3. *Rotary-drum driers* are not good for crystal solutions. Water is removed by heat conduction over a thin film of solution on the steam-heated surface of the rotating drum. The dried product is scraped from the drum with the aid of a knife at the discharge point.

4. *Spray dryers* employ atomization and spraying of product solution into a heated chamber through a nozzle. Hot gas inside the chamber provides the necessary heat for evaporation of the liquid. Dried particles are separated from hot gases using cyclones. Spray dryers are expensive to purchase but are the preferred method for heat-sensitive materials.

5. *Pneumatic conveyor driers* use a hot air stream to suspend and transport particles. The retention time of a particle in the gas stream is short, usually a few seconds. Such systems work well when surface drying is critical, but do not provide sufficient exposure times to dry large porous particles where water removal is diffusion controlled. Pneumatic conveyor systems are well suited for heat-sensitive and easily oxidized materials.

11.6. INTEGRATION OF REACTION AND SEPARATION

The separate optimization of fermentation and recovery does not necessarily yield the optimal process. Traditionally, the strain development, fermentation, and recovery experts worked essentially in isolation. A lack of formal training of engineers in basic biological concepts and of biologists in engineering and process concepts has made an integrated view of the bioprocesses difficult to obtain. Systems have been optimized sequentially, with only modest feedback from downstream to upstream. With improved training of both engineers and biologists, it has become possible to begin to build better processes.

One form of this integration is to try to couple some aspects of recovery and purification with the bioreactor. The motivation for such approaches has come initially from a desire to relieve product inhibition and thereby increase reaction rates and/or allow the use of a more concentrated feed. However, other advantages may accrue, such as improved selectivity for the product of interest, conversion of a primarily intracellular to an extracellular product, and protection of a product from degradation.

Isolated examples of attempts at bioprocess integration stretch back to the beginning of modern biochemical engineering. In the 1970s the energy crisis and interest in ethanol production spawned many suggested approaches to bioprocess integration. Some early approaches to such process integration include vacuum fermentation, membrane processes, addition of solid adsorbents, and liquid extractants.

One example is a reactor system that allows the use of tributylphosphate (TBP) as an extractant of ethanol from fermentation broths. Although TBP had been recognized as

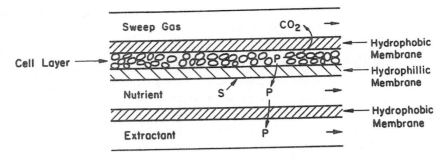

Figure 11.33. Example of a multimember reactor. The extractant fills the pores of the hydrophobic membrane but cannot pass into the nutrient solution if the pressure on the nutrient side is higher than on the extractant side. S is the substrate and P is the product. (With permission, from T. Cho and M. L. Shuler, Biotechnol. Prog. 2:53–60, 1986.)

an attractive solvent for this purpose, its direct use for *in situ* extraction was thought to be impossible due to toxicity. However, further work led to the recognition that a solvent could be toxic due to two features: (1) a chemical or molecular level toxicity due to solvent dissolved in aqueous broth, and (2) a physical or phase toxicity due to direct interaction of yeast with droplets of emulsified solvent. It turned out that TBP's toxicity was solely due to phase toxicity. One way to circumvent this is to immobilize cells in a manner that allows entry of nutrients and exit of product while excluding droplets of solvent. One device that can do this is a *multimembrane bioreactor* (Fig. 11.33).

With a hydrophobic membrane, the solvent will readily fill the pores of the membrane. If the aqueous solution is at a pressure higher than the solvent, but less than the critical entry pressure of the aqueous solution into the membrane pores, then the solvent is effectively immobilized in the membrane. Such a system allows *in situ* solvent extraction of product while preventing the emulsification of solvent and phase toxicity. The primary advantages for this type of bioreactor are the use of more concentrated feeds, relief of feedback inhibition, and reduced distillation or separation costs (e.g., for ethanol recovery).

Not only can this integrated view be expressed through the addition of extracting agents, but it can be expressed at the molecular level. Several groups have worked on protein excretion into the extracellular compartment from *E. coli* using appropriately constructed plasmids. The motivation for this approach has been to greatly simplify recovery and purification, rather than any potential increases in reactor productivity.

It is important for the bioprocess engineer to consider the whole system. The design or choice of culture, reactor configuration, and separation and purification train must be made carefully. Small changes in the upstream process can either greatly simplify or complicate the design of the downstream process.

11.7. SUMMARY

The recovery and purification of products at the end of fermentation processes is an essential step. Product separation can be accomplished either integrated with or following the fermentation step. The major categories of separation of fermentation products are

(1) separation of biomass or other insoluble products, (2) concentration or primary isolation of product, and (3) purification.

When the product is the biomass, the separation strategy is relatively simple and involves filtration or centrifugation following coagulation–flocculation of the cells. Drying is required for a marketable biomass, such as a single-cell protein (SCP) source.

For soluble products such as antibiotics, enzymes, and organic acids, the separation strategy depends on whether the products are extracellular or intracellular. Some products can be both. For intracellular products, the cells need to be disrupted for the release of the products. Mechanical, chemical, and enzymatic methods can be used for cell disruption. Products need to be separated from the other cell and medium components after cell disruption. Depending on the type and the molecular weight of the product, separation methods involve the removal of cell debris by centrifugation or filtration, followed by extraction or precipitation (for protein products) of the product. The separation strategy for extracellular products is simpler than that for intracellular products, since cell disruption is not required. After the separation of the biomass from the fermentation medium, the product is separated from the broth by one or a combination of the following methods: liquid–liquid extraction (antibiotics), two-aqueous-phase extraction (enzymes), adsorption (antibiotics), or precipitation (proteins).

Protein (enzymes, hormones) separations usually require a special strategy. Proteins are usually precipitated from the media either by salting-out or by the addition of organic solvents. The desired protein is crudely separated from other protein components, usually by ultrafiltration on the basis of its molecular weight. Then some chromatographic methods, such as ion-exchange or affinity chromatography, need to be used for further purification. After elution of the protein from the chromatographic column, the product must be dried and crystallized, if necessary, before packaging. Depending on the chemical nature of the desired protein and the composition of the protein mixture, sometimes special protein separation methods, such as isoelectric focusing or electrophoresis (for charged proteins), may need to be used.

Simultaneous separation and fermentation schemes offer special advantages over consecutive schemes, since they may overcome product inhibition and provide a more compact and economical alternative. Membrane, adsorptive, or extractive separation schemes can be utilized for the simultaneous separation of products during fermentation.

SUGGESTIONS FOR FURTHER READING

AIBA, S., A. E. HUMPHREY, AND N. F. MILLIS, *Biochemical Engineering*, 2d ed., Academic Press, New York, 1973.

ATKINSON, B., AND F. MAVITUNA, *Biochemical Engineering and Biotechnology Handbook*, 2d ed., Stockton Press, New York, 1991.

BAILEY, J. E., AND D. F. OLLIS, *Biochemical Engineering Fundamentals*, 2d ed., McGraw-Hill, New York, 1986.

BELTER, P. A., E. L. CUSSLER, AND W. S. HU, *Bioseparations*, John Wiley & Sons, New York, 1988.

BLANCH, H. W., AND D. S. CLARK, *Biochemical Engineering,* Marcel Dekker, New York, 1996.

BOWEN, R., Understanding Flux Patterns in Membrane Processing of Protein Solutions and Suspensions, *Trends Biotechnol.* 11; 451–460, 1993.

HAMEL, J., J. B. HUNTER, AND S. K. SIKDAR, *Downstream Processing and Bioseparation: Recovery and Purification of Biological Products*, American Chemical Society, Washington, DC, 1990.

HARRIS, E. L. V., AND S. ANGAL, *Protein Purification Methods: A Practical Approach*, Oxford University Press, New York, 1989.

FOSTER, D., Cell Disruption: Breaking Up Is Hard to Do, *Bio/Technology 10*: 1539–1541, 1992.

JAIN, S. M., AND P. B. REED, "Electrodialysis," in M. Moo-Young, ed., *Comprehensive Biotechnology*, Vol. 2, Pergamon Press, Elmsford, NY, pp. 575–590, 1985.

KUNDU, A., K. A. BARNTHOUSE, AND S. M. CRAMER, Selective Displacement Chromatography of Proteins, *Biotechnol. Bioeng.* 56:119–129, 1997.

LADISH, M. R., *Bioseparations Engineering: Principles, Practice, and Economics*. Wiley-Interscience, New York, 2001.

TUTUNJIAN, R. S., "Ultrafiltration Processes in Biotechnology," in M. Moo-Young, ed., *Comprehensive Biotechnology*, Vol. 2, Pergamon Press, Elmsford, NY, pp. 411–438, 1985.

WANG, D. I. C., AND OTHERS, *Fermentation and Enzyme Technology*, John Wiley & Sons, New York, 1979.

PROBLEMS

11.1. Yeast cells are recovered from a fermentation broth by using a tubular centrifuge. Sixty percent (60%) of the cells are recovered at a flow rate of 12 l/min with a rotational speed of 4000 rpm. Recovery is inversely proportional to flow rate.
   a. To increase the recovery of cells to 95% at the same flow rate, what should be the rpm of the centrifuge?
   b. At a constant rpm of 4000 rpm, what should be the flow rate to result in 95% cell recovery?

11.2. Gentamycin crystals are filtered through a small test filter medium with a negligible resistance. The following data were obtained:

t (sec)	10	20	30	40
V (l)	0.6	0.78	0.95	1.1

The pressure drop in this test run was 1.8 times that when water was used with a filter area of 100 cm^2. The concentration of gentamycin in solution is 5 g/l. How long would it take to filter 5000 l of gentamycin solution through a filter of 1.5 m^2, assuming the pressure drop is constant and $\mu = 1.2$ centipoise?

11.3. Streptomycin is extracted from the fermentation broth using an organic solvent in a countercurrent staged extraction unit. The distribution coefficient of streptomycin at pH = 4 is $K_D = Y_i/X_i = 40$, and the flow rate of the aqueous (H) phase is $H = 150$ l/min. If only five extraction units are available to reduce the streptomycin concentration from 10 g/l in the aqueous phase to 0.2 g/l, determine the required flow rate of the organic phase (L) in the extraction unit.

11.4. A new antibiotic is separated from a fermentation broth by adsorption on resin beads in a fixed bed. The bed is 5 cm in diameter and contains 0.75 cm^3 resin/cm bed. The overall mass transfer coefficient is 12 h^{-1}. If the antibiotic concentration in the feed is 4 g/l and is desired

umn. The equilibrium relationship is $C_S^* = 20(C_L^*)^{1/2}$, and the operating-line relationship is $C_S = 5C_L$, where C_S is g solute/l resin and C_L is g solute/l solution.

11.5. In a cross-flow ultrafiltration unit, a protein of MW = 3×10^5 da is separated from the fermentation broth by using a UF membrane. The flow rate of liquid through a tube of diameter $d = 2$ cm and length $L = 50$ cm is $Q = 2$ l/min. The flow regime is turbulent, $f = 0.0005$, and $C_4 = 2$ [atm (s/cm)^2]. The inlet pressure is $P_i = 2$ atm. Protein concentrations in the solution and on gel film are $C_B = 30$ mg/l and $C_G = 100$ g/l, respectively.
   a. Determine the exit pressure (P_0).
   b. Determine the transmembrane pressure drop (ΔP_M).
   c. If the mass transfer coefficient (k) for protein flux is $k = 5$ cm/s, determine the flux of liquid through the UF membrane (J).
   d. If the resistance of the filter is $R_M = 0.002$ atm $\cdot$ cm^2 $\cdot$ s/cm^3, determine the cake resistance, R_G.

11.6. Components A and B of a binary mixture are to be separated in a chromatographic column. The adsorption isotherms of these compounds are given by the following equations:

$$m_A = f_A(c) = \frac{k_1 C_A}{k_2 + C_A}$$

$$m_B = f_B(c) = \frac{k_1' C_B}{k_2' + C_B}$$

where $k_1 = 0.2$ mg solute A absorbed/mg adsorbent
   $k_2 = 0.1$ mg solute/ml liquid
   $k_1' = 0.05$ mg solute B adsorbed/mg adsorbent
   $k_2' = 0.02$ mg solute/ml liquid

The bed contains 3 g of very fine support particles. The bed volume is 150 ml, bed porosity is $\varepsilon = 0.35$, and the cross-sectional area of the bed is $A = 6$ cm^2. If the volume of the mixture added is $\Delta V = 50$ ml, determine the following:
   a. Position of each band A and B in the column, L_A and L_B (or ΔX_A and ΔX_B).
   b. L_A/L_B; $R_{fA} = L_A/L_c$; $R_{fB} = L_B/L_c$ when $C_A = 10^{-1}$ mg/ml and $C_B = 0.05$ mg/ml in liquid phase at equilibrium.

11.7. Consider the use of gel chromatography to separate two proteins A and B. The partition coefficient (K_D) for A is 0.5 and for B is 0.15. V_o, the void volume in the column, is 20 cm^3. V_i, the void volume within the gel particles, is 30 cm^3. The total volume of the column is 60 cm^3. The flow rate of elutant is 100 cm^3/h. Ignoring dispersion and other effects, how long will it take for A to exit the column? How long for B?

11.8. Biomass present in a fermentation broth is to be separated by vacuum filtration. Filter and broth characteristics are given below.

$$A = 50 \text{ m}^2, \qquad \Delta P = 0.01 \text{ N/m}^2, \qquad C = 15 \text{ kg/m}^3$$

$$\mu = 0.003 \text{ kg/m-s}, \qquad \alpha = 2 \text{ m/kg}$$

   a. If rate of filtration has a constant value of $dV/dt = 50$ l/min, determine the cake and filter resistances at $t = 30$ min.
   b. Determine the filter surface area (A) required to filter 5000 l broth within 60 min with the same pressure drop across the filter.

11.9. A fermentation broth with a protein concentration of $C_0 = 100$ mg/l and flow rate of $Q = 4$ m³/h is passed through two downflow adsorption columns connected in series. The adsorption isotherm is: $q = 4C^{1/4}$. Assuming the system is in equilibrium, calculate the following:

    a. Minimum amount of activated carbon required for two days of operation if removal efficiency for the column is $E = 50\%$.

    b. Protein concentration in the effluent of the second column is desired to be $C = 0.5$ mg/L. Determine the minimum amount of activated carbon required for two days of operation.

11.10. In a cross-flow ultrafiltration system used for filtration of proteins from a fermentation broth, gel resistance increases with protein concentration according to the following equation:

$$R_G = 0.5 + 0.01(C), \text{ where } C \text{ is in mg/l.}$$

Pressure at the entrance of the system is $P_i = 6$ atm and at the exit is $P_0 = 2$ atm. The shell side of the filter is open to the atmosphere, resulting in $P_f = 1$ atm. The membrane resistance is $R_M = 0.5$ atm/(mg/m² · h), and protein concentration in the broth is $C = 100$ mg/l. Determine:

    a. The pressure drop across the membrane.

    b. Filtration flux.

    c. Rejection coefficient of the membrane for effluent protein concentration of $C_f = 5$ mg/l.

11.11. A solute protein is to be separated from a liquid phase in a chromatographic column. The adsorption isotherm is given by the following equation:

$$C_S = kC_L^2$$

where C_S is the solute concentration in solid phase (mg solute/mg adsorbent) and C_L is the liquid phase concentration of solute (mg solute/ml liquid). Use the following information:

$$k = 0.4, \qquad \varepsilon = 0.3, \qquad A = 25 \text{ cm}^2,$$

$$\mu = 10 \text{ g ads}/100 \text{ ml column} = 100 \text{ mg/ml}$$

    a. For $V = 400$ ml and $X = 25$ cm determine the equilibrium solute concentrations in liquid and solid phases.

    b. Determine the ratio of travel distances of solute to solvent, R_f.

11.12. Consider the scale-up of a chromatography column for purification of a protein. A column of 40 cm length is used with a superficial velocity of 40 cm/h. The peak concentration of the target protein exits at a time of 100 min. The standard deviation of the peak is 14 min.

    a. How long must you wait to collect 90% of the protein?

    b. If the same column is used, but velocity is increased to 60 cm/h and external or Taylor dispersion controls, what will be the value of σ?

    c. If the column is lengthened to 60 cm while the velocity is at 60 cm/h, how will σ change? Will the peak be sharper or broader than at 40 cm/h with a 40-cm column?

12

Bioprocess
Considerations in Using
Animal Cell Cultures

Animal cells have become important catalysts for many bioprocesses. They are capable of making a vast array of important therapeutic proteins. With recombinant DNA techniques, it is possible to achieve significant levels of production of compounds made only in minute amounts in native cells. Animal cells possess the machinery to do complex post-translational modifications to proteins. This capacity is essential in some cases to the formation of a clinically useful product. Let us consider some of the important constraints in using these cells.

12.1. STRUCTURE AND BIOCHEMISTRY OF ANIMAL CELLS

Animal cells vary in size (10 to 30 μm) and shape (spherical, ellipsoidal). In terms of intracellular structures, animal cells are typical eucaryotes, and we discussed their cellular structure in Chapter 2. Figure 12.1 summarizes the structure of a typical animal cell.

Animal cells do not have a cell wall, but are surrounded by a thin and fragile plasma membrane that is composed of protein, lipid, and carbohydrate. This structure results in significant shear sensitivity. In some cells, a portion of the plasma membrane is modified to form a number of projections called *microvilli*. The microvilli increase the surface area of the cell and provide more effective passage of materials across the plasma membrane. The composition of plasma membrane is not uniform and varies in different regions of the

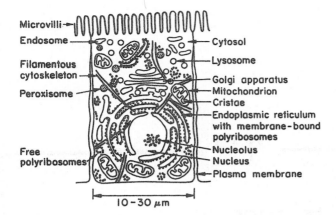

Microvilli
Endosome
Filamentous cytoskeleton
Peroxisome
Free polyribosomes

Cytosol
Lysosome
Golgi apparatus
Mitochondrion
Cristae
Endoplasmic reticulum with membrane-bound polyribosomes
Nucleolus
Nucleus
Plasma membrane

10-30 μm

Figure 12.1. Generalized animal cell. The cytosol, endoplasmic reticulum, Golgi apparatus, endosome, nucleus, lysosome, mitochondrion, and peroxisome are distinct compartments isolated from the rest of the cell by selectively permeable membranes.

membrane. The surface of an animal cell is negatively charged, and cells tend to grow on positively charged surfaces, such as Sephadex or collagen (anchorage-dependent cells). Many cells possess specific cell surface receptors that adhere to ligands on the surface. For example, the binding to collagen may be nonspecific or may be mediated by specific cell surface receptors. The degree of cell adhesiveness is usually greater if attachment is receptor mediated. Some animal cells such as hybridomas are nonanchorage dependent and grow in suspension culture.

Inside the cytoplasm of most animal cells is an extensive network of membrane-bounded channels called the *endoplasmic reticulum* (ER). The membranes of the ER divide the cytoplasm into two phases: the *lumenal* phase (inside the endoplasmic reticulum) and the *cytosol* (outside the rough endoplasmic membrane). Ribosomes are usually located on the outer surface of the endoplasmic reticulum. Some ribosomes are located in cytoplasm and may be interconnected by fine filaments. The ER is critical in protein synthesis and initial stages of posttranslational processing (see Chapter 4).

Mitochondria are the powerhouse of cells where respiration takes place and the bulk of the ATPs are produced. Mitochondria are independent organelles in the cytoplasm containing DNA and are capable of independent reproduction. Each mitochondrion is surrounded by a double membrane: a smooth outer membrane and a highly folded inner membrane called the *cristae*. The mitochondrial matrix often contains crystallike inclusions.

Lysosomes are rather small cytoplasmic organelles bound by a single membrane, and they contain various hydrolytic enzymes, such as proteases, nucleases, and esterases. Lysosomes are responsible for the digestion of certain food particles ingested by the cell.

The *Golgi body* is a cytoplasmic organelle surrounded by a rather irregularly shaped membrane called the *cisternae*. The cisternae of a Golgi body are often stacked together in parallel rows, called *dictyosome*. The Golgi apparatus is responsible for the completion of complex glycosylation and for collecting and secreting extracellular proteins or directing intracellular protein traffic to other organelles.

Some cells contain small cytoplasmic organelles called *peroxisomes* and *glyoxysomes*. These organelles are bounded by a single membrane and contain a number of enzymes, including peroxidases (hydrolysis of H_2O_2) and glyoxalases (glyoxylic acid metabolism).

The nucleus is bounded by two nuclear membranes that form a nuclear envelope. At certain positions, the inner and outer membrane of nucleus fuse to form *pores*. Nuclear pores provide continuity between the cytoplasm and the inner part of the nucleus. The *perinuclear space* (the space between the two membranes of the nucleus) may have access to the outer cell. The nucleus contains chromosomal DNA and also some dark granular structures called *nucleoli*, which can be seen under an electron microscope. Nucleoli are not bound by a membrane and appear to be formed by ribosomal material.

Animal cells have a *cytoskeleton* or system of protein filaments that provide cell mechanical strength, control cell shape, and guide cell movement. These elements are often critical components in controlling cell response to mechanical forces such as from fluid flow or from attachment to surfaces. The three types of filaments are *actin filaments*, *intermediate filaments*, and *microtubules*. Actin filaments are relatively thin, while microtubules have the largest diameter. Microtubules are critical in cell division and cell movement. Cell division and movement require the polymerization and depolymerization of microtubules; the anticancer agent, paclitaxel (Taxol®), works by preventing depolymerization of microtubules and thus "freezing" the cell in a nondividing step. The *centrosome* is the primary microtubule-organizing center in the cell and is a small organelle involved in the formation of spindle poles during mitosis. Some animal cells may contain *cilia*, which are used to transport substrate across the cell surface, but not for self-locomotion. Each cilium is covered by a plasma membrane and contains a specific array of microtubules.

The metabolism of nutrients by animal cells in culture is very different from that in other types of cells. A typical growth medium of an animal cell culture contains glucose, glutamine, nonessential and essential amino acids, serum (horse or calf), and mineral salts (for example, Dulbecco's modified Eagle's media, DME). The major metabolic pathways for animal cells in culture are depicted in Fig. 12.2. Glucose is converted to pyruvate by glycolysis and also is utilized for biomass synthesis through the pentose phosphate pathway. Pyruvate is converted partly to CO_2 and H_2O by the TCA cycle, partly to lactic acid, and partly to fatty acids. Glucose is used as a carbon and energy source, as is glutamine. Part of the glutamine is deaminated to yield ammonium and glutamate, which is converted to other amino acids for biosynthesis purposes. Glutamine also enters into the TCA cycle to yield carbon skeletons for other amino acids and to yield ATP, CO_2, and H_2O. Animal cells are also capable of synthesizing glucose from pyruvate by the gluconeogenesis pathway. The release of lactate and ammonia as waste products of metabolism is a major problem in high-cell-density culture systems. Both lactate and ammonia at high levels are toxic to cells, primarily by altering intracellular and lysosomal pH.

12.2. METHODS USED FOR THE CULTIVATION OF ANIMAL CELLS

The techniques used for cultivation of animal cells differ significantly from those used with bacteria, yeasts, and fungi. Tissues excised from specific organs of animals, such as lung and kidney, under aseptic conditions are transferred into a growth medium containing serum and small amounts of antibiotics in small T-flasks. These cells form a *primary* culture. Unlike plant cells, *primary* mammalian cells do not normally form aggregates, but grow in the form of monolayers on support surfaces such as glass surfaces of flasks.

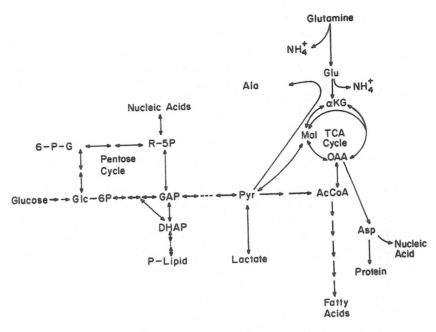

Figure 12.2. Major metabolic pathways in an animal cell. For cultured cells, glucose and glutamine are the major sources of carbon and energy. Lactate and ammonia are the primary waste products, although the release of alanine can be important. Symbols: 6-P-G is 6-P-gluconate; Glc-6P is glucose-6-phosphate; R-5P is ribose-5-phosphate; GAP is glyceraldehyde; DHAP is dihydroxyacetone; PYR is pyruvate; AcCoA is acetyl-CoA; OAA is oxaloacetate; Mal is malate; αKG is α-ketoglutarate.

Using the proteolytic enzyme trypsin, individual cells in a tissue can be separated to form single-cell cultures.

To start cultures of animal cells, excised tissues are cut into small pieces (~2 mm³) and are placed in an agitated flask containing a dilute solution of trypsin (~0.25% w/v) in buffered saline for 120 min at 37°C. The cell suspension is passed through a presterilized filter to clear the solution, and cells are washed in the centrifuge. The cells are then resuspended in growth medium and placed in T-flasks or roller bottles. Cells usually attach onto the glass surface of the bottle and grow to form a monolayer. The cells growing on support surfaces are known as *anchorage-dependent* cells. Surface attachment is necessary for these cells to assume the three-dimensional shape necessary for the alignment of internal structures in a manner allowing growth. However, some cells grow in suspension culture and are known to be *nonanchorage-dependent* cells.

The cells directly derived from excised tissues are known as the primary culture. A cell line obtained from the primary culture is known as the *secondary* culture. Cells are removed from the surfaces of flasks using a solution of EDTA, trypsin, collagenase, or pronase. The exposure time for cell removal is 5 to 30 min at 37°C. After cells are removed from surfaces, serum is added to the culture bottle. The serum-containing suspension is centrifuged, washed with buffered isotonic saline solution, and used to inoculate secondary cultures. Many secondary lines can be adapted to grow in suspension and are nonanchorage dependent.

Most differentiated mammalian cell lines (e.g., human fibroblasts such as WI-38 and MRC-5 that are licensed for human vaccine production) are *mortal*. These cell lines undergo a process called *senescence*. Essentially, the cells, for reasons that are not completely understood, will divide only for a limited number of generations (e.g., about 30 generations for the MRC-5 cells). Cells that can be propagated indefinitely are called *continuous, immortal*, or *transformed* cell lines. Cancer cells are naturally immortal. All cancerous cell lines are transformed, although it is not clear whether all transformed cell lines are cancerous.

Because of the linkage of cancer to cell transformation, the FDA had been reluctant to approve products made from transformed cells. However, transformed cells usually become attachment independent and can be propagated indefinitely in suspension culture, which is highly desirable for large-scale production in bioreactors. A little over a decade ago the FDA began to approve processes for production of products, such as the therapeutic protein, tissue plasminogen activator, from processes using immortalized cells. The approval of such processes has provided a major impetus for development of bioprocesses based on suspension cultures of animal cells.

Table 12.1 provides a summary of differences between nontransformed and transformed cells. One particular characteristic is *contact inhibition*. In two-dimensional culture on a surface nontransformed cells form only a monolayer, as cell division is inhibited when a cell's surface is in contact with other cells. Transformed cells do not "sense" the presence of other cells and keep on dividing to form multilayer structures.

Although mammalian cell lines have been the primary focus of work in animal cell culture, the two are not synonymous. Insect, fish, and crustacean cell cultures are evolving technologies. In particular, insect cell culture is unusually promising for biotechnological purposes. The baculovirus that infects insect cells is an ideal vector for genetic engineering, because it is nonpathogenic to humans and has a very strong promoter that encodes for a protein that is not essential for virus production in cell culture. The insertion of a gene under the control of this promoter can lead to high expression levels (40% of the total protein as the target protein). Most cell lines are derived from ovaries or embryonic tissue, although at least one differentiated cell line (a BTI-EAA blood cell line) has been maintained in indefinite culture for 25 years. Insect cell lines are not transformed but are naturally continuous. In contrast, senescence is observed with many fish cell lines.

TABLE 12.1 Comparison of "Normal" and "Transformed" Cells

Normal	Transformed
1. Anchorage-dependent (except blood cells)	1. Nonanchorage-dependent (i.e., suspension culture possible)
2. Mortal; finite number of divisions	2. Immortal or continuous cell lines
3. Contact inhibition; monolayer culture	3. No contact inhibition; multilayer cultures
4. Dependent on external growth factor signals for proliferation	4. May not need an external source of growth factors
5. Greater retention of differentiated cellular function	5. Typically loss of differentiated cellular function
6. Display typical cell surface receptors	6. Cell surface receptor display may be altered

Another important category is the culture of hybridomas. Hybridoma cells are obtained by fusing lymphocytes (normal blood cells that make antibodies) with myeloma (cancer) cells. Lymphocytes producing antibodies grow slowly and are mortal. After fusion with myeloma cells, hybridomas become immortal, can reproduce indefinitely, and produce antibodies. Using hybridoma cells, highly specific, monoclonal (originating from one cell) antibodies can be produced against specific antigens.

To produce hybridoma cells, animals (mice) are injected (immunized) with certain antigens (see Fig. 12.3). As a reaction to antigens, the animals produce antibodies. The antibody-producing cells (e.g., spleen B lymphocytes) are separated from blood sera and fused with certain tumor cells (e.g., myeloma cells with infinite capacity to proliferate and derived from lymphocytes). The resulting cells are hybrid cells (hybridomas), which secrete highly specific antibodies (monoclonal Ab's) against the immunizing antigen.

Mammalian cell culture to produce proteins other than antibodies usually involves genetically engineered cells. Although many host cells can be used, Chinese hamster ovary (CHO) cells are particularly popular. Most of the vectors for genetic engineering are derived from viruses (usually primate). Some inducible promoters exist. However, expression levels are relatively low, often in the range of 1% to 5% of the total cellular protein.

A typical growth medium for mammalian cells contains serum (5% to 20%), inorganic salts, nitrogen sources, carbon and energy sources, vitamins, trace elements, growth factors, and buffers in water. Serum is a cell-free liquid recovered from blood. Examples are FBS (fetal bovine serum), CS (calf serum), and HS (horse serum). The exact composition of any serum product is not known. However, serum is known to contain amino acids, growth factors, vitamins, certain proteins, hormones, lipids, and minerals. A list of major

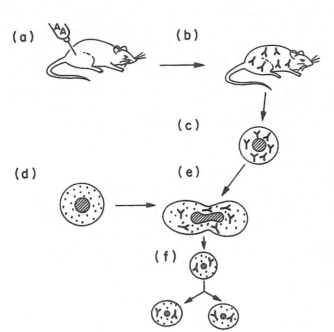

Figure 12.3. Formation of a hybridoma for making a monoclonal antibody. (a) Antigen is injected into a mouse; (b) lymphocytes in the mouse are activated to produce specific antibodies to the antigen; (c) lymphocytes are collected from the mouse; these lymphocytes grow poorly in tissue culture; (d) myeloma (cancer) cells growing in tissue culture are produced; (e) myeloma cells are fused with lymphocytes; (f) the hybrid cell grows well in tissue culture and makes a single monoclonal antibody. Progeny are called hybridomas and can be propagated indefinitely.

serum components and their functions is given in Table 12.2. The major functions of serum are as follows:

1. To stimulate cell growth and other cell activities by hormones and growth factors
2. To enhance cell attachment by certain proteins such as collagen and fibronectin
3. To provide transport proteins carrying hormones, minerals, and lipids

Serum is an expensive medium component ($100 to $500/l) and may cause further complications in the cultivation and downstream separation processes. The demand for serum is increasing rapidly and periodically exceeds supply. The presence of serum proteins and peptides greatly complicates downstream processing. Serum must be filtered sterilized, and contamination with viruses and possibly mycoplasma (a wallless bacterium) are potential problems. Contamination by prions (agents that cause diseases such as "mad cow" disease) is a real concern, and source animals cannot come from regions known to have contaminated animals. Serum-containing medium foams easily. Perhaps most disturbing from a production perspective is the intrinsic variability in serum. No one batch of serum

TABLE 12.2 Important Components of Serum and Their Probable Role in Cell Culture

Component	Probable function
Proteins	
Albumin	Osmoticum and buffer
	Lipid, hormone, mineral carrier
Fetuin	Cell attachment
Fibronectin	Cell attachment
α_2-Macroglobulin	Trypsin inhibitor
Transferrin	Binds iron
Polypeptides	
Endothelial growth factor (ECGF)	Mitogen[a]
Epidermal growth factor (EGF)	Mitogen
Fibroblast growth factor (FGF)	Mitogen
Insulin-like growth factors (IGFI and IGFII)	Mitogen
Platelet-derived growth factor (PDGF)	Mitogen and major growth factor
Hormones	
Hydrocortisone	Promotes attachment and proliferation
Insulin	Promotes uptake of glucose and amino acids
Growth hormones	Mitogen—present in fetal sera
Metabolites and nutrients	
Amino acids	Cell proliferation
Glucose	Cell proliferation
Keto-acids (e.g., pyruvate)	Cell proliferation
Lipids (e.g., cholesterol)	Membrane synthesis
Minerals	
Iron, copper, zinc, and selenium	Enzymes and other constituents
Inhibitors	
γ-globulin	
Bacterial toxins from prior contaminants	
Chalones (tissue-specific inhibitors)	

[a]A mitogen is a substance that signals mitosis to begin.

is ever identical to another. A batch of serum will last only a year before deteriorating. Changes in serum batches require extensive testing.

Because of the aforementioned disadvantages of serum-containing media, low-serum or serum-free media have been developed. Typical compositions of two defined media are listed in Table 12.3. Serum-free medium contains basal salts (inorganic salts, carbon, nitrogen compounds), vitamins, growth factors, and hormones (e.g., insulin, trans-ferrin, hydrocortisone, progesterone, fibronectin). By using a serum-free medium, we can reduce the cost of media, eliminate some potential problems in product purification, im-prove the reproducibility of results, and reduce the chance of contamination. Different cell

TABLE 12.3 Examples of Composition of Serum-Containing and Serum-Free Media

Component	Eagle's Minimum Essential Medium (mg/l)	MCDS 170 (mg/l)
Amino Acids (L-enantiomers)		
Alanine	—	8.9
Arginine-HCl	126	63.3
Asparagine	—	132.0
Aspartic acid	—	13.3
Cysteine-HCl	—	35.2
Cysteine	—	8.5
Cystine	24	—
Glutamic acid	—	14.7
Glutamine	292	292
Glycine	—	7.5
Histidine-HCl-H$_2$O	42	21.0
Isoleucine	52	13.1
Leucine	52	39.3
Lysine-HCl	73	36.6
Methionine	15	4.5
Phenylalanine	33	5.0
Proline	—	5.8
Serine	—	31.5
Threonine	48	35.7
Tryptophan	10	6.1
Tryosine	36	9.1
Valine	47	35.1
Vitamins		
Biotin	—	0.0073
D-Ca pantothenate	1.0	0.48
Choline chloride	1.0	14.0
Folic acid	1.0	—
Folinic acid	—	0.0051
i-inositol	2.0	18.0
DL-α-lipoic acid	—	0.0026
Niacinamide	—	6.1
Nicotinamide	1.0	—
Pyridoxal HCl	1.0	0.06
Riboflavin	0.1	0.11

Component	Eagle's Minimum Essential Medium (mg/l)	MCDS 170 (mg/l)
Thiamin HCl	1.0	0.34
Vitamin B12	—	0.14
Other Organics		
Adenine	—	0.135
D-glucose	1000	1440
Linoleic acid	—	0.028
Putresuine 2HCl	—	1.6×10^{-4}
Na pyruvate	—	110
Thymidine	—	0.073
Major Inorganic Salts		
$CaCl_2$	200	222
KCl	400	225
$HgSO_4$	—	180
$MgSO_4 \cdot 7H_2O$	200	—
NaCl	6800	6960
$NaH_2PO_4 \cdot H_2O$	140	—
NaH_2PO_4	—	71
Trace Elements		
$CuSO_4$	—	1.6×10^{-4}
$FeSO_4$	—	0.75
$Fe(NO_3)_3 \cdot 9H_2O$	0.10	—
H_2SeO_3	—	0.0387
$MnSO_4$	—	7.6×10^{-5}
NaS_1O_3	—	0.050
$(NH_4)_6Mo_7O_{24}$	—	0.00116
NH_4VO_3	—	5.9×10^{-4}
$NiCl_2$	—	6.5×10^{-7}
$SnCl_2$	—	9.5×10^{-7}
$ZnSO_4$	—	0.080
Buffers and Indicators		
HEPES ($C_8H_{18}N_2O_4S$)	—	7140
$NaHCO_3$	2200	—
Phenol red	10	1.31
CO_2 (Gas Phase)	5%	2%
Serum/Serum Replacements		
Serum (e.g., fetal bovine serum, FBS)	5 to 10%	—
Phosphoethanolamine	—	15.9
Ethanolamine	—	6.1
Prostaglandin E1	—	0.01
Hydrocortisone	—	0.05
EGF	—	0.01
Insulin	—	0.5
Transferrin	—	0.5
Prolactin	—	0.5
Ovine protactin	—	0.1

lines require different serum-free media composition. Not all cell lines have been adapted to serum-free media.

A number of defined media have been developed. Eagle's minimal essential medium (MEM), Dulbecco's enriched (modified) Eagle's medium (DMEM), and rather complex media, such as Ham's F12, CMRL 1066, and RPMI 1640, are commonly used. Media such as Eagle's MEM are often supplemented with 5% to 10% (by volume) of serum when used. For serum-free media formulations, often a 1:1 (v/v) mixture of DMEM (nutrient rich) and F12 (rich in trace elements and vitamins) is used. More specialized media (e.g., MCDB 170MDS) may be used for serum-free growth of specific cell lines. A simpler serum-free medium may contain insulin, transferrin, and selenium as serum replacement components, in addition to glucose, glutamine, other amino acids, and salts. Filtered whole lymph from a cow has been used as a growth medium for some mammalian cells.

Mammalian cells grow at 37°C and pH $\approx$ 7.3. Typical doubling times are 12 to 20 h. Usually, 5% CO_2-enriched air is used to buffer the medium pH around pH = 7.3. A carbonate buffer ($HCO_3^{2-}/H_2CO_3^-$) is used to control pH around 7.3. Since bicarbonate is used up by the cells, CO_2-enriched air is provided to balance carbonate equilibrium to keep pH $\approx$ 7.3. The culture medium needs to be gently aerated and agitated. Other buffers such as HEPES (N-[2-hydroxyethyl]piperazine-N'-[2-ethanesulfonic acid]) are also used to keep pH $\approx$ 7.3 $\pm$ 0.1. Insect cells grow best at about 28°C and a pH of about 6.2. Fish cell lines tolerate a wide range of temperature, although temperatures below 37°C are usually preferred (25° to 35°C). Values for pH typical of mammalian cell cultures are usually satisfactory for fish cells; pH values between 7.0 and 7.5 usually give good growth.

The kinetics of the growth of mammalian cells are similar to microbial growth. Usually, the stationary phase is relatively short, and the concentration of viable cells drops sharply thereafter, as a result of the accumulation of toxic metabolic products such as lactate and ammonium. A high level of ammonium is usually the result of glutamine metabolism, and lactate is usually a product of glucose metabolism. Cell concentration reaches a peak value within three to five days. However, product formation, such as monoclonal antibody formation by hybridoma cells, can continue under nongrowth conditions. Most of the products of mammalian cell cultures are mixed-growth associated, and product formation takes place both during the growth phase and after growth ceases. Figure 12.4 depicts a typical variation of growth, product (MAb's) formation, and glucose utilization by hybridoma cells.

The kinetics of product formation (e.g., MAb formation by hybridoma cells) can be described by a Luedeking–Piret equation:

$$\frac{1}{X}\frac{dp}{dt} = q_p = \alpha\mu + \beta \qquad (6.18)$$

where μ is the specific growth rate. The first term in eq. 6.18 is for growth-associated production and the second term is nongrowth-associated. The doubling time of mammalian cells varies between 10 and 50 h, and a typical value is 20 h. As expected, the growth rate varies depending on cell type, medium composition (including growth factors), and other environmental conditions (dissolved oxygen, carbon dioxide levels, pH, ionic strength).

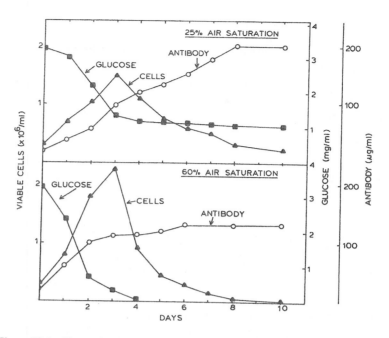

Figure 12.4. Time trajectories of hybridoma growth and monoclonal antibody production at different dissolved-oxygen concentrations in a 3-1 fermenter. (With permission, from S. Reuveay, D. Velez, J. D. Macmillian, and L. Miller, *J. Immunol. Methods* 86:53–59, 1986.)

Animal cells require 0.06 to 0.2×10^{-12} mol O_2/h · cell, which is about five times less than the oxygen requirements of many plant cells and much lower than for microbial cells. Typical animal cell concentration in suspension culture media is 0.1 to 1 g/l (5×10^5 to 5×10^6 cells/ml) for suspension cultures and 10 to 50 times higher than this in immobilized/surface culture systems. For a cell culture density of 10^6 cells/ml, the oxygen requirement is on the order of 0.1 to 0.6 mmol O_2/l · h. For shallow culture depths of 2 to 5 cm, oxygen supplied by air in the head space of the vessel should meet the oxygen requirements of cells. Forced aeration is required for submerged cultivation of denser cultures. However, animal cells are very shear sensitive, and rising air bubbles may cause shear damage to cells, particularly at the point of bubble rupture. Very small bubbles are less damaging. For this reason, special aeration and agitation systems are designed for animal cell cultures. Techniques that decouple agitation (and shear) from oxygenation are particularly attractive. Chemicals (e.g., Pluronic® F-68) can be added to provide shear protection. Since the oxygen requirements of animal cells are relatively low, the oxygen transfer coefficient need not be as high as for bacterial fermentations. Typical values of $k_L a$ for suspension cultures (10^6 cells/ml) are 5 to 25 h^{-1}. Animal cells are usually cultivated in spinner flasks of 0.5 to 10 l in the laboratory. Spinner flasks contain a magnetically driven impeller or spoonlike agitators that operate at 10 to 60 rpm. Aeration is usually by surface aeration using 5% CO_2-enriched and filtered air for mammalian cell lines. Spinner flasks are set on a magnetic stirrer plate in a CO_2 incubator.

Although cell lysis is the most obvious result of excess shear, sublytic levels of shear can be important. For a variety of cell lines, it has been shown that attached cells respond to shear by elongating and reorienting themselves to minimize shear stress. Also, the distribution and numbers of cell surface receptors can be altered. These receptors interact directly with components in the medium, such as growth factors, that regulate cell metabolism. Alterations in the production of specific proteins and the rates of DNA synthesis have been observed as responses to shear. This response to sublytic shear complicates scale-up, since the same cells exposed to the same medium, but at different levels of shear, may produce the product of interest at different rates. Also processes, such as N-linked glycosylation, can be altered by shear, resulting in protein products that differ potentially in therapeutic value.

12.3. BIOREACTOR CONSIDERATIONS FOR ANIMAL CELL CULTURE

Mammalian cells are large (10 to 20 μm diameter), slow growing ($t_d \approx 10$ to 50 h), and very shear sensitive. Moreover, some animal cells are anchorage dependent and must grow on surfaces of glass, specially treated plastics, natural polymers such as collagen, or other support materials; some are not anchorage dependent and can grow in suspension culture. Product concentration (titer) is usually very low (μg/ml), and toxic metabolites such as ammonium and lactate are produced during growth. These properties of animal cells set certain constraints on the design of animal cell bioreactors. Certain common features of these reactors are·the following:

1. The reactor should be gently aerated and agitated. Some mechanically agitated reactors operating at agitation speeds over 20 rpm and bubble-column and airlift reactors operating at high aeration rates may cause shear damage to cells. Shear sensitivity is strain dependent.
2. Well-controlled homogeneous environmental conditions (T, pH, DO, redox potential) and a supply of CO_2-enriched air need to be provided.
3. A large support material surface–volume ratio needs to be provided for anchorage-dependent cells.
4. The removal of toxic products of metabolism, such as lactic acid and ammonium, and the concentration of high-value products, such as MAb's, vaccines, and lymphokines, should be accomplished during cell cultivation.

The laboratory-scale cultivation of animal cells is carried out in (1) T-flasks (25 ml to 100 ml) for anchorage-dependent cell lines and for shallow suspension cultures, (2) spinner flasks (100 ml to 1 l) with paddle-type magnetic agitators, (3) roller bottles (50 ml to 5 l) rotating at about 1 to 5 rpm, and (4) trays containing shallow liquid suspension culture. These laboratory-scale reactors are placed in a carbon dioxide incubator at 37°C (5% CO_2-containing air) for the cultivation of mammalian cells. Different steps need to be taken for anchorage-dependent and suspension cells when scaling up animal cell cultures. An example of these steps for one cell line is presented in Fig. 12.5. Reactors with a high surface–volume ratio (microcarrier systems, hollow-fiber reactors, ceramic

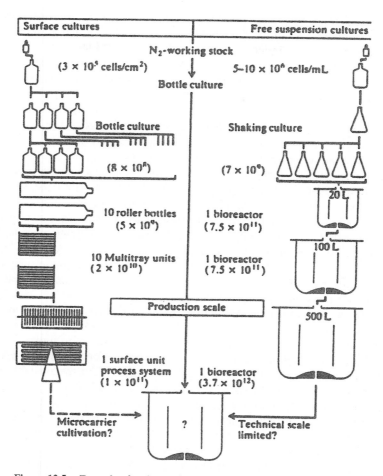

Figure 12.5. Example of scale-up of animal cell cultivation using BHK (baby hamster kidney) 21 cells. Numbers in parentheses denote total amount of cells produced at the indicated stage. (With permission, from H. W. D. Katinger and W. Scheirer, *Acta Biotechnologica* 2:3, 1982.)

matrix systems, weighted porous beads) are used for anchorage-dependent cells. However, some modified stirred reactors and airlift or bubble-column reactors are used for suspension cultures. Membrane bioreactors and microencapsulation methods have been developed for simultaneous cell cultivation, product concentration, and toxic product removal. Many of these systems behave as perfusion reactors (see Chapter 9). In essence, a perfusion system exists when cells are retained in the reactor, medium is added continuously or semicontinuously, and spent medium is removed. The removal of spent medium removes toxic metabolites.

The classic method for animal cell culture has been roller bottles, and they have been important components in the scale-up of processes using anchorage-dependent cells (Fig. 12.5). With roller bottles the liquid covers about 25% of the surface area. Bottles are rotated about the long axis (1 to 5 rpm). Cells adhere to the walls of the bottle and are exposed to liquid 25% of the time and to gas (e.g., 5% CO_2 and 95% air) 75% of the time. When exposed to the liquid, nutrients can be transported into the cell, and when in the gas phase, aeration takes place. The roller-bottle system has an advantage over T-flasks because of increased surface area, agitation of the liquid, and better aeration. Roller bottles are not typically used for large-scale production because of high labor requirements and bottle-to-bottle variability. However, a highly automated facility making extensive use of robots uses roller bottles for commercial production of erythropoietin (a therapeutic protein). Roller bottles are also used for production of some vaccines.

The use of microcarriers for the cultivation of anchorage-dependent mammalian cells is an attractive approach. For example, microcarriers such as DEAE-Sephadex beads, which provide large surface per unit volume of reactor (~70,000 cm^2/l), allow high cell concentrations in the medium (~10^7 cells/ml). The microcarrier beads with cells are suspended in a stirred bioreactor. Other microcarrier beads, such as polyacrylamide, polystyrene, cellulose fibers, hollow glass, and gelatin-coated dextran beads, have been developed. At present the most widely used microcarriers are dextran- and DEAE-based (DEAE-Sephadex, DEAE-polyacrylamide). The surfaces of these microcarrier beads can be modified by addition of compounds, such as collagen, to promote cell adhesion and enhance cell function. Cells grow on the surfaces of microcarriers, usually in the form of monolayers and sometimes as multilayers. Microcarrier culture methods provide a large surface area for growth and a rather homogeneous growth environment. Microcarrier beads can be placed in a gently agitated reactor vessel, in a fluidized bed, or in an airlift (bubble-column) fermenter. Bead-to-bead contact and abrasion of the surfaces can be a problem, and the nonporous nature of these carriers limits the available surface area. Agitation of large-scale vessels with microcarrier cultures is difficult, owing to balancing the needs for aeration and mixing against the shear sensitivity of cells attached to microcarriers. Shear forces are more harmful to cells attached to microcarriers than to suspended cells because attached cells cannot "tumble," and shear tends to "rip" them off the carrier surface. Macroporous microcarriers can be used; cells which enter the pores are shielded from shear effects. However, diffusional limitations and heterogeneous growth conditions may cause undesirable results. Because cells are easily retained in the bioreactor, microcarrier cultures are well suited for perfusion operation.

Conventional stirred-tank bioreactors have been modified to reduce shear rates on cells in suspension. Sail-type and axial-flow hydrofoil agitators have been developed and used for suspension cultures. The agitation rate in these reactors is on the order of 10 to 40 rpm, providing low shear rate. Many animal cells can be cultured as suspensions in bioreactors up to 10,000 l in size. Because cells are significantly smaller than the turbulent eddies in these bioreactors, cell lysis is minimized. Only cells that are at the interface of an eddy and another eddy or another surface (e.g., reactor wall) are likely to experience damage. Cells at the gas–liquid interface are particularly prone to damage. The breakage of air bubbles is particularly destructive to cells that accumulate at the interface of a gas bubble and medium.

Many shear protecting agents (such as serum or Pluronic F-68) work by preventing cells from accumulating at the gas–liquid interface. While gas exchange can be accomplished with membranes or silicone rubber tubing, sparging of gas is the simplest method and is acceptable with many cell lines when shear protectants are used. Cell lines differ in shear sensitivity; Chinese hamster ovary (CHO) cells, widely used for protein production, are relatively resistant to shear damage. Gross shear damage where cells lyse (*necrosis*) is obvious. More subtle effects of shear include changes in physiology and, possibly, induction of *apoptosis* or programmed cell death. These more subtle changes can alter apparent growth rates, productivity, and product quality (e.g., due to release of degradative enzymes into the medium).

Hollow-fiber reactors have also been used to provide a high growth surface–volume ratio and, therefore, high cell concentrations. Cells are immobilized on the external surfaces of hollow fibers, and nutrients pass through the tubes. Cell concentrations comparable to those found in tissues can be reached in hollow-fiber reactors. However, the control of microenvironmental conditions inside the reactor where cells are immobilized on fiber surfaces is very difficult, because there is almost no mixing (i.e., a heterogeneous environment). Also, the quantification of growth and product formation is difficult in such a system. However, it is possible to construct the reactor using fibers of known MW cutoff to control the flux of different-molecular-weight products into the effluent stream. The accumulation of some toxic products, such as lactate and ammonium, in the fiber reactors may cause high levels of inhibition. Selection of fiber material (MW cutoff) and flow regime are critical factors for the rapid removal of toxic metabolic products. Hollow-fiber reactors have been used for the production of MAb's from hybridoma cells. Antibody concentrations on the order of 5 to 50 mg/ml have been obtained with this reactor since product is retained by the membrane while medium flows in through the membrane and lower molecular wastes flow out of the cell compartment. Because of severe mass transfer and control problems, this reactor may not be suitable for large-scale production purposes. To overcome some of the difficulties with axial-flow hollow-fiber reactors, radial-flow or cross-flow hollow-fiber units have been developed. Hollow fiber reactors are well suited to perfusion operation with continuous feed. MAb production from hybridomas can be sustained for extended periods (e.g., 100 days), as stationary-phase cells still synthesize product.

The immobilization of mammalian cells in gel beads (agar, alginate, collagen, polyacrylamide) and the use of such systems in a packed- or fluidized-bed configuration are possible. Such immobilization methods and reactor configurations will reduce or eliminate the shear effects on cells. High cell densities make high volumetric productivities possible. However, the control of microenvironmental conditions inside bead particles and the accumulation of toxic metabolic products in beads are potential problems.

A tubular ceramic matrix has been used for the immobilization and cultivation of hybridoma cells. High cell and, therefore, MAb concentrations can be obtained using such systems. The quantification of cell concentration and the control of microenvironmental conditions within the heterogeneous cell culture in ceramic matrix are some of the major difficulties, although these matrices do provide well-defined flow channels. Scale-up of a ceramic matrix reactor may be difficult because of the heterogeneous nature of the system when long tubes are used.

Microencapsulation is another method used for the immobilization of animal cells. Hybridoma cells have been encapsulated within spherical membranes of polylysine–alginate for production of MAb's. Typical capsule size is 300 to 500 μm, and the molecular weight cutoff of these capsule membranes is 60 to 70 kda. Microcapsules operate like small membrane bioreactors, in which very high cell concentrations can be reached ($\sim 10^8$ cells/ml). Using the right capsule membrane with a desired MW cutoff, toxic products such as lactate and ammonium can be removed from intracapsule culture media, and high-MW products such as MAb's or lymphokines can be concentrated inside capsules. Moreover, cells are protected from hydrodynamic shear by encapsulation, providing a suitable environment for direct aeration. However, the growth and product-formation rate inside capsules may be limited by the diffusion of nutrients such as dissolved oxygen or glucose, especially at large capsule sizes. The reduction of capsule size to eliminate mass transfer limitations (down to 200 μm) and the control of pore size to concentrate desired products within the capsule are two promising approaches to solve some of the problems associated with encapsulated cells. By varying the concentration and average MW of poly-L-lysine (PPL), the pore size of the capsule membrane can be controlled within certain limits (30 to 80 kda). Monitoring and control of microenvironmental conditions within capsules (pH, DO level, nutrient–product gradient) are major problems that impose limitations on growth and product formation inside capsules. The fragile nature of the microcapsule also limits the scale-up potential of this system.

12.4. PRODUCTS OF ANIMAL CELL CULTURES

Animal cell products usually consist of high-molecular-weight proteins with or without glycosidic groups. A number of enzymes, hormones, vaccines, immunobiologicals (monoclonal antibodies, lymphokines) and anticancer agents can be produced using animal cell culture technology (see Table 12.4).

The major types of immunobiologicals produced by animal cells are (1) monoclonal antibodies, (2) immunobiological regulators (interleukins, lymphokines), and (3) virus vaccines (prophylactics).

12.4.1. Monoclonal Antibodies

Among the most important products of animal cell culture have been monoclonal antibodies (MAb's). MAb's are produced by hybridoma cells and are used in diagnostic assay systems, for therapeutic purposes, and for biological separations (e.g., affinity chromatography). MAb's have been used as diagnostic agents to determine hundreds of drugs, toxins, vitamins, and other biological compounds. MAb's are also used for chromatographic separations to purify protein molecules. Purification of interferon by affinity chromatography is an example of the use of MAb's for protein purification purposes. MAb's

TABLE 12.4 Major Categories of Products of Animal Cell Cultures

Immunobiologicals	Hormones
Monoclonal antibodies	Enzymes
Lymphokines and interleukines	Insecticides
Virus vaccines	Growth factors
Cell surface antigens	Whole cells
	Tissues
	Organs

also may serve as "magic" bullets to target toxic agents to cancer tumors. However, the large size of MAb's has limited their ability to penetrate some tumors. Antibody fragments can be used instead; these products can be made in nonmammalian cells.

12.4.2. Immunobiological Regulators

Interferon (an anticancer glycoprotein secreted by animal cells upon exposure to cancer-causing agents) is an example of an immunoregulator produced by mammalian cells. Interferon can be produced by either animal cells or recombinant (genetically engineered) bacteria. Other immunoregulators are lymphokines (hormonal proteins regulating immune responses of human body), interleukines (anticancer agents), tissue plasminogen activator (a compound preventing blood clotting), and thymosin. The production of immunoregulators is a very promising area with a great potential for growth in the near future.

12.4.3. Virus Vaccines

Various virus vaccines (prophylactics) have been produced for animal and human use. In some cases live virus is propagated in animal cells. Virus is then collected and inactivated ("killed" virus) and used as a vaccine. In other cases a weakened or *attenuated* form of the virus is used that will induce a protective response but no disease. However, most vaccines now under development are *subunit vaccines*. Typically a protein displayed on the surface of the virus particle is produced. An *epitope* or small region of the protein is recognized by the immune system and induces a protective immune response. For the epitope to be displayed properly, the protein may need to assemble into a *viruslike particle*. Such a particle is an empty capsid that contains no nucleic acid. The absence of viral DNA or RNA increases the safety of the product.

12.4.4. Hormones

Some animal hormones are large molecules (50 to 200 amino acids) and are glycosylated. These hormones can be produced by using cell cultures of the hormone-synthesizing organ.

   Some potential hormone products from animal cells are follicle-stimulating hormone, chorionic hormone, and erythropoetin. Some animal hormones are relatively small polypeptides (20 to 30 amino acids) and may be produced by chemical synthesis. Erythropoetin is a very successful commercial product useful in treating anemia in a wide range of disorders from patients on artificial kidneys to those with AIDS.

12.4.5. Enzymes

A number of enzymes can be produced by animal cell cultures. The synthesis and excretion of these enzymes are targets of genetic engineering. Glycosylation, posttranslational modification, and excretion from the host organism are potential problems. Some potential enzyme products from animal cell cultures are urokinase, rennin, asparaginase, collagenase, pepsin, trypsin, hyaluronidase, and blood-clotting compounds such as factor VII, factor VIII, and factor X. The large-scale production of tissue plasminogen activator is an important commercial process.

12.4.6. Insecticides

Animal cell cultures have been used to produce some insect viruses that are highly specific and safe to the environment. Several of the baculoviruses have federal approval for use as insecticides. Genetically engineered variants with greater virulence can now be produced. The commercial production of such viruses from tissue culture has not yet been accomplished.

12.4.7. Whole Cells and Tissue Culture

The production of differentiated cells for medical use is under early development. Artificial skin for burn patients is one example. Artificial organs and semisynthetic bone and dental structures are potentially feasible.

12.5. SUMMARY

Animal cells are well suited for the production of proteins requiring extensive and accurate posttranslational processing, organized tissues (artificial skin), certain vaccines, and viruses that can be used as pesticides.

Animal cells are more complex and fragile than bacterial, fungal, or yeast cultures. These characteristics require the development of unique strategies for cultivation.

Critical distinctions are *primary* cell lines (established directly from tissue) and *secondary* cell lines, which are established from primary cell lines. Normal cell lines are *mortal*; they can divide only a finite number of times. Other cell lines are *transformed* or *immortal* and divide indefinitely. Many cell lines will grow only when attached to a surface and are thus *anchorage dependent*. Some cells, particularly transformed cell lines, can grow in suspension. One particularly important type of suspension cell is the *hybridoma*. This cell is a hybrid of a mortal antibody-producing cell with a transformed cell, which results in an immortal, easily cultured cell line that produces a single type of antibody or a *monoclonal antibody*.

The media used to grow animal cells must provide a variety of growth factors, glucose, glutamine, amino acids, specific salts, and vitamins. In some cases, serum from animals (especially mammals) is used. However, serum is expensive, increases the probability of contamination, is undefined in composition, is variable from lot to lot, and complicates the downstream recovery of protein products. In many cases, our knowledge of cellular nutrition has allowed the development of serum-free media, and the use of such media is of increasing importance.

Many potential bioreactor configurations are commercially available for use with animal cells. The reactor design problem centers on providing sufficient oxygen and nutrients and (for anchorage-dependent cells) surface area to achieve high cell densities while minimizing exposure to liquid shear. Also, complete medium utilization and high product concentrations are sought. Designs must minimize the formation of toxic by-products (for example, lactate and ammonium) or effectively remove such metabolites from the microenvironment of the cells.

SUGGESTIONS FOR FURTHER READING

CHALMERS, J. J., "Animal cell culture. Effects of agitation and aeration on cell adaption," in *Encyclopedia of Cell Technology,* R. Spier, J. B. Griffiths, and A. H. Seragg, eds., Wiley, New York, 2000. (See also other related articles in this book.)

FRESHNEY, R. I., *Culture of Animal Cells: A Manual of Basic Technique,* 2d ed., Alan R. Liss, Inc., New York, 1987.

HO, C. S., AND D. I. C. WANG. *Animal Cell Bioreactors,* Butterworth-Heinemann Press, Stoneham, MA, 1991.

HU, W-S., AND PESHWA, M. V., Mammalian Cells for Pharmaceutical Manufacturing, *ASM News.* 59(2):65–68, 1993.

MARINO, M., C. ANGELO, A. IPPOLITO, G. CASSANI, AND G. FASSINA, Effect of bench-scale culture conditions on murine IgG heterogeneity, *Biotechnol. Bioeng. 54*: 7–25, 1997.

SHULER, M. L., H. A. WOOD, R. R. GRANADOS, AND D. A. HAMMER, *Baculovirus Expression Systems and Biopesticides*, Wiley-Liss, New York, 1995.

The series *Advances in Cell Culture* published by Academic Press, New York, is a good source of review articles.

PROBLEMS

12.1. Cite the major differences between animal, plant (see Chapter 13), and bacterial cells in terms of cell structure and physiology.

12.2. Describe the role of glutamine in animal cell metabolism.

12.3. Compare serum-containing and serum-free media in terms of their advantages and disadvantages.

12.4. What are the roles of CO_2 provided in air to animal cell cultures?

12.5. What are the common features of animal cell bioreactors?

12.6. Compare the following immobilization methods used for animal cells in terms of their relative advantages and disadvantages: microcarrier (surface) culture, porous beads, encapsulation, and gel entrapment.

12.7. What do you think is the most suitable reactor type, mode of operation (batch, continuous perfusion), and cultivation method (suspended, immobilized) for animal cell cultures, in general? Compare your choice with the other alternatives in detail.

12.8. Compare various methods for the aeration of animal cell cultures. Comment on the advantages and disadvantages of each method.

12.9. Describe the process for formation of hybridomas.

12.10. What value of k_La must be achieved to sustain a population of 1×10^7 cells/ml when the oxygen consumption is 0.1×10^{-12} mol O_2/h-cell?

12.11. Hybridoma cells immobilized on surfaces of Sephadex beads are used in a packed column for production of monoclonoal antibodies (Mab). Hybridoma concentration is approximately $X = 5$ g/l in the bed. The flow rate of the synthetic medium and glucose concentration are $Q = 2$ l/h and $S_0 = 40$ g/l, respectively. The rate constant for Mab formation is $k = 1$ gX/l-d. Assume that there are no diffusion limitations and glucose is the rate limiting nutrient.

    a. Determine the volume and the height of the packed bed for 95% glucose conversion. Bed diameter is $D_0 = 0.2$ m. Neglect the growth of the hybridomas and assume first order kinetics.

    b. If $Y_{p/s}$ is 4 mg Mab/g glu, determine the effluent Mab concentration and the productivity of the system.

13

Bioprocess Considerations in Using Plant Cell Cultures

13.1. WHY PLANT CELL CULTURES?

The plant kingdom is very diverse. Many thousands of chemicals are produced only in plants. The biosynthetic capabilities of plant cells have been little explored. Only a few percent of the world's plants have been scientifically named or described, and only a small fraction of these have been screened for the production of novel and useful compounds. The great genetic potential of plants to produce compounds of use to humans has been little exploited. In fact, the rapid destruction of forests worldwide is leading to the extinction of many plants without preservation of their genomes.

Even the small number of characterized plants have yielded many important compounds (e.g., over 120 prescription drugs). In the Western world, over 25% of pharmaceuticals are derived from extraction from whole plants, and a much greater fraction of medicinals are plant-derived in Asia and the rest of the world. In addition to uses as medicinals, plant products are of interest as dyes, food colors, food flavors, fragrances, insecticides, and herbicides. Some compounds of potential commercial interest are listed in Table 13.1.

These compounds are chemically complex and are generally nonproteins. The compounds most likely to be of commercial interest are those that are not synthesized by microbes and are of sufficient complexity that chemical synthesis is not a reasonable alternative. Also, these nonprotein products are often made by the condensation of intermediates from separate pathways. Plant genetics and physiology are poorly

405

TABLE 13.1 Examples of Plant Products of Potential Commercial Interest

Pharmaceuticals:
  Ajmalicine, atropine, berberine, camptothecine, codeine, diosqenin, digoxin, L-dopa, hypercium,
  hyoscyamine, morphine, phototoxin, sangranine, scopolamine, Taxol, ubiquinone-10, vaccines,
  vincristine, vinblastine
Food Colors or Dyes:
  Anthocyanins, betacyanins, saffron, shikonin
Flavors:
  Vanilla, strawberry, grape, onion, garlic
Fragrances:
  Jasmine, lemon, mint, rose, sandalwood, vetiver
Sweeteners:
  Miraculin, monellin, stevioside, thaumatin
Agricultural Chemicals:
  Alleopathic chemicals, azodirachtin, neriifolin, pyrethrine, rotenone, salannin, thiophene

understood in comparison to animal or bacterial systems. Because of the difficulty of finding and recovering genes from unrelated pathways, genetic engineering using more easily grown organisms to make these products is not now an attractive alternative to plant cell culture. Plant cell tissue cultures have low volumetric productivities, so higher-value products ($\geq$ \$500/kg) are reasonable targets. Thus, the chemicals of greatest commercial interest must be of intermediate to high value and unique to the plant kingdom.

A good example of bioprocess development using plant cell cultures is for the production of the important anticancer agent, paclitaxel (Taxol). Supply problems greatly impeded the clinical development of this drug. Initially paclitaxel was produced by extraction from the bark of the Pacific yew tree (*Taxus brevifolia*). It required three 100-year-old trees to supply enough paclitaxel to treat one patient. This tree is relatively uncommon, and its harvestation had significant adverse environmental impacts. Harvestation of whole trees has been replaced by a method of semisynthesis. Needles and branches of more common yews can be collected, a precursor compound extracted, and the precursor chemically converted to paclitaxel. The collection of needles and branches can be done in a way to prevent death of the plant. However, there are environmental concerns with both needle collection and disposal of solvents from the chemical processing. Production of paclitaxel in large-scale bioreactors provides a product with fewer contaminants, is highly controllable and reproducible, and is environmentally friendly. Commercial production of paclitaxel using large-scale bioreactors is now a reality. Suspension cultures in stirred tank vessels of about 30000 l are used.

The factory production of chemicals from plant cell tissue culture offers a number of important advantages:

1. Control of supply of product independent of availability of the plant itself.
2. Cultivation under controlled and optimized conditions.
3. Strain improvement with programs analogous to those used for microbial systems.
4. With the feeding of compounds analogous to natural substrates, novel compounds not present in nature can be synthesized.

The production of the dye, shikonin, further illustrates these advantages. Japan could not grow sufficient *Lithospermum* to supply its needs, nor was it assured of a stable supply to import. The plant had to grow for at least two to three years before reaching harvestable size; at harvest, about 1% to 2% of the dry weight was shikonin. Using a combination of strain selection and optimization of reactor conditions, a commercial process that could produce cells with 14 wt % shikonin in a three-week batch cultivation period was established.

However, the role of bioreactors in plant cell tissue is not limited to chemical production. Also, transgenic plant cell cultures can be used to produce proteins such as vaccines. The production of propagules or of artificial seeds may be economically important. The use of efficient submerged cultivation devices to generate elite plants may replace the current labor-intensive processes for the micropropagation of plants.

Whether the product is a chemical or a new plant, the bioprocess engineer must become familiar with some basic characteristics of plants and their implications for reactor design.

13.2. PLANT CELLS IN CULTURE COMPARED TO MICROBES

Plant cells in culture are not microbes in disguise. Aspects of plant cell structure and physiology were discussed in Chapter 2. See Fig. 13.1 for a summary of plant cell structure. The primary difference between plant cells and microbes is the ability of the cells to

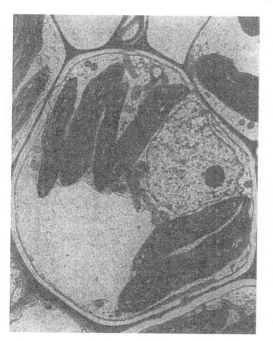
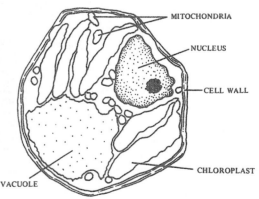

Figure 13.1. (Left) Cross section of a plant cell enlarged about 9000 times its actual size by an electron microscope. (Right) The labeled drawing identifies some of the cell's structures and organelles, those specialized cell parts that resemble and function as organs. This particular cell is a bundle sheath cell from the leaf of a *Zea mays* plant. Cells like this one are part of the veins of the leaf and completely surround the water- and food-conducting tissues (xylem and phloem) of the veins. (With permission, from Michael A. Walsh, Utah State University.)

undergo differentiation and organization even after extended culture in the undifferentiated state. The capacity to regenerate whole plants from undifferentiated cells under appropriate environmental conditions is called *totipotency*. This capacity is essential to plant micropropagation and is often associated with secondary metabolite formation. If we could do the same for animals as we can for many plants, we could remove a cell from the reader's tongue and generate millions of nearly identical clones of the reader's whole body!

Let us begin by considering the establishment of cell cultures for producing a particular chemical (see Fig. 13.2). *Callus* and *suspension cultures* have been established from hundreds of different plants. A *callus* can be formed from any portion of the whole plant containing dividing cells (Fig. 13.3). The excised plant material is placed on solidified medium containing nutrients and hormones that promote rapid cell differentiation. The callus that forms can be quite large (> 1 cm across and high) and has no organized structure. Although we speak of a callus as being *dedifferentiated tissue*, it contains a mixture of cell types. The sources of the initial tissue (root, shoot, and so on) can make some difference as well. If we are interested in maximizing the formation of a particular compound, it is advisable to initiate the callus from a plant that is known to be a high producer. Even from the same batch of seeds, there may be considerable variation from plant to plant in their biosynthetic capacity.

For both callus and suspension cultures, a chemically defined medium is used. Typically, cultures, especially suspension cultures, are maintained in the dark. While exposure to light may be used to regulate expression of specific pathways, light is rarely used solely to support growth as most cells are incapable of sustained photoautotrophic growth.

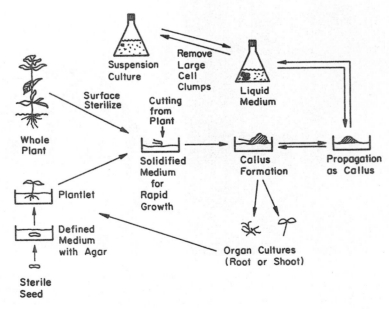

Figure 13.2. Suspension cultures can be established from many plants using this approach. For some plants, the process can be reversed to generate whole plants from suspensions.

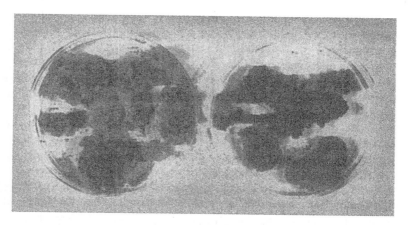

Figure 13.3. Calli of grape cultures. The plate on the left contains an unselected culture in which a few cells make high levels of reddish-purple pigments (anthocyanins). The plate on the right contains highly selected cultures in which most cells make pigment.

Typical media use a carbon/energy source such as sucrose. Inorganic nutrients, vitamins, and "hormones" (or growth regulators) are included in media. Classes of plant hormones that are growth promoters are auxins, cytokinins, and giberellins. Most media contain at least an auxin such as naphthalene acetic acid (NAA) or 2,4-Dichlorophenoxyacetic acid (2,4-D) and a cytokinin such as kinetin or benzyladenine (BA). Ethylene is a plant hormone and is typically produced by the culture itself.

The establishment of *suspension cultures* (see Fig. 13.4) from callus is generally straightforward if the callus is *friable* (easily breaks into small pieces). A piece of callus is

Figure 13.4. Suspension cultures of grape cells. The flask on the left is a culture with modest selection, while the culture on the right is highly selected for anthocyanin production.

placed in a liquid medium in a shake flask. With gentle to moderate agitation, cells or small aggregates of cells will slough off. Typical conditions would be 27°C and a pH of 5.5 in the dark. These suspended cells then replicate. After two or three weeks, the suspended cells are transferred to fresh medium, and large aggregates or residual callus are discarded. Suspensions can grow to high cell densities (similar to applesauce). All suspension cultures contain *aggregates;* however, it is possible to select for fine suspensions by discarding large aggregates at each subsequent transfer.

The aggregates are due not so much to clumping as to the failure of progeny to separate. Cell-to-cell communication is very important in plants. Normally, all cells in a whole plant are connected to one another by small pores called the *plasmodesmata.* The plasmodesmata allow the interchange of lower-molecular-weight compounds (< 800 da) from the cytoplasm of one cell to another. The plasmodesmata are formed principally during cell division. Cells in aggregates can communicate with one another through plasmodesmata and also through diffusable species. For example, plants generate ethylene, which acts as a hormone. In a large aggregate, there will be concentration gradients of such metabolic products as well as nutrients (e.g., oxygen and hormones). These gradients lead to a wide variety of microenvironments and, as a consequence, cells in different positions in the aggregates may have greatly different biochemical and morphological structure. Although plant cell cultures are pure cultures in the sense that only one species is present (ideally one genotype), it behaves very much like a mixed culture. A mix of cell types may be necessary for the formation of some products.

Cells in suspension can be made to undergo differentiation and organization if the correct environmental conditions can be found. Nutrient and hormone levels must be adjusted. Embryos, shoots, and roots can be made from aggregates in suspension. However, let us turn our attention from aggregates back to individual cells.

Plant cells can be very large, with cell diameters typically in the range of 10 to 100 μm. They grow slowly, with doubling times typically ranging from 20 to 100 h. Growth is usually nonphotosynthetic, with sucrose or glucose supplied exogenously as the carbon and energy source. Typical respiration rates are roughly 0.5 mmol O_2/h-g dry weight or about 5% to 15% of that in *E. coli*. Plant cells can often be cultivated at very high densities, up to 70% of the total reactor volume as cells. Although the respiration rate is low, high cell density requires very good oxygen-transfer capabilities in a reactor.

Plant cells contain a higher percentage of water than bacterial cells (90% to 95% versus 80%). This higher water content is due, in part, to the presence of the central vacuole, which can occupy as much as 95% of the intracellular volume in extreme cases. Plant cells tend to secrete relatively few compounds, but sequester them inside the vacuole. Many of these compounds are secondary metabolites of commercial interest; many would also be cytotoxic if not removed from the cytoplasm. Their intracellular storage will have important implications in our later discussions on bioreactor strategies.

One other biological aspect of critical importance is the role of *elicitors.* In whole plants, there are a number of defensive mechanisms that plants use to contain attacks from pathogenic fungi, bacteria, and viruses. Defense mechanisms can be activated if breakdown products of fungal or plant cell walls (oligosaccharides) are present. Other compounds, some even nonorganic, can also elicit a response. Many secondary metabolites are involved in the plant's defensive mechanisms. The exposure of suspension cultures to elicitors has led in many cases to rapid increases in the accumulation of secondary products. The use of elicitors has been an important breakthrough in improving volumetric productivities in bioreactors. The most useful elicitor is methyl jasmonate. This compound is relatively inexpensive and readily introduced into large bioreactors. It regulates expression of a wide variety of plant genes.

Recent advances in genomics and our understanding of plant molecular biology make it possible to consider genetically engineering the plant cell's biosynthetic pathways. It will soon be possible to redirect these toward desired chemicals.

13.3. BIOREACTOR CONSIDERATIONS

13.3.1. Bioreactors for Suspension Cultures

Many of the differences between plant cell cultures and microbes have direct implications for the design and scale-up of suspension culture systems (see Table 13.2). In particular, the degree of physical mixing that is desirable is important. However, the basic reactor types we discussed in Chapter 10 can usually be adapted to plant cells.

Plant cells are large, and when they are exposed to turbulent shear fields where the eddy size approaches the cell size, the cells can be exposed to a twisting motion that can damage them. Lower levels of shear appear to affect cell surface receptors and nutrient transport. Reactors with high shear must be avoided. However, plants cells can withstand far more shear than animal cells, and shear-tolerant lines can sometimes be developed. Stirred tanks designed for the culture of bacteria are not good choices, but modified stirred tanks can be suitable. Reactors up to 75000 l have been used successfully.

Plant cell cultures can achieve high cell densities and viscosities. Their reduced respiration rate, however, compensates in part for the need for vigorous agitation. Airlift reactors for low or moderate cell densities (< 20 g dry wt/l) or paddle-type or helical-ribbon impellers for high-cell-density systems have been advocated as reactors that strike a good compromise between the need for good mixing and the shear sensitivity of plant cells.

More important than actual shear damage is the role that mixing plays in the biological response of some systems to scale-up. In at least some cases, the formation of aggregates appears to be necessary to achieving the mix of cell types essential for good

TABLE 13.2 Differences between Plant Cells and Microbes and the Implications for Bioreactor Design

Differences	Implications for reactor design
Lower respiration rate	Lower O_2 transfer rates required
More shear sensitive	May require operation under low-shear conditions
Cells often grow as aggregates or clumps	May have mass transfer limitations that limit the availability of nutrients to cells within the aggregate
Degree of aggregation may be important with regard to secondary metabolism	May be an optimal aggregate size for product synthesis
Volatile compounds may be important for cell metabolism (e.g., CO_2 or ethylene)	May need to sparge gas mixtures

With permission, from G. F. Payne and others, in B. K. Lydersen, ed., *Large Scale Cell Culture Technology*, John Wiley & Sons, Inc., New York, 1987.

secondary-product formation. The degree of aggregate formation is influenced by the degree of mixing. Since the degree of mixing necessary to achieve equivalent oxygen transfer or equivalent shear changes upon scale-up, the degree of aggregation and productivity may change (see Fig. 13.5).

Mixing depends on a combination of sparging and mechanical agitation. Oversparging can be a problem for plant cell cultures. Even though photoautotrophic growth is not important in most cultures, elevated CO_2 levels can enhance productivity. Plants make at least one volatile hormone, ethylene, and its rapid removal can affect productivity. With plant cell cultures, optimizing the gas composition, sparging rate, and degree of mechanical agitation is a difficult problem.

The low growth rates of plant cells present other problems for large-scale systems. The primary problem is maintenance of aseptic conditions for the two to four weeks often necessary to complete a fermentation. Additionally, the low growth rates reduce volumetric productivities to levels that diminish the economic attractiveness of plant cell cultures.

A further problem is the apparent genetic instability of many cell lines. Often, suspension and callus cultures will produce a desired compound initially, but after three to six months the capacity to make the product often decreases considerably. This loss of productivity may be due to inadequate nutrition or may have a genetic or epigenetic basis. Artificial cell culture tends to select for cells with higher ploidy levels (tetraploid instead

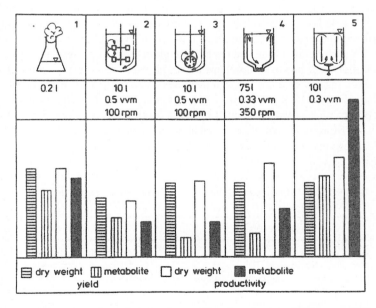

Figure 13.5. Comparison of yield (g/l, mmol/l) and productivity (g/l/day, mmol/l/day) for biomass (*Morinda citrifolia*) and anthraquinone in different culture systems: (1) shake flask; (2) flat-blade turbine; (3) perforated disc impeller; (4) draft tube reactor with Kaplan turbine; (5) airlift reactor. (With permission, from F. Wagner and H. Vogelmann in W. Barz, E. Reinhard, and M. H. Zenk, eds., *Plant Tissue Culture and Its Biotechnological Application*, Springer-Verlag, New York, 1977, p. 250.)

of diploid[†]). Cryopreservation of cell lines is not possible in all cases, and many cell lines must be maintained through routine subculture.

This loss of activity can be counteracted in many cases by the active selection of productive subclones. Selection is facilitated when the product of interest is readily visible. In cases where extensive screening and selection have been done, stable, high-producing clones have been reported. A stable, high-producing cell line is essential to any commercial process.

One strategy that has proved to be useful in increasing productivity has been the use of a two-phase culture. The first phase uses a medium optimized for growth, while a second phase uses a different medium optimized for product formation. The first commercial process with plant cell culture has been shikonin production, which utilizes two batch reactors in series.

13.3.2. Reactors Using Cell Immobilization

Further improvements in productivity may be possible due to cell immobilization. Immobilized-cell reactors inherently follow the two-phase approach. Cells are grown and then immobilized; once immobilized, conditions for product formation are optimized. Immobilized-cell systems are advantageous when continuous operation is possible and if the product is, or can be made, extracellular. Some advantages and disadvantages of immobilization are given in Table 13.3. Probably the most important advantage is that the degree of cell-to-cell interaction can be manipulated.

Plant cells will often self-immobilize by preferentially attaching to or within a porous matrix. The resulting biofilm (if on a surface) has been shown to be very effective in a number of cases. Plant cells have also been entrapped in gels or between membranes. Immobilization generates concentration gradients that alter the biosynthetic capacity of the culture (sometimes negatively and sometimes positively). The cell-to-cell contact due to immobilization or the contact of the cell surface with the surrounding gel phase may also alter cell physiology. In some cases, immobilization has improved intrinsic production rates by more than an order of magnitude.

Plant cell cultures can be cultivated at three levels of cell-to-cell communication. In a fine suspension, aggregates are small and mass transfer gradients are unimportant. Any diffusable species that might act as chemical messengers are diluted to a low concentration. Plasmodesmata interconnect only a small fraction of the cells. If cells are concentrated and entrapped, they form a pseudotissue; since the entrapping matrix is typically on the order of 1 to 10 mm, mass transfer gradients become important. Diffusable chemical species can build to high local concentrations because of mass transfer resistances. However, plasmodesmata are relatively unimportant in pseudotissues. If a few cells are immobilized (say, between two membranes or in the pockets of a foam matrix) and allowed to grow in place, they will develop a tissuelike structure. Not only can diffusable species accumulate, but cells will be interconnected through the plasmodesmata. Then immobilization (and the method used for immobilization) can be used as an engineering design parameter to alter the degree and type of cell-to-cell communication. Immobilization can be coupled with other strategies to enhance product formation.

[†]Recall that a diploid cell has two copies of each chromosome; a tetraploid cell has four copies.

TABLE 13.3 Potential Advantages and Problems of Large-scale Immobilized
Plant Cell Cultures

Potential advantages
  Continuous operation facilitated
  High cell concentrations
  Cell reuse may lead to increased efficiency
  Cells can be protected from shear
  Once immobilized, the slow growth and strain instability of plant cells are no longer problems
  Media can be easily changed, which would be important for processes that require a series of media for
    optimal production
  Continuous removal of inhibitory metabolites may enhance the overall cellular metabolism or unmask
    biochemical pathways
  May be able to better exploit the biological relationships between aggregation, morphological
    differentiation, and secondary metabolite production

Potential problems
  Large-scale aseptic immobilization procedures must be developed
  Mass transfer limitations may significantly affect cell metabolism (positively and adversely)
  Products must be produced by nongrowing cells
  Products must be released from the cell into the medium
  Experience in the scale-up of immobilized-cell systems is limited

With permission, from G. F. Payne and others, in B. K. Lydersen, ed., *Large Scale Cell Culture Technology*,
John Wiley & Sons, Inc., New York, 1987.

Consider the production of ajmalicine from periwinkle (*Catharanthus roseus*). Ajmalicine is stored primarily in the vacuole in cells in suspension culture, although a small amount (10%) is excreted. Ajmalicine has a pKa of 6.3, so that at the growth pH (5.6) much of the alkaloid is in the neutral form. A neutral resin can be used to remove ajmalicine *in situ*. The *in situ* removal of the product can enhance formation of a product by relief of feedback inhibition or protection from degradation or further conversion.

Figure 13.6 shows the results of experiments involving combinations of *in situ* adsorption, use of a fungal elicitor, immobilization in calcium alginate gels, and the use of a production medium. Using all four approaches increases extracellular concentrations almost 100-fold. The purity is high. The closely related alkaloid, serpentine, has a much higher pKa (10.8) and will not adsorb into the resin. The adsorption of ajmalicine increases its synthesis preferentially. Under some circumstances, the intracellular levels of ajmalicine and serpentine are nearly equal. In this system, the ratio of extracellular ajmalicine to serpentine was 60-fold, with serpentine accounting for a vanishingly small amount of the adsorbed alkaloid. Perhaps the most important result of such an experiment is that at least some normally intracellular compounds can be excreted from viable cells when environmental conditions are correctly manipulated.

13.3.3. Bioreactors for Organized Tissues

In many cases, neither suspension nor immobilized-cell cultures will produce satisfactory amounts of a desired metabolite. However, organ cultures from the same plant may give good yields (Table 13.4). In addition to high yields, organ cultures have a number of distinct advantages (Table 13.5).

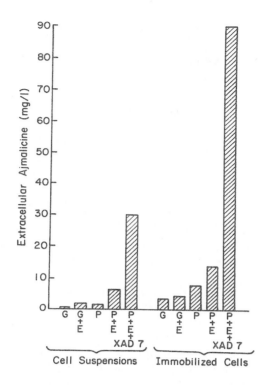

Figure 13.6. Summary of the effects of various treatments on the formation and release of ajmalicine from *C. roseus*. G stands for growth medium and P for production medium. E indicates elicitor addition (2%) of an autoclaved culture of the fungus, *Phytophthora cactorum*, for a two-day exposure time; the neutral resin, amberlite XAD-7, indicates the addition of resin. Samples were analyzed 23 days after inoculation. (With permission, from M. Asada and M. L. Shuler, *Appl. Microbiol. Biotechnol. 30*:475, 481, 1989.)

TABLE 13.4 Examples of Compounds Made in High Yields in Organ Culture But Not in Corresponding Callus or Suspension Culture

Compound	Plant	Form of culture
Cardenolides	*Digitalis lanata*	Somatic embryos
Digitoxin	*Digitalis purpurea*	Shoots
Diosgenin	*Dioscorea composita*	Shoots
Monoterpenes	*Pelargonium fragrans* (geranium)	Shoots
Morphiran alkaloids	*Papaver somniferum* (opium poppy)	Shoots
β-Peltatin	*Linum flavum*	Roots
Pyrethrins	*Chrysanthemum cinerariaefolium*	Shoots
Scopolamine	*Datura innoxia*	Roots
Tropane alkaloids	*Hyoscyamus niger*	Roots
Valepotriates	*Centranthus* sp.	Roots
Withanolides	*Withania somnifera*	Shoots

TABLE 13.5 Advantages and Disadvantages of Organ Cultures

Advantages
  Biosynthetic capacity often returns upon organogenesis.
  Product spectrum for flavors and fragrances returns to that for parent plant.
  Product secretion is enhanced in many cases (particularly root cultures).
  Genetic stability under growth conditions is greatly improved over callus or suspension culture.
  Self-immobilization; provides more optimal mix of cell types.

Disadvantages
  Growth rates may be lower than suspension cultures in some, but not all cases.
  Efficient, scalable reactors for organized tissues need to be developed.
  Control of microenvironmental conditions is often more difficult.

The primary disadvantages for organ cultures have been their apparently slow growth rates and difficulties in designing effective bioreactors to handle organized tissues. Recent advances suggest that both limitations may be circumvented, and in several cases, organ cultures that grow more rapidly than their corresponding suspension cultures have been found.

Some plants (some dicots, but no monocots) will respond to infection by *Agrobacterium rhizogenes* with rapid root proliferation. This infection leads to *hairy roots*. The best doubling times for hairy root cultures approach or exceed typical values for many suspension cultures. Even in species not susceptible to *A. rhizogenes* infection, recent results indicate that proper control of hormone content can accelerate growth rates to acceptable levels (three- to five-day doubling times). The biggest difficulty with the large-scale culture of roots is the formation of root *mats*. These mats restrict internal mass transfer, can entrap gas and float, and present significant problems in maintaining a scalable uniform environment. However, large-scale units for root culture have been built (20000 l for ginseng roots in Japan). Laboratory-scale reactors using a mist or forced convection of liquid nutrients appear promising and provide much better mass transfer in the center of root mats.

One other advantage of using organ cultures over whole plants is the possibility of using precursor feeding and elicitors. For example, with species of onion and garlic, the use of precursors can greatly enhance the formation of flavor compounds. Different precursors give different levels of enhancement to particular components of the flavor spectrum. By using combinations of chemical precursors, it may be possible to custom-make flavors for specific applications.

Shoot cultures present additional problems. Light may be required for some shoot cultures, while roots can be grown easily in the dark. Shoot cultures may be *mixotrophic*, involving the exogenous supply of sugars as well as some photosynthesis. In some cases, control of the *photoperiod* (hours of exposure to light per day) is important. However, the most crucial need for light comes from the role that light often plays in cellular regulation. Exposure to light of certain wavelengths is essential to induce synthesis of some enzymes. In at least some cases, these enzymes play a crucial role in secondary metabolism. The maintenance of uniform light intensity in a large reactor is a challenging and partially unsolved problem.

One other perceived limitation on shoot cultures has been the belief that they could not be grown under totally submerged conditions. Recent experiments have shown that tobacco shoots will grow as well in submerged culture as a standard shake flask (three-day doubling time in both cases), provided that the liquid phase is well mixed. In Israel, commercial production of plantlets using totally submerged culture has been reported.

Although organ cultures have not yet been extensively exploited, they are promising vehicles for the production of some important secondary metabolites.

13.4. ECONOMICS OF PLANT CELL TISSUE CULTURES

Far fewer commercial processes for plant cell culture have been established than for bacterial, fungal, or animal cell cultures. Currently, four large-scale systems have been constructed in Japan and South Korea (1200 to 30,000 l) and one in Germany (75,000 l). In North America, the possible commercial development of three products is under active consideration.

As a rule of thumb, most commercial fermentations yield a revenue of about 20 ¢/l-day. If the volumetric productivity of a process is known, the wholesale price necessary to achieve this level of revenue can be estimated. Table 13.6 lists the published values of the volumetric productivities reported for some of the more intensely studied cell lines. Bulk prices in excess of $220/kg would be necessary to yield 20 ¢/l-day for these products.

13.5. SUMMARY

Plants produce a wide range of commercially important compounds. For some of these chemicals, plant cell tissue culture is a potential technique for their production. In addition, the bioreactor techniques that are applied to plant cell culture will be useful in the micropropagation of plants.

TABLE 13.6 Volumetric Productivities for Some Products Made in Plant Cell Culture and Their Corresponding Value

Product	Cell line	Productivity (g/l-day)	Bulk price to give 20 ¢/l-d revenue (¢/g)
Anthocyanin	Grape (Bailey alicant A, *Vitis* hybrid)	0.15	133
Berberine	*Thalictrum minus*[a] and	0.05	400
	Coplis japonica[b]	0.60	33
Diosgenin	*Dioscorea* spp.	0.0098	2040
Paclitaxel (Taxol)	*Taxus canadensis*	0.75	27
Podoverine	*Podophyllum versipelle*	0.15	133
Rosmarinic acid	*Coleus blumei*	0.91	22
Sanguinarine	*Papaver somniferum*	0.034	588
Shikonin	Lithospermum erythrorhizon	0.15	133

[a]Excreted into the medium, where it crystallizes.

[b]Continuous-flow system with cell retention; intracellular product.

Since most secondary products are not made in rapidly growing cultures, two-phase fermentations are employed. The first phase is used to promote growth, and a second phase is optimized for product formation. Liquid suspensions can be cultured analogously to fermenters for microbes, if the special characteristics of plant cells are taken into account. Alternatively, immobilized or entrapped cell systems can be used. In either case, *elicitors* can often be used to enhance product formation. A combination of immobilization and *in situ* separation techniques can often dramatically improve production.

Organized tissues (shoots, roots, and embryos) hold promise as techniques to improve yields. However, bioreactor designs that supply homogeneous environments for organ cultures are only now being developed.

SUGGESTIONS FOR FURTHER READING

Overview on plant physiology

SALISBURY, F. B., AND C. W. ROSS, *Plant Physiology*, 3d ed., Wadsworth Publishing Company, Belmont, CA, 1985.

Overview on plant cell culture

DORAN, P. M., Design of Mixing Systems for Plant Cell Suspensions in Stirred Reactors, *Biotechnol. Prog. 15*:319–355, 1999.

FLORES, H. E., Plant Roots as Chemical Factories, *Chem. & Ind* (May 18):374–377, 1992.

GUNDLACH, H., M. J. MULLER, T. M. KUTCHAN, AND M. H. ZENK, Jasmonic Acid Is a Signal Transducer in Elicitor-induced Plant Cell Cultures, *Proc. Natl. Acad. Sci. (USA) 89*:2389–2393, 1992.

HSIAO, T. Y., F. T. BACANI, E. B. CARVALHO, AND W. R. CURTIS, Development of a Low Capital Investment Reactor System: Application for Plant Cell Suspension Culture, *Biotechnol. Prog. 15*:114–122, 1999.

KARGI, F., Plant Cell Bioreactors, Present States and Future Trends, *Biotechnol. Prog. 3*:1, 1987.

KETCHUM, R. E. B., D. M. GIBSON, R. B. CROTEAU, AND M. L. SHULER, The Kinetics of Taxoid Accumulation in Cell Suspension Cultures of *Taxus* following Elicitation With Methyl Jasmonate, *Biotechnol. Bioeng. 62*(1):97–105, 1999.

MACLOUGHLIN, P. F., D. M. MALONE, J. T. MURTAGH, AND P. M. KIERAN, The Effect of Turbulent Jet Flows on Plant Cell Suspension Cultures, *Biotechnol. Bioeng. 58*:595–604, 1998.

PAYNE, G. F., V. BRINGI, C. L. PRINCE, AND M. L. SHULER, *Plant Cell and Tissue Culture in Liquid Systems,* Hanser Publ., New York, (available through Wiley), 1992.

ROBERTS, S. C., AND M. L. SHULER, Large Scale Plant Cell Culture, *Current Opin. Biotechnol. 8*/2:154–159, 1997.

SCHRIPSEMA, J., AND R. VERPORTE, Search for Factors Related to the Indole Alkaloid Production in Cell Suspension Cultures of *Tabernaemontana divaricata, Planta Med. 58*:245–249, 1992.

PROBLEMS

13.1. The uptake of the auxin, indole acetic acid (or IAA), by suspension cultures of *Parthenocissus* sp. is nearly zero order at 1 nmol/mg dry cell weight-min. The diffusivity of IAA in water is 5×10^{-6} cm^2/s. Beads of calcium alginate are most conveniently made as spheres with a 4-mm diameter. Assume the beads are made 25% by volume of plant cells. Assume the plant cells are 90% water and that the diffusivity of IAA in the gel is the same as in water. If the external concentration is maintained at 2 μmol, will IAA penetrate to the center of the bead?

13.2. Gel-immobilized cells of *Papaver somniferum* (opium poppy) can make codeine from codeinone. The rate of codeinone uptake is first order, with a rate constant of 3.3×10^{-8} l/g cells dry weight-s. The diffusivity of codeinone in the gel is 0.2×10^{-9} m^2/s. For a gel particle of 4-mm diameter with a 25% volume loading of cells (95% water), what will be the effectiveness factor?

13.3. The $k_L a$ of a small bubble column (2 l) has been measured as 20 h^{-1} at an airflow of 4 l/m in. If the rate of oxygen uptake by a culture of *Catharanthus roseus* is 0.2 mmol O$_2$/g dry weight-h and if the critical oxygen concentration must be above 10% of saturation (about 8 mg/l), what is the maximum concentration of cells that can be maintained in the reactor?

13.4. *C. roseus* cells immobilized in Ca-alginate beads of diameter 0.5 mm are used for production of indole alkaloids (IA) in a fluidized-bed bioreactor. The rate limiting nutrient is glucose and no intraparticle diffusion limitations exist. Use the following data: Flow rate of the feed: Q = 1 l/h, Glucose in the feed: S_o = 30 g/l, Plant Cell Concentration: X = 6 g/l reac. The rate constant for IA formation: k = 5 d^{-1} (g/l)$^{-1}$ K_s = 0.4 g/l, Column diameter: D_o = 0.15 m. Growth is negligible and Monod kinetics is valid. Determine the following:

a. For 95% glucose conversion determine required hydraulic residence time, volume, and the height of the column

b. If $Y_{p/s}$ is 0.02 g IA/g glu, determine IA concentration in the effluent and the productivity.

14

Utilizing Genetically Engineered Organisms

14.1. INTRODUCTION

In Chapter 8 we discussed how cells could be genetically engineered. The techniques of genetic engineering are fairly straightforward; the design of the best production system is not. The choice of host cell, the details of the construction of the vector, and the choice of promoter must all fit into a processing strategy. That strategy includes plans not only for efficient production but also for how a product is to be recovered and purified. The development of processes for making products from genetically engineered organisms requires that many choices be made. Any choice at the molecular level imposes processing constraints. As indicated in Fig. 14.1, an interdisciplinary team approach to process development is necessary to make sure that a well-designed process results.

In this chapter we will discuss some of the questions that must be considered in building processes using genetically engineered cells.

14.2. HOW THE PRODUCT INFLUENCES PROCESS DECISIONS

Genetically engineered cells can be used to make two major classes of products: proteins and nonproteins. Nonprotein products can be made by metabolically engineering cells, inserting DNA-encoding enzymes that generate new pathways or pathways with an

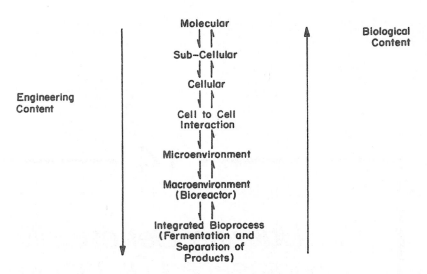

Figure 14.1. Biocatalyst and bioreactor design are strongly interconnected. Choices made at the molecular level can have profound effects on bioreactor design. A multidisciplinary approach is necessary to ensure that a well-designed process is developed.

enhanced capacity to process the precursors to a desired metabolite. However, most current industrial emphasis has been on proteins. Table 14.1 lists examples. The majority of these proteins are human therapeutics, but proteins that can be used in animal husbandry, in food processing, or as industrial catalysts are of interest. Sales in the United States for these products are forecasted to be $32 billion in 2006. Over 200 other proteins are in clinical trials.

With therapeutic proteins that are injectable, the prime concern is the clinical efficiency of the product. Such products must be highly pure, since strong immunogenic reactions by patients or other side effects can be disastrous. The authenticity of the product is often critical. Correct or near correct posttranslational processing of the protein (e.g., glycosylation or phosphorylation) is sometimes essential to its therapeutic action. Any variant forms of the protein (e.g., the modification of side groups on amino acids) are highly undesirable and present very difficult purification problems.

The processing challenges in making therapeutic proteins are to ensure product quality and safety. Process efficiency to reduce manufacturing cost, although important, is of less concern, since these products are required in relatively small amounts, because they can command high prices, and because the selling price is much more determined by the costs of process development and regulatory approvals, particularly clinical trials. Thus, for these products the choice of biological system and processing equipment is dictated by the need to produce highly purified material in an absolutely consistent manner.

Other protein products are purchased strictly on an economic basis, and manufacturing costs play a much more critical role in the viability of a proposed process. In this case, regulatory demands are of lesser importance than in the production of therapeutic proteins. For animal vaccines or animal hormones, the products must be very pure and render favorable cost ratios. For example, the use of bovine somatropin to increase

TABLE 14.1 Key Biopharmaceutical Products

Product	End use
A. Hormones and peptide factors	
Human insulin	Diabetes
Factor VIII-C	Hemophilla
Human growth hormone	Growth deficiency
Erythropoietin	Anemia, chronic renal failure
Interferon-alpha 2a	Hairy cell leukemia, AIDS-related cancer
Interferon-alpha 2b	Hairy cell leukemia, Herpes
Interferon-beta	Cancer
Interferon-gamma	Cancer, venereal warts, infectious disease
Interleukin-2	Cancer immunotherapy, AIDS
Muromonab-CD3	Acute kidney transplant rejection
Granulocyte colony stimulating factor (G-CSF)	Chemotherapy effects, AIDS
Granulocyte macrophage-colony stimulating factor (GM-CSF)	Autologous bone marrow transplant
B. Enzymes	
Tissue plasminogen activator	Acute myocardial infarction, stroke
Prourokinase/urokinase	Heart attack
DNase	Pulmonary treatment
Glucocerebrosidase	Gaucher
C. Vaccines	
Hepatitis B	Hepatitis B vaccine
Herpes	Herpes
D. Monoclonal antibodies	Wide range of different antibodies for diagnostics
	Prevention of blood clots
	Breast cancer
	Lung cancer

milk production requires that the increased value of the milk produced be substantially greater than the cost of the hormone and any increase in feed costs due to increased milk production.

For food use, product safety is important, but purity requirements are less stringent than for an injectable. The volume of the required product is often substantial (several metric tons per year), and price is critical because alternative products from natural sources may be available. Proteins used as specialty chemicals (e.g., adhesives and enzymes for industrial processes) usually can tolerate the presence of contaminating proteins and compounds. The manufacturing costs for such proteins will greatly influence the market penetration.

For nonprotein products based on metabolically engineered cells, processing costs compared to costs of other routes of manufacture (usually nonbiological) will determine the success of such a process.

As the reader has certainly noted, the constraints on production can vary widely from one product class to another. These constraints determine which host cells, vectors, genetic constructions, processing equipment, and processing strategies are chosen.

14.3. GUIDELINES FOR CHOOSING HOST–VECTOR SYSTEMS

14.3.1. Overview

The success or failure of a process often depends on the initial choice of host organism and expression system. Such choices must be made in the context of a processing strategy.

Table 14.2 summarizes many of the salient features of common host systems. The most important initial judgment must be whether posttranslational modifications of the product are necessary. If they are, then an animal cell host system must be chosen. If some simple posttranslational processing is required (e.g., some forms of glycosylation), yeast or fungi may be acceptable. Whether posttranslational modifications are necessary for proper activity of a therapeutic protein cannot always be predicted with certainty, and clinical trials may be necessary.

Another important consideration is whether the product will be used in foods. For example, some yeasts (e.g., *S. cerevisiae*) are on the FDA GRAS list (*generally regarded as safe*), which would greatly simplify obtaining regulatory approval for a given product. In some cases edible portions of transgenic plants can be used to deliver vaccines or therapeutic proteins.

14.3.2. *Escherichia coli*

If posttranslational modifications are unnecessary, *E. coli* is most often chosen as the initial host. The main reason for the popularity of *E. coli* is the broad knowledge base for it. *E. coli* physiology and its genetics are probably far better understood than for any other

TABLE 14.2 Characteristics of Selected Host Systems for Protein Production from Recombinant DNA

| Characteristic | Organism | | | | |
	E. coli	*S. cerevisiae*	*P. pastoris*	Insect	Mammalian
High growth rate	E[a]	VG	VG	P–F	P–F
Availability of genetic systems	E	G	F	F–G	F–G
Expression levels	E	VG	E	G–E	P–G
Low-cost media available	E	E	E	P	P
Protein folding	F	F–G	F–G	VG–E	E
Simple glycosylation	No	Yes	Yes	Yes	Yes
Complex glycosylation	No	No	No	Yes[b]	Yes
Low levels of proteolytic degradation	F–G	G	G	VG	VG
Excretion or secretion	P normally VG in special cases	VG	VG	VG	E
Safety	VG	E	VG	E	G

[a]E, excellent; VG, very good; G, good; F, fair; P, poor.

[b]Glycosylation patterns differ from mammalian cells.

living organism. A wide range of host backgrounds (i.e., combinations of specific mutations) is available, as well as vectors and promoters. This large knowledge base greatly facilitates sophisticated genetic manipulations. The well-defined vectors and promoters greatly speed the development of an appropriate biological catalyst.

The relatively high growth rates for *E. coli* coupled with the ability to grow *E. coli* to high cell concentrations (> 50 g dry wt/l) and with the high expression levels possible from specific vector–promoter combinations (about 25% to 50% or more of total protein) can lead to extremely high volumetric productivities. Also, *E. coli* will grow on simple and inexpensive media. These factors give *E. coli* many economic advantages.

An important engineering contribution was the development of strategies to grow cultures of *E. coli* to high cell densities. The buildup of acetate and other metabolic by-products can significantly inhibit growth. Controlled feeding of glucose so as to prevent the accumulation of large amounts of glucose in the medium prevents overflow metabolism and the formation of acetate. Glucose feeding can be coupled to consumption rate if the consumption rate can be estimated on-line or predicted.

However, *E. coli* is not a perfect host. The major problems result from the fact *E. coli* does not normally secrete proteins. When proteins are retained intracellularly and produced at high levels, the amount of soluble active protein present is usually limited due to either proteolytic degradation or insolubilization into *inclusion bodies.*

The production of large amounts of foreign protein may trigger a *heat-shock* response. One response of the heat-shock regulon is increased proteolytic activity. In some cases, intracellular proteolytic activity results in product degradation at a rate nearly equal to the rate of production.

More often, the target protein forms an inclusion body. Although the heterologous protein predominates in an inclusion body, other cellular material is also often included. The protein in the inclusion body is misfolded. The misfolded protein has no biological activity and is worthless. If the inclusion bodies are recovered from the culture, the inclusion bodies can be resolubilized and activity (and value) restored. Resolubilization can vary tremendously in difficulty from one protein to another. When resolubilization is straightforward and recoveries are high, the formation of inclusion bodies can be advantageous, as it simplifies the initial steps of recovery and purification. It is important that during resolubilization the protein be checked by several analytical methods to ensure that no chemical modifications have occurred. Even slight changes in a side group can alter the effectiveness of the product.

Other consequences of cytoplasmic protein production can be important. The intracellular environment in *E. coli* might not allow the formation of disulfide bridges. Also the protein will usually start with a methionine, whereas that methionine would have been removed in normal posttranslational processing in the natural host cell. If the product is retained intracellularly, then the cell must be lysed (broken) during recovery. Lysis usually results in the release of endotoxins (or pyrogens) from *E. coli*. Endotoxins are lipopolysaccharides (found in the outer membrane) and can result in undesirable side effects (e.g., high fevers) and death. Thus, purification is an important consideration.

Many of the limitations on *E. coli* can be circumvented with protein secretion and excretion. *Secretion* is defined here as the translocation of a protein across the inner membrane of *E. coli*. *Excretion* is defined as release of the protein into the extracellular compartment.

About 20% of all protein in *E. coli* is translocated across the inner membrane into the periplasmic space or incorporated into the outer membrane. As the reader will recall from Chapter 4, secreted proteins are made with a signal or leader sequence. The presence of a *signal sequence* is a necessary (but not sufficient) condition for secretion. The signal sequence is a sequence of amino acids attached to the mature protein, and the signal sequence is cleaved during secretion.

Many benefits are possible if a protein is secreted. Secretion eliminates an undesired methionine from the beginning of the protein. Secretion also often offers some protection from proteolysis. Periplasmic proteases exist in *E. coli*, but usually at a low level. They are most active at alkaline pH values, and pH control can be used to reduce target protein degradation. The environment in the periplasmic space promotes the correct protein folding in some cases (including the formation of disulfide bridges). Proteins in the periplasmic space can be released by gentle osmotic shock, so that fewer contaminating proteins are present than if the whole cell were lysed.

Even more attractive would be the extracellular release of target proteins. Normally, *E. coli* does not excrete protein (colicin and haemolysin are the two exceptions), but a variety of schemes to obtain excretion in *E. coli* are being developed. Strategies usually involve either trying to disrupt the structure of the outer membrane or attempting to use the colicin or haemolysin excretion systems by constructing a fusion of the target protein with components of these excretion systems. Excretion without cell lysis can simplify recovery and purification even more than secretion alone, while achieving the same advantages as secretion with respect to protein processing. Excretion also facilitates the potential use of continuous immobilized cell systems.

Excretion of normally cytoplasmic or human-designed proteins is problematic. Two preliminary reports for the excretion of normally cytoplasmic proteins have been made, but the general principles for the extension of excretion to cytoplasmic proteins are still being developed. At this time, the excretion of normally secreted proteins can be obtained in *E. coli* (and other cells) even when the protein is derived from animal cells. However, extension to nonsecreted proteins is difficult.

The lack of established excretion systems in *E. coli* has led to interest in alternative expression systems. Also, in some cases patent considerations may require the use of alternative hosts.

14.3.3. Gram-positive Bacteria

The gram-positive bacterium, *Bacillus subtilis,* is the best studied bacterial alternative to *E. coli.* Since it is gram positive, it has no outer membrane, and it is a very effective excreter of proteins. Many of these proteins, amylases and proteases, are produced commercially using *B. subtilis.* If heterologous proteins could be excreted as efficiently from *B. subtilis,* then *B. subtilis* would be a very attractive production system.

However, *B. subtilis* has a number of problems that have hindered its commercial adoption. A primary concern has been that *B. subtilis* produces a large amount and variety of proteases. These proteases can degrade the product very rapidly. Mutants with greatly reduced protease activity have become available, but even these mutants may have sufficient amounts of minor proteases to be troublesome. *B. subtilis* is also much more difficult to manipulate genetically than *E. coli* because of a limited range of vectors and promoters.

Also, the genetic instability of plasmids is more of a problem in *B. subtilis* than in *E. coli*. Finally, the high levels of excretion that have been observed with native *B. subtilis* proteins have not yet been obtainable with heterologous protein (i.e., foreign proteins produced from recombinant DNA).

Other gram-positive bacteria that have been considered as hosts include *Streptococcus cremoris* and *Streptomyces* sp. These other systems are less well characterized than *B. subtilis*.

Both gram-negative and gram-positive bacteria have limitations on protein processing that can be circumvented with eucaryotic cells.

14.3.4. Lower Eucaryotic Cells

The yeast, *Saccharomyces cerevisiae,* has been used extensively in food and industrial fermentations and is among the first organisms harnessed by humans. It can grow to high cell densities and at a reasonable rate (about 25% of the maximum growth rate of *E. coli* in similar medium). Yeast are larger than most bacteria and can be recovered more easily from a fermentation broth.

Further advantages include the capacity to do simple glycosylation of proteins and to secrete proteins. However, *S. cerevisiae* tends to hyperglycosylate proteins, adding large numbers of mannose units. In some cases the hyperglycosylated protein may be inactive. These organisms are also on the GRAS list, which simplifies regulatory approval and makes yeast particularly well suited to production of food-related proteins.

Generally, the limitations on *S. cerevisiae* are the difficulties of achieving high protein expression levels, hyperglycosylation, and good excretion. Although the genetics of ·*S. cerevisiae* are better known than for any other eucaryotic cell, the range of genetic systems is limited, and stable high protein expression levels are more difficult to achieve than in *E. coli*. Also, the normal capacity of the secretion pathways in *S. cerevisiae* is limited and can provide a bottleneck on excreted protein production, even when high expression levels are achieved.

The methylotrophic yeasts, *Pichia pastoris* and *Hansenula polymorpha,* are very attractive hosts for some proteins. These yeasts can grow on methanol as a carbon-energy source; methanol is also an inducer for the AOX 1 promoter, which is typically used to control expression of the target protein. Very high cell densities (e.g., up to 100 g/l) can be obtained. Due to high densities and, for some proteins, high expression levels, the volumetric productivities of these cultures can be higher than with *E. coli*. Protein folding and secretion are, also, often better than in *E. coli*. These yeasts do simple glycosylation and are less likely to hyperglycosylate than *S. cerevisiae*. Like many host systems, their effectiveness is often a function of the target protein. The disadvantages of the methylotrophic yeast are due to the high cell density and rate of metabolism, which creates high levels of metabolic heat that must be removed and high oxygen demand. Effective induction of expression, while maintaining cell activity, requires very good process control due to methanol's dual rate as growth substrate and inducer. Further, high levels of methanol are inhibitory (i.e., substrate inhibition), which also demands good process control. Scale-up to large reactors often is very challenging, since heat removal, oxygen supply, and process control are typically more difficult in large reactors with longer mixing times (see Chapter 10). Also, methanol is flammable, and handling large volumes of methanol is a safety concern. Nonetheless, these methylotrophic yeasts are of increasing importance.

Fungi, such as *Aspergillus nidulans* and *Trichoderma reesei,* are also potentially important hosts. They generally have greater intrinsic capacity for protein secretion than *S. cerevisiae.* Their filamentous growth makes large-scale cultivation somewhat difficult. However, commercial enzyme production from these fungi is well established, and the scale-up problems have been addressed. The major limitation has been the construction of expression and secretion systems that can produce as large amounts of extracellular heterologous proteins as some of the native proteins. A better understanding of the secretion pathway and its interaction with protein structure will be critical for this system to reach its potential.

All these lower eucaryotic systems are inappropriate when complex glycosylation and posttranslational modifications are necessary. In such cases, animal cell tissue culture has been employed.

14.3.5. Mammalian Cells

Mammalian cell culture is chosen when the virtual authenticity of the product protein must be complete. Authenticity implies not only the correct arrangement of all amino acids, but also that all posttranslational processing is identical to that in the whole animal. In some cases, the cells in culture may not do the posttranslation modifications identically to those done by the same cell while in the body. But for bioreactor processes, mammalian cell tissue culture will provide the product closest to its natural counterpart. Another advantage is that most proteins of commercial interest are readily excreted.

Slow growth, expensive media, and low protein expression levels all make mammalian cell tissue culture very expensive. As discussed in Chapter 12, a wide variety of reactor systems are being used with animal cell cultures. Although many of these can improve efficiency significantly, processes based on mammalian cells remain very expensive.

Several cell lines have been used as hosts for the production of proteins using recombinant DNA. The most popular hosts are probably lines of CHO (Chinese hamster ovary) cells.

In addition to the cost of production, mammalian cells face other severe constraints. Normal cells from animals are capable of dividing only a few times; these cell lines are *mortal.* Some cells are *immortal* or *continuous* and can divide continuously, just as a bacterium can. Continuous cell lines are *transformed* cells. Cancer cells are transformed also (i.e., have lost the inhibition of cell replication). The theoretical possibility that a cancer-promoting substance could be injected along with the desired product necessitates extreme care in the purification process. It is particularly important to exclude nucleic acids from the product. The use of transformed cells also requires cautions to ensure worker safety.

Additionally, the vectors commonly used with mammalian cell cultures have been derived from primate viruses. Again, there is concern about the reversion of such vectors back to a form that could be pathogenic in humans.

Most of these vectors cannot give high expression levels of the target protein in common host cells (usually < 5% of total protein). However, higher levels of expression can be obtained (e.g., > 100 mg/l of secreted, active protein) through amplification of number of gene copies. It may take six months with a CHO cell line to achieve stable, high-level expression. It is often easier to obtain high titers (or product concentration) when producing monoclonal antibodies from hybridoma cultures.

While the quality of the protein product may change upon scale-up with any system, this issue is particularly important with animal cell cultures. This contention is due, in part, to the fact that animal cell cultures are used primarily because authenticity of the protein product is a major concern. Since culture conditions (shear, glucose, amino sugars, DO, etc.) can change upon scale-up, the efficiency of cellular protein processing can change, altering the level of posttranslational processing. Further, it has been shown that protein quality may change with harvest time in batch cultures. This change may be due to alterations in intracellular machinery, but often it is due to release of proteases and siladase (an enzyme that removes the silalic acid cap from glycosylated proteins) from dead cells. Also, excessive levels of protein production may saturate the intracellular protein-processing organelles (i.e., ER and Golgi), leading to incompletely processed proteins. These problems can significantly impact process strategy. One well-known company has been forced to harvest 24 h early to maintain the silalic acid/protein ratio specified for the product. Early harvest resulted in a 30% loss in protein concentration (as compared with delayed harvest).

Strategies to reduce such problems include selection of cell lines or genetic manipulation of cell lines with reduced levels of siladase production or enhanced protein processing capacity. Redesign of medium can be beneficial; chemicals that inhibit undesirable extracellular enzyme activity can be added or precursors added (e.g., amino sugars) to improve processing. Cell lysis can be reduced by adding genes (e.g., *bcl* 2) to the host cell that reduce apotosis. An engineering solution is to remove the product from the medium as it is formed. For example, perfusion systems with an integrated product capture step can be used.

14.3.6. Insect Cell–Baculovirus System

A popular alternative for protein production at small (< 100 l) or laboratory scale is the insect cell–baculovirus system. This system is particularly attractive for rapidly obtaining biologically active protein for characterization studies. Typical host cell lines come from the fall armyworm (*Spodoptera frugiperda*) and the cabbage looper (*Trichoplusia ni*). The baculovirus, *Autographa californica* nuclear polyhedrosis virus (Ac NPV), is used as a vector for insertion of recombinant DNA into the host cell.

This virus has an unusual biphasic replication cycle in nature. An insect ingests the occluded form, in which multiple virus particles are embedded in a protein matrix. The protein matrix protects the virus when it is on a leaf from environmental stresses (e.g., UV radiation). This protein matrix is from the polyhederin protein. In the mid-gut of the insect, the matrix dissolves, allowing the virus to attack the cells lining the insect's gut; this is the primary infection. These infected cells release a second type of virus; it is nonoccluded (no polyhederin matrix) and buds through the cell envelope. The nonoccluded virus (NOV) infects other cells throughout the insect (secondary infection).

In insect cell culture, only NOVs are infectious, and the polyhederin gene is unnecessary. The polyhederin promoter is the strongest known animal promoter and is expressed late in the infection cycle. Replacing the polyhederin structural gene with the gene for a target protein allows high-level target protein production (up to 50% of cellular protein). Proteins that are secreted and glycosylated are often made at much lower levels than nonsecreted proteins.

In addition to high expression levels, the insect–baculovirus system offers safety advantages over mammal–retrovirus systems. The insect cell lines derived from ovaries or embryos are continuous but not transformed. The baculovirus is not pathogenic toward either plants or mammals. Thus, the insect–baculovirus system offers potential safety advantages. Another important advantage is that the molecular biology and high-level expression of correctly folded proteins can be achieved in less than a month.

This system also has the cellular machinery to do almost all the complex posttranslational modifications that mammalian cells do. However, even when the machinery is present, at least some of the proteins produced in the insect cell–baculovirus system are not processed identically to the native protein. In some cases, their slight variations may be beneficial (e.g., increased antigenic response in the development of an AIDS vaccine), while in others they may be undesirable. While complex glycoforms (including silaic acid) have been made, it is more common to observe only simple glycoforms. Production of complex forms requires special host cell lines and is sensitive to culture conditions.

The insect cell–baculovirus system is a good system to illustrate a holistic perspective on heterologous protein production. Any bioprocess for protein production is complex, consisting of the nonlinear interaction of many subcomponents. Thus the optimal process is not simply the sum of individually optimized steps. Figure 14.2 presents a holistic view for the insect cell–baculovirus system. Because of the viral component, this system is even more complex than most other bioprocesses, as the infection process and resulting protein expression kinetics must be considered. One factor is the ratio of infectious particles to cells (e.g., multiplicity of infection or MOI), which alters the synchrony of infection and the resulting protein expression kinetics. Another is the genetic design of the virus (which shares many of the general features of vector design). Also, the quality of the virus stock is important; if the virus stock is maintained incorrectly, mutant virus can form. One example is the formation of *defective interfering particles* (DIPs) that reduce protein expression in the culture by 90% when high MOIs are used.

You should work from the bottom of Fig. 14.2 toward the top. What is the desired product quality? What is the product worth? These questions then guide selection of bioprocess strategies to achieve the cost and quality desired. To develop that strategy requires an understanding of the basic kinetics and capabilities of the biological system. Understanding these requirements guides selection of the specific host cell line, the medium, and the molecular design of the virus. For example, the addition of serum to the medium of some *Ti ni* cell lines results in production of proteins with complex N-linked glycosylation, including a sialic acid cap, which may be a requirement for product quality. However, the use of serum alters growth kinetics, expression levels (often less), and the difficulty of purification, which may alter cost. Such trade-offs need to be considered with respect to alternative approaches (e.g., development of a genetically engineered host that could perform the same reactions, but in serum-free medium).

14.3.7. Transgenic Animals

In some cases proteins with necessary biological activity cannot be made in animal cell culture. While posttranslational protein processes, such as N-linked glycosylation, can be done in cell culture, other more subtle forms of posttranslational processing may not be done satisfactorily. An alternative to cell culture is the use of transgenic animals. Animals

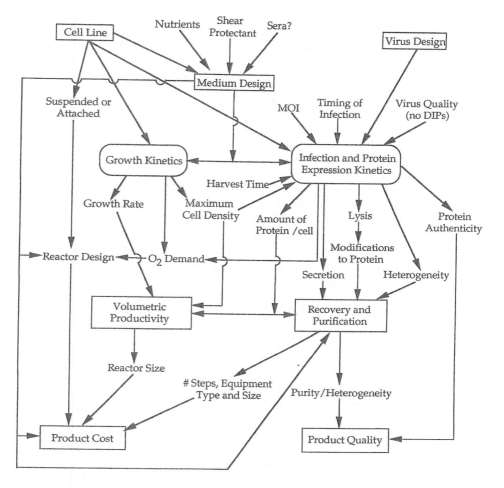

Figure 14.2. Holistic perspective on heterologous protein production with the baculovirus expression system.

are engineered to express the protein and release it into specific fluids, such as milk or urine. High concentrations of complex proteins can be achieved, and such approaches can be cost effective for complex proteins. In these cases the role for the bioprocess engineer is in protein recovery and purification, although significant issues exist for agricultural engineers and animal scientists in devising appropriate systems to obtain the protein-containing fluid (e.g., pig milking stations). While transgenic animals can be developed from many mammals, sheep, goats, and pigs are the primary species used commercially.

There are significant limitations on this technology. In some cases, the protein of interest will cause adverse health problems in the producing animal. The use of animals raises safety concerns with respect to virus or prion transmission. The process to generate and screen for high-producing animals is inefficient and costly (e.g., $100,000 for a goat and $500,000 for a cow). Perhaps surprisingly, not all of the complex posttranslational processing steps necessary to achieve the desired product occur when the protein is expressed in milk, urine, or blood. Nonetheless, transgenic animals will be critical to production of some proteins.

14.3.8. Transgenic Plants and Plant Cell Culture

Proteins, including many complex protein assemblies, such as antibodies and virus-like particles (as vaccines), can be made inexpensively in plants. Transgenic plants offer many potential advantages in addition to cost. Since plant viruses are not infective for humans, there are no safety concerns with respect to endogenous viruses or prions. Scale-up is readily accomplished by planting more acreage. The protein can be targeted for sterile, edible compartments, either reducing the need for rigorous purification or making it an ideal vehicle for oral delivery of a therapeutic protein. Indeed, development of edible vaccines for use in developing countries is being actively pursued.

The disadvantages of transgenic plants are that expression levels are often low (1% of total soluble protein is considered good), N-linked glycosylation is incomplete, and some other mammalian posttranslational processing is missing. While inexpensive, with easy scale-up, it takes 30 months to test and produce sufficient seed for unlimited commercial use. Such long lead times are undesirable. Further, environmental control on field-grown crops is difficult, so the amount (and possibly quality) of the product can vary from time to time and place to place.

While many crops could be used, much of the commercial interest centers on transgenic corn. Some corn products are used in medicinals, so there exist some FDA guidelines (e.g., contamination with herbicides and mycotoxins), and there is considerable processing experience. Production costs vary with degree of desired purity. For high-purity material (95% pure), a cost of about $4 to $8/g can be estimated. For higher-purity material (99%) with full quality assurance and control, a cost of about $20 to $30/g is reasonable. At least two enzymes are being produced commercially from transgenic plants. Large-scale production of monoclonal antibodies (e.g., 500 kg/yr) for topical uses is being considered.

The use of plant cell cultures is also being explored. The primary advantage of such cultures over transgenic plants is the much higher level of control that can be exercised over the process. Plant cell cultures, compared to animal cell cultures, grow to very high cell density, use defined media, and are intrinsically safer.

14.3.9. Comparison of Strategies

The choice of host–vector system is complicated. The characteristics desired in the protein product and the cost are the critical factors in the choice. The dominant systems for commercial production are *E. coli* and CHO cell cultures. An interesting study of the process economics of these two systems for production of tissue plasminogen activator, tPA, has been published by Datar, Cartwright, and Rosen.[†] Their analysis was for plants making 11 kg/yr of product. The CHO cell process was assumed to produce 33.5 mg/l of product, while *E. coli* made 460 mg/l. The CHO cell product was correctly folded, biologically

[†]*Bio/Technology 11*:349 (1993).

active, and released into the medium. The *E. coli* product was primarily in the form of inclusion bodies, and thus biologically inactive, misfolded, and insoluble. The process to resolubilize and refold the *E. coli* product into active material requires extra steps. The recovery process for the CHO cell material requires five steps, while 16 steps are required for the *E. coli* process. The larger the number of steps, the greater the possibility of yield loss. Total recovery of 47% with the CHO-produced material was possible compared to only 2.8% for the *E. coli*-produced material.

The extra steps in the *E. coli* process are for cell recovery, cell breakage, recovery of inclusion bodies, resolubilization of inclusion bodies, concentration, sulfination, refolding, and concentration of the renatured protein. The difficulty of these processes depends on the nature of the protein; tPA is particularly difficult. With tPA the concentration of tPA had to be maintained at 2.5 mg/l or less, and refolding is slow, requiring 48 hours. A 20% efficiency for renaturation was achieved. Many proteins can be refolded at higher concentrations (up to 1 g/l) and much more quickly. For tPA the result is unacceptably large tanks and very high chemical usage. In this case, five tons of urea and 26 tons of guanidine would be necessary to produce only 11 kg of active tPA.

For tPA the required bioreactor volumes were 14,000 l for the CHO process and 17,300 l for the *E. coli* process. The capital costs were $11.1 million for the CHO process and $70.9 million for the *E. coli* process, with 75% of that capital cost being required for the refolding tanks. Under these conditions, the unit production costs are $10,660/g for the CHO process versus $22,000/g for the *E. coli* process. The rate of return on investment (ROI) for the CHO process was 130% versus only 8% for the *E. coli* process. However, if the refolding step yield were 90% instead of 20%, the overall yield would improve to 15.4%, and the unit production cost would fall to $7,530 with an ROI of 85% for the improved *E. coli* process at production of 11 kg/yr. If the *E. coli* plant remained the original size (17,300 l fermentor) so as to produce 61.3 kg/yr, the unit production cost would drop to $4,400/g. The cost of tPA from the CHO process is very sensitive to cost of serum in the medium. If the price of media dropped from $10.5/l to $2/l (e.g., 10% to 2% serum in the medium), the cost of the CHO cell product would drop to $6,500/g.

A primary lesson from this exercise is the difficulty of making choices of host–vector systems without a fairly complete analysis. The price will depend on the protein, its characteristics, and intended use. Changes in process technology (e.g., low serum medium for CHO cells or protein secretion systems in *E. coli*) can have dramatic effects on manufacturing costs and choice of the host–vector system.

14.4. PROCESS CONSTRAINTS: GENETIC INSTABILITY

There is a tension between the goal of maximal target-protein production and the maintenance of a vigorous culture. The formation of large amounts of foreign protein is always detrimental to the host cell, often lethal. Cells that lose the capacity to make the target protein often grow much more quickly and can displace the original, more productive strain. This leads to *genetic instability* (see Fig. 14.3).

Genetic instability can occur due to *segregational loss, structural instability, host cell regulatory mutations,* and the *growth-rate ratio* of plasmid-free or altered cells to

Genetic Instability

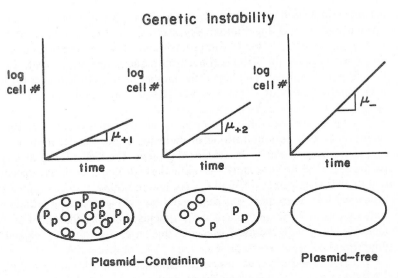

Figure 14.3. Cells that contain plasmids (O) actively making protein (p) must direct many of their cellular resources away from growth and hence grow more slowly than plasmid-free cells (that is, $\mu_{+1} < \mu_{+2} < \mu_{-}$).

plasmid-containing unaltered cells. Genetic instability can occur in any expression system. We illustrate this problem by considering gene expression from plasmids in bacteria.

14.4.1. Segregational Loss

Segregational loss occurs when a cell divides such that one of the daughter cells receives no plasmids (see Fig. 14.4). Plasmids can be described as *high-copy-number plasmids* (> 20 copies per cell) and *low-copy-number plasmids* (sometimes as low as one or two copies per cell). Low-copy-number plasmids usually have specific mechanisms to ensure their equal distribution among daughter cells. High-copy-number plasmids are usually distributed randomly (or nearly randomly) among daughter cells following a binomial distribution. For high copy numbers, almost all the daughter cells receive some plasmids, but even if the possibility of forming a plasmid-free cell is low (one per million cell divisions), a large reactor contains so many cells that some plasmid-free cells will be present (e.g., 1000 l with 10^9 cells/ml yields 10^{15} cells and 10^9 plasmid-free cells being formed every cell generation).

The segregational loss of plasmid can be influenced by many environmental factors, such as dissolved oxygen, temperature, medium composition, and dilution rate in a chemostat. Many plasmids will also form *multimers,* which are multiple copies of the same plasmid attached to each other to form a single unit. The process of multimerization involves using host cell recombination systems. A *dimer* is a replicative unit in which two separate plasmids have been joined, and a *tetramer* is a single unit consisting of four separate monomers fused together.

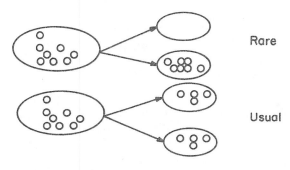

Rare

Usual

**Random Distribution
(binomial)**

Figure 14.4. Segregational instability results when a dividing cell donates all its plasmids to one progeny and none to the other.

Example 14.1.

What fraction of the cells undergoing division will generate a plasmid-free cell if:
  a. All cells have 40 plasmids at division?
  b. All cells have enough plasmid DNA for 40 copies, but one-half of the plasmid DNA is in the form of dimers and one-fifth in the form of tetramers?
  c. Half of the cells have 10 copies of the plasmid and half have 70 copies (the average copy number is 40 as in case a)?

Solution

a. If we assume a random distribution at division, the probability of forming a plasmid-free cell is

$$P = 2^{(1-Z)} \tag{14.1}$$

where Z is the number of plasmid replicative units. For $Z = 40$, $P = 1.8 \times 10^{-12}$ plasmid-free cells per division.

b. If the total amount of plasmid DNA is equivalent to 40 single copies, we can determine the plasmid distribution from

$$M + D + T = 40 \text{ monomer plasmid equivalents} \tag{14.2a}$$

$$D = \frac{1}{2}(40) = 20 \text{ monomer plasmid equivalents} \tag{14.2b}$$

$$T = \frac{1}{5}(40) = 8 \text{ monomer plasmid equivalents} \tag{14.2c}$$

which implies that

$$M = 40 - 20 - 8 = 12 \text{ monomer plasmid equivalents} \tag{14.3}$$

The number of copies of replicative units is then

$$M + D/2 + T/4 = \text{total replicative units} \tag{14.4a}$$

$$12 + 10 + 2 = 24 \text{ total replicative units} \tag{14.4b}$$

since a dimer consists of two monomer equivalents and a tetramer of four monomer equivalents.

Equation 14.1 can then be applied to yield

$$P = 2^{(1-24)} = 1.2 \times 10^{-7} \qquad (14.5)$$

c. When plasmids are not evenly distributed, the probability of forming a plasmid-free cell changes. In this case,

$$\frac{1}{2}P_{10} + \frac{1}{2}P_{70} = \frac{1}{2}[2^{(1-10)}] + \frac{1}{2}[2^{(1-70)}]$$
$$= 9.8 \times 10^{-4} + 8.5 \times 10^{-22} = 9.8 \times 10^{-4} \qquad (14.6)$$

Note that although this population has the same average copy number as case a, the probability of forming a plasmid-free cell is 5.4×10^8 greater.

14.4.2. Plasmid Structural Instability

In addition to the problems of segregational instability, some cells retain plasmids but alter the plasmid so as to reduce its harmful effects on the cell (structural instability). For example, the plasmid may encode both for antibiotic resistance and for a foreign protein. The foreign protein drains cellular resources away from growth toward an end product of no benefit to the cell. However, if the investigator has added antibiotics to the medium, the cell will benefit from retaining the gene encoding the antibiotic resistance. Normal mutations will result in some altered plasmids that retain the capacity to encode for desirable functions (for example, antibiotic resistance) while no longer making the foreign protein. In other cases, cellular recombination systems will integrate the gene for antibiotic resistance into the chromosome. Cells containing structurally altered plasmids can normally grow much more quickly than cells with the original plasmids. A culture having undergone a change in which the population is dominated by cells with an altered plasmid has undergone structural instability.

14.4.3. Host Cell Mutations

Mutations in host cells can also occur that make them far less useful as production systems for a given product. These mutations often alter cellular regulation and result in reduced target-protein synthesis. For example, if the promoter controlling expression of the foreign protein utilizes a host cell factor (e.g., a repressor), then modification of the host cell factor may greatly modulate the level of production of the desired plasmid-encoded protein. The lac promoter (see our discussion in Chapter 4 of the lac operon) can be induced by adding chemicals; for a lac promoter lactose or a chemical analog of lactose (e.g., IPTG or isopropyl-β-D-thiogalactoside) can be used. Such promoters are often used in plasmid construction to control the synthesis of a plasmid-encoded protein. If induction of plasmid-encoded protein synthesis from this promoter reduces cellular growth rates (as in Fig. 14.2), then a mutation that inactivates lac permease would prevent protein induction in that mutant cell. The lac permease protein is necessary for the rapid uptake of the inducer. Thus, the mutant cell would grow faster than the desired strain. Alternatively, a host cell mutation in the repressor, so that it would not recognize the inducer, would make induction impossible.

The key feature of this category of genetic instability is that a host cell mutation imparts a growth advantage to the mutant, so that it will eventually dominate the culture. In this case the mutant cell will contain unaltered plasmids but will make very little of the target, plasmid-encoded protein.

14.4.4. Growth-rate-dominated Instability

The importance of all three of these factors (segregational loss of plasmid, structural alterations of the plasmid, and host cell mutations) depends on the growth-rate differential of the changed cell-plasmid system to the original host–vector system. If the altered host–vector system has a distinct growth advantage over the original host–vector system, the altered system will eventually dominate (i.e., genetic instability will occur).

The terms used to describe the cause of genetic instability are based on fairly subtle distinctions. For example, if genetic instability is due to segregational instability, we would infer that the rate of formation of plasmid-free cells is high. In this case, the number of plasmid-free cells would be high irrespective of whether the plasmid-free cells had a growth-rate advantage. If, on the other hand, we claimed that the genetic instability is growth-rate dependent, we would imply that the rate of formation of plasmid-free cells is low, but the plasmid-free cells have such a large growth-rate advantage that they outgrow the original host–vector system. In most cases, growth-rate dependent instability and one of the other factors (segregational loss, structural changes in the plasmid, or host cell mutations) are important.

The growth-rate ratio can be manipulated to some extent by the choice of medium (e.g., the use of *selective pressure* such as antibiotic supplementation to kill plasmid-free cells) and the use of production systems that do not allow significant target-protein production during most of the culture period. For example, an inducible promoter can be turned on only at the end of a batch growth cycle when only one or two more cell doublings may normally occur. Before induction, the *metabolic burden* imposed by the formation of the target protein is nil, and the growth ratio of the altered to the original host–vector system is close to 1 (or less if selective pressure is also applied). This *two-phase fermentation* can be done as a modified batch system, or a multistage chemostat could be used. In a two-stage system, the first stage is optimized to produce viable plasmid-containing cells, and production formation is induced in the second stage. The continual resupply of fresh cells to the second stage ensures that many unaltered cells will be present.

The problem of genetic instability is a more significant in commercial operations than in laboratory-scale experiments. The primary reason is that the culture must go through many more generations to reach a density of 10^{10} cells/ml in a 10,000 l tank than in a shake flask with 25 ml. Also, the use of antibiotics as selective agents may not be desirable in the large-scale system, owing either to cost or to regulatory constraints on product quality.

In the next section we will discuss some implications of these process constraints on plasmid design. In the section following, we will discuss how simple mathematical models of genetic instability can be constructed.

14.5. CONSIDERATIONS IN PLASMID DESIGN TO AVOID PROCESS PROBLEMS

When we design vectors for genetic engineering, we are concerned with elements that control plasmid copy number, the level of target-gene expression, and the nature of the gene product, and we must also allow for the application of selective pressure (e.g., antibiotic resistance). The vector must also be designed to be compatible with the host cell.

Different *origins of replication* exist for various plasmids. The origin often contains transcripts that regulate copy number. Different mutations in these regulatory transcripts will yield greatly different copy numbers. In some cases, these transcripts have temperature-sensitive mutations, and temperature shifts can lead to *runaway replication* in which plasmid copy number increases until cell death occurs.

Total protein production depends on both the number of gene copies (e.g., the number of plasmids) and the strength of the promoter used to control transcription from these promoters. Increasing copy number while maintaining a fixed promoter strength increases protein production in a saturable manner. Typically, doubling copy numbers from 25 to 50 will increase protein production twofold, but an increase from 50 to 100 will increase protein production less than twofold. If the number of replicating units is above 50, pure segregational plasmid loss is fairly minimal. Most useful cloning vectors in *E. coli* have stable copy numbers from 25 to 250.

Many promoters exist. Some of the important ones for use in *E. coli* are listed in Table 14.3. An ideal promoter would be both very strong and tightly regulated. A zero basal level of protein production is desirable, particularly if the target protein is toxic to the host cell. A rapid response to induction is desirable, and the inducer should be cheap and safe. Although temperature induction is often used on a small scale, thermal lags in a large fermentation vessel can be problematic. Increased temperatures may also activate a heat-shock response and increased levels of proteolytic enzymes. Many chemical inducers are expensive or might cause health concerns if not removed from the product. Some promoters respond to starvation for a nutrient (e.g., phosphate, oxygen, and energy), but the control of induction with such promoters can be difficult to do precisely. The recent isolation of a promoter induced by oxygen depletion may prove useful, because oxygen levels can be controlled relatively easily in fermenters.

Anytime a strong promoter is used, a strong transcriptional terminator should be used in the construction. Recall from Chapter 4 that a terminator facilitates the release of RNA polymerase after a gene or operon is read. Without a strong terminator, the RNA polymerase may not disengage. If the RNA polymerase reads through, it may transcribe undesirable genes or may disturb the elements controlling plasmid copy numbers. In extreme cases, this might cause runaway replication and cell death.

The nature of the protein and its localization are important considerations in achieving a good process. To prevent proteolytic destruction of the target protein, a hybrid gene for the production of a fusion protein can be made. Typically, a small part of a protein native to the host cell is fused to the sequence for the target protein with a linkage that can be easily cleaved during downstream processing. Also, fusion proteins may be constructed to facilitate downstream recovery by providing a "handle" or "tail" that adheres easily to a particular chromatographic medium.

TABLE 14.3 Strong *E. coli* Promoters

Promoter	Induction method	Characteristics[a]
lacUV5	Addition of IPTG	(about 5%)
tac	As above	Induction results in cell death (> 30%)
Ipp-OmpA	As above	Suitable for secreted proteins (20%)
Ipp^p-5	As above	Strongest *E. coli* promoter (47%)
trp	Tryptophan starvation	Relatively weak (around 10%)
λp$_L$	Growth at 42°C	See text (> 30%)
λp$_L$/cl$_{trp}$	Addition of tryptophan	Easily inducible in large-scale production (24%)
att-nutL-p-att-N	10-Min incubation at 42°C	No product is synthesized before induction (on/off promoter)
T7 promoter	Addition of IPTG or viral infection	As above (> 35%), low basal levels
T4 promoter	Viral infection	Method of induction inhibits product degradation
phoA	Phosphate starvation	Induction in large-scale production is complicated

[a]Typical values of accumulated product as a percent of the total protein of induced cells are given in parentheses.

With permission, from G. Georgiou, *AIChE J. 34:*1233 (1988).

Another approach to preventing intracellular proteolysis is to develop a secretion vector in which a signal sequence is coupled to the target protein. If the protein is secreted in one host, it will usually be excreted in another, at least if the right signal sequence is used. Replacement of the protein's natural signal sequence (e.g., from an eukaryotic protein) with a host-specific signal sequence can often improve secretion.

The secretion process is complicated, and the fusion of a signal sequence with a normally nonsecreted protein (e.g., cytoplasmic) does not ensure secretion, although several cases of secretion of normally cytoplasmic proteins have been reported. Apparently, the mature form of the protein contains the "information" necessary in the secretion process, but no general rules are available to specify when coupling a signal sequence to a normally cytoplasmic protein will lead to secretion.

To ensure the genetic stability of any construct and to aid in the selection of the desired host–vector combination, the vector should be developed to survive under selective growth conditions. The most common strategy is to include genes for antibiotic resistance. The common cloning plasmid, pBR322, contains both ampicillin and tetracycline resistance. Multiple resistance genes are an aid in selecting for human-designed modifications of the plasmid.

Another strategy for selection is to place on the vector the genes necessary to make an essential metabolite (e.g., an amino acid). If the vector is placed in a host that is auxotrophic for that amino acid, then the vector complements the host.[†] In a medium without that amino acid, only plasmid-containing cells should be able to grow. Because the genes for the synthesis of the auxotrophic factors can be integrated into the

[†]Recall that an auxotrophic mutant would be unable to synthesize an essential compound on its own.

chromosome or because of reversions on the parental chromosome, double auxotrophs are often used to reduce the probability of nonplasmid-containing cells outgrowing the desired construction.

One weakness in both of these strategies is that, even when the cell loses the plasmid, the plasmid-free cell will retain for several divisions enough gene product to provide antibiotic resistance or the production of an auxotrophic factor (see Fig. 14.5). Thus, cells that will not form viable colonies on selective plates (about 25 generations are required to form a colony) can still be present and dividing in a large-scale system. These plasmid-free cells consume resources without making product.

Another related problem, particularly in large-scale systems, is that plasmid-containing cells may protect plasmid-free cells from the selective agent. For example, auxotrophic cells with a plasmid may leak sufficient levels of the auxotrophic factor that plasmid-free cells can grow. With an antibiotic, the enzyme responsible for antibiotic degradation may leak into the medium. Also, the enzyme may be so effective, even when retained intracellularly, that all the antibiotic is destroyed quickly in a high-density culture, reducing the extracellular concentration to zero. Although genes allowing the placement of selective pressure on a culture are essential in vector development, the engineer should be aware of the limitations of selective pressure in commercial-scale systems.

The other useful addition to plasmid construction is the addition of elements that improve plasmid segregation. Examples are the so-called *par* and *cer* loci. These elements act positively to ensure more even distribution of plasmids. The mechanisms behind these elements are incompletely understood, although they may involve promoting plasmid-membrane complexes (the *par* locus) or decreasing the net level of multimerization (the *cer* locus). Recall from Example 14.1 that multimerization decreases the number of independent, inheritable units, thus increasing the probability of forming a plasmid-free cell.

Any choice of vector construction must consider host cell characteristics. Proteolytic degradation may not be critical if the host cell has been mutated to inactivate all

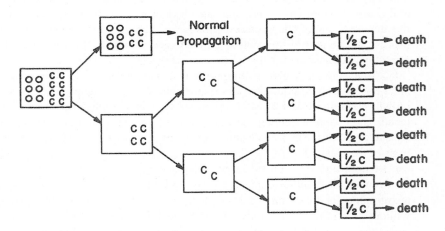

Figure 14.5. Newly born plasmid-free cells usually contain sufficient complementing factor (C) to withstand killing by a selective agent or starvation from the lack of a growth factor. In this case, the plasmid-free cell undergoes three divisions before the complementing factor is reduced to an ineffective level.

known proteases. Multimerization can be reduced by choosing a host with a defective recombination system. However, host cells with a defective recombination system tend to grow poorly. Many other possible host cell modifications enter into considerations of how to best construct a vector for a commercial operation.

These qualitative ideas allow us to anticipate to some extent what problems may arise in the maintenance of genetic stability and net protein expression. However, a good deal of research has been done on predicting genetic instability.

14.6. PREDICTING HOST–VECTOR INTERACTIONS AND GENETIC INSTABILITY

Many of the structured mathematical models we discussed previously can be extended to include component models for plasmid replication. Such models can then predict how plasmid-encoded functions interact with the host cell. The quantitative prediction of the growth-rate ratio and the development of plasmid-free cells due to segregational losses can be readily made. The most sophisticated models will predict the distribution of plasmids within a population and even the effects of multimerization on genetic stability. These models are too complex to warrant discussion in an introductory course.

We will consider some simple models that mimic many of the characteristics we discussed with models of mixed cultures. A number of simple models for plasmid-bearing cells have been proposed. The key parameters in such models are the relative growth rates of plasmid-free and plasmid-containing cells and the rate of generation of plasmid-free cells (i.e., segregational loss). These parameters can be determined experimentally or even predicted for more sophisticated models of host–vector interactions.

Let's consider how a simple model may be constructed and how the parameters of interest may be determined experimentally.[†]

The simplest model considers only two cell types: plasmid free (n_-) and plasmid containing (n_+), where n_- and n_+ are the number concentrations of plasmid-free and plasmid-containing cells, respectively. The model assumes that all plasmid-containing cells are identical in growth rate and in the probability of plasmid loss. This assumption is the same as assuming that all cells have exactly the same copy number. As we showed in Example 14.1, the actual distribution of copy numbers can make a significant difference on plasmid loss. Also, plasmid-encoded protein production is not a linear function of copy number, so assuming that all cells have the same copy number may lead to incorrect estimates of the growth rate of plasmid-bearing cells. The assumption of a single type of plasmid-bearing cell is a weak assumption, but other assumptions result in a level of complexity inconsistent with this book's purpose. However, models that recognize the segregated nature of the plasmid population are available in the research literature.

Let us further restrict our initial considerations to a single-stage chemostat. Then

$$dn_+ / dt = \mu_+ n_+ - D n_+ - R n_+ \tag{14.7}$$

$$dn_- / dt = \mu_- n_- - D n_- + R n_+ \tag{14.8}$$

[†]This analysis is adapted from the paper of N. S. Cooper, M. E. Brown, and C. A. Cauleott, *J. Gen. Microbiol. 133*:1871 (1987).

$$dS/dt = [S_0 - S]D - \frac{\mu_+ n_+}{Ys_+} - \frac{\mu_- n_-}{Ys_-} \qquad (14.9)$$

where R is the rate of generation of plasmid-free cells from plasmid-containing cells and Ys_+ and Ys_- are cell-number yield coefficients (i.e., the number of cells formed per unit mass of limiting nutrient consumed). R can be represented by

$$R = P\mu_+ \qquad (14.10)$$

where P = probability of forming a plasmid-free cell. P can be estimated by eq. 14.1 if the copy number is known or can be predicted with a more sophisticated structured–segregated model. A value for P could be estimated from an experimentally determined copy-number distribution as in Example 14.1c, which would be more realistic than assuming a monocopy number. As we will soon see, R can be determined experimentally without a knowledge of copy number.

Equations 14.7 through 14.9 assume only simple competition between plasmid-containing and plasmid-free cells. No selective agents are present, and the production of complementing factors from the plasmid is neglected. The simplest assumption for cellular kinetics is

$$\mu_+ = \mu_{+\max} \frac{S}{K_{S_+} + S} \qquad (14.11a)$$

$$\mu_- = \mu_{-\max} \frac{S}{K_{S_-} + S} \qquad (14.11b)$$

The situation can be simplified even more if we assume that after a few generations in a chemostat with constant operating conditions the total number of cells (N^t) is constant. This approximation will be acceptable in many cases as long as the metabolic burden imposed by plasmid-encoded functions is not too drastic and D is less than 80% of either $\mu_{+\max}$ or $\mu_{-\max}$. For allowable dilution rates, these assumptions allow us to decouple the substrate balance (eq. 14.9) from immediate consideration. If we then add eqs. 14.7 and 14.8, we have

$$dn_+/dt + dn_-/dt = \mu_+ n_+ + \mu_- n_- - D(n_+ + n_-) \qquad (14.12)$$

Since N^t is constant and

$$N^t = n_+ + n_- \qquad (14.13)$$

at quasi-steady-state, eq. 14.12 becomes

$$0 = \mu_+ n_+ + \mu_- n_- - D(N^t) \qquad (14.14a)$$

or

$$0 = \mu_+ f_+ + \mu_- f_- - D \qquad (14.14b)$$

where f_+ is the fraction of the total population that is plasmid-containing cells and f_- is the fraction of plasmid-free cells. Since $f_+ + f_- = 1$, then

$$D = \mu_+ f_+ + \mu_- (1 - f_+) \tag{14.15}$$

Substituting eq. 14.15 into eq. 14.7 after eq. 14.7 is divided by N^t yields

$$df_+ / dt = \mu_+ f_+ - f_+ [(\mu_+ f_+) + \mu_- (1 - f_+)] - Rf_+ \tag{14.16}$$

and, after rearrangement,

$$df_+ / dt = f_+^2 \Delta\mu - f_+ (\Delta\mu + R) \tag{14.17}$$

where

$$\Delta\mu = \mu_- - \mu_+ \tag{14.18}$$

Equation 14.17 is a Bernoulli's equation. It can be solved in terms of a dummy variable $v = 1/f^+$, assuming a constant $\Delta\mu$ and R. A constant $\Delta\mu$ and R would be achieved only if the copy number distribution of plasmid-containing cells remained constant during the experiment. Also, strictly speaking, the assumption of constant $\Delta\mu$ is inconsistent with constant N^t. This analysis is best applied to situations where $\Delta\mu$ is not extremely large. The solution is

$$\frac{1}{f_+} = \frac{1}{1 - f_-} = \frac{\Delta\mu}{\Delta\mu + R} + ce^{(\Delta\mu + R)t} \tag{14.19}$$

where c is the constant of integration. The initial condition is $f_- = f_{-0}$ at $t = 0$. Then c must be

$$c = \left(\left[\frac{1}{1 - f_{-0}} \right] - \frac{\Delta\mu}{\Delta\mu + R} \right) \tag{14.20}$$

Once c is evaluated, eq. 14.19 can be rearranged to yield

$$f_- = \frac{(f_{-0}\Delta\mu + R)e^{(\Delta\mu + R)t} - R(1 - f_{-0})}{(f_{-0}\Delta\mu + R)e^{(\Delta\mu + R)t} + \Delta\mu(1 - f_{-0})} \tag{14.21}$$

Since f_{-0} is usually $\ll 1$, eq. 14.21 becomes

$$f_- = \frac{(f_{-0}\Delta\mu + R)e^{(\Delta\mu + R)t} - R}{(f_{-0}\Delta\mu + R)e^{(\Delta\mu + R)t} + \Delta\mu} \tag{14.22}$$

Further simplification is possible by considering three limiting cases:

1. $\Delta\mu \gg R$ (growth-rate-dependent instability dominant)
2. $\Delta\mu \leq R$ (segregational instability dominant)
3. $\Delta\mu < 0$ and $|\Delta\mu| \gg R$ (effective selective pressure)

These limiting cases yield:

Case 1: Where t is sufficiently small so that $\Delta\mu \gg (f_{-0}\Delta\mu + R)e^{(\Delta\mu t)}$, then

$$f_- \approx (f_{-0} + R/\Delta\mu)e^{\Delta\mu t} \tag{14.23}$$

Case 2: Where $f_{-0}\,\Delta\mu$ will be $<< R$ and using a binomial expansion,

$$f_{-} \approx 1 - (1 + \Delta\mu / R)e^{-(\Delta\mu + R)t} + \Delta\mu / R(1 + \Delta\mu / R)e^{-2(\Delta\mu + R)t} \tag{14.24}$$

Case 3: The denominator is $\approx \Delta\mu$, since $|\Delta\mu| >> R$ and $f_{-0} << 1$, and we denote $\Delta\mu = -|\Delta\mu|$:

$$f_{-} \approx \frac{R}{|\Delta\mu|} + \left(f_{-0} - \frac{R}{|\Delta\mu|} \right)e^{(R-|\Delta\mu|)t} \tag{14.25}$$

For each of these cases, a straight-line portion of the data will allow estimates of $\Delta\mu$ and R if the basic assumptions for each case are met.

Figure 14.6 shows the behavior that would be expected for each case in a chemostat. The appropriate plots to estimate the parameters $\Delta\mu$ and R are also given (see Fig. 14.6).

In case 1 for $\Delta\mu t > 1$,

$$\ln f_{-} = \ln(f_{-0} + R/\Delta\mu) + \Delta\mu t \tag{14.26}$$

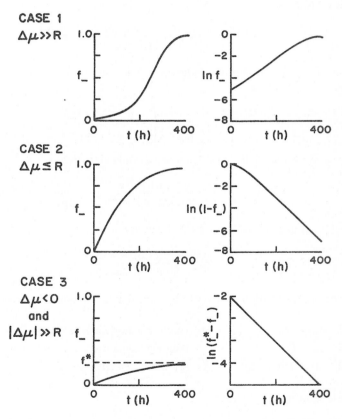

Figure 14.6. The shape of a plot of f_{-} versus t is diagnostic for limiting cases of eq. 14.21. For each case, a plot with a substantial linear region can be constructed and used to evaluate $\Delta\mu$ and R.

and thus the slope, m, gives $\Delta\mu$, and R is given by the intercept as

$$R = m(f_{-i} - f_{-0}) = \Delta\mu(f_{-i} - f_{-0}) \tag{14.27}$$

where f_{-i} is the value of f_- at the intercept.

Equation 14.27 has interesting implications in helping to quantify the conditions for which eq. 14.26 is valid. That is, t must be sufficiently small that

$$\Delta\mu \gg (f_{-0}\,\Delta\mu + R)e^{\,\Delta\mu t}$$

when $\Delta\mu \gg R$. Thus,

$$\Delta\mu \gg (f_{-0}\Delta\mu + \Delta\mu f_{-i} - \Delta\mu f_{-0})e^{\Delta\mu t} \tag{14.28}$$

or

$$1 \gg f_{-i}\,e^{\Delta\mu t} \tag{14.29}$$

or

$$0 \gg \ln f_{-i} + \Delta\mu t \tag{14.30}$$

or

$$t \ll \frac{-\ln f_{-i}}{\Delta\mu} \tag{14.31}$$

The analysis also assumes that $\Delta\mu t > 1$; thus, the linear region suitable for use for case 1 situations is

$$\frac{1}{m} < t < \frac{-\ln f_{-i}}{m} \tag{14.32}$$

In case 2, for $t > 1/(\Delta\mu + R)$,

$$f_- \approx 1 - (1 + \Delta\mu/R)e^{-(\Delta\mu + R)t} \tag{14.33a}$$

$$\ln f_+ = \ln(1 - f_-) = \ln(1 + \Delta\mu/R) - (\Delta\mu + R)t \tag{14.33b}$$

in which case the intercept is used to evaluate f_{+i} and the slope $m = -(\Delta\mu + R)$:

$$\Delta\mu = m(f_{+i} - 1) = mf_{-i} \tag{14.34}$$

$$R = -mf_{+i} = -m(1 - f_{-i}) \tag{14.35}$$

In case 3, we note that f_- assumes a constant value between 0 and 1, which we denote as f_-^*. Equation 14.25 yields f_-^* by allowing $t \to \infty$, or

$$f_-^* \approx R/\Delta\mu \tag{14.36}$$

A plot of $\ln(f_-^* - f_-)$ versus time will give a slope, m, such that

$$m = R - |\Delta\mu| \tag{14.37}$$

Equations 14.36 and 14.37 can be combined to give

$$\Delta\mu = \frac{m}{f_-^* - 1} \tag{14.38}$$

$$R = \frac{-m f_-^*}{1 - f^*} \tag{14.39}$$

This analysis provides a method to estimate experimentally the parameters of importance in predicting genetic instability. Once these parameters are known, eq. 14.21 can be used to predict chemostat performance. However, the reader should recall the large number of assumptions that went into these expressions. Case 1 is the one relevant to most commercial systems, where high expression levels are coupled with high-copy-number plasmids. Case 2 will occur only for low-copy-number plasmids that do not have a par locus or other stabilizing features. The results for case 3 must be applied very cautiously. For μ_- to be less than μ_+, selective pressure usually must be applied (e.g., antibiotic resistance or auxotrophic hosts). Equations 14.7 and 14.8 do not recognize the possibility of leakage of nutrients or enzymes, which would degrade the antibiotics. If there were no leakage and the antibiotic-to-cell ratio were high enough to leave an effective residual level of antibiotic, then the analysis for case 3 could be applied if μ_- were interpreted correctly. For a newly born plasmid-free cell, there will be a carry-over of the complementing factor, which will be gradually reduced by turnover and dilution (recall Fig. 14.5). Thus, in the first generation after plasmid loss, μ_- may even be greater than μ_+, but μ_- will decrease from generation to generation until it becomes zero. The analysis from case 3 will give an effective average value of μ_-.

Removal of many of these assumptions to yield a more realistic analysis increases complexity and makes the development of simple analytical expressions difficult. Let us consider how this simple analysis may be used and how it can be extended.

Example 14.2.

The data in Table 14.4 were obtained for *E. coli* B/r-pDW17 at two different dilution rates in glucose-limited chemostats. The average plasmid copy number for pDW17 is about 40 to 50 copies per cell. About 12% of the total protein synthesized is due to the plasmid. The proteins are retained intracellularly in soluble form. Use these data to estimate $\Delta\mu$, R, and P.

Solution Note that data on plasmid stability are usually given in terms of the number of generations, which is usually calculated as $(\ln 2)/D = 1$ generation. Clearly, the trend of the data in Table 14.4 is a sigmoidal dependence of f_- on time, which corresponds to growth-rate-dependent instability or a case 1 situation (see Fig. 14.6). Thus, we need to plot the data as $\ln f_-$ versus time. Such a plot is given as Fig. 14.7. Recall that eq. 14.32, which gives the bounds on t, suggests that only data at intermediate times can be used to estimate $\Delta\mu$ and R. In this case, data for $0.01 < f_- < 0.40$ were used as a basis for the initial estimates of values of $\Delta\mu$ and R. Equation 14.32 can be used later to check the appropriate data range.

For $D = 0.3 \ h^{-1}$ the slope, m, is 0.15 gen^{-1} or 0.066 h^{-1}, and the intercept is -5.3, implying $f_{-1} = 0.005$. For $D = 0.67 \ h^{-1}$, $m = 0.12 \ gen^{-1}$ or 0.11 h^{-1}, an intercept of -5.6 gives $f_{-1} = 0.0037$. Recall that, for case 1, $m = \Delta\mu$ and $R = m(f_{-1} - f_{-0})$. The value of f_{-0} presents a problem, as an accurate value is difficult to determine experimentally when it is low (many colonies must be plated to obtain any statistical accuracy). R will be maximal if f_{-0} is assumed to be zero. If a culture is developed under strong selective pressure, with plasmid-

TABLE 14.4 Experimental Values for Plasmid Stability in a Glucose-limited Chemostat

$D = 0.30\ h^{-1}$				$D = 0.67\ h^{-1}$		
Time, h	No. of generations	f_-		Time, h	No. of generations	f_-
0	0	≤ 0.002		0	0	≤ 0.002
11.6	5	0.010		10.3	10	0.010
23.1	10	0.030		20.7	20	0.035
46.2	20	0.13		31.0	30	0.15
57.8	25	0.22		41.4	40	0.34
69.3	30	0.39		51.7	50	0.65
80.9	35	0.76		62.1	60	0.81
92.4	40	0.88		72.4	70	0.93
104.0	45	0.97		82.8	80	0.98
115.5	50	1.00		93.1	90	1.0
138.6	60	1.00		103.5	100	1.0

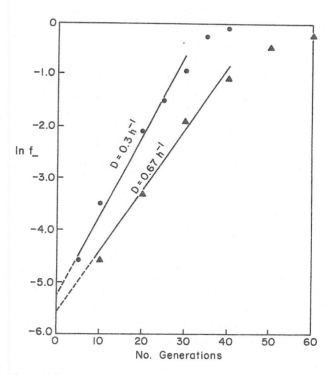

Figure 14.7. Determination of $\Delta\mu$ and R using data from Example 14.2.

encoded protein synthesis uninduced before the beginning of the experiment, then f_{-0} will approach zero.

Thus,

$m = \Delta\mu$	$D = 0.3\ h^{-1}$ $0.15\ gen^{-1}$	$D = 0.67\ h^{-1}$ $0.12\ gen^{-1}$
$R\ max\ (f_{-0} = 0)$	$7.5 \times 10^{-4}\ gen^{-1}$	$4.4 \times 10^{-4}\ gen^{-1}$
R if $f_{-0} = 0.002$	$4.5 \times 10^{-4}\ gen^{-1}$	$2.0 \times 10^{-4}\ gen^{-1}$
R if $f_{-0} = 0.003$	$3.0 \times 10^{-4}\ gen^{-1}$	$8.4 \times 10^{-5}\ gen^{-1}$
R if $f_{-0} = 0.0037$	$2.0 \times 10^{-4}\ gen^{-1}$	$0\ gen^{-1}$

To determine P, recall that $R = P\,\mu_+$ (eq. 14.10). A value for μ_+ can be estimated from eq. 14.14b after some rearrangement to give

$$D = \mu_+ + \Delta\mu f_-$$ (14.40)

Values of μ_+ and μ_- vary slightly during culture as a consequence of the assumption that $dN'/dt = 0$.[†] An average value of μ_+ is

$$\mu_+ = D - \Delta\mu/2$$ (14.41)

R, in units of gen^{-1}, is changed to h^{-1} by multiplying by $D/\ln 2$. Thus,

D	μ_+	R_{max}	P_{max}
$0.3\ h^{-1}$	$0.267\ h^{-1}$	$7.5 \times 10^{-4}\ gen^{-1}$ or $3.2 \times 10^{-4}\ h^{-1}$	1.2×10^{-3}
$0.67\ h^{-1}$	$0.615\ h^{-1}$	$4.4 \times 10^{-4}\ gen^{-1}$ or $4.2 \times 10^{-4}\ h^{-1}$	0.7×10^{-3}

A change in the probability of segregational losses as a function of dilution rate corresponds to the copy number, which also changes with dilution rate. The pDW17 plasmid has an *average* copy number of about 40 to 50 in exponential growth in minimal medium.

Consider eq. 14.1, which predicts the probability of plasmid loss by random segregation:

$$P = 2^{(1-Z)}$$ (14.1)

for $Z = 40$ and $P = 1.8 \times 10^{-12}$, which is certainly less than the maximum value calculated in these experiments. Note from Example 14.1, however, that if the average copy number were distributed so that half the cells had 10 plasmids and half had 70, then the probability of plasmid loss would be 9.8×10^{-4}, which is very close to the maximum probabilities allowed by these experimental data.

[†]Physically, these changes in μ_+ and μ_- could occur due to small changes in residual substrate concentration. This observation implies that the assumption of constant $\Delta\mu$ and R used in integrating eq. 14.17 would fail if $\Delta\mu$ were very large.

Example 14.3.

Derive equations to describe the dynamics of a plasmid-containing population when the plasmid-free host is auxotrophic for metabolite M. The metabolite is made and released from a plasmid-containing cell into the medium in a chemostat.

Solution This situation is similar to that in eqs. 14.7, 14.8, and 14.9, except for the production and release of M. The only modification is to add a mass balance for M and to alter the equations for μ_-. Let M be the concentration of M. Then

$$dn_+/dt = \mu_+ n_+ - Dn_+ - Rn_+ \tag{14.7}$$

$$dn_-/dt = Rn_+ + \mu_- n_- - Dn_- \tag{14.8}$$

$$dS/dt = D(S_0 - S) - \frac{\mu_+ n_+}{Y_{S_+}} - \frac{\mu_- n_-}{Y_{S_-}} \tag{14.9}$$

$$\frac{dM}{dt} = \delta \mu_+ n_+ - \frac{\mu_- n_-}{Y_{M_-}} - DM \tag{14.42}$$

$$\mu_+ = \mu_{+max} \frac{S}{K_{S_+} + S} \tag{14.11a}$$

$$\mu_- = \mu_{-max} \frac{S}{K_{S_-} + S} \frac{M}{K_{M_-} + M} \tag{14.43}$$

where δ is the stoichiometric coefficient relating the production of M to the growth of n_+.

This set of equations would have to be solved numerically. However, the type of analysis used in Example 14.2 could be applied if M were supplied in the medium such that the residual level of $M \gg K_{M_-}$.

Our discussion and examples so far have been for continuous reactors. For most industrial applications batch or fed-batch operations will be used. While continuous reactors are particularly sensitive to genetic instability, genetic instability can be a significant limitation for batch systems.

For a batch reactor, the cell-number balances are:

$$dn_+/dt = \mu_+ n_+ - P\mu_+ n_+ \tag{14.44}$$

$$dn_-/dt = P\mu_+ n_+ + \mu_- n_- \tag{14.45}$$

with the initial conditions of

$$@\ t = 0, \qquad n_{+0} = N_{+0}, \qquad n_{-0} = 0$$

Solving eq. 14.44 with these initial conditions yields

$$n_+ = N_{+0} e^{(1-P)\mu_+ t} \tag{14.46}$$

and eq. 14.45 using eq. 14.46 yields

$$n_- = \frac{P\mu_+ N_{+0}}{(1-P)\mu_+ - \mu_-} \left[e^{(1-P)\mu_+ t} - e^{\mu_- t} \right] \tag{14.47}$$

Since $1 - f_- = f_+ = \dfrac{n_+}{n_+ + n_-}$, then

$$f_+ = \frac{e^{(1-P)\mu_+ t}}{e^{(1-P)\mu_+ t} + \dfrac{P\mu_+}{(1-P)\mu_+ - \mu_-}\left[e^{(1-P)\mu_+ t} - e^{\mu_- t}\right]} \qquad (14.48)$$

Equation 14.48 is usually recast in terms of number of generations of plasmid-containing cells (n_g), a dimensionless growth-rate ratio (α), and the probability of forming a plasmid-free cell upon division of a plasmid-containing cell (P). The mathematical definitions of n_g and α are:

$$n_g = \frac{\mu_+ t}{\ln 2} \qquad (14.49)$$

$$\alpha = \frac{\mu_-}{\mu_+} \qquad (14.50)$$

With these definitions, eq. 14.48 becomes:

$$f_+ = \frac{1 - \alpha - P}{1 - \alpha - P \cdot 2^{n_g(\alpha + P - 1)}} \qquad (14.51)$$

The use of this approach is illustrated in the following example.

Example 14.4.

Estimate the fraction of plasmid-containing cells in a batch culture under the following circumstances. Cells are maintained at constant, maximal growth rate of 0.693 h^{-1} during scale-up from shake flask through seed fermenters into production fermenters. The total time for this process is 25 h. Assume that the inoculum for the shake flask was 100% plasmid-containing cells. It is known that the growth rate for a plasmid-free cell is 0.97 h^{-1}. The value of P is 0.001.

Solution   First we calculate α and n_g.

$$\alpha = 0.97/0.693 = 1.4$$

and

$$n_g = \frac{0.693 \text{ h}^{-1} \cdot 25 \text{ h}}{\ln 2} = 25 \text{ gen.}$$

Substituting into eq. 14.51

$$f_+ = \frac{1 - 1.4 - 0.001}{1 - 1.4 - 0.001 \cdot 2^{25(1.4 + .001 - 1)}}$$

$$f_+ = 0.27$$

Consequently, 73% of the cells are plasmid-free, which would result in considerable loss of productivity.

14.7. REGULATORY CONSTRAINTS ON GENETIC PROCESSES

When genetic engineering was first introduced, there was a great deal of concern over whether the release of genetically modified cells could have undesirable ecological consequences. Reports in the popular press led to fears of "genetic monsters" growing in our sewers, on our farm lands, or elsewhere. Consequently, the use of genetic engineering technology is strictly regulated.

The degree of regulatory constraint varies with the nature of the host, vector, and target protein. For example, consider a scenario where serious harm might arise. The gene for a highly toxic protein is cloned into *E. coli* to obtain enough protein to study that protein's biochemistry. Assume that a plasmid that is *promiscuous* (i.e., the plasmid will shuttle across species lines) is used. Also assume that laboratory hygiene is not adequate and a small flying insect enters the laboratory and comes into contact with a colony on a plate awaiting destruction. If that insect leaves the laboratory and returns to its natural habitat, then the target gene is accidentally released into the environment. Laboratory strains of *E. coli* are fragile and usually will not survive long in a natural environment. However, a very small probability exists that the plasmid could cross over species lines and become incorporated into a more hardy soil bacterium (e.g., *Pseudomonas* sp.). The plasmid would most certainly contain antibiotic resistance factors as well. The newly transformed soil bacterium could replicate. Many soil bacteria are opportunistic pathogens. If they enter the body through a wound, they can multiply and cause an infection. If, in addition, the bacterium makes a toxic protein, the person or animal that was infected could die from the toxic protein before the infection was controlled. If the plasmid also confers antibiotic resistance, the infection would not respond to treatment by the corresponding antibiotic, further complicating control of the spread of the gene for the toxin.

This scenario requires that several highly improbable events occur. No case of significant harm to humans or the environment due to the release of genetically modified cells has been documented. However, the potential for harm is real.

Regulations controlling genetic engineering concentrate on preventing the accidental release of genetically engineered organisms. The deliberate release of genetically engineered cells is possible, but an elaborate procedure must be followed to obtain permission for such experiments.

The degree of containment required depends on (1) the ability for the host to survive if released, (2) the ability for the vector to cross species lines or for a cell to be transformed by a piece of naked DNA and then have it incorporated into the chromosome via recombination, and (3) the nature of the genes and gene products being engineered. Experiments involving overproduction of *E. coli* proteins in *E. coli* using plasmids derived from wild populations are readily approved and do not require elaborate containment procedures (see Table 14.5). Experiments that would move the capacity to produce a toxin from a higher organism into bacteria or yeast would be subjected to a much more thorough evaluation, and more elaborate control facilities and procedures would be required.

The National Institutes of Health (NIH) have issued guidelines that regulate the use of recombinant DNA technology. Special regulations apply to large-scale systems (defined as > 10 l). There are three different levels of containment: BL1-LS, BL2-LS, and BL3-LS. BL1-LS (biosafety level 1, large scale) is the least stringent. Table 14.6 compares the requirements for the three containment levels.

TABLE 14.5 Minimum Laboratory Containment Standards for Working with Cells with Recombinant DNA

Biosafety Level 1 (BL1)
A. *Standard Microbiological Practices*
   1. Access to the laboratory is limited or restricted at the discretion of the laboratory director when experiments are in progress.
   2. Work surfaces are decontaminated once a day and after any spill of viable material.
   3. All contaminated liquid or solid wastes are decontaminated before disposal.
   4. Mechanical pipetting devices are used; mouth pipetting is prohibited.
   5. Eating, drinking, smoking, and applying cosmetics are not permitted in the work area. Food may be stored in cabinets or refrigerators designated and used for this purpose only.
   6. Persons wash their hands after they handle materials involving organisms containing recombinant DNA molecules, and animals, and before leaving the laboratory.
   7. All procedures are performed carefully to minimize the creation of aerosols.
   8. It is recommended that laboratory coats, gowns, or uniforms be worn to prevent contamination or soiling of street clothes.
B. *Special Practices*
   1. Contaminated materials that are to be decontaminated at a site away from the laboratory are placed in a durable, leakproof container that is closed before being removed from the laboratory.
   2. An insect and rodent control program is in effect.
C. *Containment Equipment*
   1. Special containment equipment is generally not required for manipulations of agents assigned to Biosafety Level 1.
D. *Laboratory Facilities*
   1. The laboratory is designed so that it can be easily cleaned.
   2. Bench tops are impervious to water and resistant to acids, alkalis, organic solvents, and moderate heat.
   3. Laboratory furniture is sturdy. Spaces between benches, cabinets, and equipment are accessible for cleaning.
   4. Each laboratory contains a sink for hand-washing.
   5. If the laboratory has windows that open, they are fitted with fly screens.

In all cases, no viable organisms can be purposely released. Gas vented from the fermenter must be filtered and sterilized. All cells in the liquid effluent must be killed. The latter can present some operating issues, since the inactivation of the host cell must be done in such a way as not to harm what are often fragile products. Emergency plans and devices must be on hand to handle any accidental spill or loss of fluid in the fermentation area. These extra precautions increase manufacturing costs.

The issue of the regulation of cells and recombinant DNA will undoubtedly undergo further refinement with time. Both laboratory and manufacturing personnel need to keep abreast of any such changes.

14.8. METABOLIC ENGINEERING

Metabolic or pathway engineering uses the tools of genetic engineering to endow an organism to make a totally new pathway, amplify an existing pathway, disable an undesired pathway, or alter the regulation of a pathway. The principle motivations for metabolic engineering are the production of specialty chemicals (e.g., indigo, biotin, and amino acids), utilization of alternative substrates (e.g., pentose sugars from hemicellulose), or degrada-

TABLE 14.6 Physical Containment Requirements for Large-scale Fermentations Using Organisms Containing Recombinant DNA Molecules

Item no.	Description	BL1-LS	BL2-LS	BL3-LS
1	Closed vessel	×	×	×
2	Inactivation of cultures by validated procedure before removing from the closed system	×	×	×
3	Sample collection and addition of material in a closed system	×	×	×
4	Exhaust gases sterilized by filters before leaving the closed system	×	×	×
5	Sterilization by validated procedures before opening for maintenance or other purposes	×	×	×
6	Emergency plans and procedures for handling large losses	×	×	×
7	No leakage of viable organisms from rotating seals and other mechanical devices		×	×
8	Integrity evaluation procedure: monitors and sensors		×	×
9	Containment evaluation with the host organism before introduction of viable organisms		×	×
10	Permanent identification of closed system (fermenter) and identification to be used in all records		×	×
11	Posting of universal biohazard sign on each closed system and containment equipment when working with a viable organism		×	×
12	Posting of universal biohazard sign on entry doors when work is in progress			×
13	Operations to be in a controlled area:			×
	Separate specified entry			×
	Double doors, air locks			×
	Walls, ceiling, and floors to permit ready cleaning and decontamination			×
	Utilities and services (process piping and wiring) to be protected against contamination			×
	Handwashing facilities and shower facilities in close proximity			×
	Area designed to preclude release of culture fluids outside the controlled area in the event of an accident			×
	Ventilation: movement of air, filtration of air			×

With permission, from R. J. Giorgiou and J. J. Wu, *Trends in Biotechnology 4:*198 (1986).

tion of hazardous wastes such as benzoates or trichloroethylene (see Fig. 14.8 for an example). The same concepts form the basis for gene therapy.

The reader may ask why genetically engineered organisms should be used instead of natural isolates. The potential advantages over natural isolates are as follows:

- Can put an "odd-ball" pathway under the control of a regulated promoter. The investigator can turn on the pathway in situations where the pathway might normally be suppressed (e.g., degradation of a hazardous compound to a concentration lower than necessary to induce the pathway in the natural isolate).
- High levels of enzymes in desired pathways can be obtained with strong promoters; only low activity levels may be present in the natural isolate.
- Pathways moved from lower eucaryotes to bacteria can be controlled by a single promoter; in lower eucaryotes, each protein has a separate promoter.

Sec. 14.8   Metabolic Engineering

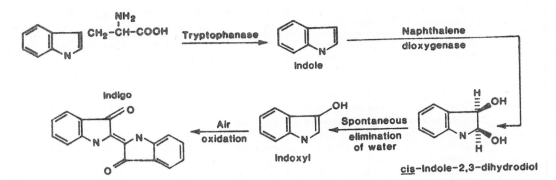

Figure 14.8. Proposed pathway for indigo biosynthesis in a recombinant strain of *E. coli*. Indole is formed from tryptophan by tryptophanase, a natural enzyme in *E. coli*. Naphthalene dioxygenase formed by the expression of the cloned *Pseudomonas* DNA oxidizes indole to indigo; *cis*-2,3-dihydroxy-2,3-dihydroindole and indoxyl have not been isolated. Their inclusion is based on the known activities of aromatic hydrocarbon dioxygenases and established mechanisms for the chemical synthesis of indigo. (With permission, from B. D. Ensley et al., *Science 222:*167, 1983, and American Association for the Advancement of Science.)

- Several pathways can be combined in a single organism by recruiting enzymes from more than one organism.
- Can move a pathway from an organism that grows poorly to one that can be more easily cultured.
- The genetically engineered cell can be proprietary property.

Cells that have engineered pathways face many of the same limitations that cells engineered to produce proteins face. Two issues that perhaps assume greater importance with metabolic engineering are stability and regulatory constraints.

Protein products are of high value and can be made in batch culture. Instability is avoided by inducing overproduction only at the end of the culture cycle. The productive phase is too short for nonproducers to grow to a significant level, and cells are not reused. With metabolically engineered cells, the same strategy is untenable. Lower product values necessitate cell reuse or, at least, extended use. The use of antibiotics as selective agents may be undesirable because of contamination of product or cost. For a culture with a 1.5-h doubling time and a 20-h batch cycle, a continuous system has a 14-fold advantage in productivity over a batch system. Although the levels of protein overproduction are lower in metabolically engineered cells, they can experience as high a level of metabolic burden as "protein producers" because of the diversion of cellular building blocks to nonessential metabolites. Also, if the cells are used to treat hazardous compounds, the genetically engineered cells will face competition from a natural flora.

In addition to the need for extra efforts in engineering design to ensure genetic stability, regulatory approval may be more difficult. If a genetically engineered cell is to be used to treat hazardous wastes, containment of the engineered organism is difficult or impossible. Pump-and-treat scenarios for leachates allow the possibility of control. *In situ* use of such organisms would have to satisfy the constraints for deliberate release.

In addition to these two problems, engineering pathways require good quantitative information on the flow of metabolites in a cell. Structured models of cells can be used in conjunction with experiments to optimize the design of new pathways. Metabolic engineering has been a fertile area for engineering contributions.

A goal of the engineering approach has been to develop rational design techniques for metabolical engineering. Due to the highly nonlinear, dynamic nature of a cell and its metabolism, uninformed changes intended to improve production of a specific compound often fail to give the desired result. The problem of metabolic engineering is to express appropriate genes in the "right" amount. Too much of a key enzyme may result in little change, if it is not a rate-influencing step, as there is insufficient reactant to increase the flow of substrate (or *flux*) into the reaction product. If it is rate-influencing, the increased reaction of the substrate will alter fluxes of precursors in other pathways; sometimes these changes compromise the cell's ability to grow or to provide co-reactants when needed. Metabolic engineering (and gene therapy) require a delicate balance of activities and are a problem of quantative optimization rather than simple maximization of expression of a gene.

Metabolic control theory and the closely allied activity of *metabolic flux analysis* are mathematical tools that can be applied to such problems. A flux balance equation consists of the product of a stoichiometric matrix and an intracellular flux vector to yield an overall rate vector. Typically such equations are underdetermined and require assumptions (e.g., on energy stoichiometry) that may not be justified for a metabolically engineered cell. However, improved experimental techniques to measure the intracellular flux vector (e.g., mass spectrometry) coupled with genomic/proteomic information may provide the data to allow complete solution of the flux balance equations. One caution is that analysis of pathways in isolation can be misleading. An assumption of such analysis is that the products of the pathway do not influence inputs into that pathway. Given the complexity of a cell, it is difficult to assure that such an assumption is satisfied.

Metabolic control theory is based on a sensitivity analysis to calculate the response of a pathway to changes in the individual steps in a pathway. This approach allows calculations of flux control coefficients, which are defined as the fractional change of flux expected for a fractional change in the amount of each enzyme. This process involves a linearization, so that flux coefficients can vary significantly if growth conditions and the cell's physiological step change. An important result of the application of this theory to many cases is that it is rare for a single enzyme step to be rate-controlling. Typically several enzymatic steps influence rate. Maximization of flux through a particular pathway would require several enzyme activities to be altered simultaneously. The potential of such analysis to contribute to rational design of microbes is clear, but not yet routinely applicable.

Real progress has been made toward the industrial use of metabolically engineered cells. Processes to convert glucose from cornstarch into 1,3-propanediol, an important monomer in the polymer, polytrimethylene terepthalate, is poised for commercialization. A process to make 2-keto-L-gluconic acid as a precursor for vitamin C production is in the pilot stage (30,000 l), using metabolically engineered bacteria with glucose as a substrate. Other products from metabolically engineered cells may be new polyketides, modified polyhydroxyalkanoates (as biodegradable polymers), indgo, xylitol, and hybrid antibiotics.

14.9. PROTEIN ENGINEERING

Not only can cells be engineered to make high levels of naturally occurring proteins or to introduce new pathways, but we can also make novel proteins. It is possible to make synthetic genes encoding for totally new proteins. We are beginning to understand the rules by which a protein's primary structure is converted into its three-dimensional form. We are just learning how to relate a protein's shape to its functional properties, stability, and catalytic activity. In the future, it may become possible to customize protein design to a particular well-defined purpose.

Protein engineering at present mainly involves the modification of existing proteins to improve their stability, substrate and inhibitor affinity and specificity, and catalytic rate. Generally, the protein structure must be known from x-ray crystallography. Key amino acids in the structure are selected for alteration based on computer modeling, on interactions of the protein with substrates, or by analogy to proteins of related structure. The technique used to generate genes encoding the desired changes in protein structure is called *site-directed mutagenesis* (see Fig. 14.9). Using this approach, any desired amino acid can be inserted precisely into the desired position.

The reader may wonder why site-directed mutagenesis would be preferred to simple mutation-selection procedures. One reason may be that mutation followed by selection for particular properties may be difficult when the alterations in protein properties are subtle and confer no advantages or disadvantages on the mutant cell. A second reason is that site-directed mutagenesis can be used to generate the insertion of an amino acid in a particular location, while a random mutation giving the same result would occur so infrequently as to be unobtainable.

To make this point more evident, consider the degenerate nature of the genetic code. Each codon consists of three letters. The odds for mutation in one of these three letters is about 10^{-8} per generation. The odds that two letters would simultaneously be altered is much lower (order of 10^{-16}). The codon UAC (for tyrosine) could be altered by single-letter substitutions to give AAC (asparagine), GAC (aspartate), CAC (histidine), UCC (serine), UUC (phenylalanine), UGC (cysteine), UAA (stop signal), UAG (stop signal), and UAU (tyrosine). Random mutants in this case are very unlikely to carry substitutions for 13 of the 20 amino acids. Thus, most of the potential insertions can be generated reliably only by using site-directed mutagenesis.

The above approaches are directed toward the rational design of proteins. An alternative, and often complementary, approach is that of *directed evolution*. This process is based on random mutagenesis of a gene and the subsequent selection of proteins with desired properties. Large libraries of mutant genes must be made so that the rare beneficial forms are present. A rapid screen or selection must be available to select those mutants with the desired function or characteristics. One technique to generate mutants is the use of "error-prone" PCR. Another approach is DNA shuffling, which requires genes from homologous proteins. Segments from the genes are recombined randomly to form chimeric genes. Proteins are typically selected for improved stability, binding strength, catalytic activity, or solubility. In some cases, the screen is for new activities based on ability to bind other molecules not normally bound by the native protein. These techniques combined with improved biochemical methods are leading to a better understanding of the relation of protein structure to function. Also, these techniques support the extension of protein

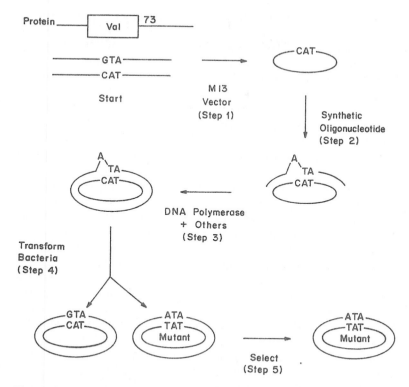

Figure 14.9. The general approach to site-directed mutagenesis is depicted. For this example the amino acid, valine, at position 73 of a particular protein, is to be replaced with isoleucine. The first step is to clone the gene into the single-stranded DNA vector, M13, and collect a working amount of gene. In the second step, an oligonucleotide essentially complementary to that region of the gene to be mutated is synthesized chemically. A typical length for such an oligonucleotide is 18 residues. However, this synthetic oligonucleotide contains one mismatched nucleotide so as to cause the desired mutation. In the third step, enzymes are added that complete the *in vitro* synthesis of a double-stranded vector. In the fourth step, the amount of double-stranded vector is amplified. In the last step, cells with the mutant gene are detected by hybridization with the synthetic oligonucleotide.

catalysis to unusual environments, such as in organic media rather than in aqueous solutions.

14.10. SUMMARY

The application of recombinant DNA technology at the commercial level requires a judicious choice of the proper host–vector system. *E. coli* greatly facilitates sophisticated genetic manipulations, but process or product considerations may suggest alternative hosts. *S. cerevisiae* is easy to culture and is already on the GRAS list, simplifying regulatory approval, although productivities are low with some proteins and hyperglycosylation is a problem. *P. pichia* can produce very high concentrations of proteins, but the use of methanol presents challenges in reactor control and safety. *Bacillus* and the lower fungi may have well-developed secretion systems that would be attractive if they can be harnessed. Animal cell culture is required when posttranslational modifications are essential. Mammalian cells offer the highest degree of fidelity to the authentic natural product. Insect cell systems potentially offer high expression levels and greater safety than mammalian cells at the cost of a potential decrease in the fidelity of posttranslational processing. Transgenic animals are good alternatives for complex proteins requiring extensive posttranslational processing. Transgenic plants and plant cell cultures are emerging as production systems for high-volume protein products. Plants are well suited to produce proteins for oral or topical delivery.

A host–vector system is useful only if it can persist long enough to make commercially important product quantities. *Genetic instability* is the loss of the genetic information to make the target protein. *Segregational instability* arises when a plasmid-free cell is formed during cell division. *Structured instability* results when the cell loses the capacity to make the target protein in significant quantities due to changes in plasmid structure. *Host-cell derived instability* occurs when chromosomal mutations decrease effective plasmid-encoded protein production while the cell retains the plasmid. *Growth-rate-dependent instability* is a function of the growth-rate differential between plasmid-free and plasmid-containing cells; it is usually a determining factor in deciding how long a fermentation can be usefully maintained. Simple models to evaluate genetic instability are available.

The vector must be designed to optimize a desired process. Factors to be considered include vector copy number, promoter strength and regulation, the use of fusion proteins, signal sequences and secretion, genes providing selective pressure, and elements enhancing the accuracy of vector partitioning.

The engineer must be aware of the regulatory constraints on the release of cells with recombinant DNA. These are particularly relevant in plant design, where guidelines for physical containment must be met. Deliberate release of genetically modified cells is possible, but extensive documentation will be required.

Two increasingly important applications of genetic engineering are metabolic or pathway engineering for the production or destruction of nonproteins and protein engineering for the production of novel or specifically modified proteins.

SUGGESTIONS FOR FURTHER READING

Overviews on genetic engineering for protein production

CREGG, J. M., AND D. R. HIGGINS, Production of Foreign Proteins in the Yeast *Pichia pastoris, Can. J. Bot. 73 (suppl 1):*5891–5897, 1995.

DATAR, R. V., T. CARTWRIGHT, AND C-G. ROSEN, Process Economics of Animal Cell and Bacterial Fermentations: A Case Study Analysis of Tissue Plasminogen Activator, *Bio/Technology 11:*349, 1993.

EVANGELISTA, R. L., A. R. KUSNADI, J. A. HOWARD, AND Z. L. NIKOLOV, Process and Economic Evaluation of the Extraction and Purification of Recombinant β-glucuronidase from Transgenic Corn, *Biotechnol. Prog. 14:*607–614, 1998.

FERNANDEZ, J. M., AND J. P. HOEFFLER, *Gene Expression Systems,* Academic Press, New York, 1999.

Utilizing Genetically Engineered Organisms   Chap. 14

Georgiou, G., Optimizing the Production of Recombinant Proteins in Microorganisms, *AIChE J.* *34:*1233, 1988.

Goochee, C. F., and T. Monica, Environmental Effects on Protein Glycosylation, *Bio/Technology* *8:*421, 1990.

Hooker, A. D., M. H. Goldman, N. H. Markham, D. C. James, A. P. Ison, A. T. Bull, P. G. Strange, I. Salmon, A. J. Baines, and N. Jenkins, N-Glycans of Recombinant Human Interferon-γ Change during Batch Culture of Chinese Hamster Ovary Cells, *Biotechnol. Bioeng.* *48:*639, 1995.

Kunkel, J. P., D. C. H. Jan, M. Butler, and J. C. Jamieson, Comparisons of the Glycosylation of a Monoclonal Antibody Produced under Nominally Identical Cell Culture Conditions in Two Different Bioreactors, *Biotechnol. Prog. 16:*462–470, 2000.

Lillie, H., E. Schwarz, and R. Rudolph, Advances in Refolding Proteins Produced in *E. coli, Curr. Opin. Biotechnol. 9:*497–501, 1998.

Shuler, M. L., H. A. Wood, R. R. Granados, and D. A. Hammer, *Baculovirus Expression Systems and Biopesticides,* Wiley-Liss, New York, 1995.

Velander, W. H., H. Lubon, and W. N. Drohan, Transgenic Livestock as Drug Factories, *Sci. Am. 276:*70–74, 1997.

Overviews that include aspects of metabolic engineering

Bailey, J. E., Towards a Science of Metabolic Engineering, *Science 252:*1668–1675, 1991.

Lee, S. Y., and E. T. Papoutsakis, *Metabolic Engineering,* Marcel Dekker, Inc., New York, 1999.

Stephanopoulos, G., J. Nielsen, and A. Aristidou, *Metabolic Engineering. Principles and Methodologies,* Academic Press, 1998.

Overviews on protein engineering

——— Biocatalysis: Synthesis Methods that Exploit Enzymatic Activities. *Nature Insight* 409, #6817, 2001. This issue contains excellent review articles on protein engineering and the industrial use of such enzymes.

Blundell, T. L., Problems and Solutions in Protein Engineering—Towards Rational Design, *Trends in Biotechnol. 12:*145–148, 1994.

Klibanov, A. M., Enzyme Catalysis Without Water, *R & D Innovator* 2(4):1–7, 1993.

Kuchner, K., and F. H. Arnold, Directed Evolution of Enzyme Catalysts, *Trends in Biotechnol. 15:*523–530, 1997.

Overviews on regulatory issues

Giorgiou, R. J., and J. J. Wu, Design of Large Scale Containment Facilities for Recombinant DNA Fermentation, *Trends in Biotechnol. 4:*60, 1986.

National Academy of Sciences, *Introduction of Recombinant DNA-Engineered Organisms into the Environment: Key Issue,* National Academy Press, Washington, DC, 1987.

National Institutes of Health, Guidelines for Research Involving Recombinant DNA Molecules, *Federal Register 51* (May 7) 16958, 1987.

Examples of specific articles on modeling genetic instability

Aiba, S., and T. Imanaka, A Perspective on the Application of Genetic Engineering: Stability of Recombinant Plasmid, *Ann. N.Y. Acad. Sci. 369:*1, 1981.

Cooper, N. S., M. E. Brown, and C. A. Caulcott, A Mathematical Method for Analyzing Plasmid Stability in Microorganisms, *J. Gen. Microbiol. 133:*1871, 1987.

PROBLEMS

14.1. Given the following information, calculate the probability of forming a plasmid-free cell due to random segregation for a cell with 50 plasmid monomer equivalents:

    a. 40% of the total plasmid DNA is in dimers and 16% in tetramers.

    b. The distribution of copy numbers per cell is as follows, assuming monomers only:

≤ 3 plasmids	~ 0%
4 to 8	4%
9 to 13	10%
14 to 18	25%
19 to 23	25%
24 to 28	20%
29 to 33	12%
34 to 38	4%
≥ 39	~ 0%

14.2. Assume that all plasmid-containing cells have eight plasmids; that an antibiotic is present in the medium, and the plasmid-containing cells are totally resistant; and that a newly born, plasmid-free cell has sufficient enzyme to protect a cell and its progeny for three generations. Estimate the fraction of plasmid-containing cells in the population in a batch reactor starting with only plasmid-containing cells after five generations.

14.3. Consider an industrial-scale batch fermentation. A 10,000 l fermenter with 5×10^{10} cells/ml is the desired scale-up operation. Inoculum for the large tank is brought through a series of seed tanks and flasks, beginning with a single pure colony growing on an agar slant. Assume that a colony (10^6 plasmid-containing cells) is picked and placed in a test tube with 1 ml of medium. Calculate how many generations will be required to achieve the required cell density in the 10,000 l fermenter. What fraction of the total population will be plasmid-free cells if $\mu_+ = 1.0$ h^{-1}, $\mu_- = 1.2$ h^{-1}, and $P = 0.0005$?

14.4. Assume that you have been assigned to a team to produce human epidermal growth factor (hEGF). A small peptide, hEGF speeds wound healing and may be useful in treating ulcers. A market size of 50 to 500 kg/yr has been estimated. Posttranslational processing is not essential to the value of the product. It is a secreted product in the natural host cell. Discuss what recommendations you would make to the molecular-biology team leader for the choice of host cell and the design of a reactor. Make your recommendations from the perspective of what is desirable to make an effective process. You should point out any potential problems with the solution you have proposed, as well as defend why your approach should be advantageous.

14.5. Develop a model to describe the stability of a chemostat culture for a plasmid-containing culture. For some cultures, plasmids make a protein product (e.g., colicin in *E. coli*) that kills plasmid-free cells but does not act on plasmid-containing cells. Assume that the rate of killing by colicin is $k_T C n_-$, where k_T is the rate constant for the killing and C is the colicin concentration. Assume that the colicin production is first order with respect to n_+.

14.6. Consider the following data for *E. coli* B/r-pDW17 grown in a minimal medium supplemented with amino acids. Estimate $\Delta \mu$ and R. Compare the stability of this system to one with a glucose-minimal medium (Example 14.2).

D = 0.3 h^{-1}			D = 0.67 h^{-1}		
Time, h	Generation	f_-	Time, h	Generation	f_-
0	0	0.003	0	0	0.003
4.6	2	0.010	5.2	5	0.010
11.5	5	0.04	10.3	10	0.015
16.0	7	0.08	15.5	15	0.05
23.1	10	0.30	20.6	20	0.13
27.7	12	0.43	31.0	30	0.34
34.7	15	0.55	41.4	40	0.68
46.2	20	0.96	51.7	50	0.95

14.7. It has been claimed that gel immobilization stabilizes a plasmid-containing population. A factor suggested to be responsible for the stabilization is compartmentalization of the population into very small pockets. For example, the pocket may start with an individual cell and grow to a level of 200 cells per cavity. Develop a mathematical formula to compare the number of plasmid-free cells in a gel to that in a large, well-mixed tank.

14.8. You must design an operating strategy to allow an *E. coli* fermentation to achieve a high cell density (> 50 g/l) in a fed-batch system. You have access to an off-gas analyzer that will measure the pCO_2 in the exit gas. The glucose concentration must be less than 100 mg/l to avoid the formation of acetate and other inhibitory products. Develop an approach to control the glucose feed rate so as to maintain the glucose level at 100 ± 20 mg/l. What equations would you use and what assumptions would you make?

14.9. Develop a simple model for a population in which plasmids are present at division with copy numbers 2, 4, 6, 8, or 10. The model should be developed in analogy to eqs. 14.7 through 14.9. You can assume that dividing cells either segregate plasmids perfectly or generate a plasmid-free cell.

14.10. Assume you have an inoculum with 95% plasmid-containing cells and 5% plasmid-free cells in a 2 l reactor with a total cell population of 2×10^{10} cells/ml. You use this inoculum for a 1000 l reactor and achieve a final population of 4×10^{10} cells/ml. Assuming $\mu_+ = 0.69$ h^{-1}, $\mu_- = 1.0$ h^{-1}, and $P = 0.0002$, predict the fraction of plasmid-containing cells.

14.11. Assume you scale up from 1 l of 1×10^{10} cells/ml of 100% plasmid-containing cells to 20,000 l of 5×10^9 cells/ml, at which point overproduction of the target protein is induced. You harvest six hours after induction. The value of P is 0.0005. Before induction $\mu_+ = 0.95$ h^{-1} and $\mu_- = 1.0$ h^{-1}. After induction μ_+ is 0.15 h^{-1}. What is the fraction of plasmid-containing cells at induction? What is the fraction of plasmid-containing cells at harvest?

$$\boxed{15}$$

Medical Applications
of Bioprocess
Engineering

15.1. INTRODUCTION

With increasing knowledge of cellular and molecular biology, the boundary between traditional bioprocess engineering and biomedical engineering has become increasingly fuzzy. In this chapter we will consider areas in which bioprocess principles are critical to solution of medical problems. Two important areas are tissue engineering and gene therapy using viral vectors. These are examples; other medical applications exist, and their number will undoubtedly increase. Indeed, we have already mentioned the use of transgenic plants as a source for edible vaccines and the techniques of microfabrication applied to make miniature process facilities for rapid DNA analysis to allow genome analysis in a physician's office.

15.2. TISSUE ENGINEERING

15.2.1. What Is Tissue Engineering?

Tissue engineering has a primary focus on developing *in vitro* bioartificial tissues, typically based on cells derived from donor tissue. These tissues can be used as transplants to improve biological function in the recipient. Commercial examples include living skin tissue and chondrocytes implanted in a damaged knee for production of hyaline-like carti-lage. An extracorporeal (outside of the body) artificial liver employing pig liver cells in a hollow fiber reactor is being clinically tested.

Artificial tissues/organs for transplantation that are under active development in-clude liver, pancreas, kidney, fat (for reconstructive surgery), blood vessel, bone marrow, bone, and neurotransmitter-secreting cell constructs. An alternative form of tissue engi-neering is *in vivo* alteration of cell growth and function. An example would be the use of implanted polymeric tubes with a controlled surface chemistry to encourage and guide re-connection of damaged nerves. Another use of artificial tissue constructs is for toxicologi-cal and pharmacological testing of potential new drugs. In this case the artificial tissue or combination of tissues acts as a surrogate, reducing the need to use animals for such testing.

The primary difference between tissue engineering and protein production from mammalian cells lies in the constraints on cell selection. For protein production we prefer continuous, transformed cell lines. A single cell type is desirable. For tissue engineering the goal is to replicate the response of a living tissue. Cells removed from a tissue and cul-tured as a homogeneous cell type in two dimensions (e.g., on a solid surface) often lose their authentic *in vivo* response. Since cell transformation and cancer are closely related, transformed cells cannot be used for a product for transplantation. Reconstruction of arti-ficial tissues requires a deep understanding of the interactions of one cell with another, control of cellular differentiation processes, and knowledge of how cells interact with sur-faces to which they attach. Most often a polymer scaffold is used to guide and organize tissue growth. Maintaining the correct ratio of cell types can be difficult, since some cell types, such as fibroblasts, can "outgrow" others. The appropriate cell types must organize themselves into the appropriate three-dimensional configuration. With such tissues, func-tion requires appropriate structure. Ideally the cells can be multiplied from donor tissue for at least 10 passages and then assume the fully differentiated phenotype when the correct stimulus is applied. A major challenge is the routine, reproducible culture of such cells, and bioprocess engineers are in an excellent position to contribute to this technology.

Of particular importance to many tissue-engineered constructs is the formation of extracellular matrix, the interaction of cells with one another, and the interaction with an artificial surface. Anchorage-dependent cells must attach and spread on a substrate surface to proliferate and function. Cell adhesion is mediated by extracellular matrix proteins such as fibronectin and collagen. Synthetic substrates can be modified by adding synthetic peptide sequences (3 to 6 amino acids) to a surface at the end of a synthetic polymer. The difference in strength of cell–substrate and cell–cell adhesion can greatly alter the three-dimensional organization of the cells.

The use of simulated microgravity reactors coupled with polymer scaffolds has been useful in some cases in the development of certain tissue constructs. The use of mi-crofabrication techniques, where the investigator has control over placement of individual cells, is an intriguing technology for more authentic tissue constructs. An increasingly large array of tools are being developed for controlling the formation of tissue constructs with improved performance.

Development of effective tissue constructs that are biologically complex, such as the liver, is a very difficult problem, and commercial production of transplantable organs is in the distant future. However, commercial production of simple tissues has been accomplished.

15.2.2. Commercial Tissue Culture Processes

The two examples of commercial production of tissue-engineered products are for skin and cartilage.

15.2.2.1. Tissue-Engineered Skin Replacements.
Skin replacements are needed for patients with burns, with pressure ulcers (usually due to diabetes), and with reconstructive surgery. Cadaver skin is undesirable because of potential disease transmission, possible immunological rejection, inflammation of the wound bed, limited availability, and high cost due to labor-intensive procurement. Animal products would be problematic due to immunological reactions. An "artificial" human skin replacement that can be taken out of the freezer and used by physicians is highly desirable. TransCyte™, a human fibroblast-derived temporary skin substitute, is such a product.

An artificial human skin is made from human neonatal foreskin fibroblasts which are obtained from routine circumcision discards. These fibroblast cells are easily available and not subject to rejection by the body; they are also capable of replication, and numbers can be expanded in a straightforward manner. Donor tissue is enzymatically digested and seeded onto a three-dimensional bioresorbable polymer scaffold. The cell-seeded scaffold is placed in a bioreactor that provides a physiologically similar environment to promote cell growth and secretion of proteins and extracellular matrix material. The cells and extracellular matrix material result in a three-dimensional tissue construct with metabolically active cells that is functionally similar to natural skin and promotes wound healing.

The manufacturing process is centered on a bioreactor system that mimics *in vivo* conditions through control of temperature, pH, oxygen level, nutrient supply, waste removal, and fluid hydrodynamics. It must produce thousands of individual products of uniform and reproducible quality. The manufacturing system requires a bioreactor for product growth that also serves as final packaging. The system is sealed after assembly and remains sealed to maintain sterility. Figure 15.1 is a schematic of the process. The bioreactor is made of plastic with inlet and outlet ports. It is flat and just large enough to hold a polymer mesh about 13 by 19 cm. Twelve individual bioreactors are manifolded together. Groups of twelve are then operated in parallel to provide the desired lot size. Once the system is sterilized, it is opened only for cell seeding and nutrient replenishment. About three weeks are required for growth. At the end of this growth period, cryopreservative is pumped into the bioreactor module. After sealing inlet and outlet tubes, the bioreactor is packaged, frozen, and then stored or shipped. The product is thawed and used by the physician.

This manufacturing process is similar to many we have discussed before, but differs in that the product is discrete and not in solution. The process must meet all of the FDA and GMP requirements discussed earlier.

15.2.2.2. Chondrocyte Culture for Cartilage Replacement.
Articulate or hyaline cartilage is a thin layer of tissue found at the ends of bones. Damage to articulate cartilage in the knee is a common health problem. Cartilage consists of three primary

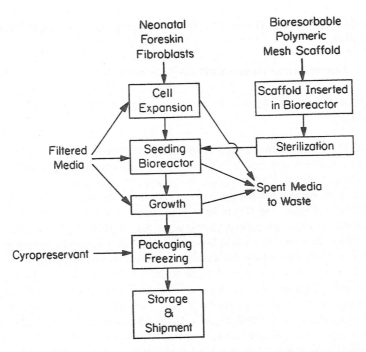

Figure 15.1. Conceptual process diagram for production of a human skin substitute.

components: collagen fibrils, proteoglycans, and cells known as chondrocytes. These cells are present in low numbers (1% by volume) but are responsible for synthesis and release of these other compounds, forming cartilage. When chondrocytes are cultured as two-dimensional (monolayer) cultures, the cells are no longer differentiated and do not make normal hyaline matrix proteins. However, if cultured in three dimensions, such as a suspension in agarose, or *in vivo,* the cells redifferentiate and begin to manufacture hyaline-like matrix.

Because of immunological responses it is advantageous for patients to supply their own cells for expansion in number before being reinjected into the knee. A manufacturing process for individual cell growth is needed. The supply and reuse of cells from an individual is termed *autologous implantation.*

The basic procedure entails biopsy of a patient to obtain a small number of chondrocytes, monolayer culture of primary chondrocytes and expansion of cell numbers, release of cells from monolayer and into suspension, assembly of cells in three dimensions, and injection into the patient's own knee. This product is known as Carticel™. The implanted cells produce hyalinelike cartilage and fill in defects in the patient's knee. Patients can often reenter vigorous physical activity in twelve months.

The primary manufacturing challenge here is the need for separate culture for each individual. To accomplish this for mass production is a challenge to any manufacturing process. There is expected to be a demand for other therapies (e.g., cancer treatment), where cells from an individual will need to be efficiently cultured and returned to the individual.

15.3. GENE THERAPY USING VIRAL VECTORS

Gene therapy is the transfer of one or more genes into cells for a therapeutic effect. It can be done *ex vivo* (outside the body) in tissues that are transplanted back into the patient or *in vivo,* where the genes, usually in a delivery vehicle such as a virus, are injected into the patient.

Gene therapy is intellectually connected to metabolic engineering. However, the complexity of humans presents an even greater challenge than the metabolic engineering of a single cell. Gene therapy is a quantitative problem that makes good use of the quantitative skills of engineers. Basically the right gene needs to be delivered to only the right tissue target in the right amount, with the gene products being expressed at the right level at the right time for the right length of time. For gene therapy to be effective many things have to go right. Clinical success with gene therapy has been minimal, as approaches have been qualitative and trial-and-error in nature. A rational analysis would be a useful tool.

Many methods can be used for gene delivery. These involve viral vectors, use of naked DNA, and liposome or particle-mediated gene delivery. In this chapter we will focus only on viral systems. Bioprocess technology is necessary for production of the viral vectors, and analyses arising from bioprocess studies are applicable to gene therapy.

15.3.1. Models of Viral Infection

The three primary virus vectors for gene therapy are retroviruses (derived from a wild-type virus that infects mice), adenoviruses, and adeno-associated viruses. Retroviruses are enveloped viruses, because they are encapsulated in a lipid bilayer membrane. The adenovirus and adeno-associated viruses are nonenveloped viruses. The model of viral trafficking that we discuss below is for enveloped viruses.

Dee et al.[†] proposed a model for the viral trafficking of Semliki Forest virus (SFV), an enveloped RNA virus that has been considered as a vector for large-scale production of heterologous proteins. However, this analysis, which was motivated by a bioprocess application, is applicable to retrovirus vectors for gene therapy.

As indicated in Fig. 15.2, enveloped RNA viruses can enter cells through receptor-mediated endocytosis. The virus binds to specific receptor molecules on the cell's surface. It is assumed (for this analysis) that attachment is irreversible and that the number of receptor molecules is much greater than the number of virus particles present. Under this circumstance:

$$dV_{ex}/dt = -k_a C V_{ex} \tag{15.1}$$

where V_{ex} is the number of extracellular viruses per cell, C is the cell concentration, and k_a is the attachment-rate constant. The value of k_a can be estimated as:

$$k_a = k_f (\alpha R) \tag{15.2}$$

where k_f is the intrinsic forward rate constant for the binding of a single viral attachment protein to a receptor, α is the number of viral attachment proteins per virus, and R is the

---

[†]K. U. Dee, D. A. Hammer, and M. L. Shuler, *Biotechnology Bioengineering* 46: 485–496 (1995).

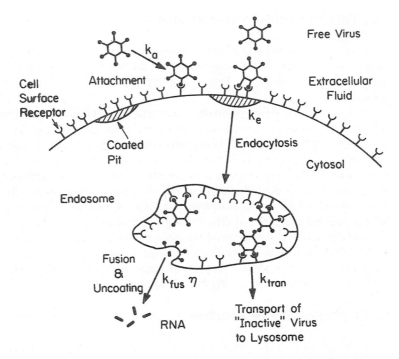

Figure 15.2. The trafficking of an enveloped RNA virus is depicted. A mathematical model of these processes is given in the text.

number of available receptors per cell. If the ratio of R to V_{ex} is large, R is approximately the total number of receptors.

Once the virus is attached, it randomly associates with "coated pits," which internalize the virus through the process of *endocytosis*. Endocytosis is a process of invagination of the plasma membrane to form a vesicle in which the receptors and anything attached to the receptors are captured within the vesicle. Such vesicles are known as *endosomes*. The rate of endocytosis is assumed to be:

$$dV_i/dt = k_e V_s \qquad (15.3)$$

where V_i is the number per cell of internalized virus, V_s is the surface concentration of virus (number per cell), and k_e is the endocytosis rate constant.

The amount of surface virus can be determined from combining eqs. 15.1 and 15.3 to give:

$$dV_s/dt = k_a C V_{ex} - k_e V_s \qquad (15.4)$$

Endosomes are intermediates in recycle of plasma membrane components and transport of internalized materials to lysosomes for degradation. For the virus to replicate successfully it must escape from the endosome before it is delivered to a lysosome. Endosomes undergo a biphasic pH change. In early endosomes pH drops from about 7.5 to 6.0 within 5 to 10 min. In the second phase pH drops slowly to about 5.2 in another 30 to 60 min. The pH drop is due to membrane proteins that pump hydrogen ions into the endosome.

The virus escapes through fusion of its membrane envelope with the endosome membrane. The fusion process is pH dependent. For SFV the threshold pH for fusion is 6.2. This threshold is reached in early endosomes. The fusion process is quite rapid. However, not all virus undergoes fusion. The virus balance in the endosome is:

$$dV_{ene}/dt = k_e V_s - k_{fus}\eta V_{ene} - k_{tran}V_{ene} \tag{15.5}$$

where V_{ene} is the number of virus in the early endosomes, k_{fus} is the observed fusion-rate constant, η is the fraction of virus that can successfully fuse with endosomal membrane, and k_{tran} is the rate constant for movement of inactive virus to the lysosome and degradation. The main cause of inactive virus is the fusion of virus with membrane fragments within the endosome instead of the endosomal membrane.

Virus that successfully fuses with the endosomal membrane is released into the cytosol. The protein coat is removed rapidly, and for this analysis we assume that effectively all cytoplasmic virus is uncoated, so that the RNA is released and can be replicated.

The rate equation for uncoating is

$$dV_{cyt}/dt = k_{fus}\eta V_{ene} \tag{15.6}$$

where V_{cyt} is the cytoplasmic virus.

The amount of RNA synthesized per cell follows a saturation type response, so that:

$$[RNA]_{cell} = \frac{k_m V_{cyt}}{K_{sv} + V_{cyt}} \tag{15.7}$$

where k_m and K_{sv} are parameters dependent on the host cell. Equation 15.7 applies only to the early parts of the infection cycle. Virus that enters after RNA replication has been completed for early entry virus probably no longer contributes to RNA synthesis, as host cell function begins to be lost.

Equations 15.1 to 15.6 allow for an analytical solution. If we assume that only free virus is present initially ($t = 0$), then the initial conditions are: $V_{ex} = V_{exo}$; $V_s = 0$; $V_i = 0$; $V_{ene} = 0$; $V_{cyt} = 0$. The solutions in this case are:

$$V_{ex} = V_{exo}e^{-k_a Ct} \tag{15.8}$$

$$V_s = \frac{k_a C V_{exo}}{k_e - k_a C}(e^{-k_a Ct} - e^{-k_e t}) \tag{15.9}$$

$$V_i = \frac{V_{exo}}{k_e - k_a C}[k_a C(e^{-k_e t} - 1) - k_e(e^{-k_a Ct} - 1)] \tag{15.10}$$

$$V_{ene} = \frac{k_e k_a C V_{exo}}{k_e - k_a C}\left[\frac{1}{k_{fus}\eta + k_{tran} - k_a C}(e^{-k_a Ct} - e^{-(k_{fus}\eta + k_{tran})t}) \right.$$
$$\left. - \frac{1}{k_{fus}\eta + k_{tran} - k_e}(e^{-k_e t} - e^{-(k_{fus}\eta + k_{tran})t})\right] \tag{15.11}$$

$$V_{cyt} = \frac{k_{fus}\eta k_e k_a C V_{exo}}{k_e - k_a C} \left\{ \frac{1}{k_{fus}\eta + k_{tran} - k_a C} \left[\frac{1}{k_a C}(1 - e^{-k_a Ct}) + \frac{1}{k_{fus}\eta + k_{tran}}(e^{-(k_{fus}\eta + k_{tran})t} - 1) \right] \right.$$

$$\left. - \frac{1}{k_{fus}\eta + k_{tran} - k_e} \left[\frac{1}{k_e}(1 - e^{-k_e t}) + \frac{1}{k_{fus}\eta + k_{tran}}(e^{-(k_{fus}\eta + k_{tran})t} - 1) \right] \right\} \tag{15.12}$$

For wild-type SFV infecting baby hamster kidney (BHK) cells the rate parameters have been estimated as: $k_a = 5.2 \times 10^{-9}$ ml/cell-min.; $k_e = 0.039$ min^{-1}; $k_{fus}\eta = k_{tran} = 0.14$ min^{-1} (based on 50% of internalized virus being uncoated). Such a model with these parameters fits time-course measurements from actual experiments and can also be used to make estimates of response if mutant virus forms are used.

These same equations can easily be applied to an *ex vivo* system for gene therapy. For *in vivo* use they apply in principle, but the presence of various natural compounds in the bloodstream can alter k_a (and potentially other parameters), and multiple cell types are present which may bind, but not endocytose virus. A pharmacokinetic model (to predict time-dependent virus distribution at different positions in the body) would need to be coupled with virus/cell interaction models (of the type developed here) for each major cell type. The reader can easily see why *ex vivo* therapy would be much easier to design than an *in vivo* protocol. For either the *ex vivo* or *in vivo* therapies to be successful, there must be a method to efficiently produce virus.

15.3.2. Mass Production of Retrovirus

The efficient production of a highly concentrated solution of retroviruses that have been genetically modified to deliver a gene is difficult. Genetically engineered retroviruses, like all recombinant viruses, are produced by a two-part system consisting of a packaging cell line and the recombinant vector. This two-part system is necessary because the recombinant virus is unable to replicate itself. The recombinant virus is derived from a wild-type virus in which essential viral genes have been deleted and replaced with the therapeutic gene of interest. The enzyme reverse transcriptase converts RNA-encoded genes into DNA, which is transported into the nucleus and integrated into the cell's chromosomal DNA.

The packaging cell line is genetically engineered to produce the essential, structural viral genes that have been deleted from the viral genome. Because the viral genes are encoded in the packaging cell's chromosome, the resulting viral particles are incapable of causing disease, but act as carriers for the desired, therapeutic genes. The resulting retrovirus vector can be used only with dividing cells, since cells must undergo mitosis before gene integration can occur. This feature limits *in vivo* use to cases such as cancer suppression but is not a limitation on *ex vivo* systems.

Another limitation on the retrovirus system is a bioprocess limitation; the production of high-titer virus and subsequent purification and concentration without loss of infectivity have been difficult. A major limitation to the effectiveness of many gene-therapy approaches, including the use of retroviruses, is that the average number of genes delivered to target cells is too low to achieve a beneficial therapeutic effect. This low efficiency is due to the bioprocess limitation on production of a concentrated, highly infectious retrovirus preparation.

The problems in producing a highly infectious preparation derive from two factors:[†] rapid decay of virus and inhibition of viral infectivity by proteoglycans released by the packaging cell line. The proteoglycans have very high molecular weight. If the virus is concentrated by ultrafiltration, the proteoglycan is also concentrated by nearly the same factor. While the high virus concentration would be expected to increase the number of infected cells, the presence of concentrated proteoglycan leads to increased inhibition of infection, so that the number of cells infected does not increase significantly and in some cases even decreases with the concentrated viral preparation.

Another approach to increase viral titer is to produce virus at a reduced temperature (e.g., 28°C versus 37°C). While both the rate of virus production and viral decay decrease with temperature, the rate of decay is more sensitive to temperature. Thus the amount of potentially active virus is severalfold higher when produced at 28°C (e.g., 2×10^5 cfu/ml vs. 6×10^4 cfu/ml; cfu is colony-forming units and measures the number of active viruses in a dilute solution). However, the transduction efficiency did not increase, presumably due to changes in proteoglycan concentration.

These results should inspire the reader to consider bioreactor options beyond batch growth followed by ultrafiltration. Clearly, the reader would want to consider a bioreactor configuration in which the virus is removed as soon as it is produced and then purified and stored at low temperature to reduce decay. Further, a method of concentration of virus needs to be developed that does not also concentrate proteoglycan. Enzymatic digestion of proteoglycans has been suggested, but due to multiplicity of proteoglycan structures such a strategy is difficult to implement. Selective adsorption and desorption of virus is attractive, but the sensitivity of the virus to these processes is a concern. A packaging cell altered in its capability to produce proteoglycan is another possible strategy.

The primary point is that bioprocess technology is intimately connected to production of agents that can be effective in gene therapy. The solution to the biomedical problem requires a solution to the bioprocess problem.

15.4. BIOREACTORS

Mass production of cells for transplantation or the use of bioreactors as artificial hybrid organs are subjects of intense development in medicine. These are issues in which bioprocess engineers may make significant contributions.

15.4.1. Stem Cells and Hematopoiesis

Some animal cells are capable of extensive replication and self-renewal. Others are highly differentiated and perform specific functions; typically differentiated cells cannot replicate (at least, replication is very limited). A *stem cell* is an undifferentiated cell capable of continuous self-renewal that can also produce large numbers of differentiated progeny, depending on extracellular factors. The best studied system is the hematopoietic system. There are eight major types of fully mature blood cells in the human circulatory system. A hematopoietic stem cell gives rise to two types of *progenitor cells*. The progenitor cells

---

[†] J. M. LeDoux, H. E. Davis, J. R. Morgan, and M. L. Yarmush, *Biotechnol. Bioeng.* *63*:654 (1999).

are capable of self-replication but have a more restricted range of cells into which they may differentiate. In the hematopoietic system one type of progenitor cell can give rise to the myeloid cells (e.g., red blood cells or erythrocytes, platelets, macrophages, etc.) and the other to lymphoid cells (e.g., T-cells and B-cells). The progression of differentiation depends on a large number of hematopoietic growth factors. Hematopoiesis, the process of generating these blood cells, takes place in the bone marrow. *Ex vivo* hematopoiesis is an alternative or supplement to bone marrow transplants. Hematopoietic stem and progenitor cells can be recovered from human blood, particularly umbilical cord blood.

There are many challenges to generating commercial-scale systems for hematopoiesis. Coculture with stroma cells from the bone marrow is necessary to generate necessary growth factors, so bioreactors must accommodate adherent cell growth. Three basic reactor types are under development. Fluidized bed reactors with macroporous supports mimic structures much like that in bone. However, there are challenges in controlling surface chemistry and maintaining the appropriate mix of cell types. Cell sampling and control of reactor conditions can be problematic.

Flatbed reactors (e.g., modified T-flasks with continuous flow possible) can carry stroma and facilitate analysis and design, since the geometry is well defined. Direct microscopic observation of cells is possible. Automated flatbed systems have been used to generate cells for human clinical trials.

Membrane-based units, such as hollow-fiber reactors, are potential solutions. This is an efficient design that is fairly easy to characterize. However, cell observation and harvesting are considered problematic. Another alternative is the possible use of spheroid cultures (a natural aggregation of a mixed population of cells), which can be done in suspension-type bioreactors.

Bioreactors for production of other tissue types from other stem cells are a real possibility. There is far less experience with bioreactors for other stem cells/tissues, but the principles the reader has learned in this book should be applicable to these challenges.

15.4.2. Extracorporeal Artificial Liver

Liver failure is a major medical problem. The liver performs many metabolic functions (carbohydrate, fat, and vitamin metabolism, production of plasma proteins, conjugation of bile acids, and detoxification), of which detoxification is the most critical. In some cases, a failing liver may recover if the metabolic and detoxification demands on it can be reduced; an artificial liver may provide the respite necessary for self-repair of a liver. In other cases an artificial liver may serve as a bridge to a transplant.

Due to the spatial and metabolic complexity of the liver, an implantable artificial liver is a distant objective. However, an extracorporeal device to serve as temporary assist device is realistic (such a system is in clinical trials). A promising design is a hollow-fiber system using porcine (pig) hepatocytes. Such cells are relatively easy to obtain in large quantities and maintain a satisfactory level of differentiated cellular activity in regard to detoxification. A disadvantage is due to limited lifespan (and proliferative ability). The membrane that separates these cells from the blood provides protection against adverse immune reactions. The issues in the design of such hollow-fiber reactors are similar, whether the reactor is to be used in a bioprocess operation or in a biomedical application.

15.5. SUMMARY

The principles from bioprocess engineering are NOT restricted to use for biomanufacture of chemicals, medicinals, or other biological compounds. These same principles have many applications in medicine. Indeed, as biomedical engineering has become more oriented to molecular and cellular systems and bioprocess engineering more toward animal cells, the boundary between these two activities has become very porous.

Important concepts in this chapter include ideas of how to produce tissue constructs on a commercial scale. Also the efficient production of virus for use in gene therapy is an unresolved bioprocess problem. The basic understanding of the dynamics of viral infection of cells is important both in design of biomanufacturing processes and for gene therapy. Quantitative approaches to describing such processes can lead to rational design both for the biomanufacturing process and for therapeutic strategies. Bioreactor design is also an important component of design of artificial organs and reactors to generate functional tissue from stem cells.

The list of examples of the intersection of bioprocessing and medicine is a growing one that will present many readers of this text with exciting career possibilities.

SUGGESTIONS FOR FURTHER READING

A good source for most topics in biomedical engineering is

BRONZINO, J. D., Editor-in-Chief, *The Biomedical Engineering Handbook* (2d ed., 2 vols.), CRC Press LLC, Boca Raton, FL, 2000.
Chapters 97, 106, 109, 116, 120, 124, and 133 were used in preparation of this chapter.

Other useful sources include

DEE, K. U., D. A. HAMMER, AND M. L. SHULER, A Model of the Binding, Entry, Uncoating, and RNA Synthesis of Semliki Forest Virus in Baby Hamster Kidney (BHK-21) Cells, *Biotechnol. Bioeng. 46*:485–496, 1995.

LE DOUX, J. M., H. E. DAVIS, J. R. MORGAN, AND M. L. YARMUSH, Kinetics of Retrovirus Production and Decay, *Biotechnol. Bioeng. 63*:654–662, 1999.

NAUGHTON, G. K., Skin: The First Tissue-Engineered Products, The Advanced Tissue Sciences Story, *Sci. Am. 280* (4):84–85, 1999.

TUBO, R., Restoring Wounded Knees, *Sci. & Med. 6*(2):6–7, 1999.

PROBLEMS

15.1. Many experiments for virus trafficking are done with prebound virus. Virus attaches at low temperature to suppress endocytosis, and unbound virus is washed away. Cells are then warmed to 37°C to initiate endocytosis. Derive equations in this case for the time-dependent concentration of internalized virus and virus that enters the cytosol and uncoats.

15.2. A mutant SFV has values of $k_{fus}\eta = 8.3 \times 10^{-3}$ min^{-1} and $k_{tran} = 1.8 \times 10^{-2}$ min^{-1}. The wild-type virus has values of $k_{fus}\eta = 6.4 \times 10^{-2}$ min^{-1} and $k_{tran} = 1.1 \times 10^{-1}$ min^{-1}. For both wild-type and mutant SFV the value of $k_e = 0.074$ min^{-1}. Prebound virus were used (see Problem 15.1) and V_{s0} was the same for both cases. At 20 min what is the ratio of uncoated virus for the wild-type virus to the mutant.

15.3. A value of $k_a C$ is measured as 0.021 min^{-1}. Assume $k_{fus}\eta = k_{tran} = 0.14$ min^{-1} and $k_e = 0.04$ min^{-1}. The initial inoculum was 500 viruses/cell. At two hours after inoculation unbound virus was removed. Calculate V_s, V_{cyt}, and V_{ene}.

15.4. You wish to produce active retrovirus, and you are investigating the effect of temperature on the process. Active virus is subject to decay with a rate constant, $k_d = 2.2$ day^{-1} at 37°C and 0.76 day^{-1} at 31°C. The production rate of virus from a packaging cell line is $k_p = 3.3$ virus/cell-day at 37°C and $k_p = 2.9$ virus/cell-day at 31°C. Assume that there are 1×10^6 cells and the volume of liquid medium is 2 ml. The initial number of virus in solution is zero. How many viruses are there per ml 12 hours after initiation of virus production?

15.5. Mass transfer limitations are often critical in design of devices such as the hollow-fiber bioartificial liver for use as an extracorporeal assist device. Consider removal (detoxification) of a slightly hydrophobic compound in a patient's blood. Draw a diagram and describe potential mass transfer limitations on the rate of detoxification by intracellular enzymes.

16

Mixed Cultures

16.1. INTRODUCTION

The dynamics of mixed cultures are important considerations in some commercial fermentations. They are critical to understanding the response of many ecological systems to stress. The use of organisms with recombinant DNA has added another dimension to our consideration of how cells within a population interact with each other.

Many food fermentations, such as cheese manufacture, depend on multiple interacting species. The biological treatment of waste waters relies on an undefined complex mixture of microorganisms. The ratio of various species in the treatment process is critical; sudden shifts in the composition of the population can lead to failure of the unit to meet its objectives.

In all natural environments, cells exist in potentially mixed populations. Understanding how these cells interact with one another is critical to understanding the natural cycles for the elements (e.g., C, N, and S), the response of ecosystems to environmental challenges (for example, acid rain), and the rate and extent of degradation of chemicals introduced into such ecosystems.

As discussed in Chapter 14, many populations with recombinant DNA behave as mixed-culture systems. Some members of the population will lose or modify the inserted gene (often carried on a plasmid). Although a single species is present, the formation of mutant or plasmidless cells leads to a distinct subpopulation. The interaction of the subpopulation with the original population follows the principles we will develop in this chapter for mixed populations.

16.2. MAJOR CLASSES OF INTERACTIONS IN MIXED CULTURES

The major interactions between two organisms in a mixed culture are competition, neutralism, mutualism, commensalism, amensalism, and prey–predator interactions. Table 16.1 summarizes these interactions.

TABLE 16.1 Scheme of Classification of Binary Population Interactions. The Roles of A and B May Be Reversed. Top, Indirect Interactions; Bottom, Direct Interactions

Effect of presence of B on growth rate of A	Effect of presence of A on growth rate of B	Qualifying remarks	Name of interaction
	Indirect Interactions		
−	−	Negative effects caused by removal of resources	Competition
−	O		
−	−	Negative effects caused by production of toxins or inhibitors	Antagonism
−	O		Amensalism
−	+	Negative effects caused by production of lytic agents; positive effects caused by solubilization of biomass	Eccrinolysis
+	O	Positive effects caused by production by B (host) of a stimulus for growth of A (commensal) or by removal by B of an inhibitor for growth of A	Commensalism
+	+	See remarks for commensalism. Also, presence of both populations is not necessary for growth of both	Protocooperation
+	+	See remarks for commensalism. Also, presence of both propulations is necessary for growth of either	Mutualism
−	+	B feeds on A	Feeding (includes predation and suspension feeding)
	Direct Interactions		
−	+	The parasite (B) penetrates the body of its host (A) and converts the host's biomaterial or activities into its own	Parasitism
+	+ (or perhaps O)	A and B are in physical contact; interaction highly specific	Symbiosis
−	−	Competition for space	Crowding

With permission, from A. G. Fredrickson, in *Foundations of Biochemical Engineering,* American Chemical Society Symposium Series, Vol. 207, pp. 201–228, 1983.

Competition is an indirect interaction between two populations that has negative effects on both. In competition, each population competes for the same substrate. Two populations or microorganisms with similar nutrient requirements usually compete for a number of common, required nutrients when grown together.

The outcome of competition between two species for the *same growth-limiting substrate* in an open system (e.g., a chemostat) is determined by the specific growth-rate-limiting substrate concentration relationship. Two different cases can be distinguished in a mixed culture of two competing species (Fig. 16.1):

1. μ_a is always greater than μ_b. The organisms with the fastest growth rate will displace the others from the culture. This is known as the *exclusion principle*.

2. Crossover in μ-S relationship. In this case, the faster-growing organism is determined by the dilution rate. Depending on the dilution rate, three different cases may be identified:

   a. At the crossover point $D = \mu_x$; $S = S_x$, two species could be maintained in a chemostat at $D = \mu_x$. However, this is an unstable operating point.

   b. If $D > \mu_x$, then $\mu_a > \mu_b$, and B will be washed out; A will dominate.

   c. If $D < \mu_x$, then $\mu_b > \mu_a$, and A will be washed out; B will dominate.

In a batch system, both species would exist in culture media. The ratio of number density of species at a given time will be determined by the relative magnitudes of the specific growth rates and the initial concentrations of cells.

Neutralism is an interaction where neither population is affected by the presence of the other. That is, there is no change in the growth rate of either organism due to the presence of the other. Neutralism is relatively rare. One example of neutralism is the growth of yogurt starter strains of *Streptococcus* and *Lactobacillus* in a chemostat. The total counts of these two species at a dilution rate of 0.4 h^{-1} are quite similar whether the populations are cultured separately or together. Neutralism may occur in special environments where each species consumes different limiting substrates and neither species is affected by the end products of the other.

Mutualism and *protocooperation* are more common than neutralism and may involve different mechanisms. In both cases, the presence of each population has a positive effect on the other. For mutualism, the interaction is essential to the survival of both species. In protocooperation, the interaction is nonessential. One mechanism is the mutual exchange of required substances or the removal of toxic end products by each organism.

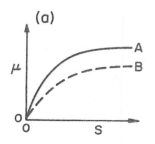

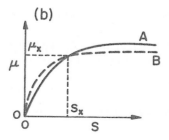

Figure 16.1. μ-S relationship for two competing species (A and B) in a mixed culture. Case (a) corresponds to $\mu_{mA} > \mu_{mB}$ and $K_{SA} \leq K_{SB}$, while case (b) corresponds to a case where $\mu_{mA} > \mu_{mB}$, but $K_{SB} < K_{SA}$. In case (b), the growth curves cross and A and B could potentially coexist at $D = \mu_x$ and $S = S_x$ in a chemostat.

The metabolisms of partner populations must be complementary to yield a mutualistic interaction. An example is the growth of a phenylalanine-requiring strain of *Lactobacillus* and a folic-acid-requiring strain of *Streptococcus* in a mixed culture. Exchange of the growth factors phenylalanine and folic acid produced by partner organisms helps each organism to grow in a mixed culture, while separate pure cultures exhibit no growth. Another example of mutualistic interaction exists between aerobic bacteria and photosynthetic algae. Bacteria use oxygen and carbohydrate for growth and produce CO_2 and H_2O. The algae convert CO_2 to carbohydrates and liberate oxygen in the presence of sunlight.

The reader should note that *symbiosis* and mutualism are not the same. Symbiosis implies a relationship when two organisms live together. A symbiotic relationship may be mutualistic, but it may also be neutralistic, parasitic, commensalistic, and so on.

Commensalism is an interaction in which one population is positively affected by the presence of the other. However, the second population is not affected by the presence of the first population. Various mechanisms may yield a commensal interaction. Two common mechanisms are the following:

1. The second population produces a required nutrient or growth factor for the first population.
2. The second population removes a substance from the medium that is toxic to the first population.

An example of the first type of commensal interaction is the production of H_2S by *Desulfovibrio* (through the reduction of SO_4^{2-}), which is used as an energy source by sulfur bacteria.

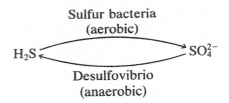

Sulfur bacteria
(aerobic)

H_2S ⇄ SO_4^{2-}

Desulfovibrio
(anaerobic)

An example of the second type of commensal interaction is the removal of lactic acid by the fungus *Geotrichium candidum,* which allows the growth of *Streptococcus lactis.* This interaction is utilized in cheese making using *S. lactis.* Lactic acid produced by *S. lactis* inhibits the growth of the bacteria. The fungus metabolizes lactic acid and improves the growth conditions for *S. lactis.*

Amensalism is the opposite of commensalism. In amensalism, population *A* is negatively affected by the presence of the other population (*B*). However, population *B* is not affected by the presence of population *A*. Various amensal interaction mechanisms are possible. Two common mechanisms are the following:

1. Population *B* produces a toxic substance that inhibits the growth of population *A*.
2. Population *B* removes essential nutrients from the media, thus negatively affecting the growth of population *A*.

One example of the first type of amensal interaction is the production of antibiotics by certain molds to inhibit the growth of others. Some microbes excrete enzymes that decompose cell-wall polymers. Such organisms destroy their competitors and also utilize the nutrients released by the lysed cells. The microbial synthesis of organic acids reduces pH and inhibits the growth of other organisms.

Predation and *parasitism* are interactions in which one population benefits at the expense of the other. These two interactions are distinguished by the relative size of organisms and the mechanisms involved. Predation involves the ingestion of prey by the predator organism. A good example of prey–predator interaction is the ingestion of bacteria by protozoa. This interaction is common in aerobic waste-treatment reactors such as activated sludge units. In parasitism, the host, which is usually the larger organism, is damaged by the parasite. The parasite benefits from utilization of nutrients from the host. A common example of parasitism is the destruction of microorganisms by microphages. Although the physical mechanisms in predation and parasitism differ, these two phenomena have many common features in their conceptual and mathematical descriptions.

In an open system, such as a chemostat where predator–prey interactions take place, the populations of predator and prey do not necessarily reach steady state but can oscillate at certain dilution rates. At the beginning of the operation prey concentration is high, but predator concentration is low. As the predators consume prey, the number of predators increases and the prey concentration decreases. After a while, a small prey population cannot support the large predator population, and the predator population decreases while prey population increases. Depending on the dilution rate and feed substrate concentration, these oscillations may be sustained or damped or may not exist.

Finally, note that these interactions can, and often do, exist in combination. For example, *A* and *B* may compete for glucose as a nutrient, but *A* requires a growth factor from *B* to grow. In such a case, both competition and commensalism would be present.

16.3. SIMPLE MODELS DESCRIBING MIXED-CULTURE INTERACTIONS

Multiple interacting species can give rise to very complex behavior. In some cases, coexistence of species is prohibited; in others, complex sustained oscillatory behavior may be observed. Multiple steady states are possible.

Writing the appropriate equations to describe mixed populations follows the principles we discussed in Chapter 6. In each case, a balance must be written for each species (organism, rate-limiting substrate, or product), and these balances will be the same as we have used previously. Although we may write chemically structured models for each organism, we will confine our discussions to the use of unstructured models for each species. Even so, the population model is still structured in the sense that the whole biomass is divided into distinct subpopulations.

These equations are often applied to chemostats that mimic many ecosystems. When these equations are solved, they often yield multiple steady states. For each steady state we need to analyze its stability. An unstable steady state will never be realized in practice; using phase-plane portraits, we can explore the approach to the steady state and the determination of the presence or absence of limit cycles. It is often easier to test a steady state for its local stability (linearized stability analysis) than for its global stability.

With this background, we will consider some representative examples.

Example 16.1.

Competition of two species for the same growth-rate-limiting substrate is common. Determine when the two organisms may stably coexist if both A and B follow Monod kinetics.

Solution   For this situation, the following equations describe the dynamic situation:

$$\frac{dX_A}{dt} = -DX_A + \frac{\mu_{mA}S}{K_{SA} + S} X_A \tag{16.1}$$

$$\frac{dX_B}{dt} = -DX_B + \frac{\mu_{mB}S}{K_{SB} + S} X_B \tag{16.2}$$

$$\frac{dS}{dt} = D(S_0 - S) - \frac{1}{Y_{X_{A/S}}} \frac{\mu_{mA}SX_A}{K_{SA} + S} - \frac{1}{Y_{X_{B/S}}} \frac{\mu_{mB}S}{K_{SB} + S} X_B \tag{16.3}$$

If both A and B coexist, we would require from eqs. 16.1 and 16.2 that

$$D = \frac{\mu_{mA}S}{K_{SA} + S} = \frac{\mu_{mB}S}{K_{SB} + S} \tag{16.4}$$

Equation 16.4 can be solved to yield

$$S = \frac{\mu_{mB}K_{SA} - \mu_{mA}K_{SB}}{\mu_{mA} - \mu_{mB}} \tag{16.5}$$

Equation 16.5 is meaningful only if $S \geq 0$. Consider the two cases in Fig. 16.1. In case (a), $\mu_{mA}K_{SB} > \mu_{mB}K_{SA}$ and $\mu_{mA} > \mu_{mB}$; consequently, S is always less than zero, and coexistence is impossible. In case (b), however, a value of S can be found from eq. 16.5 that will allow the populations to coexist. The corresponding D for the crossover point is D_c.

Although this coexistence is mathematically obtainable, it is not physically attainable in real systems. In real systems, the dilution rate will vary slightly with time, and, in fact, the variation will show a bias. Using an analysis more sophisticated than appropriate for this text, it can be demonstrated that one competitor will be excluded from the chemostat if the intensity of the "noise" (random fluctuations) in D and the bias of the mean of D away from D_c are not both zero. Also, it is possible for either competitor to be excluded, depending on how D varies.[†]

Two competitors can coexist if we modify the conditions of the experiments. Examples of such modifications include allowing spatial heterogeneity (the system is no longer well mixed or wall growth is present) or another level of interaction is added (e.g., adding a predator or interchange of metabolites). Also, operation of the chemostat in a dynamic mode (D is a function of time) can sometimes lead to coexistence. It is also interesting to note that the use of other rate expressions (e.g., substrate inhibition) can lead to multiple crossover points and potentially multiple steady states.

[†]G. N. Stephanopoulos, R. Aris, and A. G. Fredrickson, *Math Biosci. 45:*99 (1979).

Example 16.2.

Consider a case of mutualistic growth in a chemostat. How would you write the appropriate equations for this system? Determine the effects of the addition of competition to mutualism. **Solution** The physical model is for A and B as two separate species. A produces P_A as a by-product of growth and B produces P_B. Organism B requires P_A to grow, while A requires P_B. The feed to a chemostat contains all essential nutrients except for P_A and P_B, and A and B may compete for substrate, S, in the feed.

For this case the most general description is

$$\frac{dX_A}{dt} = -DX_A + \mu_A X_A \tag{16.6}$$

$$\frac{dX_B}{dt} = -DX_B + \mu_B X_B \tag{16.7}$$

$$\frac{dP_A}{dt} = -DP_A + Y_{P_A}\mu_A X_A - \frac{1}{Y_{X_B/P_A}}\mu_B X_B \tag{16.8}$$

$$\frac{dP_B}{dt} = -DP_B + Y_{P_B}\mu_B X_B - \frac{1}{Y_{X_A/P_B}}\mu_A X_A \tag{16.9}$$

And if the growth of either A or B is limited by S, then we need to consider

$$\frac{dS}{dt} = D(S_0 - S) - \frac{1}{Y_{X_A/S}}\mu_A X_A - \frac{1}{Y_{X_B/S}}\mu_B X_B \tag{16.10}$$

Note that Y_{X_B/P_A} is the biomass yield of B using P_A as substrate, and Y_{P_A} is the amount of P_A made per unit mass of A. Similar definitions apply to Y_{X_A/P_B} and Y_{P_B}. If we consider the pure mutualistic state, then we ignore eq. 16.10. For a coexistent state to exist, $D = \mu_A = \mu_B$. It is also clear that the rate of production of P_A and P_B must exceed their consumption. Thus

$$Y_{P_A}X_A > \frac{X_B}{Y_{X_B/P_A}} \tag{16.11}$$

and

$$Y_{P_B}X_B > \frac{X_A}{Y_{X_A/P_B}} \tag{16.12}$$

It then follows that

$$Y_{P_A}Y_{P_B}X_A X_B > \frac{1}{Y_{X_B/P_A}} \cdot \frac{1}{Y_{X_A/P_B}} X_A X_B \tag{16.13a}$$

or

$$Y_{P_A}Y_{P_B} > \frac{1}{Y_{X_B/P_A}} \cdot \frac{1}{Y_{X_A/P_B}} \tag{16.13b}$$

It is also clear that

$$D < \min(\mu_{mA}, \mu_{mB}) \tag{16.14}$$

Equations 16.13b and 16.14 determine whether eqs. 16.6 to 16.9 allow the potential existence of a purely mutualistic steady state.

The stability of such a coexistent state has been examined, where μ_A and μ_B were represented by various growth functions. Using a linear stability analysis, it can be shown that this pure mutualistic state results in a saddle point, and the system is unstable for all physically accessible values of D. If, however, the growth-rate-limiting substrate for either A or B is S, then a stable coexistent state can be found.

Example 16.3.

Consider the growth of a protozoa (predator) on bacteria (prey) in a chemostat. Write appropriate equations for this system.

Solution   Here bacteria constitute a substrate for protozoa. In a chemostat culture, the following balances can be written for substrate, prey, and predator.

$$\text{Substrate:} \qquad \frac{dS}{dt} = D(S_0 - S) - \frac{1}{Y_{X_b/S}} \frac{\mu_{mb}S}{K_b + S} X_b \qquad (16.15)$$

$$\text{Prey:} \qquad \frac{dX_b}{dt} = -DX_b + \frac{\mu_{mb}S}{K_b + S} X_b - \frac{1}{Y_{X_p/b}} \frac{\mu_{mp}X_b X_p}{K_p + X_b} \qquad (16.16)$$

$$\text{Predator:} \qquad \frac{dX_p}{dt} = -DX_p + \frac{\mu_{mp}X_b X_p}{K_p + X_b} \qquad (16.17)$$

where $Y_{X_b/S}$ and $Y_{X_p/b}$ are the yield coefficients for the growth of prey on substrate and the growth of predator on prey, respectively.

This model was used to describe the behavior of *Dictyostelium discoideum* and *E. coli* in a chemostat culture and was found to predict experimental results quite well. A variety of types of coexistence behavior have been revealed by stability analysis. These include no oscillations, damped oscillations, and sustained oscillations (see Fig. 16.2).

A classical model that describes oscillations in a prey–predator system is the Lotka–Volterra model, in which growth rates are expressed by the following equations:

$$\text{Prey:} \qquad \frac{dX_b}{dt} = \mu'_b X_b - \frac{\mu'_p X_b X_p}{Y_{p/b}} \qquad (16.18)$$

$$\text{Predator:} \qquad \frac{dX_p}{dt} = -k'_d X_p + \mu'_p X_b X_p \qquad (16.19)$$

The first term in eq. 16.18 describes the growth of prey on substrate and the second the consumption of prey by predators. The first and second terms in eq. 16.19 describe the death of predator in the absence of prey and the growth of predator on prey, respectively. $Y_{p/b}$ is the yield of predators on prey (g/g), μ'_b is the specific growth rate of prey on a soluble substrate (h^{-1}), μ'_p is the growth rate of predator on prey (l/g-h), and k'_d is the specific death rate of the predator (h^{-1}).

Equations 16.18 and 16.19 allow a steady-state solution for batch growth, where dX_b/dt and $dX_p/dt = 0$. Under these conditions $X_{bF} = k'_d/\mu'_p$ and $X_{pF} = \mu'_b Y_{p/b}/\mu'_p$.

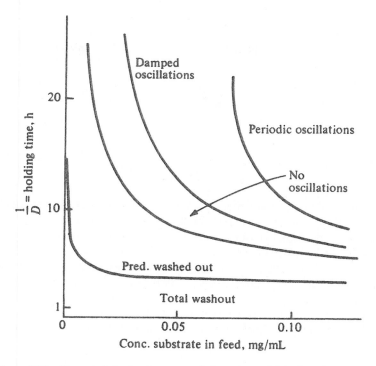

Figure 16.2. Dynamic behavior that can result from the model developed in Example 16.3. (With permission, from H. M. Tsuchiya and others, Predator-Prey Interactions of *Dictyostelium discoideum* and *Escherichia coli* in Continuous Culture, *J. Bacteriol.*, vol. 110, p. 1147, 1972.)

By defining dimensionless variables such as

$$\overline{X}_b = \frac{X_b}{X_{bF}}, \quad \overline{X}_p = \frac{X_p}{X_{pF}} \tag{16.20}$$

we can express eqs. 16.18 and 16.19 in terms of X_b and X_p.

$$\frac{d\overline{X}_b}{dt} = \mu_b'(1 - \overline{X}_p)\overline{X}_b \tag{16.21}$$

$$\frac{d\overline{X}_p}{dt} = -k_d'(1 - \overline{X}_b)\overline{X}_p \tag{16.22}$$

Equations 16.21 and 16.22 can be solved with the initial conditions of $\overline{X}_b(0) = \overline{X}_{b_0}$ and $\overline{X}_p(0) = \overline{X}_{p0}$.

The relationship between $\overline{X}_b$ and $\overline{X}_p$ can be determined by dividing eq. 16.21 by eq. 16.22.

$$\frac{d\overline{X}_b}{d\overline{X}_p} = \frac{\mu_b' \overline{X}_b (1 - \overline{X}_p)}{-k_d' \overline{X}_p (1 - \overline{X}_b)}$$

(16.23)

Integration of eq. 16.23 yields

$$\mu_b' \ln \overline{X}_p - \mu_b' \overline{X}_p + k_d' \ln \overline{X}_b - k_d' \overline{X}_b = K$$

(16.24a)

or

$$\left(\frac{\overline{X}_p}{e^{\overline{X}_p}}\right)^{\mu_b'} \left(\frac{\overline{X}_b}{e^{\overline{X}_p}}\right)^{k_d'} = e^K$$

(16.24b)

where K is an integration constant that is a function of the initial population sizes.

The phase-plane analysis of the system can be made using eq. 16.24. Figure 16.3 describes the limit cycles (oscillatory trajectories) of prey–predator populations for different initial population levels.

The Lotka–Volterra model considers the exponential growth of prey species in the absence of predator and neglects the utilization of substrate by prey species according to Monod form. The Lotka–Volterra oscillations depend on initial conditions and change their amplitude and frequency in the presence of an external disturbance. These types of oscillation are called soft oscillations. The other model based on Monod rate expressions

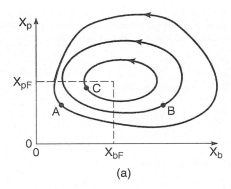

(a)

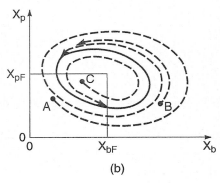

(b)

Figure 16.3. Both diagrams are schematics of phase-plane portraits for prey–predator interactions. (a) Limit cycles predicted by Lotka–Volterra (soft oscillations where initial conditions determine the dynamic behavior). (b) Limit-cycle prediction using the model developed in Example 16.3 (hard oscillations where the limit cycle is independent of initial conditions). The predicted steady-state point is defined by X_{pF} and X_{bF}. A, B, and C represent different initial conditions.

(eqs. 16.15–16.17) explains the more stable and sustained oscillations observed in nature, which are independent of initial conditions (that is, hard oscillations).

16.4. MIXED CULTURES IN NATURE

Mixed cultures of organisms are common in natural ecological systems. Microorganisms are involved in the natural cycles of most elements (e.g., carbon, nitrogen, oxygen, and sulfur). Simplified diagrams of the carbon and nitrogen cycles are presented in Figs. 16.4 and 16.5. Organisms living in soil and aquatic environments actively participate in carbon and nitrogen cycles. For example, certain organisms fix atmospheric CO_2 to form carbohydrates, while others degrade carbohydrates and release CO_2 into the atmosphere. Similarly, some organisms fix atmospheric nitrogen (N_2) to form ammonium and proteins, while others convert ammonium into nitrite and nitrate (nitrification), and others reduce nitrate into atmospheric nitrogen (denitrification). Sulfur-oxidizing organisms convert reduced sulfur compounds (sulfur and sulfide) into sulfate, and sulfate-reducing organisms reduce sulfate into hydrogen sulfide.

The aforementioned interactions among different species take place in natural systems in a more complicated manner. The complexity of such a system is depicted in

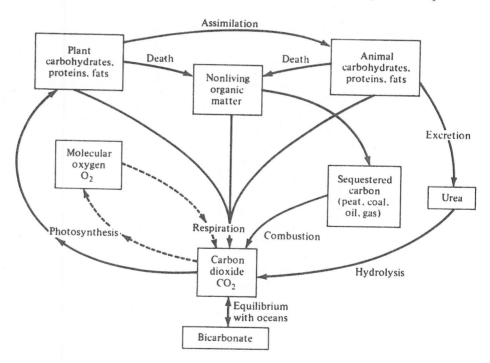

Figure 16.4. Simplified diagram of the carbon cycle. Also shown (dashed lines) is the major component of the oxygen cycle, which is closely linked to the cycle of carbon. (With permission, from J. E. Bailey and D. F. Ollis, *Biochemical Engineering Fundamentals,* 2d ed., McGraw-Hill Book Co., New York, 1986, p. 914.)

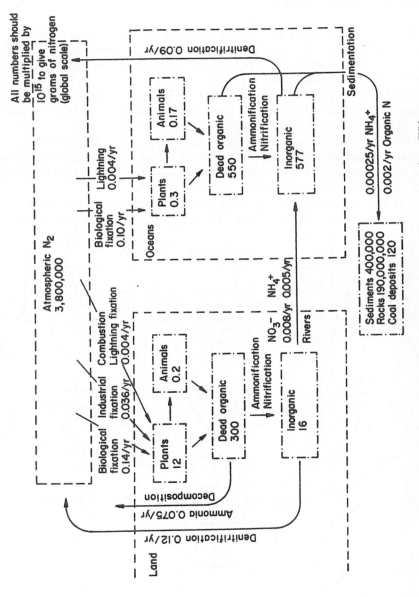

Figure 16.5. The global nitrogen cycle. Minor compartments and fluxes are not given. (With permission, from T. D. Brock, D. W. Smith, and M. T. Madigan, *Biology of Microorganisms*, 4th ed., Pearson Education, Upper Saddle River, NJ, 1984, p. 421.)

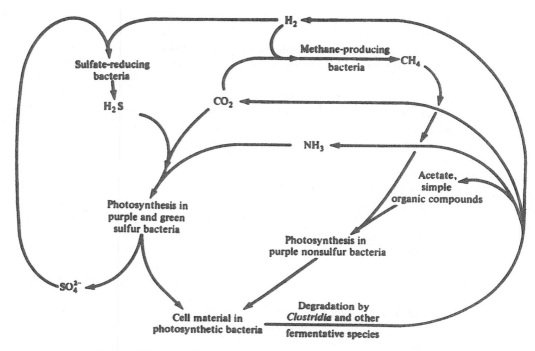

Figure 16.6. Simplified schematic of the cycles of matter in an anaerobic environment. (With permission, from J. E. Bailey and D. F. Ollis, *Biochemical Engineering Fundamentals*, 2d ed., McGraw-Hill Book Co., New York, 1986, p. 918.)

Fig. 16.6 for an anaerobic environment. Understanding such complex cycles is important to understanding the capacity of natural ecosystems to degrade organic pollutants.

16.5. INDUSTRIAL UTILIZATION OF MIXED CULTURES

Defined mixed microbial populations are commonly used in cheese making, a good example of using mixed cultures in food production. Cheeses of various types are produced by inoculating pasteurized fresh milk with appropriate lactic acid organisms. The bacteria used for lactic acid production are various species of *Streptococcus* and *Lactobacillus* in a mixed culture. Other organisms are used to develop flavor and aroma. Among these are *Brevibacterium linens*, *Propionibacterium shermanii*, *Leuconostoc* sp., and *Streptococcus diacetilactis*. After inoculation of pasteurized milk, a protein-rich curd is precipitated by the acidity of the medium, and the liquid is drained off. The precipitated curd is allowed to age by action of bacteria or mold. Some molds used in cheese making are *Penicillum camemberti* and *Penicillum roqueforti*.

Lactic acid bacteria are also used in whiskey manufacture. *Lactobacillus* added to the yeast reduces pH and, therefore, the chance of contamination. *Lactobacillus* also contributes to the flavor and aroma of whiskey. A favorable interaction between yeast and lactic acid bacteria exists in ginger-beer fermentation.

The utilization of undefined mixed microbial cultures in waste-treatment processes is typical and unavoidable. Waste-water treatment constitutes one of the largest-scale uses of bioprocesses. Mixed cultures are also utilized in the anaerobic digestion of waste materials. Cellulase producers, acid formers, and methane producers are typical organisms involved in the anaerobic digestion of cellulosic wastes. However, attempts to encourage the growth of a particular species on waste materials have been made.

The Symba process was developed in Sweden for treating starchy wastes, particularly those from potato processing. This process utilizes *Endomycopsis fibuligera* for amylase production and a yeast, *Candida utilis,* for the utilization of sugar molecules produced from the hydrolysis of starch. Single-cell protein (SCP) is produced simultaneously with potato waste treatment.

Corn and pea wastes are also treated by a mixed culture of *Trichoderma viride* and *Geotrichium* sp. *T. viride* produces cellulase to break down cellulose into reduced sugar molecules, and *Geotrichium* produces amylases to break down starch into reduced sugar molecules. Both organisms utilize reduced sugar molecules for growth.

A mixed culture of *Candida lipolytica* and *Candida tropicalis* has been grown on hydrocarbons, *n*-paraffins, or gas oil for single-cell protein (SCP) production purposes in both laboratory and pilot-scale operations. The utilization of a mixed culture of yeasts was proved to yield better product quality as compared to pure yeast strains.

Gaseous hydrocarbon substrates like methane can be utilized by certain bacteria to produce SCP. Several experimental studies have shown that mixed cultures of methane-utilizing organisms grow faster than pure cultures.

Certain methane-utilizing species of *Pseudomonas* oxidize methane to methanol. However, *Pseudomonas* is inhibited by the end product, methanol. Inclusion of a methanol-utilizing bacteria such as *Hyphomicrobium* into the growth medium eliminates the problem of methanol inhibition. This relationship is mutualistic in the sense that *Pseudomonas* supplies carbon source (CH_3OH) for *Hyphomicrobium,* and *Hyphomicrobium* removes the growth inhibitor (methanol) of *Pseudomonas.*

$$CH_4 \xrightarrow{\textit{Pseudomonas}} CH_3OH \xrightarrow{\textit{Hyphomicrobium}} CO_2 + H_2O$$

$$\text{Feedback inhibition}$$

(16.25)

16.6. BIOLOGICAL WASTE TREATMENT: AN EXAMPLE OF THE INDUSTRIAL UTILIZATION OF MIXED CULTURES

16.6.1. Overview

Waste materials generated in a society can be classified in three major categories:

1. Industrial wastes are produced by various industries, and waste characteristics vary greatly from one industry to another. Industrial wastes usually contain hydrocarbons, carbohydrates, alcohols, lipids, and aromatic organics. Industrial wastes are rich in carbon compounds and usually deficient in nitrogen (high C/N ratio);

therefore, the biological treatment of industrial wastes usually requires supplemental addition of nitrogen compounds and other nutrients. The presence of potentially toxic compounds must be carefully considered in devising a treatment strategy.

2. Domestic wastes are treated by municipalities and derive from humans and their daily activities. They include ground garbage, laundry water, excrement, and often some industrial wastes that have been sewered into the municipal system. Domestic waste varies significantly with time in terms of flow and composition due to the periodic nature of human activity (e.g., flow decreases at night when most people sleep).

3. Agricultural wastes are produced by farm animals (e.g., manure) and include waste plants, such as straws. Agricultural wastes are usually carbon rich because of high cellulosic material content, although some wastes, such as poultry manure, are high in nitrogen.

Each of these waste materials has its own characteristics, and treatment methods vary depending on these characteristics.

Three major waste treatment methods are the following:

1. *Physical treatment* includes screening, flocculation, sedimentation, filtration, and flotation, which are usually used for the removal of insoluble materials.

2. *Chemical treatment* includes chemical oxidations (chlorination, ozonation) and chemical precipitation using $CaCl_2$, $FeCl_3$, $Ca(OH)_2$, or $Al_2(SO_4)_3$.

3. *Biological treatment* includes the aerobic and anaerobic treatment of waste water by a mixed culture of microorganisms.

Certain characteristics of waste water need to be known before treatment. Among them are (1) physical characteristics, such as color, odor, pH, temperature, and solids contents (suspended and dissolved solids), and (2) chemical characteristics, such as organic and inorganic compounds. Major carbon compounds in a typical industrial waste are carbohydrates, lipids–oils, hydrocarbons, and proteins. Other compounds, such as phenols, surfactants, herbicides, pesticides, and aromatic compounds, are usually in relatively small concentrations (<1 g/l) but are difficult to degrade by biological means. Among inorganic compounds present in waste water are nitrogenous compounds (NH_4^+, NO_3^-), sulfur compounds (SO_4^{2-}, S^{2-}, S^0, S_3^{2-}), phosphorus compounds (PO_4^{3-}, HPO_4^{2-}, $H_2PO_4^{1-}$), heavy metals (Ni^{2+}, Pb^{2+}, Cd^{2+}, Fe^{2+}, Cu^{2+}, Zn^{2+}, Hg^{2+}), and dissolved gases, such as H_2S, NH_3, and CH_4.

The carbon content (strength) of a waste-water sample can be expressed in several ways: *biological oxygen demand* (BOD), *chemical oxygen demand* (COD), and *total organic carbon* (TOC). Normally, a 5-day BOD value is reported. The BOD_5 is the amount of dissolved oxygen (DO) consumed when a waste-water sample is seeded with active bacteria and incubated at 20°C for 5 days. Since the amount of oxygen consumed is stoichiometrically related primarily to the organic content of waste water, BOD is a measure of the strength of waste water. This stoichiometric coefficient is not always known, since the composition of the organics is usually unknown. Also, some nitrogen-containing or inorganic compounds will exert an oxygen demand. If the only organic compound is glucose, oxygen consumption can be easily related to the carbon content of waste water under aerobic conditions.

$$C_6H_{12}O_6 + 6\ O_2 \rightarrow 6\ CO_2 + 6\ H_2O \qquad (16.26)$$

According to the stoichiometry of this reaction 1.07 g of oxygen is required for the oxidation of 1 g of glucose.

Samples of waste water need to be properly diluted to obtain an accurate BOD_5 measurement, seeded with active bacteria, and incubated at 20°C for 5 days along with an unseeded blank. BOD_5 is calculated using the following equation:

$$BOD_5 = [(DO)_{t=0} - (DO)_{t=5}]_{sample} - [(DO)_{t=0} - (DO)_{t=5}]_{blank} \qquad (16.27)$$

BOD measurements have some shortcomings. This method is applicable only to biodegradable, soluble organics and requires a high concentration of active bacteria preadapted to this type of waste. Moreover, if organic compounds are refractory, 5 days of incubation may not suffice, and 20 days of incubation (BOD_{20}) may be required.

COD is a measure of the concentration of chemically oxidizable organic compounds present in waste water. Organic compounds are oxidized by a strong chemical oxidant, and using the reaction stoichiometry, the organic content is calculated. Almost all organic compounds present in waste water are oxidized by certain strong chemical oxidants. Therefore, the COD content of a waste-water sample usually exceeds the measured BOD ($COD > BOD_5$).

A typical chemical oxidation reaction (unbalanced) is

$$C_aH_bO_c + Cr_2O_7^{2-} + H^+ \xrightarrow{\ heat\ } Cr^{3+} + CO_2 + H_2O \qquad (16.28)$$

Dichromate may be used as an oxidizing agent, and by a redox balance, the amount of oxygen required to oxidize organic compounds can be calculated. This method is faster (order of 3 hours), easier, and less expensive than BOD measurements.

The TOC content of waste-water samples can be determined by using a TOC analyzer. After proper dilutions, samples are injected into a high-temperature (900° to 950°F) furnace and all organic carbon compounds are oxidized to CO_2, which is measured by an infrared analyzer. To determine the TOC content, waste-water samples should be acidified to remove inorganic carbon compounds (mainly carbonates). The total carbon content of waste water can be determined before and after acidification, and the difference is inorganic carbon content.

The nitrogen content of waste-water samples is usually measured by total Kjeldahl nitrogen (TKN) determination. Other key nutrient concentrations, such as phosphate, sulfur, and toxic compounds, should be determined before waste streams are treated.

The concentration of biomass in a waste treatment system using suspended cells is measured as mixed-liquor volatile suspended solids (MLVSS). Basically waste water of known volume is filtered and the collected solids dried and weighed to give mixed-liquor suspended solids (MLSS). This material is then "volatilized" by burning in air at 600°C. The weight of the remaining noncombustible, inorganic material is the "fixed" solids. The difference between the original mass prior to combustion and the fixed solids is the "volatile" portion. The volatile portion or MLVSS is assumed to be primarily microbes, although carbonaceous particles are included in the measurement.

A typical waste-treatment operation employing biological treatment includes the following steps:

1. Primary treatment includes the removal of coarse solids and suspended matter (screening, sedimentation, filtration) and conditioning of the waste-water stream by pH adjustment and nutrient additions (e.g., PO_4^{3-}, NH_4^+).

2. Secondary treatment is the major step in biological treatment; it includes biological oxidation or anaerobic treatment of soluble and insoluble organic compounds. Organic compounds are oxidized to CO_2 and H_2O by organisms under aerobic conditions. Unoxidized organic compounds and solids from aerobic treatment (e.g., cell wall material, lipids–fats) are decomposed to a mixture of CH_4, CO_2, and H_2S under anaerobic conditions. A *sludge* of undecomposed material must be purged from either system.

3. Tertiary treatment includes the removal of the remaining inorganic compounds (phosphate, sulfate, ammonium) and other refractory organic compounds by one or more physical separation methods, such as carbon adsorption, deep-bed filtration, and in some cases membrane-based techniques, such as reverse osmosis or electrodialysis.

16.6.2. Biological Waste Treatment Processes

Biological waste-water treatment usually employs a mixed culture of organisms whose makeup varies with the nature of the waste. Biological treatment may be aerobic or anaerobic. The major aerobic processes (or reactor types) used in waste-water treatment are (1) activated sludge, (2) trickling filter, (3) rotating biological contractors, and (4) oxidation ponds.

Activated-sludge processes include a well-agitated and aerated continuous-flow reactor and a settling tank. Depending on the physical design of the tank and how feed is introduced into the tank, it may approximate either a PFR or CFSTR. A long narrow tank with single feed approaches PFR behavior; circular tanks approach CFSTR. The concentrated cells from the settling tank are recycled back to the stirred-tank reactor. Figure 16.7 is a schematic of a typical activated-sludge process. Usually, a mixed culture of organisms is utilized in the bioreactor. Some of these organisms may produce polymeric materials (polysaccharides), which help the organisms to agglomerate. Floc formation is a common phenomenon encountered in activated-sludge processes, which may impose some mass transfer limitations on the biological oxidation of soluble organics; but good floc formation is essential to good performance of the system, since large dense flocs are required in the sedimentation step. Cell recycle from the sedimentation unit improves the volumetric rate of biological oxidation (i.e., high-density culture) and therefore reduces the residence time or volume of the sludge reactor for a given feed rate. The recycle ratio needs to be controlled to maximize BOD removal rate.

The selection of aerator and agitators is a critical factor in the design of activated-sludge processes. The aeration requirements vary depending on the strength of the waste water and cell concentration. Oxygen requirements for a typical activated-sludge process are about 30 to 60 m^3 O_2/kg of BOD removed. Various aeration devices with and without

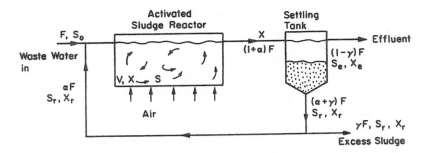

Figure 16.7. Schematic diagram of an activated-sludge unit.

mechanical agitation can be used in activated-sludge units. Mechanical surface aerators are widely used for shallow activated-sludge units. Surface aerators consist of partially submerged impellers attached to motors mounted on fixed structures. Surface aerators spray liquid and create rapid changes at the air–water interface to enhance oxygen transfer. Pure oxygen may be used for high-strength waste-water treatment. Also, stagewise operation with pure oxygen has been found to be a very effective method of waste-water treatment. The UNOX Process, first developed by Union Carbide, is based on this concept. Other forms of aeration include bubble aerators and fixed turbines similar to those we discussed in Chapter 10.

The activated-sludge system faces many uncontrolled disturbances in input parameters, such as waste flow and composition. Such disturbances can lead to system failure (less-than-adequate treatment of the waste stream). One type of disturbance is referred to as *shock loading*. A shock load indicates the sudden input (pulse) of a high concentration of a toxic compound. A CFSTR design, of course, is less affected by such inputs than a PFR design.

One response to disturbances is sludge *bulking*. A *bulking sludge* has flocs that do not settle well, and consequently cell mass is not recycled. *Bulking* sludge often results from a change in the composition of the microbial population in the treatment unit. For example, filamentous bacteria may dominate the normal floc-forming cells, leading to small, light flocs.

Various modifications of activated-sludge processes have been developed. Two examples are the following:

1. *Step feed process:* The feed stream is distributed along the length of the reactor. Such a configuration converts a conventional PFR system to more CFSTR-like behavior, which provides greater stability and more effective distribution and utilization of oxygen and oxygen transfer.

2. Solids reaeration (*contact stabilization*): In the conventional activated sludge process, the dissolved organics are quickly adsorbed onto (or into) the flocs, while the actual conversion to CO_2 and H_2O proceeds much more slowly. In contact stabilization, two tanks are used: one (about 1 h residence time) is used to promote the uptake of the organics, and the second (3–6 h residence time) is used for reaeration and the final conversion of the organic material. By concentrating the sludge before

oxidation, the total required tank volume for aeration is reduced by 50% in comparison to the conventional system.

Using rate expressions for microbial growth and substrate utilization and material balances for biomass and substrate, we can determine the required volume of an activated-sludge tank for a certain degree of BOD removal. Since an activated-sludge tank contains a mixed culture of organisms, the actual kinetics of BOD removal are complicated. Usually, interactions among various species are not known. The following analysis is based on pure-culture kinetics, which is only an approximation. It is also assumed that the specific growth-rate expression is given by the Monod equation, with a death rate (or endogenous respiration rate) term

$$\mu_{net} = \frac{\mu_m S}{K_s + S} - k_d \tag{16.29}$$

Steady-state material balances for biomass and rate-limiting substrates in an activated-sludge tank are (Fig. 16.7)

Biomass: $$\left(\frac{\mu_m S}{K_s + S} - k_d \right) XV + \alpha FX_r = (1 + \alpha) FX \tag{16.30}$$

Substrate: $$FS_0 + \alpha FS_r = \frac{1}{Y_{X/S}^M} \left(\frac{\mu_m S}{K_s + S} \right) XV + (1 + \alpha) FS \tag{16.31}$$

These equations are very similar to the case of the chemostat with recycle that we discussed in Chapter 9. They differ in that we now include the endogenous respiration term.

Assuming no substrate utilization and cell growth in the settling tank (short residence times), material balances around the settling tank yield

Biomass: $$(1 + \alpha) FX = (1 - \gamma) FX_e + (\alpha + \gamma) FX_r \tag{16.32}$$

Substrate: $$(1 + \alpha) FS = (1 - \gamma) FS_e + (\alpha + \gamma) FS_r \tag{16.33}$$

where α is the ratio of sludge recycle flow rate to feed flow rate and γ is the ratio of excess sludge flow to feed flow rate.

Assuming that the substrate is not separated in the settling tank (that is, $S = S_e = S_r$), eq. 16.33 can be eliminated. By rearranging eq. 16.32, we can obtain

$$(1 + \alpha) FX - \alpha FX_r = (1 - \gamma) FX_e + \gamma FX_r \tag{16.34}$$

Substituting eqs. 16.34 and 16.29 into eq. 16.30 yields

$$\mu_{net} VX = (1 - \gamma) FX_e + \gamma FX_r \tag{16.35}$$

Defining $\mu_{net} = 1/\theta_c$, where θ_c is the cells' (solids') residence time, we obtain

$$\theta_c = \frac{1}{\mu_{net}} = \frac{VX}{F(1 - \gamma) X_e + \gamma FX_r} \tag{16.36}$$

which is used to calculate the cellular (solids') residence time in the sludge tank. The value of θ_c is controlled by operator choice of recycle flow rates. Hydraulic (liquid) residence time is

$$\theta_H = \frac{V}{F} = \frac{\theta_c[(1-\gamma)X_e + \gamma X_r]}{X} = \frac{(1-\gamma)X_e + \gamma X_r}{\mu_{net} X} \tag{16.37}$$

Substituting $S_r = S$ in eq. 16.31 results in

$$F(S_0 - S) = \frac{1}{Y_{X/S}^M} \mu_g XV \tag{16.38}$$

or

$$V = \frac{Y_{X/S}^M F(S_0 - S)}{\mu_g X} = \frac{Y_{X/S}^M \theta_c F(S_0 - S)}{X(1 + k_d \theta_c)} \tag{16.39}$$

Equation 16.39 is used to calculate the required volume of the sludge tank for a certain degree of BOD removal $(S_0 - S)$.

The resulting reactor volume can be expressed in terms of the recycle ratio by substituting eq. 16.34 into eq. 16.37 to yield

$$V = F\theta_c \left(1 + \alpha - \alpha \frac{X_r}{X}\right) \tag{16.40}$$

The kinetic parameters of the active organisms in the sludge tank need to be known (i.e., μ_m, K_s, $Y_{X/S}^M$ and k_d), and other physical parameters (V, F, α, X_r, X, and S_0) need to be determined for the design (sizing, aeration requirement) of activated-sludge units. Typical hydraulic residence time for the aeration tank is 4 to 12 hours and typical sludge age is 3–10 days.

A summary of typical values for the kinetic parameters in biological waste treatment is presented in Table 16.2. Despite all simplifying assumptions, the pure-culture model seems to fit steady-state experimental data reasonably well, although it does not predict the dynamic performance very well.

Example 16.4.

An industrial waste with an inlet BOD_5 of 800 mg/l must be treated to reduce the exit BOD_5 level to $\leq$ 20 mg/l. The inlet flow rate is 400 m^3/h. Kinetic parameters have been estimated for waste as $\mu_m = 0.20$ h^{-1}, $K_s = 50$ mg/l of BOD_5, $Y_{X/S}^M = 0.5$ mg MLVSS/mg BOD_5, and $k_d = 0.005$ h^{-1}. A waste treatment unit of 3200 m^3 is available. Assume a recycle ratio of 0.40 and $X_e = 0$. If you operate at a value of $\theta_c = 120$ h, find S and determine if sufficient BOD_5 removal is attained in a well-mixed activated-sludge process to meet specifications. What will be X and the sludge production rate from this process?

Solution Equation 16.29 can be used to estimate S.

$$\mu_{net} = \frac{1}{\theta_c} = \frac{\mu_m S}{K_s + S} - k_d \tag{16.29}$$

$$S = \frac{K_s(1 + k_d \theta_c)}{\theta_c(\mu_m - k_d) - 1} \tag{16.29a}$$

$$S = \frac{(50 \ \text{mg/l})(1 + 0.005 \ h^{-1} \cdot 120 \ h)}{120 \ h \ (0.20 \ h^{-1} - 0.005 \ h^{-1}) - 1} = 3.57 \ \text{mg/l}$$

TABLE 16.2 Typical Monod Model Parameters for the Activated-sludge Process

Waste-water composition	μ_m (h^{-1})	K_s (mg/l)	$Y_{X/S}^M \left(\dfrac{\text{mg M}}{\text{mg waste}} \right)$	k_d (h^{-1})	Basis
Domestic	0.4–0.55	50–120	0.5–0.67	2.0–3.0×10^{-3}	BOD$_5$
Shellfish processing	0.43	96	0.58	5.8×10^{-2}	BOD$_5$
Yeast processing	0.038	680	0.88	3.3×10^{-3}	BOD$_5$
Phenol	0.46	1.66 $K_i = 380$	0.85		Phenol
Plastic processing	0.83	167	0.30	3.3×10^{-3}	COD

With permission, from D. W. Sundstrom and H. E. Klei, *Wastewater Treatment*, Pearson Education, Upper Saddle River, NJ, 1979, p. 146.

Thus, these operating parameters provide more than adequate BOD$_5$ removal, as 3.57 mg/l is significantly less than 20 mg/l.

A value for X can be found from eq. 16.39.

$$V = \frac{Y_{X/S}^M \theta_c F(S_0 - S)}{X(1 + k_d \theta_c)} \tag{16.39}$$

$$X = \frac{Y_{X/S}^M \theta_c F(S_0 - S)}{V(1 + k_d \theta_c)}$$

$$X = \frac{(0.5)(120 \text{ h})(400 \text{ m}^3/\text{h})(800 - 3.57) \text{ mg/l}}{3200 \text{ m}^3 (1 + 0.005 \text{ h}^{-1} \, 120 \text{ h})}$$

$$X = 3733 \text{ mg/l}$$

The amount of sludge produced is, from either eq. 16.36 or 16.37,

$$\theta_H = V/F = \frac{(1 - \gamma)X_e + \gamma X_r}{\mu_{\text{net}} X} \tag{16.37}$$

Since $1/\mu_{\text{net}} = \theta_c$ and $X_e = 0$,

$$F\gamma X_r = \frac{VX}{\theta_c} = \text{sludge production rate}$$

$$F\gamma X_r = \frac{(3200 \text{ m}^3)(3733 \text{ mg/l})(1000 \, \text{l/1 m}^3)}{120 \text{ h}}$$

or

$$F\gamma X_r = 9.95 \times 10^7 \text{ mg/h}$$
$$= 99.5 \text{ kg/h}$$

Although not required by the question, we could use eq. 16.40 to get a value of X_r, since $\alpha = 0.40$. Once X_r is known, γ can then be calculated if needed.

Trickling biological filters consist of a packed bed of inert support particles (sand or plastics) covered with a mixed culture of microorganisms in the form of a film or slime layer and cell aggregates. The column is loosely filled with packing material to yield a void fraction of 0.4 to 0.5. The filter bed is usually arranged in the form of a rather shallow bed with a high diameter-to-height ratio ($D/H \approx 3$) to avoid possible clogging problems and axial variations. Waste-liquid is fed to the top of the bed using rotary liquid distributors. The waste-liquid flow rate should be low enough to avoid creating shear forces that would remove biofilms from the surfaces of support particles. Air enters the bed from the bottom and moves upward by natural convection. The driving force for air circulation is the temperature difference created by heat released by biological oxidations. The reaction medium is highly heterogeneous, having temperature, pH, dissolved oxygen, and nutrient profiles throughout the column. The thickness of the biofilm and the composition of the organisms in the biofilm may vary with the length of the column. Dissolved-oxygen limitations are likely because of the high density of cells, unfavorable hydrodynamic conditions, and diffusion barriers within the film. Preaeration of the feed waste-water stream (to saturate with oxygen) and high liquid circulation rates may partly alleviate oxygen transfer problems.

As the liquid flows downward in the column on the surface of microbial films, organic compounds (substrates) diffuse through the microbial film and are utilized by organisms simultaneously. The liquid film over the biomass film should be thin enough to allow adequate aeration. A typical liquid film thickness is on the order of 0.01 mm, and biofilm thickness is 0.25 mm. Typical hydraulic residence times in trickling biological filters (TBF) are 0.5 to 4 h for high-rate filters using recirculation of effluent. Trickling filters are more stable against shock loads than are activated-sludge units. Trickling filters also entail lower operating costs and often give better effluent clarity compared to activated-sludge units. However, capital cost and space requirements for trickling filters are higher than for activated-sludge units, and trickling filters remove a smaller fraction of the soluble organics. Also, the maximum concentration of BOD in the influent is more constrained in trickling filters than in activated-sludge systems. A major problem with the operation of trickling filter units is poor control of conditions (T, pH, DO) due to the heterogeneous nature of the system. A comparison of trickle biological filters (TBF) with activated-sludge units (ASU) is presented in Table 16.3.

The liquid effluent of TBFs is usually recycled to obtain more complete removal of BOD from waste-water streams. Figure 16.8 is a schematic of a trickling biological filter.

The substrate balance on a differential height of a trickling biological filter yields

$$-F \, dS_0 = N_s a A \, dz \qquad (16.41)$$

where F is liquid flow rate (l/h), N_s is the substrate flux or rate of substrate utilization per unit surface area of biofilm (mg S/cm^2 film h), a is biofilm surface area per unit volume of the bed (cm^2 film/cm^3), and A (cm^2) and z (cm) are the cross-sectional area and the height of the bed, respectively.

The substrate flux or rate of substrate consumption is

$$N_s = -D_e \left. \frac{dS}{dy} \right|_{y=L} = \eta \frac{r_m S_0}{K_s + S_0} L \qquad (16.42)$$

TABLE 16.3 Comparison between Biological Trickling-filter and Activated-sludge Waste-water Treatment Processes

Item	Filter	Sludge tank
Capital costs	High	Low
Operating costs	Low	High
Space requirements	High	Low
Aeration control	Partial except in enclosed forced-draft types	Good
Temperature control	Difficult	Possible; heat losses small
Sensitivity to variations in applied feed concentrations	Fairly insensitive but slow to recover if upset	More sensitive, but recovery is rapid
Clarity of final effluent	Good	Not as good
Fraction of soluble organics removed	70%–90%	85%–95%
Typical θ_H		
Low rate	6 to 48 h	4 to 6 h
High rate	0.75 to 4 h	
Fly and odor nuisance	High	Low

where η is the effectiveness factor, L is the thickness of the biofilm (cm), r_m is the maximum rate of removal of substrate (mg $S/l \cdot s$), and S_0 is the substrate concentration in the bulk liquid (mg/cm^3). Typical values for r_m are 0.2 to 0.5 mg $S/l \cdot s$. Substituting eq. 16.42 into eq. 16.41 yields

$$-F\frac{dS_0}{dz} = \eta\frac{r_m S_0}{K_s + S_0} LaA \tag{16.43}$$

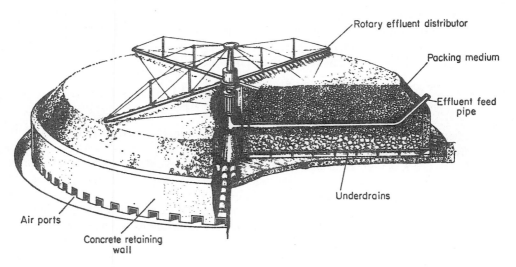

Figure 16.8. Typical trickling biological filter. (With permission, from J. W. Abson and K. H. Todhunter, in N. Blakebrough, ed., *Biochemical and Biological Engineering Science,* Vol. 1, Academic Press, New York, 1967, p. 326.)

Equation 16.43 can be integrated to yield a substrate concentration profile throughout the column. Usually, the substrate concentration in bulk liquid is very low, and the biological rate expression can be approximated to first order. That is,

$$-F\frac{dS_0}{dz} \cong \eta \frac{r_m}{K_s} LaAS_0 \qquad (16.44)$$

Integration of eq. 16.44 from $z = 0$ to $z = H$ yields

$$\frac{S_0}{S_{0i}} = \exp\left[-\frac{\eta r_m LaAH}{FK_s}\right] \qquad (16.45)$$

where S_{0i} is the inlet substrate concentration, and S_0 is concentration of substrate in the liquid phase at a distance z from the inlet. Experimental data of $\ln S_0/S_{0i}$ plotted versus z yield a straight line (in at least some cases), supporting the validity of this analysis.

The effectiveness factor can be calculated by using the following expressions:

$$\eta = 1 - \frac{\tanh\phi}{\phi}\left(\frac{\omega}{\tanh\omega} - 1\right), \qquad \text{for } \omega \leq 1 \qquad (16.46)$$

$$\eta = \frac{1}{\omega} - \frac{\tanh\phi}{\phi}\left(\frac{\omega}{\tanh\omega} - 1\right), \qquad \text{for } \omega \geq 1 \qquad (16.47)$$

where

$$\omega = \frac{\phi(S_0/K_s)}{\sqrt{2}\left(1 + S_0/K_s\right)}\left[\frac{S_0}{K_s} - \ln\left(\frac{S_0}{K_s}\right)\right]^{-1/2} \qquad (16.48)$$

and $\phi = L\sqrt{r_m/D_e K_s}$ and $r_m = \mu_m X/Y_{X/S}$. D_e is the effective diffusivity of the substrate within the biofilm (cm^2/s), and $Y_{X/S}$ is the yield coefficient (g cells/g substrate).

A similar, but more empirical design equation that has been used to determine waste-treatment parameters for biological filter is

$$\frac{S_0}{S_{0i}} = \exp\left[-Ka^{1+m}\left(\frac{A}{F}\right)^n\right] \qquad (16.49)$$

where a is the specific surface area of the inert support material (cm^2/cm^3), A is the cross-sectional area of the bed, F is liquid flow rate, and K is the apparent rate coefficient. In practice K is used as an adjustable parameter in a curve-fitting procedure. Exponents m and n vary depending on system characteristics such as geometry, hydrodynamics, biological systems, and waste-water characteristics.

Rotating biological contactors (RBC) contain rotating disks that come into contact with waste water periodically as they rotate. The disks are made of polystyrene or polyethylene, and their diameters range from 2 to 4 m. Figure 16.9 shows rotating disks used in biological waste treatment. A biological film forms on the surface of the disks. As the disks come into contact with waste water, nutrients (organics and dissolved oxygen) diffuse through the biofilm and are utilized by a mixed culture of organisms within the

Figure 16.9. Styrofoam disks used in RBC units (Biosystems Division, Autotrol Corporation). (With permission, from E. D. Schroeder, *Water and Wastewater Treatment,* McGraw-Hill Book Co., New York, 1977, p. 306.)

biofilm. The design equations for RBCs are similar to those used for trickling biological filters (TBF). RBCs are more compact and efficient than trickling-bed systems.

Oxidation ponds provide another inexpensive alternative to activated-sludge and trickling biological filter operations. Oxidation ponds are shallow (2 to 4 ft deep) waste-treatment reactors closely resembling natural aquatic ecosystems. Bacteria and algae grow in the same pond in a symbiotic relationship. Bacteria oxidize organic compounds by utilizing oxygen produced by algae and produce CO_2; algae utilize CO_2 produced by bacteria and produce oxygen by photosynthesis for bacterial consumption. Such ponds require large land areas, are less efficient than many other techniques, and may have adverse environmental side effects. Toxic or hazardous materials may collect in the sediment without degradation, creating a long-term problem.

Anaerobic digestion (or biological treatment) is usually used to treat solid wastes and excess sludge produced in aerobic waste-treatment processes. Particulate waste material removed by screening or sedimentation in primary waste treatment and biomass (concentrated sludge) produced in activated-sludge units are degraded under anaerobic conditions to produce methane. Anaerobic digestion is a slow process compared to aerobic processes; typical residence times are 30 to 60 days. The microbiology and biochemistry of anaerobic digestion are very complicated. However, the major steps involved in this process are as follows:

1. *Solubilization of insoluble organics:* Waste material may contain large amounts of solids made of cellulosics (papers, agricultural wastes), starches (potato waste), and other complex insoluble organic chemicals. Solubilization of these compounds by acid or enzymatic hydrolysis (cellulases, amylases, glucoamylases, lipases, proteases) is the first step in anaerobic digestion. These compounds are not readily utilizable by microorganisms, and their hydrolysis is essential for effective microbial digestion.

2. *Formation of volatile acids:* Solubilized organic compounds are metabolized by anaerobic bacteria to produce volatile organic acids, such as acetic, butyric, formic, and propionic acids, and short-chain fatty acids. Acid-producing organisms are a mixture of facultative anaerobes, such as enteric bacteria and clostridial species, which are called *acid formers.* Alcohol formation (butanol, propanol, ethanol) also takes place to a lesser extent. The optimal temperature and pH values for this step are $T = 35°C$ and pH = 4–6. The partial pressure of H_2 in the reactor can greatly influence metabolism.

3. *Formation of methane:* Volatile acids and alcohols produced in the second stage are converted to methane and CO_2 by methanogenic bacteria, which are strictly anaerobic. Among methanogenic bacteria used for this purpose are *Methanobacterium* (nonspore-forming rods), *Methanobacillus* (spore-forming rods), and *Methanococcus* and *Methanosarcina* (a cocci group growing in cubes of eight cells). The optimal temperature and pH range for methanogenic bacteria are $T = 35°$ to $40°C$ and pH = 7 to 7.8.

The major biological reaction steps involved in a typical anaerobic digestion process can be represented as follows:

$$\text{insoluble organics} \xrightarrow[\substack{\text{enzymatic or} \\ \text{acid}}]{\text{Hydrolysis}} \text{soluble organics} \xrightarrow[\text{bacteria}]{\text{Acid-producing}}$$

$$\text{volatile acids} + \text{alcohols} + H_2 + CO_2 \xrightarrow{\text{Methanogens}}$$
$$CH_4 + CO_2 + H_2S$$

Depending on the composition of organic waste, either enzymatic hydrolysis of insolubles (e.g., cellulosics) or methane production from volatile acids is the rate-limiting step in this reaction sequence. The concentration of H_2 is often critical in determining the rates of these processes. Hydrolysis reactions are usually carried out as a separate step. However, acid-formation and methane-formation steps can be achieved in the same reactor. Single- and two-reactor systems have been developed for anaerobic digestion. The two-step reactor scheme has been reported to result in higher reaction rates and methane yields. The operating conditions in a single-reactor system are pH = 6 to 7 and $T = 35°$ to $40°C$, which result in reasonable rates of acid formation and methane generation. Typical solids residence time in an anaerobic digester is 10 to 30 days.

Cell yields in anaerobic digesters are low, and a typical value is 0.05 g cells/g COD. Cell yield also varies with substrate. Yield on carbohydrates is larger than yield on the other carbon compounds.

Changes in temperature affect the methane-formation rate more than the acid-formation rate. Although the optimal temperature is reported to be 35° to 40°C, in some cases higher removal rates are obtained at higher temperatures ($T = 55°C$), depending on the composition of the organisms. The system does not function satisfactorily at temperatures below 35°C, and the temperature of the reactors needs to be controlled for stable methane generation.

The rate of methane formation is considerably lower than the rate of acid formation and is rate-limiting. Sudden increases in organic load concentrations may cause volatile acid (VA) accumulation in the medium, decreasing pH and altering H_2 levels, which further depresses methane production. Typical soluble organics removal rates are on the order of 0.2 g COD/g cells · day. Certain metal ions (Na^+, K^+, Ca^{2+}, Mg^{2+}) are known to be toxic to organisms of anaerobic digesters at high concentrations (> 1000 mg/l), although they are stimulatory at low levels (< 100 mg/l).

The composition of the product gas mixture varies, depending on the composition of waste material and other environmental factors. A typical gas composition is 70% to 75% CH_4, 20% to 25% CO_2, and 5% H_2S and other gases (NO_2, H_2, CO). Part of the methane produced in an anaerobic digestion unit is used to heat the digesters to fermentation temperature. The digester gas has a heating value of 24 to 28 million J/m^3, and the yield is 0.75 to 1.2 m^3 (std)/kg of organic material decomposed, resulting in an energy generation of 18 to 33 million J/kg of organic material. One problem associated with the use of digester gas is the H_2S content, which can cause foul odors and corrosion and needs to be removed before use.

Anaerobic digesters were originally designed as closed stationary tanks. In newer designs, reactor contents are mixed and the reactor is heated. Mixing is accomplished either by recycling a portion of the gas or by mechanical stirring. The reaction mixture is highly heterogeneous and viscous and, therefore, difficult to mix. Figure 16.10 is a schematic diagram of an anaerobic digester. Anaerobic digestion can be achieved in packed beds containing porous solid support particles. Bacteria grow on support surfaces and inside pores, and liquid waste may be trickled through the bed. Solids recycling schemes have been developed for use with conventional digesters that simulate activated-sludge units. Solids residence times are considerably higher than hydraulic residence times in these systems, which results in biomass accumulation and higher reaction rates. Continuous plug flow reactors operated in upflow mode are commonly used as anaerobic digesters.

The solids content of sludge (waste) is reduced by about 50% to 60% at the end of anaerobic digestion. Digested sludge is much less biodegradable than raw sludge and is easier to dewater. The digester effluent is dewatered by vacuum filtration and is dried. The dried solids are either incinerated or spread on land as fertilizer.

16.6.3. Advanced Waste-water Treatment Systems

Advanced waste-water treatment systems are used for removal of residual nitrogen and phosphorus along with refractory carbonaceous compounds. Examples are for nitrification and denitrification, phosphate removal processes (e.g., PhoStrip and A/O processes), and combined nitrogen and phosphate removal processes (e.g., A^2O and Bardenpho processes).

Nitrogen-containing organic compounds (e.g., amino acids and proteins) are oxidized biologically to ammonium, which is further oxidized to nitrite and nitrate by *Nitrosomonas* and *Nitrobacter*, respectively. The conversion of ammonium to nitrate by microbial catalysis is called *nitrification*.

$$NH_4^+ + \frac{3}{2}O_2 \xrightarrow{\ Nitrosomonas\ } NO_2^- + H_2O + 2H^+$$

$$(16.50a)$$

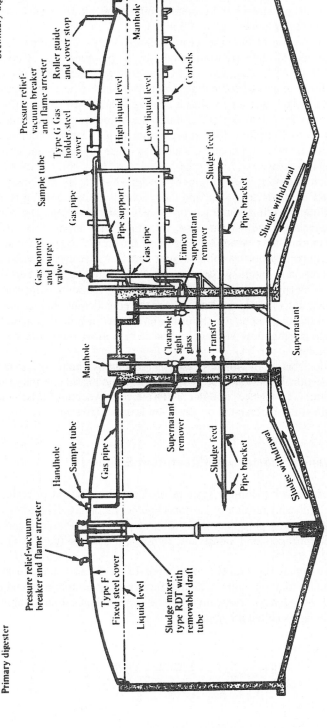

Primary digester

Pressure relief-vacuum
breaker and flame arrester

Type F
Fixed steel cover

Liquid level

Sludge mixer,
type RDT with
removable draft
tube

Handhole

Sample tube

Gas pipe

Supernatant
remover

Sludge feed

Pipe bracket

Sludge withdrawal

Manhole

Cleanable
sight glass

Transfer

Supernatant

Secondary digester

Pressure relief-
vacuum breaker
and flame arrester

Sample tube

Gas bonnet
and purge
valve

Gas pipe

Pipe support

Gas pipe

Eimco
supernatant
remover

Roller guide
and cover stop

Manhole

Type G Gas
holder steel
cover

High liquid level

Low liquid level

Corbels

Sludge feed

Pipe bracket

Sludge withdrawal

Figure 16.10. A schematic of an anaerobic digestion unit (Eimco Processing Machinery Division, Envirotech Corporation). (With permission, from E. D. Schroeder, *Water and Wastewater Treatment,* McGraw-Hill Book Co., New York, 1979, p. 327.)

502

$$NO_2 + \frac{1}{2}O_2 \xrightarrow{\text{\textit{Nitrobacter}}} NO_3^- \tag{16.50b}$$

Overall,

$$NH_4^+ + 2O_2 \rightarrow NO_3^- + 2H^+ + H_2O \tag{16.51}$$

Both of the aforementioned organisms fix carbon dioxide to be used for biosynthesis. Aerobic nitrification can take place during BOD removal in the presence of these organisms and increases the oxygen requirements in aerobic treatment.

These reactions are energy yielding, and the first step (eq. 16.50a) is the rate-limiting one. Nitrification processes can be accomplished in a single stage with BOD removal or as a two-stage system with separate BOD removal and nitrification units. The choice between the one-stage and two-stage processes typically depends on the ratio of carbon to nitrogen in the feed stream. Typically this ratio is expressed as BOD_5/TKN. The growth yield of nitrifying bacteria is low (0.2 g dry wt per g N) and the fraction of the population that is nitrifying bacteria can be low. For example, at $BOD_5/TKN = 10$ the fraction of nitrifying bacteria is 0.02; at $BOD_5/TKN = 0.5$ the fraction of nitrifying bacteria rises to 0.35. For a single-stage unit with combined BOD removal and nitrification the BOD_5/TKN ratio should be greater than 5. When it is less than 3, the two-stage process is preferred.

Nitrification is best accomplished with suspended cells, typically using a modification of the activated-sludge process. However, fixed-film processes that are well aerated can yield nitrification efficiencies over 90%. In either system DO levels should be high (> 2 mg/l), and slightly alkaline solutions (7.5–8.5) at 20° to 30°C work best.

Under anaerobic conditions, some bacteria utilize nitrate as the final electron acceptor instead of oxygen. This process is known as denitrification. Two types of nitrate reduction are possible: assimilatory and dissimilatory reduction. In assimilatory reduction, nitrate is reduced to ammonia and is incorporated into cell biomass. However, in dissimilatory reduction, nitrate is reduced to nitrite and further to elemental nitrogen. An external carbon source is needed; this source must be inexpensive (e.g., waste starch, methanol, or molasses).

Assimilatory reduction:

$$NO_3^- + H_2O \longrightarrow NH_3 \tag{16.52}$$

Dissimilatory reduction:

$$NO_3^- + C \text{ compound} \longrightarrow NO_2^- + CO_2 + \text{cells} \tag{16.53}$$

$$NO_2^- + C \text{ compound} \longrightarrow N_2 + CO_2 + \text{cells} \tag{16.54}$$

Using an average biomass composition of $C_5H_7NO_2$ and assuming that methanol is used as the carbon source, the overall balance for denitrification can be written as

$$NO_3^- + 1.08 \ CH_3OH + H^+ \longrightarrow 0.065 \ C_5H_7NO_2 + 0.47 \ N_2 + 0.76 \ CO_2 + 2.44 \ H_2O \tag{16.55}$$

Clearly, equations such as 16.55 can be used to predict the amount of carbon source (e.g., methanol) required to remove a known amount of NO_3^-. A number of microbes can act as denitrifiers; among these are species of the genera *Pseudomonas, Acaligenes, Arthobacter,* and *Corynebacter.* Such bacteria grow slowly with $\mu_g = 0.5$ d^{-1} and with a yield coefficient of 0.8 g biomass per g nitrogen removed. Saturation constants are low (e.g., 0.1 mg/l $< K_5 < 1$ mg/l) for both nitrification and denitrification, so these reactions can be approximated as zero order for nitrogen concentrations above about 2 mg/l.

Activated-sludge or packed-bed biofilm type reactors may be used for anaerobic denitrification. Upflow packed-bed reactors utilizing biofilms on porous support material are reported to result in high nitrate removal rates.

A typical two-stage nitrification/denitrification process is depicted in Figure 16.11. Optimal pH and temperature values for denitrification are pH = 6.5–7 and T = 20°–30°C.

Phosphate removal from waste waters by biological means can be realized either by assimilation or by "luxury" phosphate uptake. Assimilation of phosphate occurs in all aerobic/anaerobic processes since nearly 3% of cell mass is made of phosphorus. Luxury phosphate uptake is accomplished by organisms belonging to the genera *Acinetobacter,* among which *Acinetobacter calcoaceticus* is the most widely known. *Acinetobacter* species utilize carbohydrates as carbon and energy source by the Entner–Doudoroff pathway. Those organisms utilize acetate and fatty acids under anaerobic conditions to synthesize polyhydroxybutyrate (PHB). Energy required for PHB synthesis is obtained from ATP molecules generated by breakdown of polyphosphates. Under aerobic conditions in the absence of carbon source, the organisms utilize PHB molecules as carbon/energy source to synthesize polyphosphates and store them in form of polyphosphate granules inside the cell.

Aerobic polyphosphate synthesis:

$$(PO_4)_n + ATP \longrightarrow (PO_4)_{n+1} + ADP \tag{16.56}$$

Anaerobic polyphosphate depolymerization:

$$(PO_4)_n + H_2O \longrightarrow (PO_4)_{n-1} + H_2PO_4^{-1} \tag{16.57}$$

Consequently, the organisms remove phosphate from liquid media to synthesize and store polyphosphates under aerobic conditions and release phosphate into the media under anaerobic/anoxic conditions. *Acinetobacter* species can store polyphosphates up to 30% of their dry weight.

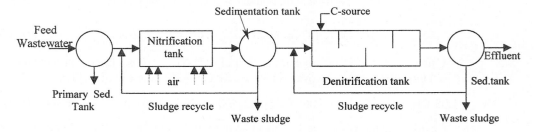

Figure 16.11 A schematic of the two-stage nitrification–denitrification process.

In phosphate removal bioprocesses, organisms are exposed to alternating anaerobic and aerobic conditions. The two major processes developed to promote luxury phosphate uptake are the A/O (Anaerobic/Oxic) and PhoStrip processes.

The A/O process is used for combined BOD and phosphate removal. It is a two-stage process incorporating anaerobic and oxic (aerobic) steps in sequence. In the anaerobic stage, the phosphate in the recycled sludge is released into liquid media. The released phosphate is taken up by the cells under aerobic conditions along with phosphate in the feed waste-water stream and is stored as polyphosphate granules. Polyphosphate is removed from the system by the waste sludge, which has a high value as fertilizer. When the BOD/P ratio in the feed wastewater is greater then 10, the effluent phosphate levels can be lower than 1 mg/l (Fig. 16.12).

The PhoStrip process consists of an aerobic stage and an anaerobic one on the sludge recycle stream, as shown in Fig. 16.12. The waste-water stream is fed to the aerobic stage for polyphosphate synthesis, the effluent of which is fed to a sedimentation tank. The underflow of the sedimentation tank is fed to the anaerobic stage for phosphate release. The released phosphate is precipitated in a sedimentation tank with the addition of lime in form of calcium phosphate. Part of the sedimented sludge and anaerobic reactor sludge is recyled back to the aerobic stage. Phosphate is removed from the process in the

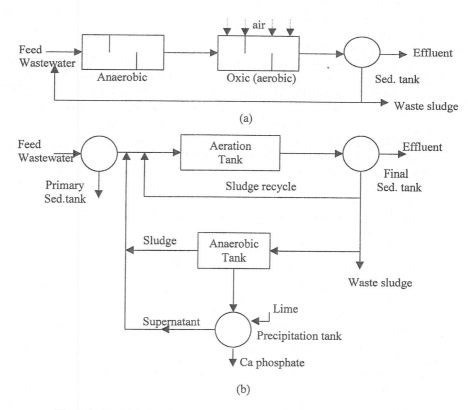

Figure 16.12 Phosphate removal processes: (a) A/O Process, (b) PhoStrip Process.

form of calcium phosphate. The effluent water stream would contain less then 1.5 mg/l phosphate in a well-operating PhoStrip process.

Combined nitrogen and phosphate removal processes employ combinations of anaerobic, anoxic, and aerobic zones in different orders which accomplish simultaneous BOD removal also. The most commonly used processes for this purpose are the A^2/O and the five-stage Bardenpho processes. Among other processes are the UCT (The University of Cape Town) and the VIP (Virginia Initiative Plant) processes.

The A^2/O process includes an anoxic zone for denitrification in addition to A/O process (Fig. 16.13). Hydraulic residence times in anaerobic and anoxic zones are less then 1.5 h, and in the aerobic zone 4–6 h. The effluent of the aerobic stage is recycled back to the anoxic zone for denitrification purposes. Phosphate release takes place in the anaerobic zone. Denitrification and BOD removal are the major functions of the anoxic zone. Phosphate removal in the form of polyphosphates, nitrification, and some BOD removal take place in aerobic zone. Phosphate is removed from the system in form of waste sludge; nitrogen is released in form of N_2 (gas); and BOD is converted to CO_2 and H_2O.

The five-stage Bardenpho process includes two additional anoxic and aerobic zones as compared to the A^2/O process (Fig. 16.13). This process is known as the $A^2/O/A/O$ process. The effluent of the aerobic zone is partly recycled back to the anoxic zone for denitrification purposes. The last two zones are for additional denitrification and polyphosphate removal to further reduce nitrate and phosphate levels in the effluent. Hydraulic residence times in anaerobic and anoxic zones are less than 4 h and in the first aerobic zone 4–12 h.

The UCT and VIP processes employ similar zones and recycle schemes in different orders, resulting in low nitrogen, phosphate, and BOD levels in the effluent.

In some cases sulfur removal from waste water is needed. Elemental sulfur in waste water can be oxidized to sulfate by *Thiobacillus* species. Sulfate can be further reduced to sulfides by anaerobic sulfate-reducing bacteria, such as *Desulfovibrio*, and sulfides can be precipitated out of waste water in the presence of certain metal ions, such as Fe^{2+}, Zn^{2+}, and Pb^{2+}. Sulfates can also be precipitated with the addition of limestone ($CaCO_3$) and $Ca(OH)_2$ and can be filtered out of waste water.

16.6.4 Conversion of Waste Water to Useful Products

Some waste materials containing starch and cellulose or other easily utilizable carbonaceous compounds can easily be converted to useful products such as high-protein feedstuff or single-cell protein (SCP), ethanol, organic acids (acetic, butyric, propiyonic), methane, and methanol. Agricultural wastes, waste paper, and waste wood constitute major cellulosic wastes. Major starch-containing wastes are some agicultural wastes (wheat, corn, rice, potato), some domestic solid wastes (vegetable/fruit), and some industrial wastes (food industry).

Cellulosic and starch-containing wastes are first hydrolyzed into sugar molecules, which are further converted to high-protein feedstuff (SCP) under aerobic conditions or ethanol/organic acids under anaerobic conditions. While many organisms can be used, yeasts, particularly *Saccharomyces,* are well suited to protein production, since they already have FDA approval for use in foods. The waste must be free of toxic contaminants.

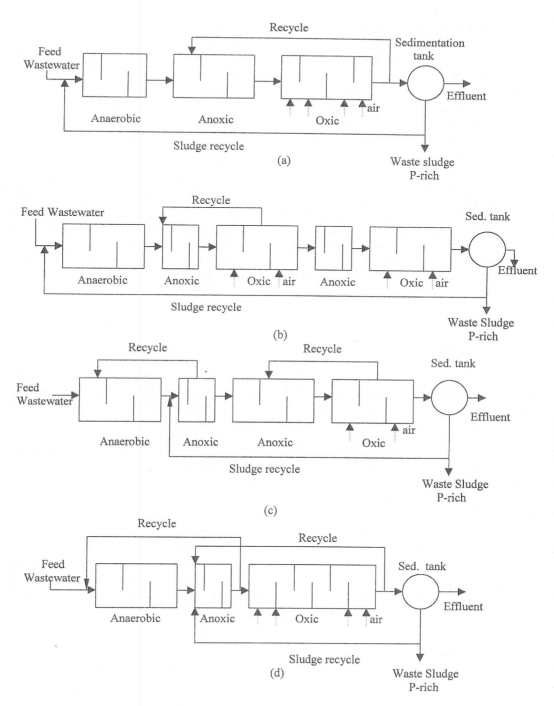

Figure 16.13 Combined biological nitrogen and phosphate removal processes: (a) A²/O process, (b) five-stage Bardenpho process, (c) UCT process, (d) VIP process.

To insure complete utilization of mineral nutrients may need to be added to give a C/N/P ratio of 100/15/3.

Two widely known processes for conversion of wastes to SCP are the SYMBA and the PEKILO processes. The SYMBA process converts potato wastes to SCP and consists of two stages. The first stage is for enzymatic hydrolysis of starch by *Endomycopsis* sp. and the second is for SCP production on glucose by *Saccharomyces* sp. PEKILO process is based on growth of *Paecilomyces* sp. on waste sulfite liquor.

Such waste materials can also be converted to chemicals such as ethanol for use as a fuel, solvents such as acetone/butanol, or monomers for synthesis of polymers (lactic acid). Conversion of starches to glucose and fermentable sugars is a well-established industry that uses large quantities of enzymes such as α-amylase, β-amylase, and glucoamylase. The economically feasible conversion of cellulosic wastes into fermentable sugars remains a bioprocess challenge. Many types of cellulases made by microbes can be used; new processes using genetically engineered microbes (e.g., *Trichoderma* sp., *Thermomonospora* sp., and some *Clostrida* sp.) may lead to less expensive cellulases. Acid hydrolysis can be used instead of enzymatic processes but is less environmentally friendly. The economical conversion of cellulosic wastes into products, such as ethanol, would be a major contribution to human welfare.

16.7 SUMMARY

Populations containing multiple species are important in natural ecosystems, well-defined processes, waste-water treatment, and systems using genetically modified cells. Some examples of interactions among these species are competition, neutralism, mutualism, protocooperation, commensalism, amensalism, predation, and parasitism. In real systems, several modes of interaction may be present. Mathematical analyses can be used to show that neither pure competition nor pure mutualism gives a stable steady state in a chemostat. However, spatial heterogeneity, dynamic fluctuations, and the addition of other interactions can lead to the sustained coexistence of species with competitive or mutualistic interactions.

One of the major process uses of mixed cultures is waste-water treatment. The activated-sludge system is commonly employed in treating waste waters. Such a system can be considered a chemostat with cell recycle under aerobic conditions. Alternative methods of aerobic treatment include trickling-bed filters, rotating biological discs, and oxidation ponds. Anaerobic digestion can be used to directly treat waste waters or, more commonly, to further degrade sludges from the primary and secondary treatment levels of a waste-treatment plant.

SUGGESTIONS FOR FURTHER READING

BULL, A. T., "Mixed Culture and Mixed Substrate Systems," M. Moo-Young, ed., *Comprehensive Biotechnology*, Vol. 1, Pergamon Press, Elmsford, NY, pp. 281–300, 1985.

FREDRICKSON, A. G., Behavior of Mixed Cultures of Microorganisms, *Ann. Rev. Microbial* *31*:63–87, 1977.

HAMMER, M. J., AND M. J. HAMMER, JR., *Water and Wastewater Technology,* 4th ed., Prentice Hall, Upper Saddle River, NJ, 2001.

SCHROEDER, E. D., *Water and Wastewater Treatment,* McGraw-Hill Book Co., New York, 1977.

SUNDSTROM, D. W., AND H. E. KLEI, *Wastewater Treatment,* Prentice Hall, Englewood Cliffs, NJ, 1979.

TEHOBANOGLOUS, G., *Wastewater Engineering Treatment, Disposal and Reuse* (Metcalf & Eddy, Inc.), 3d ed., McGraw-Hill Book Co., New York, 1991.

VIESSMAN, W., JR., AND M. J. HAMMER, *Water Supply and Pollution Control,* 6th ed., Prentice Hall, Upper Saddle River, NJ, 1998.

PROBLEMS

16.1. A batch fermenter receives 1 l of medium with 5 g/l of glucose, which is the growth-rate-limiting nutrient for a mixed population of two bacteria (a strain of *E. coli* and *Azotobacter vinelandii*). *A. vinelandii* is five times larger than *E. coli*. The replication rates for the two organisms are:

$$\mu_{EC} = \frac{1.0 \ \text{h}^{-1} \ \text{s}}{0.01 \ \text{g/l} + \text{s}} - 0.05 \ \text{h}^{-1}$$

and

$$\mu_{AV} = \frac{1.5 \ \text{h}^{-1} \ \text{s}}{0.02 \ \text{g/l} + \text{s}} - 0.10 \ \text{h}^{-1}$$

The yield coefficients are:

$$Y_{EC} = 0.5 \ \text{g dw/g glucose}$$

$$Y_{AV} = 0.35 \ \text{g dw/g glucose}$$

The inoculum for the fermenter is 0.03 g dw/l of *E. coli* (1×10^8 cells/ml) and 0.15 g dw/l of *A. vinelandii* (1×10^8 cells/ml).

What will be the ratio of *A. vinelandii* to *E. coli* at the time when all of the glucose is consumed?

16.2. Consider Example 16.1, where we demonstrated that two bacteria competing for a single nutrient in a chemostat (well-mixed) could not coexist. Consider the situation where B can adhere to a surface but A cannot. Redo the balance equations, where a is the surface area available per unit reactor volume and the rate of attachment is first order in X_B with a rate constant k_{aB}. The sites available for attachment will be $(X_{BM}^{At} - X_B^{At})aV$. The attached cells can detach with a first-order dependence on the attached cell concentration (X_B^{At}) with a rate constant of k_{dB}. Attached cells grow with the same kinetics as suspended cells.

   a. Without mathematical proofs, do you think coexistence may be possible? Why or why not?

   b. Consider the specific case below and solve the appropriate balance equations for $D = 0.4 \ \text{h}^{-1}$:

$$\mu_{mA} = 1.0 \text{ h}^{-1}; \quad \mu_{mB} = 0.5 \text{ h}^{-1}$$

$$K_{SA} = K_{SB} = 0.01 \text{ g/l}$$

$$Y_{A/S} = Y_{B/S} = 0.5 \text{ g/g}$$

$$S_0 = 5 \text{ g/l}, \quad a = 10 \text{ cm}^2/\text{cm}^3$$

$$X_{BM}^{At} = 1 \times 10^{-4} \text{ g/cm}^2$$

$$k_{dB} = 0.5 \text{h}^{-1}, \quad k_{aB} = 1000 \text{ cm}^3/\text{g-h}$$

16.3. Organism A grows on substrate S and produces product P, which is the only substrate that organism B can utilize. The batch kinetics are

$$\frac{dX_A}{dt} = \frac{\mu_A S X_A}{K_s + S}$$

$$\frac{dX_B}{dt} = \frac{\mu_B P X_B}{K_p + P}$$

$$\frac{dP}{dt} = Y_{P/A} \frac{\mu_A S X_A}{K_S + S} - \frac{\mu_B P X_B}{Y_{X_B/P}(K_P + P)}$$

$$\frac{dS}{dt} = -\frac{\mu_A S X_A}{Y_{X_A/S}(K_S + S)} - \frac{Y_{P/A}}{Y_{P/S}} \frac{\mu_A S X_A}{(K_S + S)}$$

Assume the following parameter values:

$$\mu_A = 0.18 \text{ hr}^{-1}, \quad K_s = 0.42 \text{ g/l}, \quad \mu_B = 0.29 \text{ hr}^{-1}$$

$$K_P = 0.30 \text{ g/l}, \quad Y_{X_A/S} = 0.3 \text{ g/g}, \quad Y_{X_B/P} = 0.5 \text{ g/g}$$

$$Y_{P/S} = 1.0 \text{ g/g}, \quad Y_{P/A} = 4.0 \text{ g/g}, \quad S_0 = 10 \text{ g/l}$$

Determine the behavior of these two organisms in a chemostat. Plot S, P, X_A, and X_B versus dilution rate. Discuss what happens to organism B as the dilution rate approaches the washout dilution rate for organism A. (Courtesy of L. Erickson, from "Collected Coursework Problems in Biochemical Engineering," compiled by H. W. Blanch for 1977 Am. Soc. Eng. Educ. Summer School.)

16.4. The BOD_5 value of a waste-water feed stream to an activated-sludge unit is $S_0 = 300$ mg/l, and the effluent is desired to be $S = 30$ mg/l. The feed flow rate is $F = 2 \times 10^7$ l/day. For the recycle ratio of $\alpha = 0.5$ and a steady-state biomass concentration of $X = 5$ g/l, calculate the following:

a. Required reactor volume (V).
b. Biomass concentration in recycle (X_r).
c. Solids (cells) residence time (θ_c).
d. Hydraulic residence time (θ_H).
e. Determine the daily oxygen requirement.

Use the following kinetic parameters:

$$\mu_m = 1.5 \text{ day}^{-1}, \qquad K_s = 400 \text{ mg/l}$$
$$Y_{X/S}^M = 0.5 \text{ g dw/g BOD}, \qquad k_d = 0.07 \text{ day}^{-1}$$

16.5. For the activated-sludge unit shown in Fig. 16.7, the specific growth rate of cells is given by

$$\mu_{net} = \frac{\mu_m S}{K_s + S} - k_d$$

The following parameter values are known: $F = 500$ 1/h, $\alpha = 0.4$, $\gamma = 0.1$, $X_e = 0$, $V = 1500$ 1, $K_s = 10$ mg/l, $\mu_m = 1$ h^{-1}, $k_d = 0.05$ h^{-1}, $S_0 = 1000$ mg/l, $Y^M_{X/S} = 0.5$ g dw/g substrate.
  a. Calculate the substrate concentration (S) in the reactor at steady state.
  b. Calculate the cell concentration(s) in the reactor.
  c. Calculate X_r and S_r in the recycle stream.

16.6. In a trickling biological filter, the BOD value of the feed stream is $S_{0i} = 500$ mg/l with a feed flow of $F = 10^3$ 1/h. The effluent BOD value is desired to be $S_0 = 10$ mg/l. The following kinetic parameters for the biocatalysts are known: $r_m = 20$ mg/S/l · h and $K_s = 200$ mg S/l. The biofilm thickness is $L = 0.1$ mm. The cross-sectional area of the filter is $A = 2$ m^2, and the biofilm surface area per unit volume of the bed is $a = 500$ cm^2/cm^3. Assume that dissolved oxygen is the rate-limiting substrate and the diffusion coefficient of oxygen is $D_{O_2} = 2 \times 10^{-5}$ cm^2/s. Determine the required height of the bed. You can assume first-order bioreaction kinetics.

16.7. An activated-sludge waste treatment system is required to reduce the amount of BOD$_5$ from 1000 mg/l to 20 mg/l at the exit. The sedimentation unit concentrates biomass by a factor of 3. Kinetic parameters are $\mu_m = 0.2$ h^{-1}, $K_s = 80$ mg/l, $k_d = 0.01$ h^{-1}, and $Y^M_{X/S} = 0.5$ g MLVSS/g BOD$_5$. The flow of waste water is 10000 l/h and the size of the treatment basin is 50,000 l.
  a. What is the value of the solids residence time (i.e., θ_c)?
  b. What value of the recycle ratio must be used?

16.8. Consider a well-mixed waste treatment system for a small-scale system. The system is operated with a reactor of 1000 l and flow rate of 100 l/h. The separator concentrates biomass by a factor of 2. The recycle ratio is 0.7. The kinetic parameters are $\mu_m = 0.5$ h^{-1}, $K_s = 0.2$ g/l, $Y^M_{X/S} = 0.5$ g/g, and $k_d = 0.05$ h^{-1}. What is the exit substrate concentration?

16.9. Redo Example 16.4 if the Contois equation for growth applies. In this case

$$\mu_{net} = \frac{\mu_m S}{K_{sx} X + S} - k_d$$

The values of μ_m and k_d are the same as for Example 16.4, but K_s no longer applies. Assume $K_{sx} = 0.02$ g BOD$_5$/g MLVSS.

17

Epilogue

Although we hope that you have learned a great deal about bioprocess engineering, you should realize that the material in this book is only the beginning. None of the topics has been covered in real depth. Most chapters would require a whole book to provide that depth. However, you should now have the vocabulary and perspective to benefit from more specialized books.

Some topics of increasing importance to biotechnology are functional genomics, cell surface receptor-mediated phenomena, high-throughput drug screening, protein engineering, metabolic engineering, and tissue engineering. A clear understanding of how cells perceive external signals through cell surface receptors promises to be a key to developing new therapies, as well as to understanding how shear effects in a bioreactor may alter cellular physiology. Insights into the interrelationship of protein structure with function promise the possibility of large-scale design of custom catalysts and drugs. Here the engineering approach to computational molecular dynamics may yield important dividends. Genetic engineering to produce proteins is beginning to mature; the development of cells with altered metabolic pathways is at a much earlier stage of development. The engineer's sense of optimization and the ability to quantitatively model intracellular interactions will be critical to the economically attractive development of this technique and its extension to gene therapy. Engineers will be needed to develop new and effective ways to culture organized tissues and artificial organs. Questions on the relationship of protein posttranslational processing to the microenvironment about cultured cells will be a key factor in more effective scale-up of mammalian cell reactors. The ability to control microenvironmental conditions in a spatially heterogeneous system will be essential. Bioprocess engineers can play key roles in developing the experimental tools and computational approaches required for functional genomics.

Improvements in the computer models of cellular populations will lead to new insights into bioreactor dynamics and appropriate control strategies. Although we have not talked much about structured and structured-segregated models, recent developments in this area promise significant rewards. Also, improved models, coupled with a better understanding of transport processes within bioreactors, suggest that rational scale-up of bioreactors may be achievable.

New ideas on the use of cells and enzymes in environments with low water activity (organic or gas phases) may open doors to wholly new bioprocessing opportunities. Substrates with low solubility in water may become increasingly important targets for bioprocesses.

In addition to the improvements in biocatalysts and our understanding of bioreactors, there will be improved understanding and development of bioseparation processes. Significant progress has already been made in affinity separation processes. New membrane materials are being rapidly developed. Ideas such as two-phase aqueous systems, reverse micelles, and gel swelling systems may become important tools for the engineer.

This discussion of future possibilities is in no way exhaustive. However, the reader must sense that the technological opportunities are expanding rapidly. The recent advances in genomics provide tools that will open new vistas for the bioprocess engineer. Much excitement exists, and rightly so. Now you have a better chance to participate in this excitement. The purpose of this book has been to give you a foundation. Build upon it and join the biological revolution.

Appendix

Traditional
Industrial Bioprocesses

Enzymes and microbial cells are used for production of chemicals, pharmaceuticals, foods, flavors and fragrances, and vitamins and for waste treatment. Each of these bioprocesses has unique characteristics in terms of the processing and separation technologies involved. Having covered the basics of bioprocess engineering, this appendix presents some examples of industrial bioprocesses and technologies used for production of various chemicals and pharmaceuticals. Waste-treatment aspects of bioprocess engineering are covered in Chapter 16.

A.1. ANAEROBIC BIOPROCESSES

A.1.1. Ethanol Production

Ethanol has many applications as a raw material, solvent, and fuel and is utilized in large quantities in the chemical, pharmaceutical, and food industries. Worldwide, four million tons of industrial ethanol are produced annually, 80% by fermentation. It is expected that the demand for ethanol as a fuel oxygenate will increase. An annual growth of U.S. ethanol consumption of 3.2% over the next 20 years has been predicted by the Energy Information Administration.

Yeasts are the preferred organisms for industrial-scale ethanol production. Different species can be utilized, depending on the composition of the raw material utilized. *S. cerevisiae*, particularly suitable for fermentation of hexoses, has been the major organism used so far. *Kluyveromyces fragilis* or *Candida* sp. can be utilized when lactose or pentoses, respectively, are the available substrates. Other alternative organisms capable of producing ethanol such as *Zymomonas mobilis, Pachysolen* sp., are not used in industrial production. However, *Zymomonas* species have significant advantages over yeast and may be used at industrial scale in the future. Other pentose and hexose fermenting organisms such as *Clostridium hermosaccharolyticum* and *Thermoanaerobacter ethanolicus* are thermophilic and provide significant advantages for ethanol fermentations and separation. However, they produce some undesirable end products and yield dilute ethanol solutions. Genetic engineering has transformed *E. coli* into a very efficient ethanol producer, reaching ethanol concentrations up to 43% (vol/vol).

Raw materials can constitute up to 70% of the cost of ethanol production. Therefore, the selection of inexpensive materials has an important impact on process economics. The material selected should be readily and economically available in the fermentation plant. In Brazil cane sugar is extensively used, while in the United States sugar prices make sugar a nonviable raw material. Instead, corn is the most utilized material in the United States for the production of industrial ethanol. Most microorganisms require readily available sugar compounds for fermentation. Such sugar compounds are present in inexpensive raw materials, as sugar cane, sweet sorghum, or sugar beet juices and molasses. Whey is also utilized in commercial fermentation. Starch- and cellulose-containing materials, such as grains, fruits, vegetables, wood, biomass, waste paper, and agricultural wastes, can also be used, but they need to be hydrolyzed before fermentation. Starch is enzymatically saccharified before fermentation. Acid hydrolysis of starch is not recommended, as it can result in nonfermentable or inhibitory by-products. Hydrolysis of cellulosic materials can be difficult, expensive, and inefficient. However, interest exists on increasing their utilization in an effort to recycle wastes.

Yeast convert hexoses to ethanol and carbon dioxide by glycolysis, as shown by the following reaction:

$$C_6H_{12}O_6 \longrightarrow 2C_2H_5OH + 2CO_2 \qquad (A.1)$$

The theoretical ethanol yield over glucose is 0.51 g/g, and the growth yield over glucose is 0.12 g/g. Usually by-products such as glycerol, succinic acid, and acetic acid are produced, and the actual yield is about 90–95% of the theoretical yield. Optimal temperature and pH values for yeast are 30° to 35°C and 4–6, respectively. For thermophilic organisms optimal temperature may range from 50° and 60°C. Ethanol production is triggered by anaerobic conditions. Trace amounts of oxygen (0.05–0.1 mm Hg) are required by yeast for lipid biosynthesis and maintenance of cellular processes.

The feed solution should be balanced in terms of nitrogen, phosphorus, minerals, and some trace elements. Glucose medium is usually supplemented with NH_4Cl, KH_2PO_4, $MgSO_4$, $CaCl_2$, and yeast extract. In industry, only some ammonium and phosphate salts need to be added to diluted molasses. Glucose concentration in feed solution has an important effect on the rate and extent of ethanol production. Glucose concentrations above 100 g/l are inhibitory for yeast.

Ethanol and some of the other by-products are inhibitory to yeast above concentrations of 5% (vol/vol). Therefore, the glucose concentration in feed solution in continuous fermentation should be less than 100 g/l, resulting in ethanol concentration in the effluent below 50 g/l. Ethanol-tolerant yeast strains are being developed to avoid ethanol inhibition. Simultaneous removal of ethanol from fermentation broth is another alternative for alleviation of ethanol inhibition.

Conventional ethanol fermentations operate in batch mode under aseptic conditions. Mechanically agitated stainless-steel reactors are used for this purpose. A reactor is filled with a nutrient medium up to 70% of its volume. After pH and temperature adjustment the reactor content is sterilized and cooled to fermentation temperature. Temperature and pH are controlled during operation, and redox potential is kept below −100 mV by using reducing agents such as Na_2S. A sterile yeast culture is prepared and used for inoculation of the reactor. A batch fermentation cycle lasts nearly 30-40 hours. Part of the yeast and aqueous medium can be recycled. Batch operation with cell recycle (the Melle–Boinot Process) results in reduced fermentation times (10 h) and improved productivities of 6 g/l h. At the end of batch operation the reactor content is emptied and yeasts are separated by filtration or centrifugation. The liquid broth is further processed for separation of ethanol by distillation.

Continuous operation in ethanol fermentations has significant advantages over batch operation. With continuous media sterilization and aseptic operation techniques, contamination problems associated with continuous operation can be eliminated. About 95% of sugar can be converted to ethanol in continuous operation with a residence time of 21 h, as compared to batch operation time of 40 h. Under optimized conditions the residence time for 95% conversion can be as low as 10 h. Continuous operation with cell recycle may increase fivefold the cell concentration in the reactor, resulting in faster conversion. Sedimentation tanks, centrifuges, or filters can be used for cell separation from the fermenter effluent. With cell recycle, the residence time for 95% conversion may be reduced to 1.6 h with a productivity of 30 g/l h for a feed glucose of 100 g/l. Multistage continuous operation with cell recycle may further improve the productivity of the process. When six fermenters in series are used without cell recycle, it is possible to obtain 95 g/l in 9 h of total residence time and a productivity of 11 g/l h.

Figure A.1 depicts the biostill process used for ethanol fermentations. This process employs continuous operation with cell recycle and a distillation column for ethanol separation. A mechanically agitated stainless-steel fermenter is used in continuous mode, and the effluent is centrifuged for yeast separation. Part of the separated yeast is recycled back to the fermenter, and the liquid medium is fed to a distillation column for separation of ethanol. Ethanol-free medum is recycled back to the fermenter. Yeast cell recycle provides high conversion rates, and liquid recycle reduces the amount of waste water generated and dilutes the feed sugar concentration down to noninhibitory levels.

Immobilization of yeast within porous or polymeric matrices results in high cell concentrations in the reactor and, therefore, high ethanol productivities. Immobilized cell reactors may be in form of packed columns or fluidized beds. Some flocculating yeast strains that settle rapidly may also be used in tower fermenters to obtain high cell concentrations.

A comparison of industrial processes used for ethanol production is presented in Table A.1.

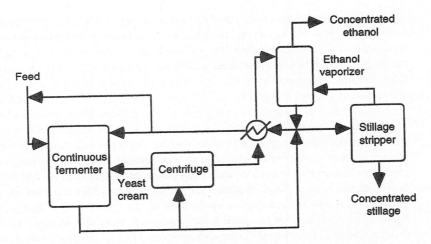

Figure A.1. Schematic for the biostill process for ethanol production.

Ethanol can be separated from the culture vessel during fermentation using low-temperature vacuum distillation, adsorption, or membrane separations. This reduces ethanol inhibition but is seldom used in industry because of operating difficulties. Separation of ethanol from fermentation broth is usually accomplished by distillation, which is energy intensive and constitutes more than 50% of the total energy consumption of the plant. The grade of industrial ethanol is usually 95% (190 proof) for chemical and pharmaceutical use, which can be obtained by using distillation columns. However, it is difficult to obtain 100% pure ethanol, since 95% ethanol in water constitutes an azeotropic mixture. A third component is added to alcohol–water mixture to break the azeotrope, but this is expensive and the third component may be toxic.

The cost of ethanol production mainly depends on the raw material and operating costs (chiefly electricity and water cooling). Total operating cost for an ethanol plant producing 190 million liters of ethanol per year is estimated to be 42 cents/l (2001 basis). Raw materials costs constitute 26 cents/l; utilities, 3.4 cents/l; labor, 6 cents/l; and fixed charges (depreciation, taxes, insurance, maintenance), 6.6 cents/l. Plant capital investment for the same size plant is estimated to be $71 million. Produced ethanol can be used as solvent, chemical intermediate (for production of acetaldehyde, acetic acid, ethylene), and fuel. In recent years there has been an increasing demand for ethanol utilization as fuel in

TABLE A.1. Comparison of Industrial Ethanol Fermentation Processes

	Total cycle time (h)	Ethanol concentration (g/l)	Productivity (g/l h)
Simple batch	36	80	2.2
Melle–Boinot	12	80	6.6
Continuous in series	8	86	11

the form of either gasohol (10% alcohol) or pure alcohol. However, the use of ethanol as fuel can be viable only if its cost is comparable to that of oil-derived fuels.

A.1.2. Lactic Acid Production

Lactic acid was first isolated from sour milk (1780) and has two optically active forms called D- and L-lactic acids. The major use of L-lactic acid is in foods (more than 50%) as an acidulant and preservative. Lactic acid is also used as a chemical intermediate to produce other chemicals and in the pharmaceutical industry. Most lactic acid is produced by fermentation.

In industry, usually a mixture of lactic acid bacteria is used to ferment a mixture of carbohydrates. A mixture of strains may result in faster fermentation rates than pure cultures. Selected organisms should grow fast, produce lactic acid with high yields and productivities, and have low nutritional requirements. Lactic acid formation is a mixed growth associated process that requires high growth rate and cell concentration. Lactic-acid-producing bacteria are classified in two major groups, homolactic (*Lactobacillus* sp., *Streptococcus* sp., *Pediococcus* sp.) and heterolactic bacteria (some *Streptococcus* sp., *Leuconostoc* sp.).

Homolactic species of *Lactobacillus* and *Streptococcus* are usually used for the industrial production of lactic acid. Homolactic bacteria use the EMP pathway to generate two moles of pyruvate from one mole of glucose that are further reduced to lactic acid, as summarized below.

$$C_6H_{12}O_6 \longrightarrow 2C_3H_6O_3 \tag{A.2}$$

The product yield over glucose is usually above 0.9 g/g. The organisms are facultative anaerobes, but generate ATP only by anaerobic fermentation. Industrially important homolactics grow at temperatures above 40°C and pH between 5 and 7. High temperature and low pH (pH < 6) reduce the risk of contamination. Homolactic bacteria can produce lactic acid from pentoses as well as hexoses other than glucose. Usually lactic and acetic acids are produced from fermentation of pentoses. Most of the lactic acid bacteria also require several B vitamins, amino acids, and phosphate. Peptides may increase cell growth rate. Fermentation yield varies, depending on substrate and the organism used.

Heterolactic fermentation is undesirable because of by-product formation. However, depending on microbial flora, energy availability, and fermentation conditions, it may occur. Heterolactic bacteria produce one mole of lactic acid, ethanol, and CO_2 from one mole of glucose as shown below.

$$C_6H_{12}O_6 \longrightarrow C_3H_6O_3 + C_2H_5OH + CO_2 \tag{A.3}$$

The theoretical lactic acid yield is 0.5 grams per gram of glucose.

Some species of *Rhizopus* (e.g., *R. oryzae*) have low nutritional requirements and can be used to produce lactic acid from carbohydrates. They have the advantage of producing stereochemically pure L-(+)-lactic acid. *Rhizopus* species can also utilize starch for the production of lactic acid.

The ideal raw material must be inexpensive, must result in high rates with high product yields and no by-product formation, should not require significant pretreatment, and should be available year around. Sucrose from sugar cane or sugar beet juice, lactose from cheese whey, and maltose or dextrose from hydrolyzed starch are used as raw materials in industry. Molasses can also be used; however, its complex nature makes separation of lactic acid problematic. Dextrose from corn starch was the most commonly used raw material in the late 1950s. Other processes based on pasteurized milk (*L. bulgaricus*), dextrose from corn, glucose (*Rhizopus*), crude sorghum extract, and potato hydrolysate have also been developed. Direct hydrolysis and fermentation of corn starch by certain *Lactobacillus* species seem to be quite promising from the economical point of view. Nitrogen sources such as malt extract, corn steep liquor, barley, and yeast extract should be added to the fermentation media to improve growth and lactic acid formation.

Industrial processes are operated batchwise (Fig. A.2). Fermenters are made of stainless steel and are equipped with heat-transfer coils. Vessels are steam-sterilized before being filled with a pasteurized medium. Slow agitation to prevent settling of calcium carbonate is provided with top-mounted mechanical stirrers. Fermentation conditions are different for each industrial producer, but are usually in the range of $T = 45–60°C$ and pH = 5–6.5 for *L. delbruckii;* $T = 43°C$, pH = 6–7 for *L. bulgaricus;* and $T = 30–50°C$, pH < 6 for *Rhizopus*. The fermentation time is 1 to 2 days for a 5% sugar source such as whey, and 2 to 6 days for a 15% glucose or sucrose source. Under optimal conditions the processing time may be reduced to 1–2 days. The rate of lactic acid formation depends on temperature, pH, sugar, nitrogen, and lactic acid concentrations. Temperature and pH control at optimum levels improves the rate of lactic acid formation. Produced lactic acid must be neutralized, usually by addition of calcium carbonate or calcium hydroxide. CO_2 is continuously released during fermentation, which creates anaerobic conditions in the fermenter. Lactic acid formation productivities are in the range of 1–3 kg/m^3 h. The yield of lactic acid at the end of fermentation is 90–95% of initial sugar concentration. Cell mass yield is usually less than 15% of initial sugar concentration; however, the yield may be as high as 30%, depending on the organism and culture conditions.

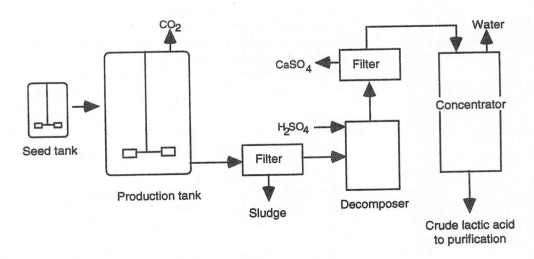

Figure A.2. Process diagram for lactic acid production.

Continuous, high cell density and immobilized cell reactors have been used for laboratory-scale lactic acid fermentations. Productivity and yield (100 kg/m^3) of lactic acid formation are higher in such reactors than in batch cultures. Continuous removal of lactic acid from cultures by dialysis membranes increases production rates.

Recovery of lactic acid from the fermentation broth constitutes a significant part of production cost. The use of pure sugar solutions with minimal amounts of nitrogen simplifies product recovery. A process diagram is shown in Fig. A.2. The first step is to increase the temperature of the fermenter to 80–100°C and the pH to 10–11. This procedure kills the organisms, coagulates the proteins, solubilizes calcium lactate, and degrades some of the residual sugars. The liquid is then filtered to remove biomass, and sulfuric acid is added to obtain lactic acid. Calcium sulfate is removed by filtration, and lactic acid is then concentrated. Alternatively, purification of lactic acid may be accomplished by calcium lactate precipitation. The fermenter broth is filtered and evaporated to 25% lactic acid. The calcium lactate is then crystallized and separated from the mother liquor. Several purification procedures can be performed to obtain lactic acid of different grades. These include bleaching with activated carbon, ion exchange, electrodialysis, solvent extraction, and esterification. Edible lactic acid is colorless and contains 50–65% lactic acid. Pharmaceutical applications require over 90% pure lactic acid. Traditionally, polymer-grade lactic acid had to be as pure as possible. However, recently polylactate has been obtained from industrial-grade lactic acid.

World demand for lactic acid is estimated as $150 million (100,000 tons). About 50% of the market is in food and beverage applications, which is a mature and stable market. However, niche applications of lactic acid are expected to increase its demand. These include biodegradable polymers, solvents, and cleaning agents. An annual growth of 8.6% of the lactic acid market is expected between 2000 and 2003.

Major producers of lactic acid are United States, Japan, Belgium, the Netherlands, Spain, and Brazil. China and India are also producers. The price of pure lactic acid is about $2.3/kg (2000). Derivatives of lactic acid such as calcium lactate sell at a higher price ($2.8/kg, 2000 basis). Assuming 20% profit margin, one can estimate the manufacturing price of lactic acid approximately as $1.8/kg.

A.1.3. Acetone–Butanol Production

Acetone is used mainly as a solvent for fats, oils, waxes, resins, lacquers, and rubber plastics. Butanol is used in the production of lacquers, rayon, detergents, and brake fluids; and as a solvent for fats, waxes, and resins. Butanol is a better fuel additive than ethanol because of its low vapor pressure, low solubility with water, and complete solubility with diesel fuel. Acetone and butanol (A/B) are produced from petrochemical industry intermediates. However, it is expected that the demand for cleaner processes and shortage of oil-derived products will require production by fermentation.

Clostridium species are used for acetone–butanol production. *C. acetobutylicum* is a strict anaerobe with a versatile metabolic capacity. It has amylolytic and saccharolytic enzymes to hydrolyze gelatinized starch to glucose and maltose. *C. acetobutylicum* can ferment a large number of carbohydrates such as glucose, lactose, fructose, galactose, xylose, sucrose, maltose, and starch. The fermentation products include acetone, butanol, and ethanol. Acetic and butyric acids, CO_2, and hydrogen are also produced in small amounts.

Starch, molasses, cheese whey, Jerusalem artichoke, and lignocellulosic hydrolyzates can be used as raw material for acetone–butanol fermentations. More than one of these carbon sources may be utilized simultaneously. The composition of culture medium determines the butanol-to-acetone ratio obtained and can be manipulated to obtain the desired ratio. Molasses can be used as carbon source (6% sucrose) with the addition of nitrogen and phosphate. Higher concentrations of sucrose in molasses increase the ratio of butanol to acetone and ethanol. Average solvent yields of 30% were reported when molasses was used. Utilizing a 5–6% starch solution derived from corn, approximately 38% solvent yield can be obtained. A butanol-to-acetone ratio of 2.75 was obtained using Jerusalem artichoke rich in inulin.

Cheese whey with nearly 6% lactose and 1% protein content can be used as carbon and nitrogen source for acteone–butanol fermentation. Somewhat different product distribution is obtained when cheese whey is used as carbon source. When corn mash was also added, the butanol-to-acetone ratio was 2. With pure glucose or pure lactose this ratio is 2.7 and 2.9, respectively. When ultrafiltrate of cheese whey was used, butanol-to-acetone ratio of 10 was obtained without any intermediate buildup of acetic and butyric acids.

Lignocellulosics hydrolyzates (wood, paper, crop residues) contain glucose, galactose, mannose, and pentose sugars, most of which are fermentable by *C. acetobutylicum* to acetone and butanol. However, since hydrolysis of lignocellulosics is difficult, this raw material is not commonly used for A/B fermentations. Alternatively, cocultures with cellulolytic clostridia can be performed.

A schematic of the process is depicted in Fig. A.3. The Weizmann process utilizes starch as raw material for A/B fermentation. *C. acetobutylicum* hydrolyzes gelatinized corn starch to glucose and maltose with amylolytic enzymes. The grain mesh is first gelatinized at 65°C for 20 min and then sterilized at 105°C for 60 min. The cooked mash is cooled down to 35°C using heat exchangers and is pumped to presterilized fermenters of 250–2000 m^3. The fermenter is inoculated with a 5% inoculum from a 24 h culture. The final butanol-to-acetone ratio increases as the inoculum age increases. In some cases, 30–40% of the total volume of the mash is provided by stillage after solvents from the previous fermentation are stripped. Batch fermentation period is usually 2–2.5 days. First, rapid growth and production of acetic/butyric acids and carbon dioxide and hydrogen occur. The initial pH of the medium drops from 6.5 to nearly 4.5 during this phase. In a second phase, growth ceases, and the organisms convert acetic and butyric acids to neutral acetone and butanol. The acidity of the medium decreases, and gas production increases. At the end of the fermentation the pH is approximately 5.

The South African process utilizes cane molasses as raw material and *C. acetobutylicum*. Stainless-steel fermenters of 90 m^3 are used in batch operation. The effluent contains 2% A/B, and the solvents are recovered by distillation. Acetone/butanol and ethanol/isopropyl alcohol are obtained as separate fractions. The biomass, rich in riboflavin and B vitamins, is concentrated, dried, and used as animal feed supplement. Due to the high price of molasses this process is operated intermittently. A process based on utilization of cheese whey for production of A/B has been developed. If the plant is placed near cheese manufacturers, this process can be more attractive than that based on molasses.

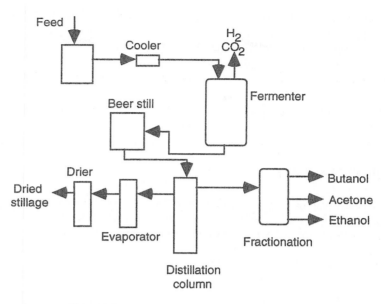

Figure A.3. Flowsheet for acetone–butanol production.

Different approaches have been developed to improve solvent yield in A/B fermentations. Phosphate limitation and addition of acetic and butyric acids improve butanol productivity. Calcium present in cheese whey complexes phosphate and also improves butanol yield. Hydrogen saturation in the medium at low agitation rates (25 rpm) and high head-space pressure improves butanol yield. Overall, butanol productivity may be improved by moderate to high agitation (300 rpm) during the acid phase followed by low agitation (25 rpm) during the solvent phase. Simultaneous removal of inhibitory products (butanol, acetone) by adsorption (activated carbon) or extraction (corn oil, paraffin oil) improves fermentation productivities. Organisms can tolerate up to 1.2% (vol/vol) of butanol.

Continuous or fed-batch operations improve solvent productivities. The possibility of controlling the growth rate and environmental and nutritional conditions improves butanol productivities (2.5 g/l h), as compared to batch cultures (0.8 g/l h). Immobilized cultures of *C. acetobutylicum* are also used for A/B fermentations with high productivities (1.2 g/l h). Biochemical and genetic manipulations of cells may further improve the solvent yield.

To recover acetone and butanol, at the end of the solvent phase the broth is transferred to a beer still that concentrates solvents (Fig. A.3). Solvents are then separated by fractionation, and the stillage is dried. Acetone and butanol prices have decreased due to a decrease in demand and excessive production capacity. It is expected that acetone demand will increase 4% per year from 2000 to 2006 to 5 million tons, while the production capacity in 2006 will be 5.3 million tons. An economic evaluation was presented for A/B fermentation from cheese whey in 1982. For a plant producing 45,000 tons of solvents per year, the total capital investment including the waste-water treatment was estimated to be $28 million and the total production cost was nearly $37 million, with an annual income of nearly $53 million and an annual profit of $16 million. Nowadays, production by fermentation is not economically attractive due to low levels of product concentrations (0.7–1.5%) and high cost of product recovery. However, fermentation may be the preferred method for production of A/B if a shortage of oil products exists, or as demands for environmentally friendly processes increase.

A.2. AEROBIC PROCESSES

Aerobic bioprocesses are widely used for the production of organic acids (citric, acetic, gluconic), vitamins, antibiotics, enzymes, flavors–fragrances, amino acids. Owing to space limitations only a few examples of aerobic bioprocesses will be presented here.

A.2.1. Citric Acid Production

Citric acid present in citrus fruits was first crystalized from lemon juice in form of calcium citrate. Later on, citric acid was synthesized from glycerol. Production of citric acid from sugar solutions by aerobic bioprocesses was first realized by using *Penicillium.* Due to low yields obtained from *Penicillium, Aspergillus niger* was utilized in subsequently developed processes.

Citric acid is used as an acidulant in food, confectionery, and beverages (75%). Pharmaceutical (10%) and industrial (15%) applications are also significant. Citric acid complexes with heavy metals such as iron and copper and can be used as a stabilizer of oil and fats. In the pharmaceutical industry, citric acid can be used in antacids, soluble aspirin preparations, and as a stabilizer of ascorbic acid. Metal salts of citric acid such as trisodium citrate are used to prevent blood clotting by complexing calcium. Trisodium citrate can be used in detergents and cleaners as a cleaning agent instead of phosphates.

Aspergillus niger is the most widely used organism for citric acid production from molasses or sugar solutions. *Candida* yeast can also be used for producing citric acid from carbohydrates or *n*-alkanes, with yields as high as 225 g/l. Beet or cane molasses can be used as source of carbohydrates for citric acid production. The concentration of heavy metals such as iron and manganese must be reduced. Typical trace-element concentrations are 0.3 ppm zinc, 1.3 ppm iron, and Mn < 0.1 ppm. High concentrations of metals can be reduced with the addition of Na- or K-ferrocyanide. Additions of nitrogen, phosphate, and other inorganic salts may not be required. Utilization of pure glucose or sucrose solutions is expensive and requires additions of nitrogen (NH_4^+), phosphate, and inorganic salts. Because of its low price and nutrient-rich nature, molasses is usually preferred to pure sugar solutions.

Citric acid production is mixed growth associated, mainly taking place under nitrogen and phosphate limitations after growth has ceased. Since citric acid is a product of primary metabolism, it is produced in high concentrations only under very specific conditions. These include restricted growth, medium deficient in one or more essential elements, high sugar concentration, high dissolved-oxygen concentration, pH below 2, and absence of trace metals. Such conditions provoke an overflow in metabolism that results in an overproduction of citric acid. Oxalic and gluconic acids are also produced if the pH is above 2. Potassium ferrocyanide is added to reduce the concentration of metals and is a growth inhibitor that promotes the production of citric acid. The metabolic imbalance that results in citric acid production also alters the morphology of fungi. The hyphae become short, stubby, and forked in small pellets (0.2–0.5 mm in diameter). Limitation of certain nutrients such as nitrogen and phosphate stimulates citric acid formation. High dissolved-oxygen concentrations must be maintained throughout the culture. Even short interruptions in oxygen provision can result in irreversible decreases in acid production rate.

Citric acid was historically produced by surface fermentation of beet molasses, and this process is still employed by some manufacturers. It is labor intensive, but power requirements are lower than for submerged fermentation. The surface process is realized on surface-aerated trays with liquid depth of 5–20 cm. Sterilized diluted beet molasses, with a sugar concentration of 150 g/l and a pH of 6, is placed on trays in a temperature-controlled (30°C) and aerated clean room. Sterilized additional nutrients and alkali ferrocyanide are added into the medium. Spores of a selected strain of *A. niger* are spread over the liquid on trays. The clean chamber is aerated with filter-sterilized air to provide oxygen to the organisms and to remove fermentation heat. Mycelium forms a layer on the surface of the medium. After 7–10 days of incubation the trays are emptied, the mycelium is removed, and the medium is transferred to the recovery section. The production of undesirable products such as gluconic and oxalic acids can be avoided by strain selection.

After the Second World War submerged fermentation processes utilizing molasses or pure sugar solutions were developed. The submerged process is realized in deep stainless-steel vessels of 100 m^3 or larger by batch or fed-batch operation. The fermenters may be mechanically agitated or aerated towers with an internal recycle draft tube. Aeration is provided to the fermenter by air sparging (0.1 to 0.4 vvm), and temperature is controlled with cooling coils. Agitation is usually gentle (50–100 rpm) to avoid shear damage on molds. Diluted molasses supplemented with other nutrients is on-line sterilized and added to the fermenter. Sterilization in the fermenter is also possible. The initial pH is adjusted at 2.5 to 3. Spores of *A. niger* are allowed to germinate in an inoculum medium before being transferred to the main fermenter. In some cases, spores are directly introduced into the fermentation media (5 to 25 × 10^6 spores/l). Since dissolved-oxygen concentration is critical for citric acid production, oxygen-enriched air may be used in some cases. About 80% of the supplied carbon is converted to citric acid in a typical fermentation. Batch operation usually results in productivities of 0.5–1 kg/m^3 h. Fed-batch operation can be performed to avoid substrate inhibition and to prolong the production phase one or two days after growth cessation. Typical volumetric yields of fed-batch processes are around 130 kg/m^3. When citric acid production stops, usually after 4–5 days, the fermenters are emptied and the biomass is separated from the broth by filtration. The liquid is transferred to the recovery section.

Precipitation is usually accomplished by addition of calcium hydroxide (lime) to the heated fermentation broth to obtain calcium citrate tetrahydrate. The precipitate is then washed and treated with dilute sulfuric acid, yielding an aqueous solution of citric acid and CaSO$_4$ (gypsum) precipitate. After bleaching and crystallization, either anhydrous or monohydrate citric acid is obtained. Solvent extraction is another option for recovery of citric acid, although it is not used commercially. Extraction avoids the use of lime and

H_2SO_4 and gypsum formation; however, it may result in extraction of some impurities present in molasses. Ketones, hydrocarbons, ethers, and esters can be used as extractants. Extraction of citric acid is realized at low temperature, and the solvent is then stripped with hot water.

Almost no information is available on process economics for citric acid production because of the proprietary nature of production schemes. The demand for citric acid in 1999 was 400,000 tons ($1,400 million). It is expected to increase 3% annually through 2003.

A.2.2. Production of Bakers' Yeast

Bakers' yeast is essential to the modern baked goods with which we are familiar. The earliest production of bakers' yeast was accomplished by a Dutch process (1781), utilizing grain mash as raw material. Later on, sugar solutions and aeration were used in Germany. The process was further improved to avoid ethanol formation and to improve biomass yield. Today, bakers' yeast worldwide production is at the level of a million tons per year. The process is based on aerobic cultivation of *Saccharomyces cerevisiae* on carbohydrates.

Today the most widely used organism for baking is *Saccharomyces cerevisiae*. In the past (prior to the nineteenth century) other strains of *Saccharomyces* such as *S. uvarum* or *S. carlsbergensis* (brewers' yeast) had been used. However, *S. cerevisiae* is superior to other species of yeast for baking purposes, primarily due to its ability to generate gas (e.g., CO_2) in the dough. The most suitable yeast strains are selected in the laboratory and are stored by freeze drying.

Molasses is the most used carbon and energy source for production of bakers' yeast. It is inexpensive, rich in nutrients (nitrogen, phosphorus, and minerals), and is available year around. Other carbohydrates such as glucose, sucrose, fructose, or hydrolyzed starch can be used. Aqueous ammonia (NH_4OH), ammonium salts, or urea may be used as nitrogen sources. Phosphoric acid may also be added as phosphate source. Addition of some Mg and Fe salts may be needed. Some yeast strains require B vitamins as well as Na, K, Mg, Ca, and sulfate ions for effective growth.

A material balance for bakers' yeast formation can be written as follows:

$$\text{Sucrose (200 g)} + NH_3 \text{ (10.3 g)} + O_2 \text{ (100.5 g)} + \text{Minerals (7.5 g)}$$
$$\longrightarrow \text{Biomass (100 g)} + CO_2 \text{ (140.g)} + H_2O \text{ (78 g)} \tag{A.4}$$

Typical yields are about 0.5 gram of cells per gram of substrate and 1 g of cells per gram of oxygen. The maximum specific growth rate of yeast is 0.6 h^{-1} (doubling time of 1.2 h). The temperature and pH of the fermentation medium are controlled at 30°C and 6–7, respectively. High dissolved-oxygen concentrations (above 2 mg/l) are required to promote biomass production. High carbohydrate concentrations may provoke ethanol production even in aerobic conditions (the Crabtree effect). Ethanol is inhibitory to cell growth and reduces biomass yield.

Growth of yeast results in a highly viscous culture broth. Consequently, mechanically agitated fermenters are preferred over airlifts or aerated columns. Typical reactors have a working capacity of 50 m^3 to 350 m^3 with a height-to-diameter ratio of 3. Vigorous aeration and agitation are required to provide oxygen for biomass production. Reactors are aerated in the bottom by perforated horizontal pipe spargers. Stirred tanks utilized for yeast production typically have oxygen transfer rates of 1 mole O_2/l h and $K_L a$ of 600 h^{-1}. Temperature control is accomplished by cooling coils, and pH is controlled by addition of acid or base. Dissolved-oxygen concentration and foam are also controlled. Some plants are computerized and automatically controlled.

To avoid the Crabtree effect, production of bakers' yeast is performed in fed-batch mode. The inoculum is prepared in smaller tanks and is added to the production fermenters. Multistage fermenters can be used for production. Diluted molasses, ammonium or urea, phosphoric acid, and mineral salts are stored in different tanks. The process starts batchwise after filling and sterilization of the fermenter content. Intermittent feeding of continuously sterilized nutrients starts when the initial charge of nutrients is depleted. The feeding rates of nutrients are adjusted to maximize growth rate. Dissolved-oxygen monitoring can be utilized to evaluate the metabolic activity of yeast and is the basis for feedback control systems for nutrient addition.

After 20 to 30 h of culture, the broth is transferred to centrifuges for separation of yeast. The cells are washed several times to remove inert solids. Centrifugal separation results in a light colored cream with up to 22% solids from yeast. The cream is stored in agitated tanks at 2–4°C, and part is used for seeding additional fermentations. Bakers' yeast can be sold in the form of cream, in compressed form (30% yeast solids), or dried (95% yeast solids). Filter press or rotary vacuum filters are used to concentrate cream. The filtered yeast may be mixed with emulsifiers prior to being extruded into yeast cakes or packaged in large paper bags. Fresh bakers' yeast may be marketed as free-flowing particles by adding ingredients such as modified starch or micronized cellulose.

To produce active dry yeast, the filtered yeast cake is extruded into particles that are dried in a hot air stream. Drying temperature, drying rate, and final moisture content of the dried yeast should be controlled during drying. The drying temperature is usually 45°C. Most of the vegetative yeast cells are killed at temperatures above 50°C. Drying time may vary from 20 min to several hours, depending on the type of drier used. The final moisture content of yeasts after drying is about 5–10%. Several types of driers can be used. The Roto–Louvre drier is an empty cylinder rotating at 1–4 rpm. Heated air at 50–60°C is blown to the cylinder containing yeast particles. The drying temperature is 45°C and the duration is 10–15 h. In fluidized bed driers, hot air (50–60°C) is blown from the bottom of the column to keep the cells suspended in air. The drying temperature and time are 45°C and 1–2 h, respectively. Spray driers are the most widely used. A suspension with 10–20% yeast is atomized into a drying chamber and dried with hot air. The drying temperature and time are 50°C and 10–20 min, respectively. Product quality depends mainly on gassing activity of yeast. Presence of certain carbohydrates such as trehalose or glycogenin increases the gassing activity of yeast. Highly stable active dry yeast can be obtained from cultures with a high protein content. Fluidized bed or spray driers yield fine yeast particles of 0.2–2 mm with high protein content and high gassing activity.

A.2.3. Production of Penicillins

Penicillin was discovered by Alexander Fleming (see Chapter 1). Different penicillins are produced by different strains of *Penicillium*. Chemical structures of penicillins G and V are given in Fig. A.4.

Sodium penicillin G

Penicillin V

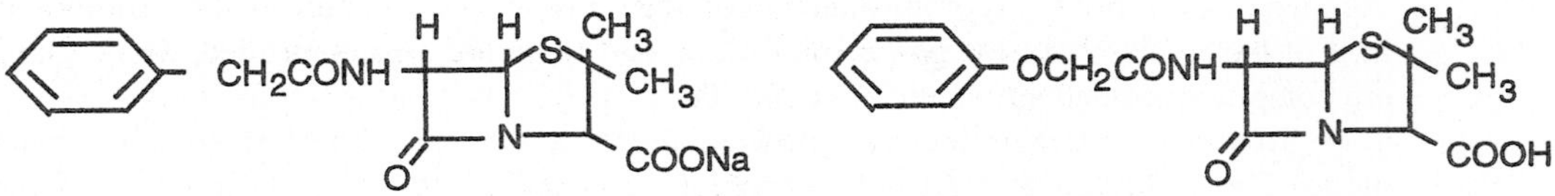

Figure A.4. Chemical structure of penicillins G and V.

Sodium penicillin G (MW = 356.4 KDa, Activity: 1,670 U/mg) is administered parenterally, as it is degraded in acid conditions; penicillin V (MW = 372.4 KDa, Activity: 1,595 U/mg) is acid stable and is orally administered. Both forms are active against Gram positive bacteria by inhibition of cell wall synthesis. Different species of the genus *Penicillium* produce different forms of penicillin. The strain used by Fleming was *P. notatum.* Later on, different strains were used, such as *P. chrysogenum,* which is the most widely used strain in industry. Before utilization in industrial scale, considerable efforts must be spent in strain selection and development to improve the yield and activity of penicillin formation. Selected strains are stored in the form of lyophilized spores. Vegetative cells may be frozen at −70°C with glycerol as a suspending medium.

P. chrysogenum can use a variety of carbohydrates and oils as carbon and energy sources. Among those are glucose, sucrose, hydrolyzed starch, lactose, and molasses. Corn oil supplemented with lactose results in fast production of highly concentrated penicillin. Medium formulation has changed significantly with new developments. The original medium (1945) contained the following compounds: lactose, 3–4%; corn steep liquor, 4%; $CaCO_3$, 1%; KH_2PO_4, 0.4%; antifoam, 0.25%. Improved media resulting in higher penicillin yields have been developed. A typical composition of such media is: glucose or molasses, 10%; corn steep liquor solids, 4–5%; phenylacetic acid (continuous feed), 0.5–0.8% total; vegetable oil-antifoam, 0.5% total.

Corn steep liquor (CSL) is used as a nitrogen source, since it results in higher penicillin yields as compared to the other nitrogen sources. Some compounds in CSL are converted to phenylacetic acid or other side-chain precursors. Cottonseed flour or soybean meal may also be used as nitrogen sources; however, they are more expensive than CSL. Continuous addition of ammonium sulfate to keep the ammonium concentration around 250–300 mg/l is required for continued synthesis of penicillin and to avoid lysis of the mycelium.

Certain precursors of the penicillin side chain need to be added into the fermentation medium. This constitutes a major cost item. Penicillin G requires 0.47 g sodium phenylacetate and penicillin V 0.5 g sodium phenoxyacetate per gram of penicillin produced. Those precursors are fed continuously to avoid possible toxic effects. More than 90% of the precursors are incorporated into the structure of penicillin. Phosphorus supplied by CSL is usually sufficient, since phosphate concentration should be limiting (200–300 mg/l) in the media for penicillin production.

Penicillin is a secondary metabolite with a nongrowth-associated production. The process to produce penicillin involves an initial batch phase in which cell growth occurs. In the first 40 h of fermentation rapid cell growth is achieved with a doubling time of nearly 6 h. After a high cell density has been obtained, nutrients (glucose and CSL) and precursors are added slowly or intermittently to reduce cell growth to $0.02 h^{-1}$ and maximize penicillin production. Oxygen, carbon, nitrogen, and phosphate concentrations should be low. A simplified diagram of the process scheme is presented in Fig. 1.3. Culture preparation starts with lyophilized spores and agar slant cultures. Vegetative cells are cultivated in shake flasks and then are transferred to seed fermenters (10–100 l). Production fermenters are agitated tanks 200–500 m^3 in volume made of stainless steel. Mechanical agitation is provided at a rate of 100 to 300 rpm. Temperature is controlled around 25–28°C (26°C optimum) by using cooling coils. Antifoam is added to reduce foam formation. Dissolved oxygen is controlled at > 2 mg/l and pH at 6.5. Vigorous aeration is supplied from the bottom of fermenters by ring or tube spargers. Due to the high viscosity of the broth, oxygen transfer is a major problem in penicillin fermentations. In some cases, strains from pellets are preferred because the medium is less viscous and oxygen supply is improved. The fermentation is stopped when the oxygen uptake rate of the culture exceeds the oxygen transfer rate of the reactor, or when 80% of the fermenter is full.

The original process for the recovery of penicillin from fermentation broth was based on adsorption on activated carbon. After washing with water, the activated carbon was eluted with 80% acetone. The penicillin was concentrated by evaporation under vacuum at 20 to 30°C. The remaining aqueous solution was cooled to 2°C, acidified to pH = 2–3, and the penicillin extracted with amyl acetate. Penicillin was crystallized from amyl acetate with excess mineral salts at pH of 7 under vacuum. This process is uneconomical because of the high cost of activated carbon.

The current recovery process includes filtration, extraction, adsorption, crystallization, and drying. Filtration is usually achieved by using high-capacity, rotary vacuum drum filters for separation of the mycelia. The mycelia are washed on the filter and disposed. The penicillin-rich filtrate is cooled to 2–4°C to avoid chemical or enzymatic degradation of the penicillin. In some early processes, the filtrate was further clarified by a second filtration with the addition of alum. In recent years, macroporous filters have been used in some plants for separation of the mycelia.

Solvent extraction is accomplished at low pH such as 2.5–3, using amyl acetate or butyl acetate as solvent. Continuous, countercurrent, multistage centrifugal extractors (Podbielniak D-36 or Alfa–Laval ABE 216) are used for this purpose. The distribution coefficient of penicillin G or V between organic and aqueous phase depends strongly on the pH of the medium. At a pH of 3 the distribution coefficient is about 20. Penicillins G and V degrade under acidic conditions with a first-order kinetics. To avoid degradation of penicillin during solvent extraction at low pH, temperature is kept around 2–4°C and filtration time is kept very short (1–2 min). Two extractors used in series result in nearly 99% penicillin recovery. Whole broth extraction (without filtration) is possible by using Podbielniak extractors in series. Due to operational difficulties, however, this approach is not used in practice. Carbon adsorption is used to remove impurities and pigments from penicillin-rich solvent after extraction. Several activated carbon columns in series can be used for this purpose.

Penicillin may be back extracted into water by addition of alkali (KOH or NaOH) or buffer at pH of 5 to 7.5. The water-to-solvent ratio in this extraction is usually between 0.1 and 0.2. A continuous, multistage, countercurrent extractor may be used for this purpose. This step is usually omitted in the new separation schemes.

Crystallization may be performed from the solvent or aqueous phase. Na, K, and penicillin concentrations, pH, and temperature need to be adjusted for crystallization. Excess amounts of Na or K are added to the penicillin-rich solvent before crystallization in an agitated vessel. The crystals are separated by a rotary vacuum filter. The crystals may be washed and predried with anhydrous butyl alcohol to remove some impurities. Large horizontal belt filters are used for collection and drying of the crystals. Usually warm air or radiant heat is used for drying.

Crystalline penicillins G or V are sold as an intermediate or converted to 6-APA (6-aminopenicillanic acid), which is used for production of semisynthetic penicillins. The enzyme penicillin acylase is used for cleavage of penicillin G or V to produce 6-APA. Some bacteria, *E. coli*, *B. megaterum*, and *P. melanogenum*, as well as some molds produce the enzyme.

Production costs for penicillin production utilizing glucose as a substrate were calculated in 1982 as $19/kg. Prices of penicillin have decreased in recent years, from $31.73/kg in 1996 to $15/kg in 2000. This has put pressure on penicillin manufacturers to reduce production costs. The most important costs in penicillin production are raw materials (35%) and utilities (14%). By using cheaper raw materials, as molasses or starch, and genetically improved strains producing higher penicillin yields, the production costs may be reduced significantly.

Worldwide, approximately 26,000 tons of penicillin G and 10,000 of penicillin V are produced annually. Penicillins for medical applications and feed have an annual demand of $4,400 million. More than 80% of the penicillin produced is utilized for the synthesis of 6-APA and other intermediates.

A.2.4. Production of High-Fructose Corn Syrup (HFCS)

High-fructose corn syrup (HFCS) is a low-calorie sweetener commonly used in beverages, desserts, and other sweet foods. Until 1935, the only syrup available was 42 DE (dextrose equivalent) acid-converted corn syrup. In 1940, enzymes were commercially available and corn starch was hydrolyzed enzymatically to produce corn syrups. In the early 1960s the first crystalline dextrose derived from corn was marketed. The commercial production of the enzyme glucose isomerase, which converts glucose to its sweeter (approx. 1.7 times) isomer fructose, was a major milestone. The first HFCS was produced in 1967 and contained 15% fructose. Further process improvements yielded 42% and 55% fructose-containing HFCS. The original conversion process was batch; however, immobilized enzyme technology was later used for the production of HFCS by continuous operation. United States sales of HFCS exceeded 9.58 million tons in 1999.

Three major HFCS products differ by their fructose content, 42%, 55%, and 90%. HFCS containing 42% fructose is mainly used in most of the food products utilizing liquid sweeteners. HFCS with higher levels of fructose (55%) are mainly used in soft drinks as a replacement for sucrose and in jams and jellies (90%) as a low-calorie sweetener.

Production of HFCS from corn starch is an enzymatic process. The process scheme may be divided into 18 steps and five major operations (Fig. A.5). Those operations are dextrose production by enzymatic hydrolysis of corn starch, primary physical and chemical treatment of dextrose syrup, isomerization of dextrose to 42% fructose, secondary

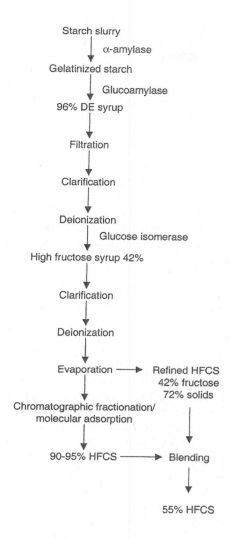

Starch slurry

↓ α-amylase

Gelatinized starch

↓ Glucoamylase

96% DE syrup

↓

Filtration

↓

Clarification

↓

Deionization

↓ Glucose isomerase

High fructose syrup 42%

↓

Clarification

↓

Deionization

↓

Evaporation ⟶ Refined HFCS
42% fructose
72% solids

↓

Chromatographic fractionation/
molecular adsorption

↓

90-95% HFCS ⟶ Blending

↓

55% HFCS

Figure A.5. Process schematic for HFCS production.

refining of fructose corn syrup, and conversion of 42% fructose- to 55% fructose-containing HFCS.

First, corn starch is gelatinized by cooking at high temperatures such as 65°C and is converted to dextrose (liquefaction) by thermostable-amylase in a two-stage continuous reactor. The product is a dextrose-rich syrup containing 10 to 15 DE. The conditions in liquefaction reactors are 105°C and a 5- to 10-min holding time for the first reactor; and 95°C and a 90- to 120-min holding time for the second reactor. The feed starch slurry contains 30–35% solids with 0.1 to 10% enzyme at a pH of 6.5. The total and soluble protein contents of the starch slurry should be lower than 0.3% and 0.03%, respectively, to avoid color formation as a result of the Maillard reaction between amino acids and sugars at high temperatures. Saccharification of liquefied starch slurry is achieved by using the enzyme glucoamylase to produce more dextrose from branched chains of the starch. In continuous saccharification reactors, glucoamylase is added to the 10–15 DE liquefied starch after temperature and pH adjustment and is fed through a number of reactors in series. The conditions for this step are 60°C, pH of 4.3, and holding time of 65 to 75 h. The feed contains 30 to 35% dry substance and 1 l of glucoamylase solution per ton of dry weight of starch. The product of this step contains 94 to 96% dextrose, 2 to 3% maltose, 1 to 2% higher saccharides, and 30 to 35% dry substance. Environmental conditions must be strictly controlled during liquefaction and saccharification to obtain the 94–96% dextrose that is required to obtain 42% fructose HFCS.

The dextrose syrup produced by the liquefaction and saccharification step is refined to remove ash, metal ions, and proteins that may interfere with isomerization. The dextrose syrup is filtered on rotary precoat vacuum filters to remove solids, proteins, and oil. The filtered liquor is then passed through several check and polish filters to remove traces of particles. The color in the filtrate is removed by granular activated carbon in columns. Carbon-treated liquor is filtered again and passed through ion-exchange columns to remove metal ions and ash. These columns are dual-pass cation–anion ion-exchange systems that also remove color. Deionized and decolorized dextrose syrup is evaporated to concentrate dextrose, and Mg ions are added to activate the isomerase.

Conversion of glucose (dextrose) to fructose by the enzyme glucose isomerize is accomplished in a packed column of immobilized enzyme. The reactor conditions are 55–65°C, pH of 7.5 to 8, and residence time of 0.5 to 4 h. The optimum temperature is 60°C. Temperatures higher than 60°C cause higher conversion rates and faster inactivation of the enzyme. The feed temperature may be as low as 55°C, resulting in slower conversion and enzyme inactivation rates. Microbial contamination may be a problem at temperatures below 55°C. The optimum pH for maximum enzyme activity is 8 and for stability is between 7 and 7.5. Therefore, the operating pH is adjusted for maximum stability and activity of the enzyme. The feed syrup contains 40 to 45% dry substance, 94–96% dextrose, 4–6% higher saccharides, and 4 mM of Mg ions as activator. The enzymatic isomerization of glucose to fructose is reversible. With 96% dextrose in the feed, the equilibrium fructose concentration in the effluent is expected to be 48%. However, the exact equilibrium is not reached with 4 h of residence time, and the effluent contains 42% fructose.

The activity of the immobilized enzyme in the column drops exponentially with time. The half-life of the enzyme is 70 to 120 days. Therefore, the residence time for maximum conversion is low for new columns with unused enzymes, and the feed flow rate should be lowered at later stages of operation to obtain 42% fructose. Usually series and parallel configurations of immobilized enzyme columns are used to compensate for activity loss of the enzyme. Parallel operation of six columns offers a good flexibility and results in good product quality. A number of process variables such as temperature, feed pH, and flow rate may be varied to obtain uniform product quality. Constant fructose levels are usually achieved by automatic back blending controlled by a polarimeter. Columns must be replaced two or three times a year. The cost of the enzyme is a major part of the operating cost. By improving the stability and activity of the enzyme, the cost of isomerization may be reduced significantly.

HFCS produced by isomerization of dextrose syrup can be further refined to remove color and ions by carbon treatment and ion exchange, respectively. The refined 42% HFCS is evaporated for shipment to yield 71% solids.

The HFCS from the isomerization step contains 42% fructose, 52% dextrose, and about 6% oligosaccharides. To obtain 55%- and 90%-fructose syrups, the fructose in 42% syrup needs to be concentrated. Fructose preferentially forms a complex with some cations such as calcium. This is used for concentrating fructose in HFCS. There are two different commercial processes for enrichment of fructose from 42% syrup. One process utilizes an inorganic resin for selective molecular adsorption of fructose. Another process employs chromatographic fractionation using organic resins. Fructose is selectively held in fractionating columns but dextrose is not. Deionized and deoxygenated water is used for the elution of fructose from the column. Usually a column packed with low cross-linked fine-mesh polystyrene sulfonate-Ca cation exchange resin is used for enrichment purpose. The enriched syrup contains nearly 90% fructose and is called very enriched fructose corn syrup (VEFCS). The VEFCS is blended with 42%-fructose syrup to obtain the desired fructose content, such as 55%. The effluent from the isomerization step may be recycled back to the feed solution to obtain 42%-fructose syrup in the effluent of the isomerization column. The raffinate stream rich in oligosaccharides is recycled back to the saccharification step. The water used as an eluent in enrichment columns should be minimized to maximize the solids content of the syrup.

Since 1972 HFCS has replaced sucrose as a low-calorie sweetener to a large extent. The price of sucrose was 31 cents/lb and the price of HFCS was 21 cents/lb, about 70% of the sucrose price in 1981. This difference has become larger with the recent developments in HFCS production technology. Consequently, HFCS has replaced sucrose and glucose as a low-calorie sweetener used in soft drinks, canned fruits, ice cream, and certain bakery products over the last twenty years. Low-calorie sweetener consumption is expected to level off at 130 lb per capita per year.

SUGGESTIONS FOR FURTHER READING

ATKINSON, B., AND MAVITUNA, F., *Biochemical Engineering and Biotechnology Handbook,* 2d ed., Stockton Press, New York, 1991.

BAILEY, J. E., AND D. F. OLLIS, *Biochemical Engineering Fundamentals,* 2d ed., McGraw-Hill, New York, 1986.

DEMAIN, A. L. Small Bugs, Big Business: The Economical Power of the Microbe, *Biotechnol. Adv.* 18:499–514, 2000.

FLICKINGER, M. C., AND DREW, S. W., eds., *Encyclopedia of Bioprocess Technology: Fermentation, Biocatalysis and Bioseparation,* John Wiley & Sons, New York, 1999.

Index

Briggs, G. E., 62
Bubble columns, 286, 323
Bubble-point tests, 322
Budding, 22
Bulk sludge, 492

C
c DNA, 232
Calcium, 51
Callus culture, 408-9
Calvin-Benson cycle, 152
Calvin cycle, 152
Cancer cells, 428
CAP (cyclic-AMP-activating protein), 120
Capsid, 14
Capsule, 17
Carbohydrates, 34-38, 54
Carbon compounds, 49
Carboxyl, 51
Carrying capacity, 180-81
Carticel, 466
Catabolism, 134
Catabolite repression, 120-21
Cell construction, 25-46
    amino acids, 26-34
    biopolymers, 25-26
    carbohydrates, 34-38
    cellular macromolecules, 26
    DNA (deoxyribonucleic acid), 40, 43-46
    fats, 39-40
    lipids, 38-39
    nucleic acids, 40-43
    proteins, 26-34
    RNA (ribonucleic acid), 40, 40-43, 46
    steroids, 40
Cell disruption, 341-43
    and heat dissipation, 341-43
    mechanical methods, 341-42
    nonmechanical methods, 342-43
Cell growth, 155-206
    batch growth, 156-75
    in continuous culture, 189-99
Cell immobilization:
    by entrapment using different support materials,
        266-67
    reactors using, 413-14
    by surface attachment, 266-67
    systems, 263-75
Cell lines, genetic instability of, 412-13

Cell lysis, 16
Cell mass concentration, determining, 158-60
Cell naming, 12-14
    archaebacteria, 13
    binary nomenclature, 12
    eubacteria, 13
    eucaryotes, 13
    genus, 12
    nomenclature, 12
    procaryotes, 13
    species, 12
    taxonomy, 12
    viruses, 13
Cell nutrients, 46-52
    growth media, 52
    macronutrients, 48, 49-50
    micronutrients, 48, 50-51
Cell separation, 231-341
    centrifugation, 332-36
    coagulation, 340-41
    filtration, 332-36
    flocculation, 340-41
Cell-to-cell communication in plants, 410
Cells, 105-32
    animal, *See* Animal cells; Animal cells cultures
    central dogma, 105-7
    construction, *See* Cell construction
    diversity of, 11-12
    DNA replication, 107-10
        origin of, 108
    and extracellular environment, 124-28
        mechanisms to transport small molecules
            across cellular membranes, 124-26
    growth of, *See* Cell growth
    immobilized cell systems, 263-75
    metabolic regulation, 119-24
        genetic-level control, 119-22
        metabolic pathway control, 123-24
    microbes, 407-11
    plant, *See* Plant cells
    replication fork, 108
    role of cell receptors in metabolism/cellular
        differentiation, 127-28
    transcription, 110-13
    translation, 113-19
        elongation, 113-15
        genetic code, 113-14
        how the machinery works, 113-14
        initiation, 113-15
        termination, 113-15

E

Eadie-Hofstee plot, 65
Eagle's minimal essential medium (MEM), 394
Eccrinolysis, 476
EDTA (ethylenediaminetetraacetic acid), 51, 67
Effectiveness factor, 88
Elder, A. L., 5
Electrodialysis (ED), 376-78
Electron transport chain, 137, 152
Electrophoresis, 234, 375-76
Electroporation, 236, 241
Elicitors, 410-11, 418
Elution chromatography, 365-66
    mobile phase (fluid phase), 365
    stationary phase, 365
Embden-Meyerhof-Parnas (EMP) pathway, 137, 152
Encapsulation, 265
End-product inhibition, 123
Endocytosis, 468-69
Endogenous metabolism, 164
Endoplasmic reticulum (ER), 21, 117, 386
Endoproteases, and proteases, 93
Endosomes, 468-69
Endospores, 17
Energy efficiency, stirred-tank systems, 291
Entner-Doudoroff (ED) pathways, 137, 149
Entrapment, physical, within porous matrices, 263
Enzyme kinetics, 60-79
    complex enzyme kinetic models, 67-75
        allosteric enzymes, 67
        inhibited enzyme kinetics, 67-75
    Hanes-Woolf plot, 74
    mechanistic models for, 61-63
        quasi-steady-state assumption, 62-63
        rapid equilibrium assumption, 61-62
    pH effects, 75-76
    temperature effects, 76-78
Enzyme-substrate interactions, 60-61
Enzymes, 57-101
    active site, 59
    as animal cell products, 401-2
    apoenzyme, 57
    classification of, 58
    coenzymes/cofactors, 59-60
    denatured, 66
    enzyme-substrate interaction, molecular aspect
        of, 59

    and free-energy change/equilibrium constant, 58-60
    function of, 58-60
    immobilized enzyme systems, 79-91
        diffusional limitations in, 84-90
        electrostatic/steric effects in, 91
        immobilization methods, 79-83
    insoluble substrates, 78-79
    isozymes, 57
    kinetics, 60-79
    large-scale production of, 91-96
        cultivation of organisms producing the
          enzyme, 91-92
        separation of cells by filtration/centrifugation, 92
    medical/industrial utilization of, 92-96
    membrane entrapment of, 80
    naming convention, 57
    orientation effect, 59
    proximity effect, 59
    ribozymes, 57
    subunits, 57
    unwinding, 108
Episomes, 228-29
Epitope, 401
Error-prone PCR, 456
Erythropoetin, 401
Estrogens, 40
Ethanol production, 515-19
Ethylenediaminetetraacetic acid (EDTA), 343
Eubacteria, 13, 15-17, 53
Eucaryotes, 13, 19-25, 53, 111-12
    cell division in, 20
    compared to procaryotes, 19
    enhancer regions, 120
    mitochondria, 20-21
    nucleus of eucaryotic cells, 20
Ex vivo gene therapy, 467
Excretion, and *E. coli*, 425
Exit-gas analysis, 308
Exocytosis, 117
Exoenzyme synthesis, 128
Extinction, probability of, 314
Extracellular enzymes, 92
Extracellular products, 17
Extracorporeal artificial liver, 472
Extremophiles, 12
Extrinsic concentrations, 184-88

F

Facilitated diffusion, 125
Facultative organisms, 12, 53
Fats, 39-40, 54
Fatty acids, 38-39
    examples of, 39
Fed-batch operation, 256-61
Feedback inhibition, 123
Feedback repression, 119
Feeding, 476
Fermentation, 137-39, 148, 152
    software overview, 312
Fermenters, *See* Bioreactors
Filamentous organism growth models, 183
Filter sterilization, 319-20, 323
Filtration, 332-36
    and cell separation, 332-36
    cross-flow, 360
    diafiltration, 364-65
    microfiltration, 358-60
    purification, 378
    separation of cells by, 92
    tangential flow filtration, 360
    ultrafiltration, 358-60
Fission, asexual reproduction by, 22
Fixed-bed columns, analysis of adsorption
        phenomena in, 353-54
Flagella, 22
Flagellates, 24-25
Flatbed reactors, 472
Fleming, Alexander, 3-4
Flocculation, 340-41
Florey, Howard, 4
Fluidized bed reactors, 472
Fluorescence, 151
Flux, 455
Foaming, 290
Folic acid, 51
Food and Drug Administration (FDA), 8-9
    drug approval process, 9
    GRAS list, 424
Fourier transform infrared (FTIR) spectra analysis,
        311
Freeze drying (lyophilization), 379
French press, 341-42
Friable callus, 409
Functional genomics, 236, 241
Fungal fermentation, and wall growth, 298
Fungi, 19-25, 22

G

G_1 and G_2 phases, eucaryote cell division, 20
Gametes, 20
Gaseous oxygen, 49
Gases, sterilization of, 320-23
Gateway sensors, 311
Gaulin-Manton press, 341-42
Gel-filtration (molecular sieving) chromatography,
        366-67, 369-70
Gelation of polymers, 264
Gene, 107
Gene bank, 234
Gene cloning, 232, 235
Gene library, 234
Gene therapy using viral vectors, 467-71
    retro virus, mass production of, 470-71
    viral infection models, 467-70
Gene transfer/rearrangement:
    conjugation, 228-29
    episomes, 228-29
    genetic recombination, 225-30
    internal gene transfer, 230
    natural mechanisms for, 225-30
    .transduction, 227-28
    transformation, 227
    transposons, 230
Generalized transduction, 227, 241
Genetic engineering, 230-36, 513
    basic elements of, 230-35
    of higher organisms, 235-36
Genetic instability, 247, 433-37
    in commercial operations vs. laboratory-scale
        experiments, 437
    growth-rate-dominated instability, 437
    host cell mutations, 436-37
    plasmid structural instability, 436
    predicting, 441-50
    segregational loss, 434-36
Genetic-level control, 119-22
Genetic processes, regulatory constraints, 451-52
Genetic recombination, 225-26, 241
Genetically engineered organisms, 421-33
    genetic instability, 433-37
        growth-rate-dominated instability, 437
        host cell mutations, 436-37
        plasmid structural instability, 436
        predicting, 441-50
        segregational loss, 434-36

Histidine, 51
Histones, 20
Hollow-fiber reactors, 399, 472
Holoenzyme, 110
Hormones, 51, 54, 128
    as animal cell products, 401
Host-cell derived instability, 458
Host cell mutations, 436-37
Host cell regulatory mutations, 433-34
Host-vector interactions, predicting, 441-50
Host-vector system selection guidelines, 424-33
    comparison of strategies, 432-33
    Escherichia coli (E. coli), 424-26
    gram-positive bacteria, 426-27
    insect cell-baculovirus system, 429-30
    lower eucaryotic cells, 427-28
    mammalian cells, 428-29
    transgenic animals, 430-43
    transgenic plants and plant cell culture, 432
Hougen-Watson kinetics, 176
Hughes press, 342
Hybrid glycoforms, 119
Hybridization, 231
Hydrocarbons, metabolism, 144-45
Hydrogen, 50
Hydrogen-ion concentration (pH), 169-70
Hydrophobic chromatography (HC), 366-67
Hypervariable regions, 34
Hyphae, 22-23

I

Immobilized cell systems, 263-75, 285
    active immobilization cells, 263-65
        adsorption of cells on inert support surfaces, 265
        encapsulation, 265
        gelation of polymers, 264
        ion-exchange gelation, 264
        macroscopic membrane-based reactors, use of, 265
        polycondensation, 264
        polymerization, 264
        precipitation of polymers, 264
    active immobilization, defined, 263
    bioreactor considerations in, 273-75
    diffusional limitations in, 268-73
Immobilized enzyme systems, 79-91
    diffusional limitations in, 84-90

enzymes immobilized in a porous matrix, 86-90
    surface-bound enzymes on nonporous support materials, 84-86
    electrostatic/steric effects in, 91
    immobilization methods, 79-83
        entrapment, 79-80
        surface immobilization, 80-83
Immobilized-metal-affinity chromatography (IMAC), 371
Immortal cell lines, 389, 428
Immune response, 31
Immunobiological regulators, as animal cell products, 401
Immunoglobulins, 31-34
    major classes of, in human blood plasma, 33-34
Impeller, 286-90
In vivo gene therapy, 467
Inclusion bodies, 425
Incompressible cake, 333-35
Indirect methods, of determining cell mass concentration, 158-60
Indirect selection, 222
Inducer exclusion, 121
Induction, 119
Industrial utilization of mixed cultures, 487-88
    example of, 488-508
Inhibited enzyme kinetics, 67-75
Initiation factors, 115
Insect cell-baculovirus system, and host-vector system selection guidelines, 429-30
Insecticides, as animal cell products, 402
Insertion elements, 230
Insoluble products, separation of, 331-41
    centrifugation, 336-40
    coagulation, 340-41
    filtration, 332-36
    flocculation, 340-41
Insoluble substrates, 78-79
Insulin, 51
Interactions in mixed cultures, 476-79
    amensalism, 476, 478-79
    antagonism, 476
    commensalism, 476, 478
    competition, 477
    crowding, 476
    eccrinolysis, 476
    feeding, 476
    mutualism, 476-78
    neutralism, 477

Passive diffusion, 125
Pasteur effect, 143
Pectinases, 94, 97
    and koji fermenter, 277
PEKILO process, 508
Pellets, and molds, 23
Penicillin, 24
    process, 3-8
        schematic of production process, 8
    production of, 527-30
Penicillin acylase, 95
Penicillinase, 94, 96
Pentose-phosphate pathway, 145, 152
Pepsin, 94
Peptide bond, 28
Peptidoglycan, 16
Peptidyl site, 115
Perfusion systems, 261
Perinuclear space, 387
Periplasmic space, 16
Permease, 120
Peroxisomes, 21, 386
Petri dishes, 156
Petroff-Hausser slide, 156
Pfizer, 5, 7
Phage displays, 235
Phage lambda, 227
Phages, 14, 53
Phophoenolpyruvate (PEP), 126
Phosphate bonds, 53
Phosphenol pyruvate (PEP), 137-39, 147
Phosphofructokinase, 142
Phosphoglycerides, 40
Phospholipids, 40, 54
Phosphorus, 50
Phosphorylation, 117
Phosphotransferase system, 126
PhoStrip process, 505-6
Photoautotrophs, 49, 152
Photoperiod, 416
Photophosphorylation, 151
Photosynthesis, 151
    dark phase, 151
    light phase, 151-52
Phototaxis, 127
Phototube, 157-58
Phycomycetes, 23
Physical environment, monitoring/control of, 308
Physical waste treatment, 489
Plant cell cultures, 405-20

bioprocess development using, example of, 406
    economics of, 417
    large-scale immobilized, potential advantages/
        disadvantages, 414
    microbes compared to plant cells in culture,
        407-11
    role of bioreactors in, 407
Plant cells, 19-25
    low growth rates of, 412
    products, examples of potential commercial
        interest, 406
Plasmids, 43, 227, 241
    design, and avoidance of process problems,
        438-41
    structural instability, 436
Plasmodesmata, 410, 413
Plate counts, 156-57
Plug flow reactor (PFR), 191
Pneumatic conveyor driers, 379
Point mutations, 241
Polyadenylation, 113
Polycondensation, 264
Polygenic messages, 112
Polyhydroxybutyrate (PHB), 40, 504
Polymerization, 264
Polynucleotides, 43
Polypeptides, 28
Polyphosphates, 51
Polysaccharides, 54
P/O ratio, 209
Pores, 387
Porous polymers, 263
Posttranslational movement of a protein, 116
Potassium, 50
Precipitation, 349-51
    adsorption zone, 353-54
    isoelectric precipitation, 350-51
    operating line, 353
    saturated zone, 353
    solvent precipitation, 350
    virgin zone, 353
Precipitation of polymers, 264
Precipitin, 31
Predation, 479
Pressure-drop-versus-flow-rate tests, 322
Primary biological waste treatment, 491
Primary culture, 387
Primary mammalian cells, 387-88
Primary metabolites, 163
Primary structure, proteins, 26, 29, 53

RNA translation, 46
Root mats, 416
Rotary-drum driers, 379
Rotary-drum fermenters, 277-78
Rotary vacuum precoat filter, 332
Rotating biological contactors (RBC), 498-99
RPMI 1640, 394
r-RNA, 46
Runaway replication, 438
Rupture with ice crystals, 342
Rushton impeller, 287-90

S

S phase, eucaryote cell division, 20
Sail-type agitators, 398
Salting-out, 349-50
Saturated zone, 353
Saturation kinetics, *See* Michaelis-Menten kinetics
Scale-down, 301-2, 323
Scale-up, 303-6, 323
Scarce enzymes, 96
Secondary biological waste treatment, 491
Secondary metabolites, 163
Secondary structure, proteins, 26, 30, 53
Secretion, and *E. coli*, 425
Segregational loss (segregational instability),
        433-36, 458
Selectable mutations, 222
Selective pressure, 437
Selenium (Se), 51
Semiconservative replication, 107
Senescence, 389
Sense strand, 110
Sephadex, 80, 386
Sequential feedback inhibition, 124
Serum-containing/serum-free media, examples of
        composition of, 392-93
Sex pilus, 229
Shear protecting agents, 398
Sheet structure, 30
Shine-Delgarno box, 115
Shock loading, 492
Shotgun cloning, 230
Sigma factor, 110
Signal sequence, 426
    proteins, 116-17
Simple glycoforms, 119
Single-cell protein (SCP), 247, 488
Single chromatographic separation process, 367

Site-directed mutagenesis, 456
Sludge, 491
Sodium (Na), 51
Solid-state fermentations (SSF), 276-78, 285
    koji process in, 276-77
        major industrial use of, 277
        rotary-drum type of koji fermenter, 277-78
        rotary-tray chamber for koji fermentations,
            278
    major advantages of, 276
    major process variables in, 277-78
Soluble products, separation of, 343-78
    adsorption, 351-55
    aqueous two-phase extraction, 348-49
    chromatography, 365-75
    cross-flow microfiltration, 360-65
    cross-flow ultrafiltration, 360-65
    dialysis, 355-56
    electrodialysis (ED), 376-78
    electrophoresis, 375-76
    liquid extraction, 343-48
    microfiltration, 358-60
    precipitation, 349-51
    reverse osmosis, 356-58
    ultrafiltration, 358-60
Solvent precipitation, 350
Sonicators, 341
Sparger, 286, 293-94
Specialized transduction, 227, 241
Species, 12
Specific growth rate:
    microbial growth, 155-56
    using unstructured nonsegregated models to
        predict, 176-83
Spirillum, 12
Spontaneous rates of mutation, 221
Spores, 17
Sporozoans, 25
Spray dryers, 379
Squibb, 5
Standard operating procedures (SOPs), 9
Steady-state method, 295
Stem cells:
    defined, 471
    and hematopoiesis, 472
Sterility, maintenance of, 247
Sterilization, 314-23, 323
    continuous, 318-19, 323
    death, 314
    disinfection, 314

Sterilization *(continued)*
    filter, 319-20, 323
    of gases, 320-23
    of liquids, 315-20
    probability of extinction, 314
    sterilization chart, 318
Steroid hormone receptors, 128
Steroids, 40
    examples of, 41
Sterols, 51
Stirred-tank bioreactors, 398
Stoichiometric calculations, 209-15
    degree of reduction, 211-15
    elemental balances, 209-11
Stokes equation, 337
Stong promoter, 110
Storage granules, 17
Streptokinase, 96
Structural instability, 433-34
Structured loss (structured instability), 458
Substrate inhibition, 71-72, 178-79
Substrate-level phosphorylation, 152
Subunit vaccines, 401
Subunits, 57
Sucrose, as disaccharide, 36-37
Sulfite method, 294-95
Sulfur, 50
Surface filters, 321
Survival curve, 315
Suspension culture, 408-9
    aggregates, 410
    bioreactors for, 411-13
    elicitors, 410-11
    establishment of, 408-10
SYMBA process, 488, 508
Symbiosis, 476, 478

T

Tangential flow filtration, 360
Taxol, 406
Taxonomy, 12
t-DNA, 235-36
Teichoic acids, 16
Temperate phage, 227, 241
Temperature activation, 77
Temperature, and cells, 169
Temperature inactivation, 78
Tertiary biological waste treatment, 491
Tertiary structure, proteins, 26, 30, 53

Tetramer, 434
Therapeutic proteins, 422
Thermal denaturation, 78
Thermoacidophiles, 19
Thermophiles, 11
Thiamine, 51
Thylakoids, 22
Thymine, 41
Tissue-engineered skin replacements, 465
Tissue engineering, 463-66
    chondrocyte culture for cartilage replacement, 465-66
    commercial tissue culture processes, 465-66
    compared to protein production from mammalian cells, 464
    defined, 463-64
    tissue-engineered skin replacements, 465
Tissue plasminogen activator (TPA), 96
Total Kjeldahl nitrogen (TKN), 490
Total organic carbon (TOC), 489
Totipotency, 408
Toxic compounds, microbial growth inhibition by, 179-80
Traditional industrial biprocesses, 515-33
    aerobic processes, 524-33
        bakers' yeast production, 526-27
        citric acid production, 524-26
        high-fructose corn syrup (HFCS) production, 530-33
        penicillin production, 527-30
    anaerobic processes, 515-24
        acetone-butanol production, 521-24
        ethanol production, 515-19
        lactic acid production, 519-21
Transcription terminator, 110
Transduction, 227-28, 241
    defined, 225
    generalized, 227
    specialized, 227
Transfection, 235
Transfer RNA (t-RNA), 46
Transformation, 227, 241
    defined, 225
Transformed cell lines, 389, 428
Transgenic animals, and host-vector system selection guidelines, 430-43
Transgenic plants and plant cell culture, and host-vector system selection guidelines, 432

Transient behavior models, 183-89
    chemically structured models, 184-89
    models with time delays, 183-84
Translation, 113-19
    elongation, 113-15
    genetic code, 113-14
    how the machinery works, 113-14
    initiation, 113-15
    termination, 113-15
Transport vesicles, 117
Transposons, 230, 241
Tricarboxylic acid (TCA), 137-39, 152
Trickling biological filters, 496-98
Triple-helix structure, 30
t-RNA, 46
Trypanosomes, 25
Trypsin, 94, 95
Turbidity measurement of culture medium, 158
Turbidostat, 190-91
Two-aqueous-phase affinity partition extraction, 349
Two-phase fermentation, 437
Tyrosine, 51

U

Ultrafiltration membranes, 364-65
Ultrafiltration (UF), 358-60
Ultrasonic vibrators, 341
Unbalanced growth, 163, 175
UNOX Process (Union Carbide), 492
Unselectable mutations, 222
Unsteady-state method, 294
Unwinding enzymes, 108
Uracil, 41
USDA Northern Regional Research Laboratory, 5

V

Vacuoles, 21
Vacuum-tray driers, 379

Viable count, 157
Viral infection models, 467-70
Virgin zone, 353
Virulence factor production, 128
Virus vaccines, as animal cell products, 401
Viruses, 13, 14-15, 53
    bacteriophages, 14
    capsid, 14
    lysogenic cycle, 14
    lytic cycle, 14
    replication of a virulent bacteriophage, 15
Viruslike particle, 401
Vitamin K, 51
Vitamins, 51
Volumetric transfer coefficient, 286, 323
Volutins, 17

W

Wall growth, 298
Waste water, conversion to useful products, 506-8
Whole cells and tissue culture, as animal cell
        products, 402
Working volume, 290

X

X-press, 342

Y

Yeasts, 22
Yield coefficients, theoretical predictions of, 215-16

Z

Zinc, 51
Zwitterion, 28
Zygotes, 20, 22